Narrative Science

Narrative Science examines the use of narrative in scientific research over the last two centuries. It brings together an international group of scholars who have engaged in intense collaboration to find and develop crucial cases of narrative in science. Motivated and coordinated by the Narrative Science Project, funded by the European Research Council, this volume offers integrated and insightful essays examining cases that run the gamut from geology to psychology, chemistry, physics, botany, mathematics, epidemiology and biological engineering. Taking in shipwrecks, human evolution, military intelligence and mass extinctions, this landmark study revises our understanding of what science is, and the roles of narrative in scientists' work. This title is also available as Open Access.

MARY S. MORGAN is the Albert O. Hirschman Professor of History and Philosophy of Economics at the London School of Economics.

KIM M. HAJEK is a postdoctoral researcher associated with the Narrative Science Project, who currently works on the 'Scholarly Vices Project' at the University of Leiden.

DOMINIC J. BERRY is Research Fellow on the Narrative Science Project, and the 'Everyday Cyborgs 2.0' project at the University of Birmingham.

Narrative Science

Reasoning, Representing and Knowing since 1800

Edited by

Mary S. Morgan
London School of Economics and Political Science

Kim M. Hajek
London School of Economics and Political Science

Dominic J. Berry
London School of Economics and Political Science

CAMBRIDGE
UNIVERSITY PRESS

CAMBRIDGE
UNIVERSITY PRESS

University Printing House, Cambridge CB2 8BS, United Kingdom

One Liberty Plaza, 20th Floor, New York, NY 10006, USA

477 Williamstown Road, Port Melbourne, VIC 3207, Australia

314–321, 3rd Floor, Plot 3, Splendor Forum, Jasola District Centre,
New Delhi – 110025, India

103 Penang Road, #05–06/07, Visioncrest Commercial, Singapore 238467

Cambridge University Press is part of the University of Cambridge.

It furthers the University's mission by disseminating knowledge in the pursuit of
education, learning, and research at the highest international levels of excellence.

www.cambridge.org
Information on this title: www.cambridge.org/9781316519004
DOI: 10.1017/9781009004329

This project has received funding from the European Research Council under the European Union's
Horizon 2020 research and innovation programme (grant agreement No. 694732): www.narrative-
science.org/

First published 2022

Printed in the United Kingdom by TJ Books Limited, Padstow Cornwall

A catalogue record for this publication is available from the British Library.

Library of Congress Cataloging-in-Publication Data
Names: Morgan, Mary S., editor.
Title: Narrative science : reasoning, representing and knowing since 1800 / edited by Mary S.
Morgan, London School of Economics and Political Science, Kim M. Hajek, London School of
Economics and Political Science, Dominic J. Berry, London School of Economics and Political
Science.
Description: Cambridge, United Kingdom ; New York, NY, USA : Cambridge University Press,
2022. | Includes bibliographical references and index.
Identifiers: LCCN 2022022798 (print) | LCCN 2022022799 (ebook) | ISBN 9781316519004
(hardback) | ISBN 9781009001991 (paperback) | ISBN 9781009004329 (ebook)
Subjects: LCSH: Communication in science. | Research – Methodology. | BISAC: TECHNOLOGY
& ENGINEERING / History
Classification: LCC Q223 .N319 2022 (print) | LCC Q223 (ebook) | DDC 501/.4–dc23/eng/20220718
LC record available at https://lccn.loc.gov/2022022798
LC ebook record available at https://lccn.loc.gov/2022022799

ISBN 978-1-316-51900-4 Hardback
ISBN 978-1-009-00199-1 Paperback

Contents

Figures

Tables

Contributors

LINE EDSLEV ANDERSEN is currently on maternity leave. She wrote her chapter for this volume while working as a postdoc at the Centre for Science Studies, Department of Mathematics, Aarhus University. Her main area of research is the philosophy of mathematical practice and the philosophy of science in practice.

JOHN BEATTY is Professor in the History and Philosophy of Science and Social and Political Philosophy at the University of British Columbia. His current research projects concern, more specifically: (1) the distinction between 'history' and 'science' and the respects in which evolutionary biology is as much like the former as it is like the latter, (2) changing views of contingency and necessity in the Darwinian Revolution, (3) the relationships between biology and 'the state', from the Manhattan Project to the Human Genome Project, and (4) issues concerning the nature of scientific 'authority'.

DOMINIC BERRY is a Research Fellow on the Narrative Science Project. His research brings together historical, philosophical and social scientific methods and analyses attending to the biological sciences in particular. He has held research fellowships at the University of Leeds and the University of Edinburgh. In 2021, he joined 'Everyday Cyborgs 2.0', a Wellcome Trust-funded project based at the University of Birmingham. In 2019, he co-founded the Biological Engineering Collaboratory (www.bioengcoll.org) and in 2021 helped create the Transformational HPS network, which supports scholars who are queering, decolonizing and centring disability within HPS (www.transformationalhps.org).

DEBJANI BHATTACHARYYA is the Professor for the History of the Anthropocene at University Zürich. She is the author of *Empire and Ecology in the Bengal Delta: The Making of Calcutta* (Cambridge University Press, 2018). Currently, she is a non-resident fellow at the Center for the Advanced Study of India (CASI), University of Pennsylvania. Most recently, she was a Shelby Cullom Davis Fellow at Princeton University. Her work has been published in *Modern Asian*

Studies, Journal of Economic and Social History of the Orient and Comparative Studies in South Asia and *Africa and the Middle East*.

SHARON L. CRASNOW is Distinguished Professor Emerita, Norco College, Southern California. Her research is on methodological issues in the social sciences with a focus on political science. She has published in *Philosophy of Social Science, Philosophy of Science, Studies in History and Philosophy of Science* and *Synthese*. She also works on feminist philosophy of science and epistemology and is co-editor (with Kristen Intemann) of *The Routledge Handbook of Feminist Philosophy of Science* (2021).

STEPHANIE DICK is an assistant professor in the School of Communication at Simon Fraser University. She holds a PhD in history of science from Harvard University. She is a historian of mathematics, computing and artificial intelligence in the twentieth-century United States. In particular, she studies the early introduction of computing and automation to American mathematics and to American policing during the Cold War.

LUKAS ENGELMANN is a Chancellor's Fellow and Senior Lecturer in the History and Sociology of Biomedicine at the University of Edinburgh. His research concerned with the history of epidemiological reasoning in the twentieth century received an ERC Starting Grant in 2020. His first book, *Mapping AIDS*, was published by Cambridge University Press in 2018 and considers the visual and medical history of AIDS/HIV. In 2020, he published a co-authored monograph on sulphuric utopias, with Christos Lynteris (open access), which tells the technological history of fumigation and the political history of maritime sanitation at the turn of the twentieth century.

DEVIN GRIFFITHS is an associate professor of English and comparative literature at the University of Southern California. His first book, *The Age of Analogy: Science and Literature between the Darwins*, explores how analogy helped shape the disciplinary formations of the life sciences and humanities. With Deanna Kreisel, he is co-editor of a forthcoming Cambridge University Press collection on the influence of Darwin's thinking on the humanities, entitled *After Darwin*. He is currently working on a new book, *The Ecology of Power*, that explores, from the perspective of the energy humanities, the historical intersection of ecological and economic theory.

ELIZABETH HAINES is Vice-Chancellor's Fellow in History at the University of Bristol. Her research applies interdisciplinary approaches to the lived practices of knowledge-making, particularly in colonial and postcolonial contexts. She has a particular interest in visual and material culture as tools within knowledge practices. Recent research collaborations include work with physical scientists, NGOs, heritage and art practitioners in the UK, Kenya, Zambia, Belgium and the United States. She is currently completing a monograph, *Illegible Territory*, that explores the use and disuse of geographical knowledge by the colonial government in Northern Rhodesia, developed from her doctoral and post-doctoral research.

KIM M. HAJEK was a postdoctoral researcher on the Narrative Science Project at the LSE, and currently works at the University of Leiden in the 'Scholarly Vices Project'. She is affiliated with the Institut des humanités en médecine, Lausanne, and serves as Associate Editor for Centaurus and on the editorial board of *Journal of the History of the Behavioral Sciences*. Kim's research spans history of science and literary studies, focusing on how textual practices inform knowledge-making in the human sciences. Her publications explore ideas of normality and case-writing in nineteenth-century French psychology, as well as science/literature intersections in the history of hypnotism.

ANDREW HOPKINS is an Honorary Research Associate in the Department of Science and Technology Studies at University College London. He also served as a Research Officer on the Narrative Science Project at the London School of Economics. A former industrial geoscientist and science educator, he is currently active in the field of history and philosophy of science, with a particular interest in understanding what enables historical sciences such as geology to reconstruct the deep past. He has a PhD in marine geology and geophysics and is a Fellow of the Geological Society of London.

BRIAN HURWITZ is Emeritus Professor of Medicine and the Arts at King's College London. He worked as an inner-city general practitioner and academic for thirty years, becoming Professor of Primary Care and General Practice at Imperial College London. In 2002, he moved to King's to set up the Centre for the Humanities and Health, a multidisciplinary research unit that also offers Master's, PhD and postdoctoral education for students of the humanities, biosciences and health professionals. Based in the English Department, his research interests include narrative studies in relation to clinical practice, ethics, law and the epistemic aspects of clinical cases.

JOHN E. HUSS is professor of philosophy at the University of Akron, where he is also on the faculty of the Integrated Bioscience PhD program and a member of the Biomimicry Research and Innovation Center. His research interests include philosophy of science, philosophy of medicine, applied ethics and philosophy of popular culture.

ELSPETH JAJDELSKA is a senior lecturer in English at the University of Strathclyde. She researches the cognitive experience of narrative fiction as well as the history of reading experience. She has published on cognition of literary fiction in *Poetics Today*, *Journal of Literary Semantics*, *Philosophical Psychology* and *Frontiers in Psychology*. Her monograph, *Speech, Print and Decorum in Britain, 1600–1750*, uses anthropological theories of verbal art and performance to explain historical changes in reading experience. She is currently working on a project on the relationship between internal scene construction and narrative experience.

NINA KRANKE studied environmental sciences and philosophy in Lüneburg and Greifswald, Germany. She started her PhD project at the University of Kassel and is now a research assistant at the Department of Philosophy at the University of Münster and a member of the interdisciplinary DFG Research Training Group 2220 'Evolutionary Processes in Adaptation and Disease' (EvoPAD). Her research interests include philosophy of science, philosophy of biology, feminist philosophy, bioethics and animal ethics.

ROBERT MEUNIER is research fellow at the Institute for the History of Medicine and Science Studies at the University of Lübeck, Germany, and part of the Cluster of Excellence 'Precision Medicine in Chronic Inflammation'. In 2012, he graduated from the PhD programme, 'Foundations of the Life Sciences and their Ethical Consequences', jointly hosted by the University of Milan and the European School of Molecular Medicine. Since then, he worked at the Max Planck Institute for the History of Science, the Institute for Cultural Inquiry Berlin, the Department of History, LMU Munich, the Department of Philosophy, University of Kassel, and the ERC-funded Narrative Science Project at the London School of Economics. From 2018 to 2021 he was Principal Investigator in the DFG-funded research project, 'Forms of Practice, Forms of Knowledge', located at the University of Kassel.

TERU MIYAKE is an associate professor and head of the philosophy programme at Nanyang Technological University in Singapore, and a former Fellow of the Radcliffe Institute for Advanced Study. His research has centred on epistemological issues in the physical sciences, particularly planetary astronomy, seismology and late nineteenth- and early twentieth-century experimental physics. He has also written on measurement, the use of models in science, scientific realism and nineteenth-century British philosophy of science.

MARY S. MORGAN is the Albert O. Hirschman Professor of History and Philosophy of Economics at the London School of Economics, an elected Fellow of the British Academy and an Overseas Fellow of the Royal Dutch

Academy of Arts and Sciences. She has published on social scientists' practices of modelling, observing, measuring and making case studies, and is especially interested in how ideas, numbers and facts are used in projects designed to change the world. Her most recent books are *How Well Do Facts Travel?* (2011) and *The World in the Model* (2012), both published by Cambridge University Press.

PAULA OLMOS is an associate professor of logic and philosophy of science at Madrid's Autonomous University. Her research interests cover argumentation theory and rhetoric, focusing especially on argumentation in science. She has published papers on these subjects in journals including *Argumentation*, *Informal Logic* and *Revista Iberoamericana de Argumentación*, and in several collective volumes of Springer's 'Argumentation Series'. She has acted as co-editor of reference works including *Compendio de lógica, Argumentación y retórica* (2011, 2012, 2013, 2016) and *De la demostración a la argumentación* (2015), and has coordinated and edited the volume of essays *Narration as Argument* (2017).

MAT PASKINS works for a charity and has a PhD in the history of science. They have written about the relations between the history of science and voluntary associations, the role of tree-planting in British politics and notions of improvement and histories of chemistry and material sciences. Mat is also interested in relations between science and literature and has co-edited two anthologies of narrative science.

ANNE TEATHER is an archaeologist who specializes in the material culture of the Neolithic period (5000–2000 BCE) in northern Europe. She is Managing Director of Past Participate CIC, a non-profit community archaeology company that provides high-quality archaeological excavation and research training. She is a visiting fellow at Bournemouth University and publishes regularly on prehistoric art and archaeological theory. She taught as a fixed-term lecturer at the University of Chester (2007–12) and is currently writing a monograph, *Power from the Periphery*, an account of the role of material culture in prehistoric societies.

M. NORTON WISE is Distinguished Research Professor (Emeritus) in the Department of History at UCLA and has published widely in the history of the physical sciences in the nineteenth and twentieth centuries. His most recent book is *Aesthetics, Industry, and Science: Hermann von Helmholtz and the Berlin Physical Society*. He has also published a variety of articles on narrative in science, especially concerning the role of computer simulations and visual narratives. With Mary Morgan, he has edited a special issue of *Studies in History and Philosophy of Science* on narratives in science.

Preface and Acknowledgements

We began the Narrative Science Project with the aim of finding and analysing narratives as they occurred in the disparate and varied sites offered across the terrain of as many natural/physical/human/social sciences as we could. We did not aim to be comprehensive, for science exists in too many guises, and nor did we assume all science was narrative. But as our project progressed we came to recognize both how surprisingly widespread narratives were in science, and how they are shown in diagrams, maps and equations as well as told in texts, protocols, books and journal articles. We did not aim to impose an account of narrative onto science, but rather to explore the different narrative formats that scientists use, and the different functions that narrative fulfils for those scientists. We did not aim to create a well-researched map, but our detailed case-work created a picture with the features of a medieval tapestry: exhibiting both the detailed texture of many individual science narratives and the amazing spread of how narrative appears in science.

Our second starting ambition was to persuade both literary communities who study narrative, and science studies communities who study science, 'to take narrative science seriously'. This caused us to walk several tightropes at once. First, we were all too conscious – from typical reactions – that putting narrative and science together is problematic, for narrative is almost automatically associated with stories, and thence with something fictional. To mitigate this, we have largely kept to the terminology 'narrative'. Second, was the tendency to assume that we were interested in narrative as a public engagement device that enabled the public to understand science, whereas we were interested in how scientists use narrative within their own communities, scientist-to-scientist, for their own purposes. Third, literary scholars sometimes assumed that either our interests were focused on factual narratives, or on the literariness of scientific prose. Narratives in science may be about facts, but may be about theories, or even both at once, and we were not primarily interested in literary qualities or the range of literary devices used by scientists. Instead, our engagement with the narrative community has been to understand how narrative functions to create joined-up accounts of things in particular domains (in our case, the various domains of science), and in what makes such relatedness 'tellable'. Fourth (although this only became clearly evident as we went along), our interests covered not just the predictable places of narrative in reporting the process of scientific research activities and outcomes, or the life histories of scientific phenomena, but the ways in which narrative plays an important role in doing science. Thus, to our constituencies in history and philosophy of science, we propose that narrative-making shows its power in making sense of a phenomenon, and in so doing becomes a part of scientific reasoning, argument and inference.

What became the 'our' in our team is a critical part of our narrative-science project. We assembled a 'home team' from history and philosophy of science and literary studies (each person also held knowledge of at least one science field) and drew extensively on the help of intellectual interlocuters from the pre-history of the project. These conversations go back to an early meeting (in 2013), generously hosted by Raine Daston at the Max Planck Institute of the History of Science, and forward to the special issue of *SHPS* (published in 2017). Then, over the five years of our project time (2016–21), 'our team' widened to embrace an incredible range of scholars who came to deliver seminars, provided papers at our specialist workshops, joined us in specially constructed symposium sessions at conferences, and who contributed cases and commentaries to our Anthologies and working papers. Together they constituted a community extending and enriching the Narrative Science Project in creating both this book and the multiple further resources on our project website www.narrative-science.org/. Without them, our narrative science tapestry would be restricted in range, thin in colour and lacking depth of conviction. We thank them all below (and apologize if our listing misses any of our wider team!), as well as the 'anonymous' reviewers of our book chapters and the book as a whole.

'Intellectutors'

Norton Wise, Jim Griesemer, Raine Daston, Sharon Crasnow, John Beatty, Brian Hurwitz, Ted Porter, Naomi Oreskes, Chiara Ambrosio, Mary Terrall, Greg Priest, Paul Roth, Roman Frigg and Sabina Leonelli.

Workshop Speakers

Puzzles and Problems of Classifying and Categorizing

Staffan Mülle-Wille, Yossi Lichtenstein, Andrea Woody, Jan-Willem Romeijn, Santi Funari and Rachel Ankeny.

Narrative Science and Its Visual Practices

Mirjam Brusius, Elizabeth Haines, Nina Kranke, Nicola Williams, Annamaria Carusi and Jonathan Gray.

Expert Narratives: Systems, Polices and Practices

Andrea Mennicken, Brendan Clarke, Hannah Roscoe, Chris Hall, Shana Vijayan, Lars Bo Henriksen and Natasha McCarthy.

Narratives as Navigation Tools

Martin Stahl, Rebecca Wilbanks, Sabine Baier, Cathal Cummins, Miguel Garcia-Sancho and Karen Polizzi.

Does Time Always Pass? Temporalities in Scientific Narratives

Norton Wise, John Beatty, Dorothea Debus, Paula Olmos, Rosa Hardt, William Matthews, John Huss, Teru Miyake, Anne Teather, Elspeth Jajdelska, Thomas Bonnin, Tirthankar Roy and Daniel Pargman.

Scientific Polyphony: How Scientific Narratives Configure Many 'Voices'

Debjani Bhattacharyya, Devin Griffiths, Isabelle Kalinowski, Birgit Lang, Harro Maas, Jill Slinger, Lotte Bontje and Rhianedd Smith.

Narrative Science in Techno-Environments

Ina Linge, Jean-Baptiste Gouyon, Amelie Bonney, Ross Brooks, Louise Coueffe, Greg Lynall, John Lidwell-Durnin, Harriet Ritvo, Saliha Bayir, Ágota Ábrán, João P. R. Joaquim, Ellie Armstrong, Aadita Chaudhury, Mauricio Nicolas Vergara, Sarah Bezan, Charlotte Sleigh, Animesh Chatterjee, sam smiley, Sarah Daw, Lachlan Fleetwood, Jon Agar, Anahita Rouyan and Alexander Hall.

Narrative and Mathematical Argument

David Corfield, Michael Friedman, Line Edslev Andersen, Mikkel Willum Johansen, Henrik Kragh Sørensen, Fenner Tanswell, Karine Chemla and Stephanie A. Dick.

Anecdotes: Little Narratives That Carry Bigger Weight

Brian Hurwitz, Martin Böhnert and Guillaume Yon.

Speakers in the Public Seminar Series

Sally Atkinson, Elisa Vecchione, Julia Sánchez-Dorado, Claudia Cristalli, Caitlin Donahue Wylie, Sigrid Leyssen, Lukas Engelmann, Sabine Baier, Sharon Crasnow, Phyllis Kirstin Illari, Ivan Flis, Adrian Currie, Alfred Nordmann, Eleanore Loiodice, Annamaria Contini, Adelene Buckland, Sarah Dillon, Vito De Lucia, Marco Tamborini, Staffan Müller-Wille, Neil Tarrant, Heieke Hartung, Sally Horrocks, Paul Merchant, Emily Hayes, Veronika Lipphardt, Will Tattersdill and Lorraine Daston.

Symposium Collaborators

Sarah Lawrence, David G. Horn, Ivan Flis, Dmitriy Myelnikov, Meria Gold, Ageliki Lefkaditou, Debjani Bhattacharyya, Nicole Edelman, Sigrid Leyssen, Robert Bud, Lijing Jiang, Tiago Saraiva, Ian Hesketh, Miriam Solomon, Anna Svensson, Greg Priest, Corinne Bloch-Mullins, Ellie Armstrong, Michael Toze, Ross Brooks, Aude Fauvel, Larry Duffy, Daniela S. Barberis, Gerald Sullivan, Jonathan Shann, Junona S. Almonaitienė, Veronika Girininkaitė, Sharman Levinson, Janella Baxter, Robert Smith, Hanna Lucia Worliczek, Caterina Schürch, Thomas Bonnin and Mathias Grote.

Anthology Contributors

Sabine Baier, Debjani Bhattacharyya, Ross Brooks, Geoffrey Cantor, Silvia De Bianchi, Federico D'Onofrio, Helena Hammond, Colin McSwiggen, Felicity Mellor, Martin Böhnert, Greg Priest, Jim Scown and Keith Tribe.

Home Team

Mary S. Morgan, Dominic J. Berry, Kim Hajek, Andrew Hopkins, Robert Meunier and Mat Paskins. (These researchers were directly funded by the ERC under the European Union's Horizon 2020 research and innovation programme [grant agreement No. 694732].)

We also thank the following for their permissions to reproduce the following images:

3.1 – Raup, D. M., & Sepkoski, J. J. (1982). 'Mass extinctions in the marine fossil record'. *Science*, 215.4539: 1501–1503.

3.2 – Reproduced with thanks to the controllers of Raup and Sepkoski's respective estates.

4.2 – Reproduced courtesy of Renegade Pictures/Channel 4.

4.3 – Amor, K., Hesselbo, S., Porcelli, D., Thackrey, S. and Parnell, J. (2008). 'A Precambrian proximal ejecta blanket from Scotland'. *Geology* 36.4:303.

4.4 – Branney, M. and Brown, R., (2011). 'Impactoclastic Density Current Emplacement of Terrestrial Meteorite-Impact Ejecta and the Formation of Dust Pellets and Accretionary Lapilli: Evidence from Stac Fada, Scotland'. *The Journal of Geology* 119.3: 275–292.

5.1 – Figure kindly provided by Dr Jeroen Ritsema.

5.2 – Suzuki, W., Aoi,S., Sekiguchi, H., and Kunugi, T. (2011). 'Rupture process of the 2011 Tohoku-Oki mega-thrust earthquake (M9.0) inverted from strong-motion data'. *Geophysical Research Letters* 38, L00G16.

7.1 – Reproduced, with permission, from John van Wyhe, ed., *The Complete Work of Charles Darwin Online*. (http://darwin-online.org.uk/converted/pdf/1880_Movement_F1325.pdf).

7.2.a – With thanks to The Rare Book and Manuscript Library, University of Illinois at Urbana-Champaign.

7.2.b – With thanks for permission to reproduce from the Whipple Library, University of Cambridge.

7.3.a – Reproduced, with permission, from John van Wyhe, ed., *The Complete Work of Charles Darwin Online*. (http://darwin-online.org.uk/converted/pdf/1880_Movement_F1325.pdf).

7.3.b – Reproduced, with permission, from John van Wyhe, ed., *The Complete Work of Charles Darwin Online*. (http://darwin-online.org.uk/converted/pdf/1880_Movement_F1325.pdf).

7.3.c – Reproduced, with permission, from John van Wyhe, ed., *The Complete Work of Charles Darwin Online*. (http://darwin-online.org.uk/converted/pdf/1880_Movement_F1325.pdf).

8.1 – Reproduced, with permission, from the British Library, London, as part of the Google Books project.

8.2.a – Made available by US National Archives.

8.2.b – Made available by US National Archives.

10.3 – Reproduced, with permission, from the Kansas Natural History Museum.

10.4 – Reproduced, with permission, from Volker Sommer original author and image maker.

13.1 – Robinson, Robert. (1917). 'LXIII.—A synthesis of tropinone'. *Journal of the Chemical Society*, Transactions 111: 762–768.

13.2 – Medley, Jonathan William, and Mohammad Movassaghi. (2013). 'Robinson's landmark synthesis of tropinone'. *Chemical Communications* 49.92: 10775–10777.

I

Prologues

1   Narrative: A General-Purpose Technology for Science

*Mary S. Morgan**

Abstract

Narrative is ubiquitous in the sciences. Whilst it might be hidden, evident only from its traces, it can be found regularly in scientists' accounts of their research, and of the natural, human and social worlds they study. Investigating the functions of narrative, it becomes clear that narrative-making provides scientists with a means of making sense of the materials in their field, that narrative provides a means of representing that knowledge and that narrative may even provide the site for scientific reasoning and knowledge claims. Narrative emerges as a 'general-purpose technology', used in many different forms in different sites of science, enabling scientists to figure out and to express their scientific knowledge. Understanding scientists' use of narrative in this way suggests that narrative functions as a bridge between the interventionist practices of science and the knowledge gained from those practices.

1.1   Introduction

Scholars of scientific life see it filled with experiments, models, theories, descriptions, observations, categories, etc. It is equally full of narratives. Yet

* The Narrative Science Project was funded by the European Research Council under the European Union's Horizon 2020 research and innovation programme (grant agreement No. 694732), whose activities are reported in great detail on the project website: www .narrative-science.org/. This chapter – especially in footnotes – refers to a number of resources on that site, particularly the reports of our workshops, and the entries in our two *Anthologies* (*Anthology I* 2019 and *Anthology II* 2022). This project grew out of an earlier collaboration with Norton Wise that resulted in a special issue (see Morgan and Wise 2017) and I am grateful for Norton's 'wise' advice throughout this current project, including on this chapter. Special thanks for their help with this chapter go to Roy Weintraub, Sarah Dillon, Tarja Knuuttila, Claudia Cristalli, William Twining, Martina King and Brian Hurwitz; to the team of postdocs on the project and the wider team of authors in this book for all they have taught me, and especially to Kim Hajek and Dominic Berry for their incredible hard work on this book.

3

the levels at which narratives work, and the kinds of things that scientists come to understand through the activities of developing their narratives, are not easily described in terms of any specific ambitions or functions. Narratives themselves may be understood as a broad class of 'epistemic genre', to use the label that Pomata (2014) developed, essential to the representation of scientific knowledge.[1] But narrative is more than just a means of representing such knowledge; rather, prior to such representations, narrative-making plays a wider role in the sciences as a means of sense-making. In contrast with Crombie's (1988) historically situated categorization of ways of doing science, an account developed further in Hacking's (1992) philosophical analysis, narrative-making is not mainly about how scientists investigate the world but rather about how they make sense out of those investigations. Narrative-making does not satisfy epistemic questions and worries in the way the interventionist and observational modes of doing science described by Hacking (and others) – such as experimenting, category making, statistical work and case-making – can do. Narrative-making and -using, by contrast, are more closely aligned with ontological questions, or, rather, scientists' claims in their 'narratives of nature' are ontological claims about the way the world is and works. The role of narratives piggybacks onto the epistemology of those other, more interventionist modes of practising science. So, while narrative usage may overlap in places with Pomata's notion of 'epistemic genres' and can be an accompaniment to Hacking's modes of doing science – narrative-making and -using fulfil other distinct roles for scientists, roles that need separate recognition.

Narrative emerges from this volume as having three functions for scientists: narrative-making operates as a means of making sense of their puzzling phenomena; it provides a means of representing that scientific knowledge; and it provides resources for reasoning about those phenomena. These three functions are related: it is because scientists often make sense of their world by making narratives that they then use those narratives to represent what they believe they know, and thence to reason with them. I propose we think of narrative as a 'technology of sense-making' that enables scientists to bridge between their interventionist activities of exploring the world and their knowledge claims about the world, that is, between their epistemic and

---

[1] Pomata (2014) labelled certain kinds of texts in science 'epistemic genres' in contrast with the genres recognized in literature. As she argued (in a historical account of changes in medical reporting), an epistemic genre: develops 'in tandem with scientific practices'; is 'deliberately cognitive in purpose'; is linked 'to the practice of knowledge-making'; and has a 'primary goal' of 'the production of knowledge'. These have certain parallels in the claims made in this chapter about the roles of narrative in science, but the functions I attribute to narrative-making have greater agency in *doing* science.

ontological realms. To label narrative a technology may seem rather strange, but we are in some interesting company here. The philosopher John Dewey argued that the notion of technology was not just about how to make things in the economy, but equally attributable to the abstract and intangible work of enquiry and deliberation involving cognitive work – just as we find for narrative in science.[2] His contemporary, the sociologist-economist Thorstein Veblen, insisted on the priority of the human element in designing, making and using a technology. While narrative-making, -using, and -reasoning start with the scientist and their community, it is worth remarking that narrative also embeds its own technical elements and attributes. These three separate but related functions of narrative, broadly understood as a technology of sense-making for scientists, may be recognized in the chapters of this book by observing whether narrative is being used as a noun, verb or adjective.

All those nouns of scientific practice – experiments, models, theories, descriptions, observations, categories – hide actions and activities: experimenting, modelling, theorizing, describing, observing, categorizing. Other elements that scientists use don't immediately convert between nouns and verbs – data has to be given its own multiple verbs ('to gather, clean, assemble and prepare'), just as laws have 'to be discovered or made'. Narrative is akin to laws and data: easily understood and effective as a noun, its scope as an activity is not quite so obvious; yet appreciating that scope is critical for understanding the broader role of narrative as a technology for scientists. The quintessential feature of narrative is that it shows how things relate together, so that constructing a narrative account in science involves figuring out how the elements of a phenomenon are related to each other. This is why narrative-making and -using are conceived here as a technology, one that enables scientists to make sense of their phenomena.

These basic usages of narrative in *noun* and *verb* forms are important, of course, but they might be still awkward, and limited, if we want to go one step further and conceive of narrative as flourishing in the knowledge-claiming activity of the sciences. In this respect, the adjectival form is more immediately useful: so, 'narrative account' and 'narrative description' might both be taken for granted. And, while 'narrative inference', 'narrative argument' and 'narrative explanation' might initially sound strange (even perhaps contradictory), it will turn out that we need these terms, for the narrative form does overlap in usage into these scientific activities of reasoning and knowledge-making. Thus, narrative as an *adjective* works as an attribute of a certain form of reasoning: giving a satisfactory narrative

[2] I thank Teru Miyake for drawing my attention to Dewey's insight, best followed in Hickman (2001).

account may go beyond sense-making into the kinds of reasoning associated with inference and explanation.

None of these uses of the term narrative – in noun, verb or adjectival forms – should be problematic if we can find ways to appreciate the active work that narrative does in our sciences, particularly if we can figure out its features and its functions, just as we have for data and laws. These grammatical labels give clues, but only clues, to the ways in which scientists develop, create and use narratives in their various fields, for various purposes and in conjunction with various other forms of scientific representation and knowledge-making activities. These language terminologies need to be filled in with examples and hardened through analyses to reveal the active work we attribute to narrative in science, and so to appreciate how narrative operates as a technology for scientists in doing science.

There are, of course, many commentaries about narrative in other domains, especially in the fields of literature, narratology and legal studies. Narrative scholars from the domain of literature typically focus on the narrative as text: its plots, its structure, temporal and spatial organization, its eventfulness and cognitive function, as well as its rhetorical and aesthetic components, and terms of affect. Narratologists tend to focus on the narrators, readers, what constructions narratives follow, and their requirements for narrative tellability. It is fair to say that with few notable exceptions, neither group focuses especially on connections of narrative with knowledge-making.[3] So, in an important chapter, Kim Hajek explores what is narrative about 'narrative science', and thus extends the relevant intersections of those fields with our agenda (Chapter 2). Discussions in the field of law about narrative range over matters of rhetoric and affect, but have an equal interest in the putting together of evidence, and the role of 'theory' – meaning both the hypothesis about what happened in a particular case, but also the concepts from law that need to be taken into account.[4] As such, these latter interests fit closest to those of this chapter. But rather than work comparatively with this legal literature I treat narrative in science on its own terms – in order to examine how it makes itself 'at home' in the scientific knowledge environment.

---

[3] Dear (1991) is a notable early work in the field (on which more later in section 1.6). Of four current books that overlap with our agenda to treat narrative in science seriously: Fludernik and Ryan (2020) attends to narratives in factual spheres (while our focus is on narratives in science, which are often, or not only, about 'facts'); Carrier, Mertens and Reinhardt (2021) are concerned with the contrasts and intersections of narratives and comparison in science; Dillon and Craig (2021) analyse how narrative can be used alongside scientific evidence in the public domain on account of the cognitive value of narratives; and Kindt and King (forthcoming) focus on narrative knowledge-making from a sample of ancient to modern texts.

[4] See, especially, Nicolson ((2019): chap. 7); Twining ((2002): chaps. 13–14); Twining ((2006): chaps. 9–13). Thanks to William Twining for introducing me to this literature and discussing it with me.

Narrative is a broad, expansive term (with many definitions in narrative theory), and the challenge has been to develop an analysis which is insightful for scientists' creation and use of narrative. Our research shows narrative to be an enabling, general-purpose technology, widely used by scientists within their own different communities to fulfil certain functions in their scientific work – even when they don't use the word or recognize that label for their activity. It is important to note the limits of this claim: narratives are not found in all aspects of all sciences. Rather, they fulfil certain kinds of function with some regularity in some sciences, or some sites of science, and in conjunction with some methods of doing science. By tracing this (sometimes hidden) narrative activity, and its locations, we can understand both what is different and what is generic in these usages in different sites, and so develop an understanding of narrative in the domains of science.

1.2   Narratives of the Field

The first challenge we address in this book is to see and locate the narratives that appear in our sciences. The most obvious narratives found in science may be those wrap-up accounts in publications resulting from the activities of scientists. In modern science, these are usually impersonal narratives,[5] cut down to the essential actions that scientists tell of how they went about their research: their 'research narratives'. Less recognizable, but still apparent, as Robert Meunier argues (Chapter 12), are their 'narratives of nature': the narratives – 'as if told' by natural, human and social life – that those scientists have tried to reveal, recover and make sensible. And, as he points out, scientists' research narratives often twine in symbiosis around their narratives of nature.[6] This has fruitful consequences: the researcher–author, in guiding the scientist–reader along the path of their activities, enables the latter to gain practical familiarity with the former's narratives of nature, particularly with any new elements and concept set in use.

A broader category of narratives can be found that seek to define and lay boundaries to new approaches for a whole field, or maybe to delineate a new interstitial field. These field-making narratives might be more or less reticent in their agenda. Grand ones are epitomized in the self-proclaimed narratives of those seeking to automate and computerize the whole of mathematics. Stephanie Dick (Chapter 15) discusses two such competing self-narratives in

---

[5] The significant exception is anthropology, where the scientist must be personally present in their narratives, and attend to the narrative text they create, to signal professional credibility (see entry on Geertz, *Anthology II*).

[6] These 'research narratives' and 'narratives of nature' are often openly related in medicine and management sciences, where scientists and their subject participants recount, and often share, their expert and experiential knowledge via narratives.

late twentieth-century American mathematics: one group sought to reformulate all mathematical knowledge into one single form, and the other to enable all mathematicians to contribute elements in their own format.[7] Their politics of control vs. pluralism were explicit. Other field narratives may be more opaque, evident only in their alignments and commitments, to be discovered by an outside reader, as Dominic Berry (Chapter 16) does in looking at how 'synthetic biologists' positioned themselves between engineering and biology in defining and growing their own field. He uses longue durée changes in history writing – from chronicles through genealogies to narratives – to argue his case. These are important categories. Chronicles report events solely based on their place in a time sequence without paying attention to any relationships between those events; *genealogies* focus on the 'family' (broadly construed) relationships between the events or objects; *narratives* provide an account of the relationships between events or objects (whether or not these relationships are tied together in a time sequence or by family connections). Among narratologists, there is a widespread view that a chronicle does not count as a proper narrative because the relational content is absent, while genealogies are just a subset form of narratives. Anne Teather (Chapter 6) adopts the same categories to show how new technologies of dating in archaeology have effectively changed narrative practices in that field. Whereas archaeologists used to tell *genealogical* accounts to frame the periods of prehistory (e.g., the Neolithic period), more recent technologies of investigation have created the more limited *chronologies* or *chronicles*.

Certainly, the narratives of nature – narratives of how the world is and how it works (whether it be the natural, human or social world) – are sometimes much harder to see than these research and field-making narratives. Narratives of nature are more likely to be found implicated with, or inside, other accounts of scientific activity. Like those sherds and trenches of Teather's archaeological sites, these traces of narrative point to the scientific activities that created them, and from which we must reconstruct the power that narrative-making and narrative-using have in such spheres.

1.3   Narrative: A Means of Scientific Representation

The core function of what narrative does is to bring and bind elements in a subject field together. Narrative-making in the sciences can be found in theorizing, in creating an adequate description of empirical materials or in marrying them to each other in ways that embed ideas and concepts,

---

[7] This chapter originated in our project workshop on narrative in mathematics. See workshop on mathematics on project website: www.narrative-science.org/events-narrative-science-project-workshops.html.

that is, in activities of sense-making and knowledge-making (examined in sections 1.4 and 1.5). Since the narratives that result from these activities express, or make evident, these connections between elements in a scientific domain, narratives can be treated as a form of scientific representation akin to other forms of representation. What are the characteristic aspects of such representations, and the implications of this way of understanding the role of narratives in science?

First, narrative representations found in science may appear as free-standing or separate pieces of verbal text – in ordinary or natural language. They might be embedded in visual representations (drawn into schemas such as diagrams of mechanisms or detailed representations of empirical matters in graphs), or even expressed in the completely formal languages of abstraction and mathematics. Wise (Chapter 22) contrasts the possibilities of natural and formal languages, and the extent to which they do different kinds of work, and say different things, and thus why narratives in the two forms are not simple translations or transpositions of each other. Depending on the science in question, the narrative form of representation will be more or less formalized, more or less abstract, and may have more or less dimensionality of elements compared to other representational forms of diagrams, equations and so forth. But, whatever their form and language, it is typically the case that they are 'community narratives', to be understood without further explanation or accompanying text only by those in the expert community who use them. Mat Paskins (Chapter 13) translates/explains, for us lay-readers, the 'chemese' of chemical reaction diagrams depicting the synthesis of particular molecules. He points out that early twentieth-century versions told a different narrative from early twenty-first-century versions of essentially the same representation: in early years, the 'equation' expressed the sequence of steps taken to synthesize a certain chemical, but in later years, such diagrams came to narrate the chemical reactions that took place. The 'cartoon' narrative shown in Andrew Hopkins's chapter (Chapter 4) relates what happens in a meteorite impact as material explodes, flows out and gradually builds up deposits on the ground. This requires, for the lay reader, a lengthy verbal narrative that lets us follow the combinations of interacting processes and outcomes from these geological events.[8] In other cases, indirect representations of nature (such as mathematical models) are manipulated to show the narratives implicit in visual schematic representations. We find such narratives in the computer visualizations from simulating snowflake growth and the processes of chemical reactions, as shown by Wise (2017); the latter offers an alternative free-standing, time-stepped,

---

[8] Another great example is found in Hopkins's analysis of three different geological diagrams depicting different theories and dimensions of the formation of the continents over long geological time (in *Anthology II*).

visualization of the chemical reaction 'equations' found in Paskins's paper (Chapter 13). Such narratives give clues to the density of knowledge that typically lies behind formal language representations.

Second, more often than free-standing independent forms, textual narratives are strongly co-dependent with other forms of scientific representation, such as charts, graphs, drawings, maps, matrices, models, formulae and so forth. Such textual narrative accompaniments might well be an essential part of the identity of those representations, whether of the evidential diagrams in graphs or of the theory-based representations found in models. The classic well-known example is Darwin's pictorial 'tree of life', which – when read alongside textual information – offers a shorthand depiction showing how evolving species branch, or die out, or survive. It is a kind of genealogy – but a conceptual tree not a report of observations. Greg Priest describes this as a 'scaffold' on which we as readers can stand to 'create narratives that enable us to understand' Darwin's account of natural evolution.[9] The infamous 'prisoner's dilemma' model from economics (which was soon transferred to other social-science and biological domains) consists of a mathematical matrix, a set of inequality conditions on those numbers, and a narrative text of the possible behaviours of the 'prisoners' given the 'dilemma' of their situation (termed by economists, 'the rules of the game'). The narrative is an essential element in identifying the game and differentiating it from others that may look similar, for the matrix and inequalities are both insufficient (see Morgan 2007). Combinations of text and drawings (keyed with numbers to each other) are found as essential partners in communicating narrative accounts of metamorphic changes in the insect world (from egg to caterpillar, larva to butterfly), as seen in Mary Terrall's (2017) discussion of eighteenth-century accounts of this phenomenon. Such matching media of visual and text narratives, in which neither is primary but each depends on the other, are also used to explore possibilities of hypothetical events as we see, for example, in D'Onofrio's account of eighteenth-century generals re-running historical battles according to geometrical lines (in *Anthology II*).

There is often a kind of bonding here, rather than co-dependency, of forms and functions. Narratives embedded in formal languages and visual representations often provide a highly efficient rendering of the materials of events. The phylogenetic trees of the evolution of the kangaroo and other marsupials discussed by Nina Kranke (Chapter 10) express a travel saga that charts their geographical and biological evolution over time and space as the species evolved while members of its ancestral population 'journeyed' from South America to Australia. As she shows us, such 'trees' exist in multiple formats – showing in succinct ways, but with distinctly different variants,

---

[9] See Priest's extract from Darwin, and his commentary, *Anthology II*.

the narratives of different kinds of family trees or genealogies. Some of these are for professional audiences, some for museums; some are plain, some 'filigreed'; some read upwards, some downwards, some sideways. There is no one convention despite the related kinds of narratives that are told by these related kinds of trees. There is surely a family tree of such trees, a genealogy of trees, going forward in evolutionary biology from Darwin's tree of life, and going back in time in a long tradition of drawing human dynastic trees.

This complementarity, and bonding, of narratives alongside and inside alternative representations show how narratives fulfil their representing functions in the sciences and how narratives do the kind of representing work they do. These kinds of co-dependency also suggest there are *no* strong reasons to privilege narratives as a text form when narratives can find their primary expression in other forms of representation. Narratives in the visual, schematic or even mathematical forms of representation may perform by *showing* as much as by *telling*; they are designed to be 'seen' by others in the same community of scientists who know how to 'read' them. For example, Martina Merz (2011) recounts how readers of a scientific paper in a particular field of physics will automatically follow the diagrams that are arranged in a clockwise fashion at the beginning of the paper – these 'show not tell' the research narrative of the salient activities, and readers follow that visual narrative before bothering to read the text of the paper. In some cases, nature's entities show their own narratives directly. Devin Griffiths (Chapter 7) tells how the Darwins set up plants so that their roots traced out their own growth narratives in scientific experiments. Starting from these visual autobiographies, Darwin constructed narratives at three different genre (i.e., generic) levels: 'micro-narratives' of individual plant life, the 'novella' of the life history of plants and the saga of biological evolution.

In sum, I argue two points: first, that narratives (like models, diagrams, equations, graphs, etc.) can be understood as a mode of representing scientific things (ideas, theories, processes, evidential records, relations, etc.); and second, that such narrative forms are quite likely to hybridize or be co-dependent with, or even entirely embedded within, those other media of scientific representation.

1.4   Narrativizing: A Means of Sense-Making

Narratives in science are not given by God, or by some other external authority, but designed and made by scientists in their research communities. Attention has to be given to the ways that they create narratives as a means of sense-making – to the active work of narrative formation in the practices of scientists, especially with respect to their narratives of nature.

There are two points here:

1. I take it that the quintessential function of narratives of nature in science lies in making, or unravelling and remaking, connections between things. The world presents many puzzles and scientists seeking to understand their phenomena in their 'narratives of nature' have to figure out how that part of their world works, and to give an account of how the bits of it fit together that makes sense. And, like other ways of making other kinds of scientific representation (such as models, schemas, diagrams, tables and category descriptions), narratives have to be developed, tried out, calibrated against other information, reconfigured and re-thought-out to fit the materials that need to be understood. I have in mind something like 'narrativizing' – which has the primary and distinctive aim of 'bringing and binding' together the heterogeneous elements associated with the phenomena in a field.

2. Following this, I ask: what relational 'grids' and scaffolds do scientists use to build their narratives? This is *not* a question about the structure of the final narrative (whether it has to have a beginning, middle and end with a change of state (see Carrier, Mertens and Reinhardt 2021), nor whether it is primarily 'tellable' in terms of sufficiently interesting events (see Ryan 1986), nor on its 'affect' and how it facilities multiple connections (see Jajdelska, Chapter 18). Rather – this is a question about the basic dimensions of relations that scientists use in building or creating or supporting their narratives; it is about the lines of relationship on which narrativizing goes on.

First: what happens in narrativizing? The basic role of narrative and its special function for scientists is to put diverse materials into relation with each other through time, or across space, or through other conceptual dimensions (such as classes in society, or elements in an ecology) in order to form a coherent account of a phenomenon. Narrativizing is a way for scientists to organize their bits of scientific knowledge to create sense out of their relations. Narrativizing serves to join things up, glue them together, express them in conjunction, triangulate, splice/integrate them together (and so forth). Yet, the need to clarify relations between things means that narrativizing sometimes means scientists have to sort things out so that their interrelations can be seen more clearly.

One term that captures the challenge that scientists face when they make narratives – explicitly or implicitly – to help them order and relate the separate elements of their scientific knowledge into coherent accounts is *configuring*. This term comes from Mink (1970 and 1978), writing about the philosophy of history, but he remained opaque about the processes involved. Two other terms, discussed in Morgan (2017), offer recipes with more content for science narrativizing. *Colligating* comes from William Whewell, who used it to refer to the process of fitting together, under an idea, items primarily from the empirical domains of science (see Cristalli 2019; Kuukkanen 2015; and Swedberg 2018). *Juxtaposing* was the other term I used then – to refer to the activity of pulling together separate elements known about a phenomenon, but that did not initially make sense together. Narrativizing, constructing a narrative, was a way to make sense of them and resolve initial puzzlement. (This followed a lead from Paul Roth (1989), again for philosophy of history, rather than for science.) Both recipes offer the possibility for creating wider narrative-based understanding or even explanations (as we will see in section 1.5). I want to press the use of Whewell's terminology of colligation for two reasons. First as a process (in verb form), colligating involves *bringing* elements together and *binding them together* just as narrative-making does; the outcome (its noun form) is equally appealing, for a narrative can be understood as a colligation. (A little care is needed here: while narrative-making in science can be understood as a process of colligating (the verb), not all colligations (nouns) necessarily come from narrative-making; for example, the elements brought together could be similar things, bound together in creating a category.)[10] Second, with these two insights of bringing and binding, it is easy to see how the process of colligation can cover the many varied ways in which scientists use narrative to bring together all sorts of different kinds of elements: empirical elements, theoretical arguments and speculative claims. These practices of colligation vary from site to site, and from science to science. Thus configuring elements into a narrative that explores a time-based, path-dependent system in nineteenth-century biology mobilizes a different mode of ordering and relating, both from the juxtaposing narratives of mid-twentieth century case studies in sociology and from the 'how possibly' puzzle narratives of modern mathematical and computer-based simulation.

These examples take us to the second point: how does this narrativizing go on? We find two main sense-making strategies, two main relational 'grids' for colligating: one based on taking a possible network of relationships as the main device for ordering materials, the other by ordering elements along space or time lines. 'Grids' are not to be understood as rigid measuring rods, but rather as a shorthand way to express the main domain upon which the process of colligation – the ordering and relating, the bringing and binding together of elements – happen. These different kinds of grid are not straightforward in use, and often they are used conjointly, because the materials that need to be knitted together in science narratives are not going to be simple connections between elements as if lined up along an individual piece of string, whether that string is a time or space string or a causal path string.

---

[10] See particularly Wise (2021), who outlines the importance of colligating for understanding the role of narrative in science; and chapter 6 of Kuukkanen (2015), which discusses colligating in the context of history, and explicitly refers to categorizing in that context.

A 'cat's cradle' offers an analogy for narrative-making for the first, causal/associational, version (a term offered in Anne Teather's chapter). It is a net made from one joined-up piece of string that can be fashioned into several different network patterns. Each network pattern will be different, for it uses the same elements of the string arranged in different ways. Each network pattern can be understood as depicting a set of relationships; the nodes and spokes of the elements may denote an ambiguous relationship, or be causal in a mechanistic kind of way, or they may indicate a much looser association. Indeed, the benefit of colligating or narrativizing on a network grid is that the resulting narrative can be opaque about the exact nature of those relations; it can allow knowledge to be uncertain; it can allow for multiple perspectives; it can enable complexity to be maintained; and it can embrace context where the cut between content and context is unclear.

Time-line and spatial relations *seem* to offer simpler grids for narrativizing. But, in practice, scientists don't rest content with creating narratives just by moving along a chronological time-line or arranging items across a spatial grid. Their use of time is not straightforward: they might use relative or absolute time; will cut time up into different units; work backwards and forwards over time, etc. And while time-based accounts in sciences may find narrative necessary, it is important to remember that time-based relations are neither a necessity for narrative, nor sufficient in themselves.[11] Of course, things happen in time, and across space, but these may not be the domains in which relationships matter. And even where either time or space may be understood as the dominant dimension for observing change (as in fields such as geology, palaeontology, evolutionary biology and parts of anthropology, sociology and social science history), there is rarely any simple time or space sequencing of events. And often, these two major kinds of grids – relational and spatio-temporal – will mix together in the narrative and will interact. Sometimes, the time–space line enables the scientist to infer subject-matter connections, at other times the subject-matter connections enable the scientist to infer the time or space relations.

It perhaps helps to draw some comparisons. The main feature of the narrative form – that it fits elements together and reveals the connections between them – contrasts with other forms and modes of making and expressing scientific knowledge. The comparison here is particularly with those activities that list

---

[11] Here is where the narrative science experience moves apart from the narratological assumptions, in which time relationships are *usually* taken as essential. See Morgan (2017) for the argument that time ordering is a subset of narrative ordering; and see Hajek's chapter for consideration of this temporal assumption (Chapter 2). Many of these issues were discussed in our project workshop on 'Temporality in Scientific Narratives'. See workshop on temporality on project website: www.narrative-science.org/events-narrative-science-project-workshops.html.

or rank elements of knowledge, those focused on activities of separating out similarities and differences between things, and the consequent listing, labelling and describing of such taxa and types. A list of fossil remains, or the table of chemical elements, or the species of natural history, organize or 'order' our knowledge according to weights, or categories.[12] They produce nuggets of knowledge, orderings of knowledge, families of like and dislike things, and whole classification systems.

So, on the one hand, the configuring and colligating of narrative-making sit in contrast to the *making of tables and categories*: the former stitching together relations between things, the latter separating out different kinds of things according to their particular characteristics. On the other hand, those alternative forms of ordering and expressing scientific knowledge are also, like narrative, more than description, for they too are a means of organizing and representing our knowledge. Both narrative-making and category-making develop our scientific knowledge and facilitate our expression of our knowledge about the world rather than being primarily technologies of intervention in the world. And, as usual, there is a caveat: the sense-making quality of narratives (the attempt to find narratives of nature) can also work *co-dependently* with those other contrasting activities and forms in science (category-making, case-making, statistical thinking and so forth). Narrative ordering and relating do not always substitute, or replace, but may complement other modes of developing knowledge in science. Narrative-making is one potential element, often an essential one, in a multifaceted network of practices that enables scientists to develop ideas and accounts of their domains.

1.4.1   Joining Things Up

The two main relational grids of narrative-making in science, as suggested above, are the spatial- or time-line and the relationship net. Narratives are widely accepted to provide the kind of glue that helps us to 'follow' a set of events through time, or across space. Free-standing, time- or space-sequenced narrative representations are found most readily in the historical sciences – natural and human/social – for these deal in matters of time and space *and* where such dimensions of ordering really matter.

Where time and/or space does matter, scales, measures of time and space, and ways of dating and locating events, may be critical to the kind of narrative made. John Huss (Chapter 3) analyses the competing narratives of the set of major mass extinctions in the natural history domain. The mass extinctions in

---

[12] It is significant here that at one of the first narrative science workshops, Lorraine Daston presented an account of lists for our consideration (the work of Jack Goody) as the comparator for narrative.

species evident in the fossil records were recorded in graphs which then had to be 'explained' by the palaeontologists – either by a narrative of a periodic event (that might possibly have an unknown astronomical cause) that repeats every 26 million years, or by individual causal narrative accounts for each individual episode. Either way, periodic or individual causes, there was a desire for 'narrative closure', the satisfaction of closing the evidential/explanatory gap between the time charts of those visual artefactual fragments of fossils and understanding the causes of the timing of these enormous events. Narrative-making here required getting a satisfactory, plausible and convincing – i.e., narratively closed – alignment of evidential remains with major events, whatever the ultimate explanation might be.

In sites such as evolutionary biology, archaeology and geology, evidential requirements from both time and space typically create narrative density and narrative complexity, as found, for example, in Anne Teather's account of narrative changes in recent archaeology. Previously, the recognition of familial relations between artefacts and their spatial distribution were used to determine the *relative time datings* of cultures, transitions and migrations, and so determine the *genealogical* periods of prehistory (the bronze age, the Neolithic period, etc.). Now, the more recent methods of dating the *absolute age* of archaeological remains (by technologies of tree ring and radio-carbon dating) determine time relations, namely the *chronologies* of those civilizations, and so have changed the nature of explanations in that field.

Narrativizing (or narrative-making) in science often relies on a kind of 'tellability' that stems, as in Ryan's analysis of 'embedded narratives' in literature, from the intersections and inter-relations of characters that prompt the events or actions that happen. That is, time and space may not be the dominant dimensions needed to follow the sequence or set of events; other relationship factors may be much more important. For example, Morgan (2017) gives an account of the narrative-making habits of social anthropologists working in American cities, where the relationships between a street gang with the police, with the political machine, and with rival gangs are all drawn through the use of narrative accounts. In such contexts where existing social/class relationships are primary, time or space as a grid has almost no value. Narrative-making does especially well in enabling accounts where causal claims contain contingency, doubt and choices, as in John Beatty's discussion of the tellability requirements for evolutionary change where the order of events is not well evidenced in a time sequence, even though they must have happened in time (2017 and Chapter 20 in this volume).

The intersection of time–space and relationship grids becomes clearer in the notion of 'process tracing', an activity discussed by Sharon Crasnow (Chapter 11) and found in many scientific fields, which involves tracing the evidence of certain relationships (in processes and between events) through time and space and other dimensions, and putting them together into a narrative account. A closely detailed narrative following political changes may be the only way to open out a full understanding of a political science phenomenon which had previously only been accounted for in a spare theorizing or model format, as she argued in studying how political scientists unravelled cases of interactions between democracies to substantiate their 'democratic peace hypothesis' (see Crasnow, 2017). We can see in such process tracing how narrative-making actually depends on both kinds of grid relationality – time–space relations and causal relations. It seems in her cases that it is the causal links between events which enable the process to be traced through the time–space events, rather than the other way around. By contrast, in Huss's mass-extinction events, the time domain is the predominant medium for tracing causes.

Relational grids sometimes function more like bridges in joining up other dimensions. For example, the genetic history in Kranke's chapter involves following materials that bridge different levels of both time and space in the processes of evolution (Chapter 10). A narrative bridge might provide the link that joins over other gaps, such as it did between different accounts of evolution by R. A. Fisher and Sewall Wright, accounts which nevertheless shared the same mathematical formulation (Rosales 2017). A bridge could offer a methodological joining up, where the research life of the scientist and their narrative of nature intersect in a joint account by the scientist, as Griffiths's chapter shows for the Darwin family's investigations into plant growth (Chapter 7). A narrative bridge could be a vehicle for familiarizing the community with the research done, by overcoming the mismatch between actual research events and the given record of events (as in Meunier's account). Or it could be the way that scientists place their own particular bit of research into a longer or wider research trajectory through 'narrative positioning' (see Berry's 2019 working paper, published in 2021).

1.4.2   Sorting Things Out to Join Them Up

It is one of the paradoxes of narrativizing that it sometimes only succeeds in joining things up by first sorting things out, perhaps in order to join them up in a different order or set of relations than they first appear. The world presents phenomena in puzzling and myriad forms. For example, the massive data sets that come from modern earthquakes have to be sorted out and re-aligned before they can be joined up into any narrative. As Teru Miyake's account (Chapter 5) of the Tohoku earthquake of 2011 tells us, each measuring instrument at each geographical point produces a data series that tells its own individual story, scaled second by second. These need to be colligated: they need to be sorted out, juxtaposed, aligned and somehow melded back together to produce an

integrated full narrative of that quake. Miyake shows how the visual representations of such earthquake data enable the scientists to sort and depict the
complexity of an earthquake in narratives that require one to follow time
evidence at different geographical points. Narrative-making does the work of
both filtering and unifying these multiple records of nature. This is a time–
space-rich narrative, in which absolute time matters absolutely, but its narrative
focus may be less evident than in Huss's mass-extinction case because it is so
strictly controlled by the technical scientific language of the field. At the same
time, in both Miyake's account of earthquake science, and Andrew Hopkins's
of geology, their analyses show us how narrative sense-making works under-
ground, within and through professional accounts.

Such categorizing, sorting out and putting back together could involve a set
of more heterogeneous observations, coming in different forms from different
observers in different places, contributing diverse information in the empirical
domain. Here the technology of colligating is more like jigsaw-making – where
grids of time or space or cause are each separately insufficient, as we see in
Lukas Engelmann's 'plague narratives' from the late nineteenth-century
(Chapter 14). As in all pandemic diseases, there are many elements that matter,
and have to be sorted out for each local account of the causes of the spread of
the disease. Here, space and time may be at least as important as the multitude
of possible causes that might be 'traced' and blamed for such disease transmission. This points us to how narrative-making proves a useful way to deal with
complex phenomena that don't divide well, don't separate well and don't
simplify or abstract easily but that have multiple elements and agencies. Just
as narrative is good for following the connection of events through time, across
space and through causal relations, narratives are good for taking all the
elements into account without trying to separate them out on the grounds that
they don't exist as separate independent elements – and the scientists' problem
is to understand their interactions. That this entanglement problem can be
'solved' through narrative-making is well shown by examples in anthropology
(e.g., Geertz, and du Bois, in *Anthology II*). Narrative-making is even at home
where there are conflicting accounts of a phenomenon, which are resolved by
understanding how these conflicts are inherent in the phenomena rather than in
the scientists' understanding, as in Hajek's examples of multiple personality
and memory confusion.[13]

1.4.3   Narrative Levels

There is one other important dimension in narrativizing – almost perhaps the
first decision for the scientist: what is the focus of the narrative gaze; at what

[13] See in *Anthology I* and Hajek (2020) and also Morgan's 'juxtaposing' account (2017).

level of interest is the scientific phenomenon; and where is the narrative perspective? The relevant 'level'[14] may range from narratives about small atoms to the whole universe, from the single individual's preferences to the market economy, from the smallest ant to the planet's ecology. Scientists' narrative-making is a reflection of these interests and decisions. We remarked earlier how Darwin constructed micro-, meso- and macro- narratives of plant life. Teather's account shows how narrative-making works on two different levels in archaeology – at the broad epoch level of the bronze age or iron age, and at the small local level of the shape of flints to create fires, and in between in the styles of causeways. So we can think of narrative-making in that field as a process of erecting scaffolds on the basis of time-datings, relative or absolute, and then using these to understand both big cultural shifts and, equally important, really specific cultural habits.

Narrative-making can operate under a kind of umbrella for understanding a general approach within a science, or even across sciences: thus narratives of complexity theory, of catastrophe theory, and so forth. Mathematicians (as we have seen) sometimes like to frame their fields in broad and deep terms – a grand narrative of 'everything' that should fit under an approach or new form of theorizing (Dick, Chapter 15).

At the other end of the scale are 'nutshell narratives'. Some of these are 'anecdotes' that capture telling examples in very particular short narratives that point to something atypical, extraordinary, unusual or exemplary.[15] They are based on individualized observations and circulate just because they pick up things that don't seem to fit together. Such juxtapositions are critical, for it is this detection of oddity that sets up the 'epistemic switch' that makes the scientist think anew about something. In one of Hurwitz's cases (Chapter 17), it is the sudden recognition that a baby being observed is 'well' which surprises the medic. In another switch, it is from 'seeing' something as just a technical fault in a lung X-ray to the removal of a bike-spoke left from a long-ago accident to the patient! In Meunier and Böhert (*Anthology II*), it is the anecdotes of dogs learning to exchange small coins for buns at the baker's door that creates new reflections about the natures of animals vs humans. Hurwitz's epistemic switches are also ontological ones.

Anecdotes come from surprising observations, but other 'small stories' come from the scientist's imagination to prompt theory-making. Stephan Hartmann (1999) tells how a small imaginative story used to launch the 'MIT Bag model' lay behind certain theoretical developments in hadron physics. Marcel Boumans (1999) tells how the little story of a child hitting a rocking horse at

---

[14] This term narrative 'level' is not to imply the same usage as in literary studies.
[15] See workshop on Anecdotes on project website: www.narrative-science.org/events-narrative-science-project-workshops.html.

random intervals with random force motivated a new model of the business cycle in the 1930s. Both of these come from metaphors that were then extended and explored through narrative – a feature discussed by Gillian Beer (1983) in a literary examination of 'Darwin's plots'. In these two cases of narrative prompts to scientific model-building, the metaphor-narratives in natural language are extended into theories expressed in formal language (again, see Wise, Chapter 22).

1.5   Narrative Reasoning and Knowledge-Making

So far, we have examined the ways in which narrative provides a means, an enabling technology, for scientists to make sense of their investigations, rather than being a means of those investigations. Yet, we have also seen that narrative-making is not a passive part of science, nor an add on at the end of work, but rather (as noted by Meunier) that scientists' research narratives are symbiotic with their narratives of nature. Our cases in this volume suggest a more ambitious claim, namely that such narrative-making and -using activate scientific understanding and explanation. Narratives appear in chains and forms of reasoning associated with direct knowledge claims, which can best be expressed in terms of 'narrative argument', 'narrative explanation' and 'narrative inference'. Once again, we see that these narrative usages do not provide a competitive path to other modes of scientific reasoning and knowledge claims, but a complementary one.

1.5.1   Narrative Inference

Narrative-making and -using act as go-betweens in inferential domains – offering the means to join together, or mediate between, theories/laws and speculations on the one hand, and data, facts and specific empirical elements on the other. Drawing inferences implies a thesis of some kind that the evidence is asked to speak to; it involves making the connection from evidence to thesis. But this is rarely (perhaps never) entirely rule-bound in any science. Rather, drawing inferences involves some leaps of commitment because the evidence rarely speaks clearly, or uniformly, or exactly, and often has gaps in the chain. Constructing plausible narratives here can play a bridging role to help scientists draw and express such inferences, sometimes preliminary ones that prompt the next step, or search for further evidence.[16]

Most often narrative comes into play in inference where the evidence is heterogeneous; and where qualitative or quantitative observations need to be

---

[16] Morgan (2021) explores the notion of narrative inference further in the context of economists' attempts to pin down the behaviour of economic cycles.

joined up. Elizabeth Haines (Chapter 9) argues that narrative-formation offers a critical resource for picking out bits of heterogeneous evidence, fitting them together and drawing inferences from them, and constitutes a 'reticulate practice'. As an example, she discusses how a scientist might go about picking out particulars from a crowded field of vision – for instance, in contexts such as the photographic evidence of terrains in order to figure out what is salient and what not in a problem of intelligence-gathering. Debjani Bhattacharyya (Chapter 8) gives an account of two sites of narrative inference. One, offering a similar kind of reticulate practice, is the legal site where the various records of shipwrecks during cyclones in the Indian Ocean – as told by captains and pilots, in ships' logs and weather reports – are spliced into narratives that draw such evidence together to determine 'the main cause', and so apportion blame. Inferences depend here on the consideration of several different narrative accounts, each of which may point to a different cause.[17] On the other side, narrative works to aggregate cases: she tells how taking the evidence from many such storms created the meteorological science of cyclones. This new scientific understanding of the behaviour of cyclonic winds and storms was then used to create 'storm cards' which contained little 'recipe narratives' telling how ships' captains should steer their ships when they found themselves in such a cyclone.

Such inferential judgements and arguments may look informal and squashy, and of course such narratives may only be partially informative. But all inference has an element of informal connections to be made. Even statistical inference, which may be strongly supported by statistical rules and criteria, needs subject-matter analysis in order to make sensible claims to answer scientists' questions. We see this in Lukas Engelmann's plague narratives. Different narrative accounts of past and current episodes of plague and its treatment based on varied sources of observation were essential to make sensible inferences from facts on the ground. Equally for the scientists seeking inferences about the causes and pattern of mass extinctions. The point here is that knowing about the statistical characteristics of plague does not give automatic entry into knowing about the statistical characteristics of fossil records – the subject matter is so different that simple rules of statistical inference have limited grip; subject matter knowledge, sometimes in narrative form, is needed to draw informative inferences.

Narrative inference may be said to have its own set of criteria for inference. Following the legal literature,[18] one might reasonably argue that the requirements for narrative inference lie in consistency (taking account of all the individual bits of evidence) and coherence (fitting all the elements together in

[17] This was one of a set of examples discussed at our Project Workshop on polyphonic narratives. See workshop on polyphony on project website: www.narrative-science.org/events-narrative-science-project-workshops.html.
[18] See MacCormick (2005) for these criteria, discussed in Morgan (2017).

a way that makes sense in the context). For legal cases, there is an additional requirement of 'agency' (e.g., of those committing a crime), which might be translated for narrative science in terms of an adequacy/ plausibility and perhaps an implicit or explicit agency in the relational claims used in the inference. Surely the most formal inferential rules for ordering and categorizing, and so transforming heterogeneous evidence into a consistent and plausible narrative, is proposed in legal scholarship: namely the Wigmore Chart method, which is designed to take into account the conjunctions among the individually separate pieces of evidence that need to be combined into legal narrative accounts.[19] It is not clear that lawyers follow such strict methods of evidence colligation, but for scientists, it is clear that the use of narratives in a scientific field comes with its own generic criteria for assessing plausibility. Andrew Hopkins recounts how geologists attempting to account for a particular rock formation in Scotland inferred, on the basis of deposited material, that the cause must have been volcanic and told a story of geological formation based on that cause. Some years later, finding a different kind of deposit, the narrative changed to blame the fall of a meteorite. In neither case was there obvious evidence of that particular cause in the presence of a volcanic vent or meteorite crater! In both changes of inference – one might argue – some crucial evidence of the 'agency', or cause, was missing when these specific event narrative accounts were constructed against an ongoing background narrative in geology of more gradual causes of erosion and deposition.

1.5.2   *Narrative Argument*

Narrative argument features in our volume where narratives are involved in making arguments about causes, and about sequences, and about causal sequences – for in practical terms, single causes are hard to come by in science. The philosophical literature arguing about causes is long-standing and wide-ranging. Narrative does not solve those arguments in any principled way. Once again it helps to return to the purpose of narrative – relational sense-making. If, long ago, the adult fish species was upright and then became flat, what evolutionary causal sequence could possibly account for this change (Beatty's paper, Chapter 20)? (And if that fish species still now begins juvenile life in upright form, and becomes flat only in adulthood, how does this work?) Simple adaptionist stories of efficiency or optimality don't work very sensibly – the argument does not grip. 'Back-stories' are needed to make sense of the adaptations, and of their order, but such arguments may still not be definitive, and it is an open question how far the narrative sequence needs to go back in

---

[19] See discussions of Wigmore Charts and their usage in Twining (2006) and Nicolson (2019).

order to make an explanation that counts as satisfactory with no questions left over.

Strangely, and despite assumptions among some narrative scholars that time is integral to narrative understanding, a given temporal sequence may be consistent with very different sets of adaptations in evolution, or very different causal relations, because, as Jajdelska (Chapter 18) makes clear, narratives have their own power to invest perceptions of causality. The aesthetic details of a narrative matter to our perception and acceptance of such causal claims as being plausible, such that the narrative must be 'performative' in this kind of sense. Jajdelska's argument is paralleled in legal analyses of narrative accounts. This is obvious in court rooms, where the performative aspect of narrative is associated with a degree of rhetoric, but much more interesting for science is that the *order* in which elements of evidence are introduced into the legal narrative affects the degree of acceptance of the narrative conclusion, just as, probably, happened in those colonial courtroom narratives of shipwrecks during cyclones.

The textual details of narratives are not just performative, but, like the diagrams and schema discussed earlier, they also embed important signals of community expertise. Line Andersen (Chapter 19) analyses how mathematicians read mathematical proofs in terms of 'scripts', a literary term denoting slim chunks of text that provide shorthand access to a set/ sequence of taken-for-granted background elements for the reader of fictional or everyday factual narratives. For mathematicians, such a script can point to a set of mathematical elements that would be habitual at that point in a proof (a set of proof steps in the background, very different from the kind of 'back-story' argument Beatty tells about going back in time). They can best be construed as the denser argument behind the shorthand maths, or the thickness of activity behind the 'chemese' found in Paskins's paper (Chapter 13). As Andersen argues, mathematicians reading a proof expect to see standard habitual moves shorthanded into these 'scripts'; they are accepted by the expert community without expanding them. But gaps in a series of such scripts, or unusual linking moves between them, alert the community to some strange move in the proof argument – a narrative gap to which they must pay attention.

1.5.3 Narrative Explanation

Traditional arguments from philosophers of history portray narrative as offering an explanation for a particular set of events. By contrast, almost in direct opposition, the traditional philosophy of science position was to understand scientific explanation as both general and valid only if it were 'covered' by 'laws' as a kind of umbrella. We can see the contrast between these two notions of explanation most vividly in Huss's account of mass extinctions. The periodic narrative 'explained' (according to philosophy) mass extinctions as a regular pattern driven by law-based behaviour elsewhere in the system and was contrasted to the 'historical narrative explanation' given for each particular historical case of mass extinction in terms of the reasons why each one happened.

More recently, philosophers of science have settled on a looser or more generous account that portrays explanations as answers to 'why?' (and perhaps 'how?') questions, but still with a presumption that scientific explanation involves a high degree of generality in its scope (although the strict 'law-based' account is now regarded as old-fashioned).[20] If we concentrate on how scientists do explain things in narrative forms, we can recognize elements of all these recipes for explanation, sometimes used at the same time.

As we have already seen, scientific narratives often embed causes for things to have happened, that is, they answer 'why' questions – so, on that definition, they are readily set up to provide explanatory accounts. Especially this applies to narratives using relational networks, for, as Olmos (Chapter 21) points out, narratives that make sense of relations (causal, associational, etc.) will double as reason-givers in persuading the reader/listener of the knowledge claims embedded in the narrative. This may account for why narrative modes of 'reason-giving explanation' work more easily than general law-based accounts in some sciences. But Olmos goes further in claiming that 'law-dependent explanations' using time relations invoke narrative as soon as they are examined and unpacked in a way that shows how those 'laws' account for real particular events. Thus, taken as an argument form, such narratives of particular events embed law-type explanations.

Olmos's analysis offers us a framework for understanding narrative explanation more broadly, for we can recognize that there are a number of ways in which narrative accounts in the sciences answer 'why' questions while making use of 'generic' claims (claims relevant for a class of phenomena) without a full-blown appeal to 'laws' (this is particularly so in mechanism-type explanations). Following Olmos's point, we can find this conjunction happening first in the considerable gap between giving more general explanations and finding the particular ones that might be needed for any specific scientific problem, and to recognize how this gap may be filled by the narrative form. Why is this so common? The 'laws' of science are in many cases 'straw men' – they are supposed to provide umbrella explanations but often do not organize scientific materials very well – they lie at too general a level to connect immediately or practically with many of the scientific problems studied. For example,

---

[20] This new account is restrictive in another way, for it is most often understood to involve offering answers that require causes, and especially a specific causal 'mechanism' (a high requirement for many scientific contexts).

scientists from several disciplines with different perspectives have general knowledge about pandemics, but for answering questions about any particular disease-class pandemic, they need to fit together knowledge about the genetic form of a virus, its transmission and medical treatment, and the social behavioural responses that might be relevant to control or eradicate it. As Engelmann's paper shows, despite widespread generic-level knowledge of the plague, all explanations in his late nineteenth-century cases had to be made local, so each area narrative was relevant to particular causes, transmission, controls and effects of the plague in that context (Chapter 14).

Another conjunction of the two bases for explanation – question-answering with an appeal to the generic level in some form – also works in reverse. It starts with narrative explanations of particulars, but then generates accounts that have more general claims. Thus, in Bhattacharyya's paper (Chapter 8), we see how, studying and aggregating the narrative accounts of many examples of cyclones, her 'hero' Piddington was able to infer the stable characteristics of the behaviour of cyclonic winds, and so set out how ships' captains should behave in such storms. An alternative mode of extending particular narratives to the generic level was explored in Morgan's (2017) account of how puzzles thrown up by the juxtaposition of evidence were resolved to answer an important 'why' question. The particular case evidence showed firms exited a failing industry in the 'wrong' order, according to the theory. The narrative account answered that puzzling 'why' question with a narrative of reasons that could be (and was) extended by the community of scientists to 'explain' a set of similar cases in similar circumstances.[21]

More often, the narratives of scientific particulars don't pretend to offer generality, or extend to a more general level, but they rarely work without some generic element (including, at the limit, the use of, or appeal to, general scientific 'laws'). My earlier account of narrative explanation (Morgan 2017) showed how narratives use conceptual elements from a science to bring together a set of examples under one conceptual roof. Cristalli (2019) has urged that such colligation is the basis for a wider notion of narrative explanation, one consistent with both philosophy of science and philosophy of history. The engineering narratives of particular accident reports rely on general claims and knowledge of the behaviour of materials and people,[22] just as the legal narratives of particular cases are set within a framework that uses the general concepts and claims of the legal system. The narratives of ant-lions catching

---

[21] The critical point for narrative science is that the explanation was exported beyond the original case to other cases; for philosophy of history, that puzzle-solving explanation works for the particular case (Roth 1989), but there remained an open problem of how that argument could be extended to science (on which, see Morgan 2017, and references therein).

[22] See the reports of the NS Workshop on 'expert narratives'. See project website: www.narrative-science.org/events-narrative-science-project-workshops.html.

their prey (Terrall 2017) can be understood as particular instantiations of a more generic predator–prey account. The narratives of marsupial evolution told by Kranke provide particular accounts of general versions of genetic evolution. Félida's narrative of multiple personality is a one-off case, but can be used for broader understandings of such cases (Hajek 2020). These all rely on some kinds of generic or conceptual framing in the narrative accounts. Specific causes can also fit easily and well with laws in a narrative account as they do in geology, where the laws might be said to lurk or police, rather than be specifically determinate (Hopkins, *Anthology I*). The appeal to a general or generic level, or the use of the conceptual level, is found somewhere in most scientific narratives – in fact, it is difficult to conceive of a narrative in science that does not do so.[23] This characteristic of narrative explanation (like the other answers given here for how narrative arguments extend their reach), does not start from an appeal to something general in the nature of historical or philosophical explanation, but looks to the practices of how scientists do reason with narrative to make their knowledge stick together with theoretical and conceptual materials and so speak beyond particulars to (some) more general kinds of knowledge.

1.6   Conclusion

Narratives are made by people, perhaps mainly for enjoyment, perhaps for enlightenment, but also in the sciences for far more utilitarian purposes; thus my labelling them an enabling, sense-making, technology for scientists. I propose we go further than this, and think of narrative as a 'general-purpose technology' (GPT). This term comes from economists and economic historians who have focused on the *use* of a technology and its histories (rather than its invention or how it is reproduced), and on two particular attributes of such technologies, both of interest for this account of narrative as a general-purpose technology for science.[24] First, GPTs are, as the term suggests, technologies with usage that is both generic in its main purpose, but gradually expands across a range of unexpected sites, fulfilling that main function in different ways, and becoming co-dependent with other technologies in the process. Steam power, electricity and computing all offer supreme examples of such GPTs: each has a general-purpose use but is harnessed in different ways for

---

[23] As before, narratives, as a form of representation in science, may well share this characteristic with other more abstract forms of scientific representation, such as models and schemas, which also embed more general conceptual claims. Even tables and graphs have concept-based labels and headings that point to generic content.

[24] The notion of general-purpose technologies was labelled in a working paper of 1992, published in 1995, by Bresnahan and Trajtenberg, following work, especially of Rosenberg (1982), on the historical development of technological interdependencies.

different specific purposes, just as narrative is harnessed across the sciences. Second, and less obviously, those scholars have noted the important role of users, and user-innovation, in the spread and development of those technologies into those multiple sites, and charted how those innovations created changes in economic and social life. For scientific life, our chapters have analysed the multiple and varied usages of narrative and shown how its general purpose of sense-making can be traced into narrative representation, reasoning and knowledge claims.

Of course, narrative is not a new technology, nor invented by scientists just for their use! It would be equally unhelpful to argue that narrative was introduced into science in a particular era, and that it became a revolutionizing GPT as it spread through the sciences in the way that steam power, electricity or the computer did for our everyday lives. Not at all. I am not claiming, nor did our project suppose, that narrative was introduced into science in the way that historians of science have argued 'the experimental method' was. Arguably, we could treat that 'method' as another possible GPT: historians have tracked how it came into the sciences in the early modern period, taking over the means of investigation and mantle of experiential knowledge, and gradually morphed into the method of controlled laboratory experimentation on the one hand, and field experiments on the other. It has now appeared under many guises (in computer simulations, in medical randomized trials, in thought experiments with models, etc.) and has grown into conjunction with other modes of investigation such as statistical methods and modelling – two other modes of doing science we might also label GPTs – to create the kinds of hybrid modes that characterize modern scientific practices.

In contrast to the laboratory experiment, narrative was surely always in science, and narrative-making and -using in science is a human activity, and so could easily become a social habit in new environments. Thus it surely has a history. Neither this chapter (nor our book) offers any serious history of the place of narrative in the sciences, although we can see a number of points salient for investigating that history of the changing roles, sites and manifestations of narrative.[25] And we can suggest the kinds of materials that would be involved. For example, Dear (1991) points to the use of narratives of individual experience in reports of actual and thought experiments in the English tradition of the seventeenth century. Holmes (1991), in response, compares that tradition with French scholars' narrative modes, which aggregated several experiments at once and which melded their experimental accounts with arguments about the nature of the materials. This is the kind

---

[25] This book sadly does not offer any kind of meta-history of narrative in science equivalent to the history of 'ways of knowing and working' (Pickstone 2011), nor the earlier multi-volumed account by Crombie (1994).

of historical point when, for a particular site of science, narrative turns from being an epistemic genre (using Pomata's terminology) into a technology that goes beyond simple reporting into something like narrative inference (while still relying on particular modes of intervention and reasoning, to use Hacking's and Crombie's ideas). Another hinge point in this historical account of narrative in science might be the one noted by Terrall (2017) in the eighteenth century, when natural historians turned from pictures, narrative texts, or one plus the other, to keying the text to the pictures. Perhaps this is one of the moments when narrative became diagrammatic? These brief remarks suggest that whereas narratives in science may have been found largely as textual free-standing accounts in earlier centuries, the production and usage of narratives appear to have become increasingly intertwined with, and adopted into, other modes of doing science and making scientific knowledge over the last two centuries. In doing so, narratives and narrative-making may have changed in form, but perhaps not changed in their fundamental knowledge-making functions.

Those three GPTs of economic life – steam, electricity and computing – are called so not just because of their flexibility in use, but because they have infiltrated the ways we humans do things in ways which add power to, and expand, our human resources. Narrative, and narrative-making, have expanded or enlivened our human abilities and intelligence as scientists – just as different modes of doing science and different epistemic genres of scientific representation have done. For narrative – as we have suggested – the general purpose that makes narrative as a technology so useful to scientists in doing science lies in narrative's sense-making possibilities: the power of narrative is to colligate elements together under conceptual frames to make sense of the phenomena that exist in the world.

References

Anthology I: Case Book of Narrative Science (2019). Edited by M. Paskins and Mary S. Morgan. London: London School of Economics and Political Science. www.narrative-science.org/anthology-of-narrative-science.html.

Anthology II: Case Book of Narrative Science (2022). Edited by A. Hopkins, Mary S. Morgan and M. Paskins. London: London School of Economics and Political Science. www.narrative-science.org/resources-narrative-science-project.html.

Beatty, J. (2017). 'Narrative Possibility and Narrative Explanation'. *Studies in History and Philosophy of Science Part A* 62: 31–41.

Beer, G. (1983). *Darwin's Plots: Evolutionary Narrative in Darwin, George Eliot and Nineteenth-Century Fiction.* Cambridge: Cambridge University Press.

Berry, Dominic J. (2019). 'Narrative Positioning'. Economic History Working Papers, 001. Narrative Science Series, Economic History Department, London School of Economics and Political Science.

(2021). 'Narrative and Epistemic Positioning: The Case of the Dandelion Pilot'. In Z. Pirtle, D. Tomblin and G. Madhaven, eds. *Engineering and Philosophy: Reimagining Technology and Social Progress*. Cham: Springer, 123–139.

Boumans, M. (1999). 'Built-In Justification'. In Mary S. Morgan and Margaret Morrison, eds. *Models as Mediators*. Cambridge: Cambridge University Press, 66–96.

Bresnahan, T. F., and M. Trajtenberg (1995). 'General Purpose Technologies "Engines of Growth?"' *Journal of Econometrics* 65.1: 83–108.

Carrier, M., R. Mertens and C. Reinhardt, eds. (2021). *Narratives and Comparisons: Adversaries or Allies in Understanding Science?* Bielefeld: Bielefeld University Press, 29–61.

Crasnow, S. (2017). 'Process Tracing in Political Science: What's the Story?' *Studies in History and Philosophy of Science Part A* 62: 6–13.

Cristalli, C. (2019). 'Narrative Explanations in Integrated History and Philosophy of Science'. In E. Herring, K. M. Jones, K. S. Kiprijanov and L. M. Sellars, eds. *The Past, Present, and Future of Integrated History and Philosophy of Science*. Abingdon: Routledge.

Crombie, A. C. (1988). 'Designed in the Mind: Western Visions of Science, Nature and Humankind'. *History of Science* 26: 1–12.

(1994). *Styles of Scientific Thinking in the European Tradition: The History of Argument and Explanation Especially in the Mathematical and Biomedical Sciences and Arts*. 3 vols. London: Duckworth.

Dear, P. (1991). 'Narratives, Anecdotes, and Experiments: Turning Experience into Science in the Seventeenth Century'. In P. Dear, ed. *The Literary Structure of Scientific Argument*. Philadelphia: University of Pennsylvania Press, 135–163.

Dillon, S., and C. Craig (2021). *Storylistening: Narrative Evidence and Public Reasoning*. London: Routledge.

Fludernik, M., and M.-L. Ryan, eds. (2020). *Narrative Factuality: A Handbook*. Berlin: De Gruyter.

Hacking, I. (1992). '"Style" for Historians and Philosophers'. *Studies in the History and Philosophy of Science A* 23.1: 1–20.

Hartmann, S. (1999). 'Models and Stories in Hadron Physics'. In Mary S. Morgan and M. Morrison, eds. *Models as Mediators*. Cambridge: Cambridge University Press, 326–346.

Hajek, K. (2020). 'Periodical Amnesia and *Dédoublement* in Case-Reasoning: Writing Psychological Cases in Late 19th-Century France'. *History of the Human Sciences* 33.3–4: 95–110.

Hickman, L. A. (2001). *Philosophical Tools for Technological Culture: Putting Pragmatism to Work*. Bloomington: Indiana University Press.

Holmes, F. L. (1991). 'Argument and Narrative in Scientific Writing'. In P. Dear, ed. *The Literary Structure of Scientific Argument*. Philadelphia: University of Pennsylvania Press, 164–181.

Kindt, T., and M. King, eds. (2022). *Narrative Structure and Narrative Knowing in Science and Medicine*. (forthcoming).

Kuukkanen, J.-M. (2015). *Postnarrativist Philosophy of Historiography*. Basingstoke: Palgrave Macmillan.

MacCormick, N. (2005). *Rhetoric and the Rule of Law: A Theory of Legal Reasoning*. Oxford: Oxford University Press.

Merz, M. (2011). 'Designed for Travel: Communicating Facts through Images'. In P. Howlett and Mary S. Morgan, eds. *How Well Do Facts Travel? The Dissemination of Reliable Knowledge*. Cambridge: Cambridge University Press, 349–375.

Mink, L. (1970). 'History and Fiction as Modes of Comprehension'. *New Literary History* 1.3: 541–558.

(1978). 'Narrative Form as a Cognitive Instrument'. In R. H. Canary and H. Kozicki, eds. *The Writing of History: Literary Form and Historical Understanding*. Madison: University of Wisconsin Press, 129–149.

Morgan, Mary S. (2007). 'The Curious Case of the Prisoner's Dilemma: Model Situation? Exemplary Narrative?' In A. Creager, M. Norton Wise and E. Lunbeck, eds. *Science without Laws: Model Systems, Cases, Exemplary Narratives*. Durham, NC: Duke University Press, 157–185.

(2017). 'Narrative Ordering and Explanation'. *Studies in History and Philosophy of Science Part A* 62: 86–97.

(2021). 'Narrative Inference With and Without Statistics: Making Sense of Economic Cycles with Malthus and Kondratiev'. *History of Political Economy* 53(S1): 113–138.

Morgan, Mary S., and M. Norton Wise (2017). 'Narrative Science and Narrative Knowing: Introduction to Special Issue on Narrative Science'. *Studies in History and Philosophy of Science Part A* 62: 1–5.

Nicolson, D. (2019). *Evidence and Proof in Scotland*. Edinburgh: Edinburgh University Press.

Pomata, G. (2014). 'The Medical Case Narrative: Distant Reading of an Epistemic Genre'. *Literature and Medicine* 32.1: 1–23.

Pickstone, J. V. (2011). 'A Brief Introduction to Ways of Knowing and Ways of Working'. *History of Science* 49: 235–245.

Rosales, A. (2017). 'Theories that Narrate the World: Ronald A. Fisher's Mass Selection and Sewall Wright's Shifting Balance'. *Studies in History and Philosophy of Science Part A* 62: 22–30.

Rosenberg, N. (1982). *Inside the Black Box: Technology and Economics*. Cambridge: Cambridge University Press.

Roth, P. A. (1989). 'How Narratives Explain'. *Social Research* 56.2: 449–478.

Ryan, M.-L. (1986). 'Embedded Narratives and Tellability'. *Style* 20.3: 319–340.

Swedberg, R. (2018). 'Colligation'. In H. Leiulfsrud and P. Sohlberg, eds. *Concepts in Action*. Leiden: Brill, 63–78.

Terrall, M. (2017). 'Narrative and Natural History in the Eighteenth Century'. *Studies in History and Philosophy of Science Part A* 62: 51–64.

Twining, W. (2002). *The Great Juristic Bazaar: Jurists' Texts and Lawyers' Stories*. Aldershot: Dartmouth Publishing.

(2006). *Rethinking Evidence*. 2nd edn. Cambridge: Cambridge University Press.

Wise, M. Norton (2017). 'On the Narrative Form of Simulations'. *Studies in History and Philosophy of Science Part A* 62: 74–85.

(2021). 'Does Narrative Matter? Engendering Belief in Electromagnetic Theory'. In M. Carrier, R. Mertens and C. Reinhardt, eds. *Narratives and Comparisons: Adversaries or Allies in Understanding Science?* Bielefeld: Bielefeld University Press, 29–61.

2   What Is Narrative in Narrative Science?
The Narrative Science Approach

Kim M. Hajek

Abstract

In current English, the term 'narrative' covers a lot of conceptual ground – from an overarching position on some big issue, to all kinds of storytelling, to a general attention to language or metaphor. This chapter argues for narrowing our conception of 'narrative' to add value to scholarship in the history and philosophy of science (HPS). This narrower Narrative Science Approach treats narrative as a distinct and complex discursive form, subject to careful technical theorizing in its own right. By using analytical categories from narrative theory, we can identify in rigorous detail how scientific narratives are put together, what might distinguish them from other narrative forms, and the questions they raise for HPS and narrative enquiry. Similarly, when scientists use narrative ways of reasoning, tools from cognitive narratology enable us to reconstruct their imaginative activity. As a reciprocal movement, our Narrative Science Approach promises to enrich narrative studies.

2.1   Introduction: Narrative and the Narrative Science Approach

What do we mean by 'narrative' in enquiry into narrative science? How does the Narrative Science (NS) Approach relate to other scholarly interest in narrative? In everyday English, we most often encounter 'narrative' used to refer to an over-arching position, or set of positions, on some issue – for example, there are competing 'narratives' of climate change,[1] while marketers for a brand develop its 'narrative' to appeal to particular consumers (see, e.g., Salmon 2008). More basically, 'narrative' serves as a synonym for 'story'. The two gather literature into their associative constellation, such that it could seem straightforward in 2010 for Laura Otis to claim a 'close affinity' between literary studies and work in the

---

[1] This is narrative in its noun form, unlike in French, for instance, where *narratif* exists only as an adjective. Cf. Elisa Vecchione's talk in the NS Public Seminar Series, 9 October 2018 (www.narrative-science.org/events-narrative-science-project-public-seminar-series.html).

history (less so philosophy) of science, due to a 'common focus on narrative' (Otis 2010: 570). With the overlapping 'linguistic' and 'narrative' turns, historians have read scientific documents 'like novels' (Carroy 1991: 22),[2] and sometimes joined literary scholars in tracing patterns of influence, shared elements or dissonances between scientific and fictional texts. These approaches have been enormously fruitful, but they disperse their analytic gaze over a wide and highly varied field of view. On the one hand, most studies in literature and science have tended to concentrate their attention on one or the other kind of text – usually novels, since most work in this domain is undertaken by literature scholars.[3] Much rarer are investigations which take full advantage of the potential for careful, detailed exploration of formal reciprocities and intersections between narrative fiction and scientific writing (Vila 1998 and Griffiths 2016 are two examples).[4] On the other hand, when it comes to scientific texts, 'narrative' stands in too often for what is primarily an attention to language or metaphor, as in Otis's 2010 reflections. When narrative appears in such broad terms, it loses its value as a distinct category of analysis. This chapter aims precisely to recover narrative as a discrete analytical category – of significance in its own right, and also as one mode of writing and thinking to be investigated alongside metaphor, themes, argument, genre, etc. in scientific texts and their literary counterparts. In promoting this 'Narrative Science Approach', I construe narrative in the specific technical terms of narratology.

Narrowing our perspective in this way has value, first, for understanding the histories and philosophies of science (HPS). As Kent Puckett (2016: 8) puts it, 'looking at and naming different aspects of [narrative] gives us the ability *to see* what is weird about almost any narrative'. Narratology (or narrative theory)[5] provides technical concepts and well-determined labels with which to discuss aspects of narrative; this chapter elucidates some fundamental narratological ideas for HPS scholars (my first set of readers) and demonstrates how these concepts help open up a peculiar set of features of scientific activity – ones we call 'narrative'. Scientific texts are my priority, as they are in this volume and the wider NS Project; novels make few appearances in these pages. The formalized, technical framework of narrative theory lets us defamiliarize aspects of standard scientific texts like experimental research articles, but also to study how diagrams or computer-simulation movies function in story-like ways – I encompass

---

[2] Already in 1885, naturalist Jules Claretie – who aimed to contribute to science – had a character in one of his novels read scientific research on hypnotism 'as I would read a novel' (see Hajek 2016b).

[3] Buckland's (2013) *Novel Science* is just one example, from an English literature scholar linked to the NS Project. Jacqueline Carroy's (1991) work stands for a primarily historical approach (www.narrative-science.org/events-narrative-science-project-publicseminar-series.html).

[4] Caroline Levine's (2015) *Forms* calls for comparative investigation of forms across disciplines.

[5] I use the two terms interchangeably, although like in all academic disciplines, individual scholars frame their disciplinary allegiance in different ways.

all of these scientific outputs under the term 'text'.[6] With tools from narratology, we can also point to imaginative processes undergirding certain forms of scientific reasoning. My analysis draws together the narratological work done in this volume and unpacks its workings, with the aim of promoting further use of rigorous narrative theory by scholars in HPS.

Such a NS Approach, secondly, has benefits for narratology more broadly, as well as for interdisciplinary research into literature and science. I thus also address this chapter to scholars in literary and narrative studies. (Indeed, bringing together a dual readership follows readily from my own interdisciplinary interests, and accords with the multidisciplinarity of the NS Project.) My analysis offers these readers an exploration of particular ways that narrative analysis plays out in historically and scientifically detailed enquiry. The contextual and technical expertise of historians and philosophers leads to perhaps surprising insights, which can, in a reciprocal movement, feed back into the work of narrative scholars. Studies of the kind in this volume provide much-needed, fine-grained analyses of non-fiction narratives in their particular historical and disciplinary contexts, for instance.[7] They also open up arenas for productive comparison of scientific and literary texts in strict formal terms. My argument, then, brings narratological endeavours – including the growing field of factual narrative – and HPS studies into dialogue, for the benefit of both areas of scholarship.[8]

What the chapter is not, is a comprehensive introduction to narratological concepts – there exist many handbooks and critical introductions for that purpose.[9] Nor do I survey all the ways narratology could inform HPS scholarship. Rather, following Morgan and Wise (2017), I concentrate on how scientists use narrative when *doing* science – as opposed to when they popularize it, or formulate an argument for a wider audience – and what narratological concepts enable us to see and say about such uses. Analyses from the NS Project serve as my principal examples; indeed, even where narratological concepts do not appear explicitly, they wind through many chapters in this volume, providing more or less implicit support to contributors' arguments. My purpose here is thus twofold. In serving as an introduction to this volume, this chapter sets out how the NS Project thinks about narrative qua narrative. One might say the chapter 'translates' commonalities in contributors' approaches into (some of) the terms of narrative theory. But, like all translations, mine is not neutral; this is *my* analysis of how narratological concepts provide an angle of entry into this

---

[6] This follows standard usage in literary and cultural studies.

[7] Monika Fludernik (2020: 63) calls for such finer examination.

[8] In this, I also respond to Herman's (1998: 383) contention that 'what is needed is a more dialectical approach to the science-narrative nexus'.

[9] Some examples are Herman, Jahn and Ryan (2010); Culler (2011); Puckett (2016); Hühn et al. (2014); Fludernik and Ryan (2020).

collection. At the same time, the chapter stands on its own as a proposal for what narratology and HPS have to offer each other as fields of enquiry, and where that kind of dialogue might lead. My argument both complements and sits as counterpoint to Mary Morgan's introduction (Chapter 1) – quite deliberately; each of us offers ways of looking at this collection of essays and at wider scholarly themes that intersect them. We just do not take quite the same angle of vision. Commentaries by Sharon Crasnow and Norton Wise do similar kinds of work at the mid- and end-points of this volume (Chapters 11 and 22).

Use of narrative spans the sciences – mathematical, natural, human and social – as Morgan outlines in detail in Chapter 1. For the purposes of this introduction, I identify two major classes of narrative knowing, each of which is particularly susceptible to investigation using a particular kind of narratological tool. In the first place, there is the '*mise en mots scientifique*' (after Acquier 2010), or the '*mise en récit*' (putting into narrative): the (re)presentation of scientific activities or findings in textual form, be that written, visual or spoken. Such texts, as material expressions of scientific work, are at once a product of scientific activity (think of a research article) and an index to the active process of narrative-making. Seen as output, the substantive '*mise en récit*' takes nominal (noun) form – as *a* narrative – and overlaps with what Morgan calls 'narrative representations'. Activity, by contrast, is verbal; what I see as the active flip side of the same '*mise en récit*' is Morgan's 'narrativizing'. But, where Morgan treats the two as separate but related functions of narrative in science, I argue that they are thoroughly, even necessarily, interdependent when seen through the lens of narrative theory. Noun and verb, narrative-as-made and narrative-making, are two sides of the same coin. Both lend themselves to analysis through the output form, the text. Concepts from classical narratology serve to unravel this doubled nature of scientific narratives, as well as to pull out ways in which the events/phenomena to be recounted might differ from the *way* they are represented – which plots are told, from whose perspective, whether there are flashbacks. Such questions ultimately relate back to the fundamental distinction in narrative theory between *story* and *discourse*; this distinction, and what it reveals about scientific activity, is the subject of section 2.3 of this chapter.

Before undertaking this work of unpacking, however, it is worth asking where scientific narratives – as output, noun, representation – sit in relation to the kinds of texts usually studied by narrative scholars. This question is the subject of section 2.2. Until recent interest in 'factual narratology' (see Fludernik and Ryan 2020), and even now, narratological categories have predominantly been applied to literary texts, which are readily accepted as being narrative in nature. The NS Approach, by contrast, does not formulate an a priori definition of what counts as a scientific narrative before asking whether we can productively employ narratological tools to unpack (some of) its functions.[10] Rather, contributors to the NS Project have examined both scientific narratives in the uncontroversial sense (like medical anecdotes or psychological case histories), and also (and more frequently) portions or characteristics of texts that might more readily be called 'reports', 'accounts' or just 'articles'.[11] (Indeed, the French term I have been using, '*récit*', encompasses both forms.) The broad features of scientific narratives that I develop in section 2.2, using Ryan's (2007) elements of narrativity, thus emerge a posteriori from the NS Approach.

This definitionally flexible approach becomes especially evident in my second class of narrative knowing in science. Similar to Morgan's notion of 'narrative reasoning' (Chapter 1), I construe this form of knowing as something that a scientist does with a scientific text. Each of us places the emphasis on a different word in the pair, however. Reasoning is privileged by Morgan under her functional approach as something scientists do *with* and *within* narrative representations – a deliberate cognitive process, distinct from imagining or affective reactions. By contrast, I understand 'narrative reasoning' as cognitively broader, involving imagination, affect and reason, in variable combinations. What matters for me is the combined result of these cognitive processes: story-like representations constructed in the mind/imagination of scientist–readers as they undertake some scientific activity (reading mathematical proofs, interpreting diagrams, framing their field).[12] The attention here is on the reader's reception of a scientific document, and how it might share cognitive features with the reading of (literary) narratives, without presuming that the document is itself *a narrative* (representation). Ideas from cognitive, or post-classical, narratology are notably helpful for examining reader responses; I discuss these in section 2.4.

Importantly, this interest in narrative modes of reasoning does not mean the NS Approach makes any broad claims about narrative as a mode of human cognition; even less do we claim epistemic priority for narrative knowing. For all our definitional flexibility, we therefore set aside the perspectives of thinkers like Paul Ricœur (e.g., Ricœur 1980) or Jerome Bruner, for whom narrative fundamentally structures one or more functions of human thought (see Crossley 2010). Asking how narrative modes might enter into human cognition in general is a valuable question; it is just not one that we find particularly helpful in the context of this project. As David Herman presciently remarked in a 1998 commentary, claiming primacy for narrative is to set up an 'idyll of

---

[10] This contrasts with the focus on questions of fact, validity, authenticity etc. often present in analyses of 'factual narratives' (see science-related articles in Fludernik and Ryan 2020).

[11] The range of scientific documents examined under the NS Approach is showcased in *Anthology I* and *Anthology II*.

[12] Although I refer in this chapter to 'readers' (of written texts), many of the arguments developed could also be extended to 'listeners'.

narrative' (1998: 385), which essentially only reverses the epistemic hierarchy present in earlier philosophers' 'myth of science as univocal rationality' (Herman 1998: 384).[13] Either hierarchization precludes fine-grained attention to the contextual nuances of science and narrative studies as historically evolving activities.

It is the evolution and intricacies of scientific activity which concern us in this volume; concomitantly, we do not take account of the historicity of narrative theory as a field of study. Rather, we make flexible use of a range of concepts from narratology and use them to interrogate the doing of science in its active sense: what in science is about narrating, constructing narratives, reading narratives? The narratological tools we employ, the places we find narrative, thus expand and contract with the contingencies of our case studies, and tend to draw from varied perspectives within narratology as a field of enquiry.[14] I reflect in my concluding remarks on what it might mean to look for narrative knowing in a historicized science of narratology.

2.2   Narrativity of Scientific Narratives

When asked, 'what is a narrative?', common usage, like some cognitive-science perspectives (Crossley 2010), holds that humans are innately able to recognize story-like configurations. Morgan, in Chapter 1, circumscribes the domain of scientific narrative along functional lines – what it does for scientists alongside or in place of tables, models, diagrams and so on. For their part, narrative scholars have long striven to develop a precise and logically coherent definition of narrative.[15] But NS contributors rarely begin with these kind of definitions, or even ask explicitly, 'is it a narrative?', about the documents or actions they propose to analyse.[16] Rather, as illustrated in this volume, contributors find it more immediately significant to plunge into examining a given document's (or action's) narrative characteristics and how those function.[17] This notably allows attention to the fragmentary or lumpy ways that narratives can appear in scientific work, which might be overlooked under too stringent an initial categorization.[18] Andrew Hopkins (Chapter 4), for instance, identifies sentence-level narrative chunks in geological research articles. These highly

---

[13] See Olmos (Chapter 21), for a detailed dissection of such philosophical claims.

[14] I thus construe our flexibility as a strength, against Herman's (1998: esp. 381) implicit desire for definitional clarity.

[15] For a nuanced account of narratology's historical development, see Puckett (2016). Ryan (2007) gives a useful overview of more recent definitional stances.

[16] As per Marie-Laure Ryan's (2007) perspective on definitions.

[17] Meunier (Chapter 12) and Berry (Chapter 16) each develop more explicit definitions of narrative.

[18] See Morgan (Chapter 1) for ways that small narrative chunks in scientific texts relate to other cognitive elements of those texts.

condensed narratives recount the transformations undergone by a rock formation, but are chiefly only recognizable as narrative by trained geologists. The
narrative lies between the textual lines of the document,[19] so to speak, a point
which emerges secondarily from Hopkins's study.

In this section, I explore several characteristics of scientific narratives that
can be identified through NS enquiry, taking narratologists' definitional frameworks and theories as a sensible starting-point. Such comparison is additionally
essential to developing a genuine dialogue between narrative theory and
science studies. My preference is for Marie-Laure Ryan's (2007: esp. 28–31)
manner of classifying narratives according to a 'fuzzy set' of conditions on
their narrativity.[20] Ryan lucidly divides the degree of narrativity of a given text
into a number of 'dimensions' and 'conditions' that span narratologists'
instincts and preoccupations regarding what narrative is. By using her scheme,
we evaluate the degree of narrativity shown by a given document, not whether
it should be ruled out (or in) as *a* narrative. Here, I work with three of Ryan's
conditions in order to interrogate some salient features of scientific narratives:
whether characters in a story are individuals with a 'mental life'; the importance of the 'temporal dimension'; and the issue of narrative 'closure'.[21] My
discussion, drawing iteratively on chapters in this volume, opens up a few
intriguing narratological features of scientific narratives – which may, in turn,
inform further categorization work on narrative.

2.2.1   Narrative Protagonists

One of Ryan's conditions on narrativity that resonates with everyday experience and literary studies is the requirement for narratives to contain some
'intelligent agents', with mental or emotional responses (Ryan 2007: 29).
That is, a text has lower narrativity if it lacks this kind of 'mental dimension'.
Hopkins's mini rock-narratives are one example; rock formations as agents
have no mental reactions. What is immediately evident from the NS Project,
therefore, is the need for a capacious approach to characters in scientific
documents, because otherwise many texts would be ruled out of consideration
as narratives. Scientific narratives very often recount transformations undergone by protagonists (main characters) that are neither human nor necessarily
anthropomorphized: in this volume, the Stac Fada Member (Hopkins,

---

[19] This is the cognitive narratologists' criticism of definitions that restrict narrative to being a text-
type, rather than (also) a cognitive style (e.g., Ryan 2007: 27–28).
[20] Norton Wise, in contrast, draws heavily upon notions of experientiality in his commentary
(Chapter 22), as he explores narrative as a style of thinking in scientific activity.
[21] Ryan's (2007: 29) four dimensions comprise the spatial, temporal, mental, and formal and
pragmatic. I set aside her 'spatial dimension' and examine one condition from each of the last
three dimensions.

Chapter 4), the Tohoku earthquake (Miyake, Chapter 5), organic molecules (Paskins, Chapter 13) and substances in the fruit fly (Meunier, Chapter 12).[22] The first two examples involve narratives about *particular individualized* protagonists; there is only one Stac Fada Member – a spatially localized rock formation – only one spatially and temporally circumscribed earth rupture process that was the Tohoku earthquake. Hopkins and Miyake do each nonetheless unpack ways that these particularized narratives inform or are informed by generalized knowledge in their fields. On the other hand, organic molecules and biological substances are already less individuated, more *generic*, narrative agents; even though the fruit fly narratives distinguish between particular substances (e.g., cn^+ or v^+), all the instances of cn^+ are held to be identical (indistinguishable) and to behave in a uniform manner across all fruit flies. When cn^+ is the protagonist in a fruit fly narrative, therefore, it stands in for a class of identical cn^+ substances, to be distinguished only from other generic character-substances (such as v^+).

Robert Meunier (Chapter 12) characterizes the narratives scientists tell about such entities as 'narratives of nature'; they relate what 'happen[s] [. . .] when no researcher is intervening or even watching'. As narratives of nature are abstracted, and become part of the acquired knowledge in a scientific discipline, the phenomena they relate also tend necessarily to become stabilized. Their narrativity correspondingly decreases, according to Ryan's schema; at the abstract, generic limit, narratives of nature tell of (what have come to be seen as) habitual physical events, undergone by generic protagonists without a mental life.[23] As such, these narratives tend archetypally to fulfil conditions of factuality (or posited factuality) in a given scientific field.[24]

By contrast, mentally reacting protagonists act in particular situations in Meunier's other category of narratives: scientists' 'research narratives'. Here, scientists appear as characters performing specific actions (like steps in an experiment), and their reasoning processes or emotional reactions are often revealed through focalized narration.[25] Ryan's condition about 'mental life' in narrativity thus plays usefully into the distinction between Meunier's two categories of scientific narratives. For again, the scientist–protagonist may

---

[22] The plant narrators discussed by Griffiths (Chapter 7) occupy an intermediate space, partly anthropomorphized through their interactions with (and explicit framing by) the Darwin family. For anthropomorphized accounts in eighteenth-century natural science, see Terrall (2017).

[23] Very many contributions to the NS Project examine narratives of regular phenomena, even when those are anthropomorphized; in Beatty's (2016) terms, the regular is narrative worthy for science through its consequences for knowledge.

[24] As 'stories of the facts', narratives of nature would seem an ideal subject for those invested in questions of factual narratology – provided they are not first ruled out *as* narratives (cf. Fludernik and Ryan 2020).

[25] For HPS readers, I will explain focalization in section 2.3. Milne (2020: 449–51) builds on some earlier work of the NS Project (Morgan and Wise 2017) to distinguish between narratives with scientist-protagonists and those involving anthropomorphized objects of study.

either be *individualized* – like Charles Darwin (see Chapter 7) – or *generic*, standing in for all scientists in a field (see Chapter 12).

Examining Meunier's categories in detail can provide insight into the way a given scientific activity functions. The prevalence of a research narrative in an experimental research article helps familiarize its reader with a new approach, especially when, as Meunier demonstrates, the scientist–protagonist becomes generic, allowing the reader to imagine herself in that place. Alternatively, that both categories of narrative are intrinsically bound together in archaeological dating practices is fundamental for Anne Teather's (Chapter 6) proposal for archaeology to become more reflexive about how research questions influence the narratives it tells about the past. Across studies from the NS Project, we mostly see that, as a field of enquiry develops, its research narratives, with their individual actors and dimension of mental life, yield place to the telling of narratives of nature. This has even led contemporary chemists to call for 'thin' narratives of nature, like chemical reaction schemes, to be 'thickened' by reinsertion of the research story (Paskins, Chapter 13). Where the two categories of narrative are less distinct is in precisely those sciences which study the human, such as anthropology or psychology. Early psychological case histories, for example, weave together narration focalized on the mental processes of both individual subject and individual scientist–observer (Hajek 2020).[26] Can (or should) we distinguish the interplay of 'research narratives' and 'narratives of nature' as psychologists start to worry about the effect of their acts and thoughts on their subjects of study?

2.2.2 Time in Scientific Narratives

For the vast majority of narrative scholars, it is an essential condition of narrativity that a text deal with events that progress in time; an account of events occurring in a single moment could not be a narrative, for instance, nor could a series of instructions. This is largely taken for granted, such that questions of time in (especially classical) narratology are chiefly a matter of differences between the story and discourse in the ordering or duration of events.[27] Many scientific narratives similarly have what we might call a 'fundamental linearity'[28] – a straightforward, and highly significant, temporal structure – particularly those of the so-called historical sciences (geology, evolutionary biology).[29] Other work in the NS Project, however, has opened up the question of the relative importance of time sequencing, in comparison

---

[26] As, necessarily, do most psychoanalytic studies, into the twenty-first century (see Scheidt and Stukenbrock 2020).

[27] I return to these questions of narrative order in section 2.3.

[28] I thank Martina King for this term.

[29] See, especially, chapters by Hopkins (Chapter 4) and Griffiths (Chapter 7).

with other kinds of ordering that make meaning in a narrative. Mary Morgan's (Chapter 1; Morgan 2017) notion of colligation privileges relations between disparate items brought together by virtue of a single framing, which may then be woven into nets of similarities and differences; here, orderings other than time are the 'grid' by which narratives structure their meaning. Both Morgan (Chapter 1) and the recent work of Carrier and colleagues mark a clear separation between such 'configurational or coherentist' narratives (Carrier, Mertens and Reinhardt 2021: 20) and their time-ordered counterparts. Certainly, the two make sense of their subject matter in different ways – according to different 'grids', to use Morgan's terms. Yet the gulf between them is precisely about differences in *function*, rather than in *narrativity*, and we should not assume that 'configurational' scientific narratives are not also situated in time. It is simply that the time dimension is more or less implicit in the length and order of their 'events', as we can see from examining how 'configurational' narratives are structured and are transposable.

Chemical reaction schemes provide one example of a scientific narrative that is structured by principles other than time. Each of the diagrams in Paskins's chapter (Chapter 13) proposes to answer the puzzle of how the molecule tropinone might be synthesized from a combination of other organic molecules. The structural formulae on the diagram are ordered under a causal logic and selected according to whether they show key stages in the transformation of the starting molecules (such as proton transfer or a rearrangement of chemical bonds). If this causal ordering is also implicitly a *sequence* in time, the *duration* of each step (between the arrows) is subordinate to consideration of which transformations take place, and which chemical substances are added to or removed from the reaction vessel (see, e.g., notations above and below the arrows).[30] Transformations, not duration, are what matters for chemists. These configurations are also of principal import for NS scholars in analysing the function of the reaction scheme as a narrative. What I want to stress is that a progression in time still underlies this kind of narrative, if only in an implicit or latent form. We can see this in two different ways.

First, the reaction scheme is a thin 'narrative of nature', Paskins argues, in the sense that the actions of chemist-researchers have been flattened onto the plane of the molecules. If we 'thicken' the narrative by reintroducing elements of the research narrative, time re-enters the account explicitly as both ordering and duration, such as in the gloss provided by Pierre Laszlo: 'let this mixture return to room temperature (rt) over four hours' (quoted in Paskins, Chapter 13). By virtue of involving human agents, a research narrative will always have some basis in time – human actions are performed in time – and, as

---

[30] Sequence or order (*ordre*) is distinguished from duration (*durée*) in narratologist Gérard Genette's treatment of narrative temporality.

Meunier (Chapter 12) demonstrates, narratives of nature are often distilled out of accounts that begin by mixing human interventions and objects' reactions.

The above logic relies on re-inserting a human actor (or at least a living agent) into the narrative; it is an external logic of time-relatedness, if you will. My second proposal for understanding 'configurational' narratives as situated in time proceeds by invoking an internal transposition of the narrative.[31] I like to think of this as similar to parameterizing the narrative in time, borrowing a term from my training as a physicist. Parameterizing is what mathematicians or physicists do when they take the movement of an object in space, like a ball thrown in an arc, and instead of writing equations showing how its vertical movement relates to its horizontal position, they break both down into how they rely on time. Time order and duration thus become explicit in the latter form, where time is only implicit in the former set of equations (vertical vs horizontal position) – the physicist chooses between them depending on what she wants to examine. Similarly, physical chemists might take a chemical reaction – expressed in the transformation-based (non-temporal) logic of the reaction scheme – and create a simulation that steps in time through the process by which molecules come together, exchange protons or create different bonds (as in Wise 2017). In other words, they might transpose the 'configurational' narrative into explicitly temporal steps – for instance to investigate which parts of a reaction occur most rapidly.[32] An analogous transposition is described by seminal French narratologist Gérard Genette (1972: 78) when he compares the temporal extension of an oral narrative – the time taken to tell the story – to that of a written narrative: the written text has an extension in space (words on a page), which we can conceive metonymically as an extension in time, in terms of the time it would take to read the text.[33] Moreover, in any number of literary texts studied routinely by narrative scholars, there is a greater symbolic or semantic significance to other linkages than the temporal (Schmid 2013). Some scientific narratives have just as low a degree of narrativity – measured along the time dimension – as many of their literary counterparts studied by narratologists, and the inverse. My point, again, is that both narrative scholars and historians and philosophers can (and do) pose more fertile questions than definitional points about time-situatedness. Chapters in this volume demonstrate other, richer analyses of time in scientific narratives: whether chronologies take a relative or

---

[31] This second form is thus more broadly applicable.

[32] The situation examined by Wise (2017) is more complex than this because it is solving a different kind of puzzle, one involving quantum-level interactions. An inverse transposition (time to space) occurs in some of the spatial diagrams of the Tohoku earthquake (see Miyake, Chapter 5). Kranke (Chapter 10) also elucidates the ways phylogenetic tree diagrams can be constructed to emphasize time-progression through evolution, or, alternatively, to draw out relationships between species.

[33] The distinction here is in terms of the '*temps du récit*'. This notion is especially crucial when it comes to comparing the 'duration' (*durée*) of events in the story and in the discourse (or written narrative) (Genette 1972: 122–124).

absolute basis (Teather, Chapter 6), or the narrative implications of adopting a periodic temporal structure (Huss, Chapter 3).

2.2.3   *Narrative Closure and Narrative Levels*

The final element in my discussion of narrativity in scientific narratives is the question of closure, which falls under Ryan's (2007: 29) 'formal and pragmatic dimension' of narrativity. Narrative closure is a matter of a reader's reception of a text on a cognitive or affective level, and is usually held to occur when a reader's expectations of the story are met, or their questions answered (Klauk, Köppe and Weskott 2016). To the extent that scientists report completed research actions or propose answers to puzzles, scientific narratives tend to be constructed explicitly as closed (or alternatively as unambiguously open – when a puzzle remains unsolved). When twentieth-century palaeobiologists proposed to account for extinction events in the fossil record (Huss, Chapter 3), their narrative of how such mass extinctions are caused by periodic extraterrestrial events comes in itself to a closed ending: it answers the puzzle question of how and why extinctions occurred.

If the periodic narrative itself, along with most scientific narratives, achieves closure in the basic sense of providing an answer, the concept remains worthy of note in narrative science for pointing to the imbrication of several narrative levels in scientific knowledge-making. Narrative closure is perhaps always a matter of multiple levels, as an individual reader's affective 'sense of an ending' is informed by that reader's cultural expectations (see Klauk, Köppe and Weskott 2016). In the case of scientific narratives, this multi-level nature of closure is additionally linked to the nature of the scientific enterprise, under which knowledge must be validated by the scholarly community. John Huss (Chapter 3) teases out these intertwined narratives with regard to the periodic extinction story. It was not sufficient for the palaeobiologists to propose this new periodic narrative as explanation; while it offered a closed answer to their question, the palaeobiologists were also impelled to search for evidence to support its claims.

On the individual level, we can consider this search as palaeobiologists' striving to reach an affective sense of properly 'scientific' completeness, in accordance with prevailing scholarly virtues and community standards for knowledge: the extraterrestrial story had to be 'filled in' with a certain level of artefactual evidence, however plausibly it accounted for mass extinctions. The palaeobiologists' search for evidence also arguably constituted a pursuit of narrative closure on the level of the story of their discipline.[34] Joseph Rouse (1990) terms this level one of narratives 'in construction', in the sense that

---

[34] Other levels can sit intermediate between these two, as Hopkins details (Chapter 4): alongside particular accounts of the formation of the Stac Fada Member, and a sense of progress in their field, geologists invoke broad (uniformitarian) narratives about the earth.

actors in a field of enquiry conceive of its past and future trajectory in narrative terms, and subscribe to a shared view that knowledge proceeds by seeking evidence for hypotheses and remaining open to revising past accounts.[35] Scientific activity, then, interweaves this shared, always open-ended narrative (of science, of a discipline) with the various closed and coherent narratives developed by scientists about their objects of study; it comprises 'an ongoing tension between narrative coherence and its threatened unravelling', in Rouse's terms (1990: 183).[36] Examining the narrative condition of closure thus brings into prominence the necessary interweaving of the *social* in scientific activity (through narratives of a field, or expectations about epistemic virtues) and *particular* scientific narrative-making by scientists. What remains to be elucidated is quite what might demarcate closure of a scientific narrative in the proper sense, linked as it is to scientists' *affective* responses, from the more general tenets of scientific enquiry as it develops through time. For it is far from clear that we should follow Rouse in considering all scientific activity as a narrative in progress – that would be to turn away from our narrower conception of narrative in the NS Approach. Exploring the affective dimension of scientific narratives – why, for instance, some seem more 'elegant' or appealing – indeed comprises a vital next step in the study of narrative science. Elspeth Jajdelska's contribution to this collection (Chapter 18) makes a start, and points the way towards the kind of collaboration between cognitive science, narrative scholarship and HPS that is needed for careful work on these borders between the formal, the affective and the social.

2.3   Formal Matters

2.3.1   Story/Discourse

Thus far, I have been using the rather unwieldy term 'narrative' – as noun, as adjective – in relation to conditions on narrativity. One of the fundamental tenets of narratology, however, provides us with the possibility of bypassing the multivalent 'narrative' (especially as we use it in English), and delineating different levels of narrative as at once both act and representation. Narratologists conceive narrative as a dynamic relation between a *story* – the events which are recounted – and a *discourse* – the way those events are

---

[35] See also Borelli (2020: 435). We might also apply this notion to the narratives that synthetic biologists construct of their field, or that mathematicians employ to explain different programming architectures (see Berry, Chapter 16, and Dick, Chapter 15, respectively).

[36] See also Levine's (2015: 40–42) account of narrative closure in novels as nonetheless 'organiz[ing] relationships into the future'. Both Rouse's and Levine's analyses have explicit political aims. Meunier (Chapter 12) examines precisely the way experimental research articles both close one research episode and open onto new questions for the field.

recounted.[37] Faced with a given narrative, we only have immediate access to the discourse, that is, to the text of the document. Let us assume that we have a fixed set of events to relate, such as the sequence of actions needed to isolate a biochemical substance.[38] We could represent those events in discourse in many different ways. For example, the story of how to synthesize tropinone could be written as a chemical reaction scheme or written in words; it might include essentially no information about the chemist's actions, or it might add in those actions and their historical context; it could pass quickly over certain steps and linger when telling others. The distinction Paskins draws (Chapter 13) between thin and thick chemical narratives therefore also emerges out of considering how much information, and of what kind, is contained in different discursive versions of a single chemical synthesis story.[39] Using terms from narrative theory adds rigour to such investigations, because we can precisely label different domains of narrative structure.

To dissect a scientific narrative into story and discourse also draws our attention to potential mismatches in the order and duration of events recounted, which in turn means we can unpack the temporal dynamics of the narrative in detail. Many scholars have noted, for instance, that scientists do not necessarily recount experiments in the same order in which they performed them in the lab (see Meunier, Chapter 12).[40] Narrative theorists like Gérard Genette (1972) have given us not only the story–discourse pair (*histoire–récit*, for Genette), but also a precise, neutral terminology for designating different temporal orderings and durations.[41] As yet, detailed analysis of the temporal workings of scientific documents remains another area to be filled in by further NS studies: for example, how might differing order and pacing (between story and discourse) be used to persuade readers, generate suspense or achieve closure? Here, I develop only several possible strands of this temporal analysis.

We know from the work of scholars like Genette (1972: esp. 78–80) that it is rare in literature for the ordering of events in the story to coincide directly with that of events as recounted in discourse. Fairy tales are perhaps one exception (Puckett 2016: 184–185). In science, short narratives of nature also tend to have the ordering of story and discourse coincide – look at examples quoted at the

---

[37] Story and discourse are the most common terms, though some narratologists employ other labels.

[38] For example, a substance in the fruit fly (Meunier, Chapter 12), or glycogen as isolated by Claude Bernard (see Hajek's case in *Anthology II*).

[39] See also Kranke (Chapter 10) on different representations of a single 'underlying' phylogenetic tree diagram.

[40] Meunier's analysis is more complex than my discussion of his chapter here, as he introduces a third domain into the narratological framework, and distinguishes the 'practice-world' from the story and the discourse.

[41] I will introduce some of Genette's terms in notes and asides here, without presenting a complete overview of his scheme.

beginning of Meunier's and Miyake's chapters (Chapters 12 and 5). More intriguing is the kind of temporal dynamic required cognitively and epistemically by historical sciences like evolutionary biology. Sharon Crasnow groups these kinds of scientific endeavours under the framework of 'process tracing' in her *Interlude* (Chapter 11) and elucidates their shared reliance on forms of evidence that intermix time and causality. These are phenomena best construed by following the effect of certain causal factors through time, through a process; what does this entail for the relative temporality of their story and discourse?

Let us take John Beatty's example (Chapter 20) of the evolution of flatfish. The narrative constructed by a biologist to explain this evolution might begin with the observation that flatfish have their eyes offset on their heads – that is, the discourse begins with an observation, which is the end event of the story of how flatfish came to have the features they do. (For the investigating biologist, it is likely a middle-term event.) The discourse would then usually jump backwards to the selected starting point of the evolutionary story – i.e., a moment when flatfish swam upright and had eyes located symmetrically on their head.[42] But, after this initial jump, for a biologist to provide a properly Darwinian account of the flatfish's evolution, they must ensure that the story unfolds each of the incremental steps in time order, leading from the fish's initial form to its form with offset eyes (Beatty, Chapter 20). Such a story is narrative worthy, according to Beatty (also 2016) precisely because of its contingency. Potential evolutionary 'branches-not-taken' might appear implicitly, embedded in the narrative,[43] but there would not be the kind of jumps backwards (or forward) in time to new sets of events that we see in a novel like *Frankenstein*, or a classic Freudian psychoanalytic case.[44] The discourse also compresses millions of years of incremental changes (in story time) into a narrative tellable in human timescales.

The epistemic conditions on such a (Darwinian) historical account require a careful temporal unfolding on the level of the *story* of evolution; by implication, we would expect this to be reflected in the *discourse*. That is, we would expect the coincidence in timing between *story* and *reasoning* about the fish's evolution to mean events must follow in sequence when scientists put such a story into narrative (the discourse), such that the crucial time-ordering of events could be conveyed to the reader. Curiously, analogous examples in this volume suggest that this is not the case. Hopkins (Chapter 4) demonstrates that geologists write very few narrative discourses into their research articles about temporally unfolding geological transformations. Similarly, political scientists

---

[42] In Genette's terms, this jump is an 'analepsis'. Although I concentrate on ordering in time here, duration is equally as important in Genette's narrative theory.

[43] See Ryan (1986) on 'embedded narratives'.

[44] I'm thinking here particularly of the case of Anna O . . . (actually written by Breuer), which has been subject to much scholarly analysis of the timing of events (see, for example, Skues 2006).

trace along processual pathways to examine, for example, whether the United States would have entered the Iraq War even had G. W. Bush not been elected president – yet their publications do not recount those processes in order from beginning to end.[45] Such a choice not to have scientific discourse recount events in their story order seems surprising. To use a frequent analogy between narratives in historical sciences and classic detective stories, it is as though Holmes never unveiled his solution to Watson, but left the reconstruction of steps in the murder to the reader. For now, I can only raise the question; it must be left to further narratological investigation to ascertain the dynamics of ordering and duration in such scientific narratives.

2.3.2   *Narration and Focalization*

Beyond a careful attention to relative timings, classic narratology also directs us to interrogate whose perspective is expressed and with what authority, at each of the story and discourse levels. It is here that we can most clearly mark the ways narrative – especially in extended, verbal format – is a complex, formal edifice, however 'natural' it might often appear.[46] Narrative theorists differentiate first between the *author* of a work and its *narrator*: the author (e.g., Mary Shelley) writes down (or draws, etc.) the narrative, while the narrator tells the story (e.g., Victor Frankenstein). Although author and narrator are often presumed to be one and the same in non-fictional ('factual') narratives, Robert Meunier (Chapter 12) argues cogently that we should consider them as separate entities, especially for multi-authored scientific texts. Having posited that distinction, what interests me here are the *narrators*, the tellers from whose point of view we receive some narrative element: whether they appear as a character in the story, and how directly they reveal their perspective. In a pure narrative of nature, for instance, the narrator tells the story, but is not a character in it; the perspective is an external one, and appears impersonal, as in the quotation which opens Chapter 5. Historians of science will be used to contrasting such an impersonal narrator with the strong, self-fashioned narrative voice typical of eighteenth-century natural science (e.g., Terrall 2017). Such an early natural-scientist narrator is also a character in his story, and often relates his actions and emotional responses in the first person.[47] But there are more than these two

---

[45] This point emerged during Crasnow's contribution to the NS Public Seminar Series (www .narrative-science.org/events-narrative-science-project-public-seminar-series.html). See also the importance of interview 'data' for the El Salvador civil war case (Crasnow, Chapter 11).

[46] Here I complicate Wise's (Chapter 22) opposition of 'formal' and 'natural' language as a framework for envisaging narrative science, and his implicit privileging of 'narratives of nature' as the instantiation of scientific narrative.

[47] In Genette's terms, the first narrator is both *heterodiegetic* (not a character) and *extradiegetic* (external perspective), while the second is *homodiegetic* and *intradiegetic* (Genette 1972: chaps. 4–5).

options present in scientific narratives, and that is precisely where using narratological tools reveals complexities we might not otherwise grasp.

We notably encounter more than one internal perspective in accounts from the human sciences, when the aim is to gain access to a human subject's mental, cognitive or emotional state.[48] Such interior views can be accessed and portrayed in a variety of different ways. In the following extract, from an experiment involving hypnotic suggestion, there is a shift in the *focus* of the narrative – it begins with the narrator–experimenters' point of view,[49] then shifts subtly to that of the hypnotized subject.

We take another coat and we pass it to M. F. . ., who puts it on; the subject, who gazes fixedly at this coat with a wondering look, sees it wave about in the air and take the form of a person. 'It is, she says, like a mannequin with nothing inside it.' (Binet and Féré 1887: 229)[50]

The hypnotic suggestion in question is that Monsieur F. will be invisible to the subject. As the extract begins, we see the narrator also present as character(s) in the story, performing actions with the coat, and then observing the subject's reaction. This reaction first consists of external features of the subject – her 'wondering look' – described from the narrator's perspective, before the text moves to portray what the subject sees, and then relate the subject's words about her vision. Throughout, the telling is done by the narrator–experimenters; they refer to the subject in the third person. But the narrative also relates information to which, logically, the narrator–experimenters do not have access, in the form of the subject's interior view; there is a shift in who 'sits behind' the words of the text, with the narrator–experimenters and the subject 'doubling up' for this part. This is an example of shifting narrative *focalization*.[51]

What I want to emphasize are the kinds of questions we can ask after noticing such a shift (or, more often, repeated shifts) in focalization in a narrative. On the one hand, the subject's perspective is stamped here with the authority of the narrator as (a pair of) scientists. The description of what the subject sees is an interpretation, based on or validated by the subject's words (which are also

---

[48] Thinking back to Ryan's (2007) conditions of narrativity, we have a 'mental dimension' here for both scientist (narrator/character) and subject (character).

[49] For this limited analysis, I set aside the complexities of the plural nature of the narrator and treat it as a single entity encompassing two experimenters.

[50] My translation. I analyse this passage in greater detail in Hajek (2016a).

[51] With shifting focalization also comes narrative polyphony, a multi-vocality present in the background to Bhattacharyya's paper (Chapter 8). Scientific polyphony was also the topic of a NS workshop, 3 June 2019 (www.narrative-science.org/events-narrative-science-project-workshops.html).

reported). On the other hand, noticing the shift in perspective – and that it occurs *before* the subject speaks – draws our attention, as readers, to the representational surface of the text – to the fact that it is *a* presentation of the story, and that there might be others. There is notably a small temporal mismatch here, since the narrator–experimenters' interpretation, which occurs first in the *discourse*, must logically follow the subject's speech on the level of the *story*. We are reminded that the immediacy of this experimental report is constructed, that writing occurred after the activity of the experiment. Did, therefore, the subject say exactly what is reported, or are the words (also) a reconstruction by the narrator–experimenters to validate their interpretation? More fundamentally, when did knowledge-making occur here – during actions, or during writing, or both? I would stress that it cannot be fixed down; narrative, (even) in its textual form, is not only an output of scientific activity, but fully and necessarily participates in the activity of knowledge-making. This is narrative as 'the expansion of a verb' (Genette 1972: 75), or the binding together of 'narrativizing' and 'narrative representation', in Morgan's terms (Chapter 1).

If, in a sense, this brings us back to the kind of arguments well known in history of science under the label of 'constructivism' (e.g., Golinski 2005), it does so from the distinct perspective of narrative. Formal narrative analysis can do more than signal that knowledge emerges from putting scientific activity into words. It can suggest different patterns of authority in narratives from sciences which study humans, compared to those which do not. My brief analysis above, for instance, points to the ways that shifting narrative focalization seems essential to the business of the human sciences around the turn of the twentieth century, but also to a concomitant trade-off in the form of a more unstable textual authority. Further work could study how textual dynamics of this kind articulate with scientists' avowed theoretical orientations; for example, do behaviourist psychologists, who eschew internalized observations, nonetheless produce focalized narratives? How do these dynamics compare with narrative focalization in accounts involving anthropomorphized (non-human) protagonists, on the one hand, or multiple interacting humans, on the other hand (as in social sciences like anthropology)? Curiously, there is narrative focalization on plant growth at multiple narrative levels in the Darwins' *Power of Movement in Plants* – not only when the Darwins narrate their story, but also when the plants themselves are (co-)narrators, as Devin Griffiths's narratological reading reveals (Chapter 7).[52]

---

[52] Griffiths further explores the implications and constraints of such non-human co-narration on genre and on narrative level (see Chapter 7, esp. Table 7.1).

2.3.3 Which Comes First?

Analysing shifts in narrator focalization prompted me to ask whether Binet and Féré's subject spoke the exact words related, or whether the experimenters filled in a plausible comment while writing their text. In story-discourse terms, this is equivalent to asking whether Binet and Féré's text – as discourse – reports a pre-existing *story* and just reorders the events, or, alternatively, whether portions of the story are only constructed (and re-constructable) through their inclusion in the *discourse*. As Kent Puckett (2016: 35) asks: 'Do events precede their representation, or does a representation somehow *produce* events as significant and thus knowable?' This 'paradox' (Puckett 2016: 215) points to a central tension in narrative theory over which of story or discourse comes first; it has been a productive force structuring the work of key narrative theorists, as Puckett sees it. NS studies also provide a particularly rich site through which to trace the dynamics of this tension, with conclusions that can feed back into theoretical work on narrative.

I am not advancing some radical constructivist view here, as if there were no reality outside of that which is '*mise en récit*' in a narrative. But, when it comes to scientific narratives, it is not always straightforward to identify what counts as story, as against the discourse, especially when we are dealing with non-human, non-anthropomorphized protagonists. Hence the richness of scientific narrative. Indeed, Meunier (Chapter 12) enunciates how both discourse and story (as events and their ordering implied in discourse) can differ from the events that took place in the experimenter's laboratory in 'reality', or the 'practice-world', as Meunier terms it – and this even for actions performed by and recorded by humans. When an archaeologist finds many Neolithic stone axes at some site, these can, on the one hand, serve as evidence or markers of story events – through some absolute dating method, for instance. On the other hand, the archaeologist might construct a narrative about popular stone quarrying sites, which might frame the axe find as a trace in a story about demand for felling trees.[53] Either way, story and discourse sit in a dynamic relation within the activity of scientific narrative-making.

The interplay of story and discourse is particularly clear in those scientific endeavours where narrative is not an end point, but where discourse-making and story-reconstruction occur iteratively.[54] In this volume, Teru Miyake's study (Chapter 5) of seismological work on the Tohoku earthquake is a salient example. Miyake's seismologists first take evidence from a single

---

[53] These examples are loosely adapted from Anne Teather's paper (Chapter 6).

[54] As noted previously, what I understand as necessarily interdependent, Morgan separates into functions of 'narrativizing' (to make the representation), 'narrative reasoning' (thinking from a narrative – closest to my discourse-making), and 'narrative explanation' (thinking within a narrative – like my story-reconstruction).

kind of sensor and configure it computationally into a time-stepped narrative simulation of how events in the earthquake occurred: the *rupture narrative*. Many rupture narratives are generated (e.g., from different types of sensors), and then compared by seismologists, who next extract details which are present in several rupture narratives; these details are treated as story-level events. Finally, 'these distilled details are strung together into a model-independent rupture narrative, which [Miyake] call[s] an *integrating narrative*'. In narratological terms, successive steps in this scientific work take each of story and discourse, respectively, as pre-existing. Rupture narratives are first configured from story points (i.e., the sensor data evidence), before a switch in perspective, which construes the discourse of the rupture narrative as a source from which to reconstruct and extract a different set of story elements (Miyake's 'distilled details'). The final step flips perspective yet again, back to the work of constructing a narrative discourse (the integrating narrative) from (the new set of) pre-selected story details. Morgan (2017; and this volume) speaks of 'narrative inference' as unravelling and reknotting sets of evidential or conceptual elements.

If these iterative steps are clearly separated in Miyake's account, we could speculate that such dynamic work of narrative configuration and reconfiguration is in play in scientific activity more widely, especially where phenomena are not directly observable. For instance, Elizabeth Haines (Chapter 9) points to a doubled way of working within visual narratives, when she shows how 'neither evidence collection nor explanatory accounts were prior' in Hugh Hamshaw Thomas's botanical and intelligence-gathering practices. Opening out from this NS work, we might ask further whether scientific narrative-making (and re-making) of this kind could serve as a useful model for broader processes of narrative-writing and narrative-reading.

2.4   Narrative Reasoning

For now, I turn to existing narratological understandings of reading practice and how they can illuminate scientific reasoning. No telling is without its implied or actual readers, and they too perform important work in narrative-making, in an interplay with the narrative as textual or visual material. In a sense, therefore, I move now from considering narrative as the dynamic relation between *story* and *discourse*, to considering an interconnecting relation between *discourse/ narration* and *reader*. It is a move which brings us into the domain of cognitive narratology – a field that combines findings from psychology and artificial intelligence to explore relations between story-text and -language, on the one hand, and human memory, perception and affect, on the other.[55] Concepts from

---

[55] See Herman (1997) and Jahn (2010) for overviews.

cognitive narratology are well suited to tracing the kinds of processes occurring in a reader's mind (or imagination) as they read a scientific text or diagram; notably, narratological concepts point us towards elements of scientific reading practices that might well be compared to ways people read fictional texts.

I construe such cognitive processes under the banner of 'narrative reasoning': they comprise story-like imaginative constructions which scientific readers generate when reading a research article or examining visual evidence.[56] If the scientific text in question has a clear narrative discursive form, narrative reasoning in the mind may not differ greatly from the logic of the narrative on the page, or it might be inferred using more classical narratological tools (of the kind discussed in the previous section).[57] Narrative reasoning is more distinctive as a component in scientific activity when story-like imaginative work is prompted by apparently non-narrative scientific texts – texts with very low narrativity (to link back to my earlier discussion). An example I have already evoked is the 'implicit' or 'covert' narratives of historical sciences like geology, which Hopkins argues only unfold as narratives to an informed reader. To interrogate narrative reasoning under my NS Approach is to examine the processes by which a scientist imaginatively replays such narratives, and, importantly, how these processes map onto particular textual elements. This explicitly adds a textual dimension to the narrative thought processes opened up by Morgan (Chapter 1). We might refer to tacit knowledge, scientists' trained judgement, or their horizon of expectations – to invoke some concepts current in HPS and narrative studies. However important, these are not enough for rigorous narrative enquiry, since they operate on a more general level: they relate texts as a whole to broad-scale expectations or knowledge in a field. With the NS Approach, we can delve into the specifics of which particular elements in a research paper or diagram activate story-like imaginative responses, as opposed to other cognitive functions. Notions like *narrative performativity* and *scripts* allow contributors to this volume to begin this work.[58] I briefly outline their findings in what follows.

As Elspeth Jajdelska emphasizes in Chapter 18, the question of who narrates a story and in what circumstances matters for its reception. Jajdelska transfers the notion of *narrative performativity* from the spoken to the written domain and, in a recursive move, elucidates its workings in a research article about cognitive science. Performative language is what early narratologists might have called properly literary language, in that it draws attention precisely to its

---

[56] This differs from Morgan's use of 'narrative reasoning' to describe reasoning *from* or *within* (pre-existing) narrative representations (Chapter 1).

[57] Meunier draws precisely these kind of interpretations in Chapter 12. See also Ryan (1986).

[58] Nina Kranke (Chapter 10) also connects elements of scientific documents – in her case, visual diagrams – to the narratives that readers construct from them, though without using particular narratological ideas.

aesthetic qualities. It thus bears a greater affective force and implicitly cues a certain imaginative worldview. The worldview thus rendered can encode assumptions or perspectives which support a researcher's explicit argumentative position, as in the article analysed by Jajdelska. Importantly, under this framework, particular textual passages, or even a few words, can be identified as corresponding to a story-like cognitive effect – one which plays a highly significant role in the knowledge claims of this scientific article.

A different kind of small-scale textual (or visual) element that produces story-like reading is the *script* (Herman 1997). In her chapter examining how mathematicians read proofs, Line Andersen deploys this concept from early cognitive narratology to argue that mathematicians read proofs similarly to how people read fictional narratives. That is, portions of the proof call up a sequence of events or actions that are expected or appropriate in the context in question. These proof-segments operate, in other words, like the scripts in literary texts for events such as 'eating in a restaurant, riding a bus, watching and playing a football game, participating in a birthday party, and so on'. As the AI researchers who developed the notion go on to say, 'These scripts are responsible for filling in the obvious information that has been left out of a story. Of course, it is obvious only to those understanders who actually know and can use the script' (Schank and Abelson 1977: 41).[59] Andersen develops the correspondences between script-activating elements of a proof and steps in mathematical understanding. Like readers of novels, the mathematical reader performs the mental action of running through steps cued by a script, but since scripts deal with expected sequences of actions, the reader's attention is particularly caught when a proof deviates from the expected background of mathematical scripts. By undertaking such narrative reasoning, mathematicians are prompted to focus on the novel, likely crucial, elements of a proof. Reciprocally, HPS analysts like Andersen can identify more precisely which elements count as most significant in mathematical reasoning and understanding, and for which kind of readers, since script-activation depends on a reader's level of understanding of an expected situation. Notions such as scripts, narrative performativity and other ideas from cognitive narratology could similarly be applied to many domains studied by HPS scholars. Wise, for instance, broadens the notion of script to several areas of scientific knowing in his *Finale* to this collection (Chapter 22). But where such an approach might bear most fruit is in combined textual and ethnographic analysis, of the kind sketched by Andersen – specific elements of a scientific text can be connected to particular narrative-like reasoning, and that mapping contrasted with scientists' own accounts, as well as analysts' reconstructions, of scientific activity.

---

[59] I thank Line Andersen for drawing my attention to this quotation. Morgan also signals the communal aspect of scientific narrative in Chapter 1.

2.5   Conclusion

Narrative theory is an extensive and complex field and, in this chapter, I have only worked through some of its key concerns and ideas as they apply to scientific narratives. My aim in doing so has been twofold. On the one hand, I have sought to encourage HPS scholars to treat narrative in the focused, technical terms of narratology, by demonstrating the analytical productivity this promotes. Such analysis – as undertaken in the NS Project and chapters in this volume – reveals that a '*mise en récit*' always involves an active component of knowledge-making or reasoning, even when a narrative (representation) is also the output of some scientific endeavour. Reciprocally, if narrative in science is always active, it is not an activity divorced from any concrete, material basis; a major part of the value of narratological tools is that they can serve to trace precise connections between narrative as *text* and narrative as mode of *reasoning*. What the NS Approach provides, then, is precision and rigour to an object of study – narrative – that otherwise risks overflowing its conceptual bounds to such an extent as to offer no meaningful basis for comparison or interpretation. NS offers exciting perspectives as an approach deployed alongside the usual epistemic resources of HPS.

On the other hand, this chapter elucidates the various ways in which work in the NS Project is informed by concepts from narratology, even where such concepts are not emphasized or delineated. As historians and philosophers of science, contributors to this volume bring a sensitivity to the theoretical and contextual constellations in which their case studies can be situated. Our studies thus bring a depth of detail to explorations of narrative in a non-literary domain – they can complement and complete narratologists' investigations in this area with much-needed science-specific expertise. Just as I hope future HPS work will be open to narratological perspectives, I similarly encourage narrative scholars to draw upon HPS expertise, as showcased in this volume, in developing their field beyond the literary. This chapter has notably pointed to some distinctive characteristics of scientific narratives – their frequent non-human, even generic, protagonists; their iterations of story-making and discourse-configuration – as well as proposing that there is less of a divide between scientific and literary narratives than often assumed, when it comes to their situatedness in time – it is just that different questions of timing might arise. And, of course, there remain many areas of enquiry where collaboration between HPS and narrative studies would be fruitful: the affective charge of scientific narratives, forms of narrative focalization and the particular interplay of ordering and duration in work in the historical sciences, to mention just some I have signalled above.

But to conclude this chapter I would like to turn briefly to the ambitions held by narratology to be considered a science, from its pre-history in Russian formalism to its more recent cognitive turn.[60] Could we apply the NS Approach to narratology itself? As Puckett (2016) stresses, narratology as a domain of enquiry is not without its own history. Where he historicizes it in terms of key political and intellectual currents, we might ask how narratology is informed by other scientific fields and what role narrative-making plays in its endeavours. If we had to classify narratology, we could place it in the category of the human or social sciences, as taking a human product – narrative – and its cultural and social imbrications, as its object of study. We might then sketch a shift in perspective from a view of narratology influenced by the model of chemistry – with stories dissected into a fixed set of re-combinable elements – to one that enacts something of a convergence with cognitive science and some branches of psychology. Early structuralist Algirdas Greimas (1983: 65), for instance, praised the language of chemistry as 'a semiotic form which must, across all kinds of language, serve to express its meaning',[61] while to read Manfred Jahn's encyclopaedia entry (2010) on cognitive narratology is to be plunged into considerations of 'preference rules and processing strategies' that would not appear out of place in a research article in computational science. By analogy with chapters in this volume, we might speculate that early structuralist narratology mobilizes 'thin narratives' of the kind identified by Paskins (Chapter 13), or that recent cognitive theories enlist strategies of 'narrative performativity' to provide imaginative support for their claims (as in the article investigated by Jajdelska in Chapter 18). What might such a transition imply for understanding the evolution or limits of narratology as a 'historically specific logic', to use Puckett's terms (2016)? When we apply a notion like the *script* to a scientific narrative, to what extent do we invoke distinctively narratological theorizing, as against ideas from the script's origins in AI? Or, is to pose such questions to descend into a methodological spiral, where narrative and science turn circularly around each other?[62]

---

[60] The term 'narratology' dates from 1969, when it was coined by Tzvetan Todorov (Puckett 2016: 234 n. 23).

[61] I thank Mat Paskins for suggesting this quotation.

[62] Many thanks to Mary Morgan, Mat Paskins, Martina King, Devin Griffiths and John P. Hajek for insightful comments and suggestions on drafts of this chapter. Working with the Narrative Science core team – Mary, Dominic, Andrew, Mat, Robert – over the last few years has been a stimulating and enriching experience, for which I thank you all. Finally, I am ever grateful to Gordon P. Jardine for promoting my interest in language, narrative and their surprising turns; I dedicate this chapter to his memory. *Narrative Science* book: This project has received funding from the European Research Council under the European Union's Horizon 2020 research and innovation programme (grant agreement No. 694732). www.narrative-science.org/.

Bibliography

Acquier, M.-L. (2010). 'Soif de mots, désir de science. Quelques aspects de la relation entre littérature et science(s)'. *Cahiers de narratologie* 18. https://doi.org/10.4000/narratologie.6116.

Anthology I: Case Book of Narrative Science (2019). Edited by M. Paskins and Mary S. Morgan. London: London School of Economics and Political Science. www.narrative-science.org/anthology-of-narrative-science.html.

Anthology II: Case Book of Narrative Science (2022). Edited by A. Hopkins, Mary S. Morgan and M. Paskins. London: London School of Economics and Political Science. www.narrative-science.org/resources-narrative-science-project.html.

Beatty, J. (2016). 'What Are Narratives Good For?' *Studies in History and Philosophy of Science Part C* 58: 33–40.

Binet, A., and C. Féré (1887). *Magnétisme animal*. Paris: F. Alcan.

Borelli, A. (2020). 'Narrative in Early Modern and Modern Science'. In M. Fludernik and M.-L. Ryan, eds. *Narrative Factuality: A Handbook*. Berlin: De Gruyter, 429–442.

Buckland, A. (2013). *Novel Science: Fiction and the Invention of Nineteenth-Century Geology*. Chicago: University of Chicago Press.

Carrier, M., R. Mertens and C. Reinhardt (2021). 'Introduction'. In *Narratives and Comparisons: Adversaries or Allies in Understanding Science?* Bielefeld: Bielefeld University Press, 7–27.

Carroy, J. (1991). *Hypnose, suggestion et psychologie: L'invention des sujets*. Paris: Presses universitaires de France.

Crossley, M. L. (2010). 'Narrative Psychology'. In D. Herman, M. Jahn and M.-L. Ryan, eds. *Routledge Encyclopedia of Narrative Theory*. London: Routledge.

Culler, J. (2011). *Literary Theory: A Very Short Introduction*. 2nd edn. Oxford: Oxford University Press.

Fludernik, M. (2020). 'Factual Narration in Narratology'. In M. Fludernik and M.-L. Ryan, eds. *Narrative Factuality: A Handbook*. Berlin: De Gruyter, 51–74.

Fludernik, M., and M.-L. Ryan, eds. (2020). *Narrative Factuality: A Handbook*. Berlin: De Gruyter.

Genette, G. (1972). 'Discours du récit'. In *Figures III*. Paris: Seuil, 65–278.

Golinski, J. (2005). *Making Natural Knowledge: Constructivism and the History of Science*. Chicago: University of Chicago Press.

Greimas, A. J. (1983). *Structural Semantics*. Trans. D. MacDowell, R. Schleifer and A. Velie. Lincoln: University of Nebraska Press.

Griffiths, D. (2016). *The Age of Analogy: Science and Literature between the Darwins*. Baltimore, MD: Johns Hopkins University Press.

Hajek, K. M. (2016a). 'Fluid Boundaries: Hypnotism as Science in Late Nineteenth-Century France'. PhD thesis, University of Queensland, Brisbane.

(2016b). '"Je lis ça comme je lirais un roman": Reading Scientific Works on Hypnotism in Late Nineteenth-Century France'. *Australian Journal of French Studies* 53: 232–245.

(2020). 'Periodical Amnesia and *Dédoublement* in Case-Reasoning: Writing Psychological Cases in Late Nineteenth-Century France'. *History of the Human Sciences* 33.3–4: 95–110.

Herman, D. (1997). 'Scripts, Sequences, and Stories: Elements of a Postclassical Narratology'. *PMLA* 112.5: 1046–1059.

(1998). 'Narrative, Science, and Narrative Science'. *Narrative Inquiry* 8.2: 379–390.

Herman, D., M. Jahn and M.-L. Ryan, eds. (2010). *Routledge Encyclopedia of Narrative Theory*. London: Routledge.

Hühn, P., J. C. Meister, J. Pier and W. Schmid, eds. (2014 [2013]). *The Living Handbook of Narratology*. 2nd edn. Hamburg: Hamburg University Press, www.lhn.uni-hamburg.de/index.html.

Jahn, M. (2010). 'Cognitive Narratology'. In D. Herman, M. Jahn and M.-L. Ryan, eds. *Routledge Encyclopedia of Narrative Theory*, London: Routledge.

Klauk, T., T. Köppe and T. Weskott (2016). 'Empirical Correlates of Narrative Closure'. *DIEGESIS* 5.1: 26–42.

Levine, C. (2015). *Forms: Whole, Rhythm, Hierarchy, Network*. Princeton, NJ: Princeton University Press.

Milne, C. (2020). 'Empiricism and the Factual'. In M. Fludernik and M.-L. Ryan, eds. *Narrative Factuality: A Handbook*. Berlin: De Gruyter, 443–452.

Morgan, Mary S. (2017). 'Narrative Ordering and Explanation'. *Studies in History and Philosophy of Science Part A* 62: 86–97.

Morgan, Mary S., and M. Norton Wise (2017). 'Narrative Science and Narrative Knowing: Introduction to Special Issue on Narrative Science'. *Studies in History and Philosophy of Science Part A* 62: 1–5.

Otis, L. (2010). 'Science Surveys and Histories of Literature: Reflections on an Uneasy Kinship'. *Isis* 101.3: 570–577.

Puckett, K. (2016). *Narrative Theory: A Critical Introduction*. Cambridge: Cambridge University Press.

Ricœur, P. (1980). 'Narrative Time'. *Critical Inquiry* 7.1: 169–180.

Rouse, J. (1990). 'The Narrative Reconstruction of Science'. *Inquiry* 33.2: 179–196.

Ryan, M.-L. (1986). 'Embedded Narratives and Tellability'. *Style* 20.3: 319–340.

(2007). 'Toward a Definition of Narrative'. In D. Herman, ed. *The Cambridge Companion to Narrative*. Cambridge: Cambridge University Press, 22–35.

Salmon, C. (2008). *Storytelling, la machine à fabriquer des histoires et à formater les esprits*. Paris: La Découverte.

Schank, R. C., and R. P. Abelson (1977). *Scripts, Plans, Goals and Understanding*. Hillsdale, NJ: L. Erlbaum Associates.

Scheidt, C. E., and A. Stukenbrock (2020). 'Factual Narratives and the Real in Therapy and Psychoanalysis'. In M. Fludernik and M.-L. Ryan, eds. *Narrative Factuality: A Handbook*. Berlin: De Gruyter, 297–312.

Schmid, W. (2013). 'Non-Temporal Linking in Narration'. In P. Hühn, J. C. Meister, J. Pier and W. Schmid, eds. *The Living Handbook of Narratology*. Hamburg: Hamburg University Press. www.lhn.uni-hamburg.de/article/non-temporal-linking-narration.

Skues, R. A. (2006). *Sigmund Freud and the History of Anna O.: Reopening a Closed Case*. Basingstoke: Palgrave Macmillan.

Terrall, M. (2017). 'Narrative and Natural History in the Eighteenth Century'. *Studies in History and Philosophy of Science Part A* 62: 51–64.

Vila, A. C. (1998). *Enlightenment and Pathology: Sensibility in the Literature and Medicine of Eighteenth-Century France*. Baltimore, MD: Johns Hopkins University Press.

Wise, M. Norton (2017). 'On the Narrative Form of Simulations'. *Studies in History and Philosophy of Science Part A* 62: 74–85.

II

Matters of Time

When time matters in the sciences, it matters in their narratives, but those narratives rarely use a simple account of time

3 Mass Extinctions and Narratives of Recurrence

John E. Huss

Abstract

A narrative of recurrent causation, the Nemesis hypothesis, holds that the Sun has a companion star, Nemesis, whose orbit perturbs comets from the Oort cloud into earth-crossing orbits leading to mass extinction by impact with a nearly clocklike periodicity. Here I discuss the pursuit of the Nemesis hypothesis as the pursuit of narrative closure. Using a framework drawing on formalist analysis of narratives that distinguishes between the ordering of events in the narrative discourse (the *syuzhet*) and in their chronological sequence (the *fabula*), I describe the processes of reading and rereading the fossil and geologic records. The resulting analysis dissolves false dichotomies between nomothetic and idiographic, and catastrophic and uniformitarian approaches in the historical sciences. It also accommodates diverse philosophical views about the nature of epistemic access to the past.

3.1   Introduction

Ever since it started to look as if the dinosaurs were done in by a nagging case of asteroids, the hypothesis has been pursued that every mass extinction has had an extraterrestrial cause, while some have expressed a strong preference for an earthly cause.[1] Here I frame the pursuit of the Nemesis hypothesis of an extraterrestrially caused periodicity in mass-extinction events as a process of 'reading' the fossil and geologic records in pursuit of narrative closure.[2] In the case of mass extinction, I am particularly keen on understanding how periodicity guides the search for evidence in pursuit of a causal narrative. In contrast

---

[1] The impact hypothesis for the extinction of the dinosaurs and other taxonomic groups at the end of the Cretaceous period was put forward by Berkeley's Alvarez group (Alvarez et al. 1979; 1980). Resistance to the idea that impact is the general cause of mass extinctions was raised by, for example, Johns Hopkins palaeontologist Steven Stanley (1987).

[2] On reading (and rereading) the fossil record, see Sepkoski (2012).

to narratives of periodic extinction stand narratives of particular mass extinctions, where the plot is driven by the specific setting, characters, and one-off events. Of course, narratives of periodicity and one-time events do not exhaust the space of possible narrative explanations, and in the end I will describe somewhat of a middle path that seems to be gaining traction.

3.2   Periodicity of Mass Extinctions

In 1979, in Gubbio, Italy, a team of researchers led by Walter Alvarez discovered an iridium anomaly in sedimentary strata dated to be of end-Cretaceous age. This worldwide temporal horizon happens to coincide with the last known fossil occurrence of a number of biological taxa, including non-avian dinosaurs, ammonites, rudist bivalves, pterosaurs, mosasaurs and large numbers of plant and bird species. In terms of severity, the Cretaceous-Tertiary (or K-T) extinction (now known as the Cretaceous-Paleogene, or K-Pg extinction) ranks among the 'big five' mass extinctions in the fossil record: the end-Ordovician, Devonian, Permian, Triassic and K-Pg.

Like many discoveries in the earth sciences, the discovery of the iridium anomaly was serendipitous (Glen 2002). The Alvarez team, assuming a statistically constant rain of meteoritic iridium throughout geologic time, thought that they could use that iridium flux to estimate elapsed time represented by sedimentary deposits. But the concentration they found was far off-scale relative to the known rate, and further lab analysis of samples confirmed that there was a 'spike' in iridium in a red boundary clay layer at the top of Cretaceous strata. Iridium concentrations in strata immediately above and below that layer fell off exponentially to zero (Alvarez et al. 1980). Because Iridium is quite rare in the earth's crust, the Alvarez team hypothesized an asteroid or comet impact.[3]

Meanwhile, as the Alvarez group pursued evidence for an asteroid or other bolide impact at the Cretaceous–Tertiary boundary, David Raup and Jack Sepkoski were independently at work analysing broad extinction patterns in a synoptic database compiled by Sepkoski, *A Compendium of Fossil Marine Families* (1982). Sepkoski had been compiling this database for years by combing the published literature for new reports of fossil occurrences, and continually updated this record of the first known and last known fossil appearances of marine families.[4] By tabulating the record of first and last appearances, a diversity curve for the entire Phanerozoic eon could be generated, and the number of families becoming extinct could be chronicled for each subdivision of geologic time. By 1982, Raup and Sepkoski's statistical analyses of Sepkoski's data resulted in

---

[3] For further discussion of the pursuit by geologists of evidence for earthly events of extraterrestrial origin, see Hopkins (Chapter 4).

[4] The family is the taxonomic level just above the genus and below the order in the Linnaean hierarchy.

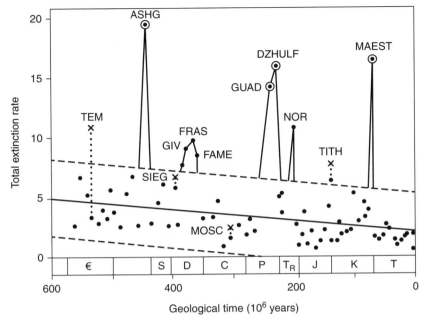

Figure 3.1 The 'big five' mass extinctions
The Ashgillian event at the close of the Ordovician, the Frasnian-Famennian event of the late Devonian, the Guadalupian-Dzhulfian event at the end of the Permian, the Norian event of the late Triassic and the Maestrichtian event at the Cretaceous–Tertiary boundary. Source: Raup and Sepkoski (1982).

a clear pattern of five large mass-extinction events – the so-called 'big five' – standing as outliers against a backdrop of smaller events (see Figure 3.1).

As Jack Sepkoski continued to compile a pen-and-ink database of first and last fossil appearance of marine families, his colleague at Chicago, David Raup, became interested in computerizing, tabulating, plotting and analysing them statistically. Whereas Sepkoski had plotted the data at the level of the stratigraphic series (e.g., upper Cretaceous), Raup decided to plot the data at a finer resolution, that of the stratigraphic stage (e.g., the Maestrichtian stage, a subdivision of the upper Cretaceous; Sepkoski Jr 1994). The gestalt they perceived was one of mass extinctions evenly spaced (Figure 3.2). Could this be a periodic array?

The stratigraphic record of the twelve largest mass-extinction events of the past 250 million years appeared to be periodic. However, two methodological constraints on the system had the potential to make the fossil record of mass extinction look periodic, regardless of whether it was or not. First, the stratigraphic record is divided into 40 stratigraphic stages (bins) of varying duration, and the dates of mass

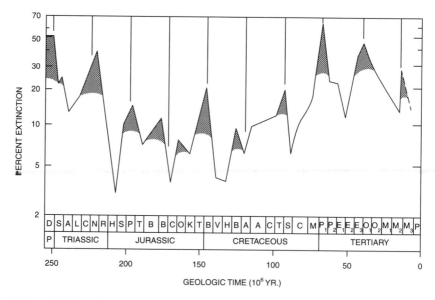

Figure 3.2 **Graph of percentage extinction of fossil marine families for each geologic stage of the past 250 million years**
With best-fit 26 million-year periodicity.
Source: Raup and Sepkoski (1984). Reproduced with thanks to the controllers of Raup and Sepkoski's respective estates.

extinctions are resolved only to the level of the stratigraphic stage. Second, extinction peaks can only be recognized if they occur in non-consecutive stages, imposing some minimum separation between events. Spurious periodicity needed to be distinguished from the real thing. The questions raised were, within these methodological constraints: (1) what periodicity best fit the data? and (2) what is the probability of obtaining such a well-fitting periodicity simply due to chance?

To answer the first question, they needed to determine the best-fit periodicity, which required a measure of goodness of fit. Raup and Sepkoski (1984) tried a range of periods from 12 million to 60 million years. For each period length they took a perfectly periodic time series and lined it up as closely as possible to the time series of mass extinctions and computed the standard deviation as a goodness-of-fit statistic. The best-fitting period came out to be 26 million years, with some standard deviation (call it *sd**) from perfect periodicity. To answer the second question, they asked how frequently such a close fit to periodicity would occur if the timescale were randomized and extinction peaks were assigned to non-adjacent stages. As it turns out, the probability of obtaining a fit of *sd** or better by chance was vanishingly small, and on this basis Raup and Sepkoski (1984) were able to argue that the periodicity of 26 million years is very unlikely to have arisen by chance and thus should be provisionally accepted.[5]

3.3   The Nemesis Affair and Narrative Closure

Raup and Sepkoski's finding of periodicity, coupled with the Alvarez group's discovery of an iridium anomaly coinciding with the mass extinction of the dinosaurs at the end of the Cretaceous period, led to the formulation of the Nemesis hypothesis (Davis, Hut and Muller 1984; Whitmire and Jackson 1984). According to the Nemesis hypothesis, the sun has a companion star, Nemesis, which every 26 million years perturbs the orbits of comets in the Oort cloud, sending some of them on an earth-crossing orbit, with the resulting impact causing a mass extinction.[6] Linking periodicity with a possible extraterrestrial cause for mass extinction altered the temporality governing palaeontological research to one based on periodicity. In addition, it set in motion a search for a cause capable of producing the extinction periodicity: an astronomical search for a companion star (Muller 1988), a statistical search for periodicity in the ages of impact craters on earth (Rampino and Stothers 1984b) and a search for indicators of impact at stratigraphic horizons corresponding with mass extinctions around the world (e.g., Claeys, Casier and Margolis 1992). In short, this new 'narrative of nature' was compelling enough to galvanize a coalition of researchers from different disciplines, and changed the nature of extinction research, setting in motion a search for narrative closure.[7] Yet alongside the search, critiques were mounted, falling into one of five categories: general scepticism about the warrant for extraterrestrial causation (e.g., Hoffman 1989), uncertainties in the ages of the dated events (e.g., Grieve et al. 1985), mismatch between timing of cause and effect, the possibility that periodicity may be spurious (e.g., Stigler and Wagner 1987) and alternative explanations for the presence of the indicator in question (e.g., Wang, Attrep and Orth 1993).

3.4   Mass Extinction as a Recurring Narrative

While it was already accepted prior to Raup and Sepkoski's finding of periodicity that there have been major mass extinctions in the history of life, there had

---

[5] Stigler and Wagner (1988) point especially to the Signor-Lipps effect (Raup 1986) and the practice of distributing coarsely resolved extinctions among adjacent stratigraphic stages as effects that act to make the empirical extinction record depart from randomness, but which are obliterated by Raup and Sepkoski's timescale randomization.

[6] See Raup (1986) and Muller (1988).

[7] On the crucial and useful distinction between a 'narrative of nature' (what happens in nature) and a 'research narrative' (the narrative of what the researchers did), see Meunier (Chapter 12).

been no reason to suspect that each of these mass extinctions had the same cause.[8] There was every reason to believe that if each mass extinction were to yield to any analysis at all, if the cause or causes were to be found, an idiographic approach was called for. Geologists and palaeontologists are highly trained in the identification of traces, in extracting information from remains, in inferring causal sequence, in arriving at consiliences of inductions and in pursuing multiple working hypotheses. In short, they are trained in reconstructing events from their available traces.[9] This may explain why, among many palaeontologists, the Nemesis hypothesis was met with suspicion. One eminent palaeontologist, Steven Stanley of Johns Hopkins, who in his 1987 book *Extinction* mounted a compelling argument that mass extinction is largely explicable in terms of well-documented changes in climate, summed up the prevailing view well:

If every peak forms part of the periodic array, then it must be attributed to the periodic agent. [. . .] *Do we really need to invoke an extraterrestrial cause for the event that occurred during the latter part of the Eocene Epoch*, for example, when we know that at this time both deep-sea waters and terrestrial climates became cold (and remained so to the present) – and when we have a potential earthly explanation for these events in the form of the isolation of Antarctica over the South Pole via the final fragmentation of a large segment of Gondwanaland?[10]

This is a paradigmatic idiographic narrative explanation. Stanley is pointing out that the elements of a narrative explanation were beginning to coalesce – approaching narrative closure – when out of nowhere, like an asteroid, comes a new narrative. Note that he is not contesting the plausibility or empirical support for the extraterrestrial narrative (although he would do so elsewhere), but rather whether, given the existence of a climatological narrative, the extraterrestrial narrative was necessary.[11]

3.5   On Rereading the Book of Nature

Historian David Sepkoski has written an account of the rise of analytical palaeobiology entitled *Rereading the Fossil Record*, focusing on the period

---

[8]   One might reasonably argue that other, competing narratives of mass extinction – volcanism, climate change, changes in sea level, ocean anoxia – posit a single recurrent cause, but each of these is better understood as a type of cause with different token instances, whereas Nemesis is understood as a single token of recurrence.

[9]   See Crasnow (Chapter 11) for an extensive discussion of evidence 'tracing' in the context of narrative construction.

[10]   Stanley (1987: 216). Emphasis mine.

[11]   Stanley (1987: 215; 1990) also questioned whether extinctions were in fact periodic, or whether their relatively even spacing was simply due to the fact that in extinctions at the global scale it takes a while for the global biota to 'rebound' from a mass extinction. Thus, even if some forcing event were to recur, a mass extinction would not occur, at least until the global biota contained a sufficient number of susceptible species. McKinney (1989) uses a mathematical model to demonstrate the plausibility of this idea.

from around 1970 to the mid-eighties. Darwin and Lyell are understood to have brought us the metaphor of the fossil record as a book from which are missing several chapters, and from the remaining chapters many pages, and from the remaining pages many words, written in a slowly changing language.[12] Sepkoski's account describes three historical phases of rereading that fossil record: literal, idealized and generalized. The literal rereading of the fossil record is exemplified by Eldredge and Gould's (1972) model of punctuated equilibria in which the absence of morphological intermediates from the fossil record is not absence of evidence so much as evidence of absence (of morphologic change in species)! The idealized rereading is exemplified by the nomothetic palaeobiology of the Marine Biological Laboratory (MBL) group, which abstracted away from species as individuals and modelled them as particles in space and time, nomothetism denoting the search for lawlike generalities among historical events.[13] The generalized rereading combines empirical and statistical analysis made possible by the painstaking compilation and digitization of taxonomic data by Sepkoski's father, Jack, with mathematical modelling undertaken for the most part with David Raup (Sepkoski 2012). During the generalized rereading phase of the rise of analytical palaeobiology emerged David Raup and Jack Sepkoski's work on mass extinctions, first as a statistical phenomenon quantitatively distinct from background extinctions and then as a recurring phenomenon registering a 26 million-year periodicity.

In coming to a better understanding of how scientists reread the fossil record, it may be helpful or at least instructive to appeal explicitly to narrative theory as it has been developed in the study of literature. Clearly this is a vast field encompassing a large body of scholarship. I would like to start with the key distinction in narrative theory, as formulated by the Russian formalists, Vladimir Propp (1895–1970) and Viktor Shklovsky (1893–1984).[14]

This is the distinction between the supposed chronological sequence of events, referred to as the *fabula*, and the way they are presented in the narrative discourse, the *syuzhet*. Notably, *fabula* and the *syuzhet* register different orderings.[15] The relationship between these two orderings of events contributes to the literary characteristics of a narrative, allowing for it to exert its effects on

---

[12] See Lyell (1833: 239; 1839: 159) and Darwin (1859: 310–311). For discussion, see Alter (1999, esp. ch. 2). For a discussion and critique of the book metaphor, see Huss (2017, esp. section 10.9, 'Closing the Book Metaphor').

[13] The MBL group consisted of David Raup, Stephen Jay Gould, Thomas J. M. Schopf, Daniel Simberloff and Jack Sepkoski, who gathered at the Marine Biological Laboratory in Woods Hole, Massachusetts, to pursue joint work in nomothetic palaeontology. See Huss (2004; 2009) and Sepkoski (2012).

[14] On narrative theory, see Hajek, Chapter 2.

[15] Gerard Genette draws a parallel distinction in his *Narrative Discourse* (1980) between *histoire* (the ordering of events as they 'actually' occurred, which we infer from the text) and *récit* (the order of presentation of the events in the text). To this he adds *narration*, the act of narrating.

a reader, and to elicit a certain aesthetic response. For example, in Dostoevsky's *Crime and Punishment*, Raskolnikov's murder of the pawnbroker is presented early in the narrative. It is only after reading for a good number of pages that we learn from Porfiry Petrovich's cross-examination that several months prior to the murder Raskolnikov had written an essay arguing that the extraordinary man is not bound by common morality. This ordering of the presentation of events between *fabula* and *syuzhet* elicits an affective response from the reader, for example a feeling of suspense over whether Raskolnikov will crack under questioning.

Crime and Punishment is rather noteworthy for its subversion of the narrative of a typical murder mystery, so, although it illustrates the difference between *fabula* and *syuzhet*, we might be better served using the more conventional genre of the 'whodunnit'. In this genre, the murder is revealed early on in the *syuzhet*, and suspense builds until the identity of the murderer is eventually revealed. I will return to this idea later.

If we take the idea of reading (or rereading) the fossil record seriously, we might regard the traces in the fossil record as forming the *syuzhet*, from which the palaeobiologist infers the *fabula*. The palaeobiologist 'reads on', and keeps rereading in a search for narrative closure. If this is so, then the narrative structure of the mass-extinction account may help explain the search for evidence as the search for closure.

It is important to acknowledge disanalogies between narrative closure in reading a work of fiction and in reading the fossil record. From the reader's point of view, in a work of fiction, the *fabula* is something inferred, and, depending on the work in question, there may not be sufficient textual evidence to adjudicate among rival *fabulae*. At first it might be tempting to think that something analogous is at work in reading the fossil record. Due to underdetermination, scientists may differ in their readings of the fossil evidence, with each reading consistent with the available evidence. In both cases, one might bring in background knowledge, theories of interpretation and the like to provide support for one reading over another. In both cases, we may have no choice but to sit pat with the situation unresolved. Yet there are at least two important disanalogies between reading a work of fiction and reading the fossil record. The first stems from the nature of fiction. It is entirely possible that an author is, to put it glibly, 'all *syuzhet* and no *fabula*'. That is to say, there need not even exist an underlying *fabula* to which the *syuzhet* refers.[16] The author may present, in whatever order, a set of events in the narrative discourse over which there could be great disagreement as to what their true chronological ordering was, and it is possible that there does not even exist any true chronological ordering: what we have are the words on the page and an argument in

---

[16] See West (2001).

favour of one reading or another. Indeed, Walsh (2001) has argued that even in conventional cases of narrative fiction, *fabula* is not ontologically prior to *syuzhet*. Rather, from the *syuzhet*, the reader is constructing – not reconstructing – a *fabula* (not the *fabula*) in an ongoing process of interpretation. *Fabula* is the reader's working version of what happened in the world of the characters – a fictional world. Yet reading the fossil record differs from this: the history of life is not a fiction. First, the palaeontologist presumes that, whether it is empirically ascertainable, there does exist an ordering of events, *wie es eigentlich gewesen*, to which the *syuzhet* (the fossil record as it is read) must in some way be connected. The *fabula* of the history of life is ontologically (and temporally) prior to the *syuzhet* (order of presentation in the fossil record that the palaeontologist is reading). It is being *reconstructed* from the record it has left behind.[17] Second, the form of reading on which the palaeontologist is embarked allows her to expand the text, to look to other stratigraphic horizons, to seek out new evidence, to read on in search of narrative closure an ever-expanding text, in which one narrative is better supported than others, at which point narrative closure will have been achieved, at least temporarily. This is not to say that the situation is *completely* unlike that of rereading a work of literature, in which other information external to the text (e.g., early drafts, memoirs by the author, inter- and extratextual references, theories of interpretation) may help to support both the existence of a *fabula* and give some notion of what it is. Indeed, in the historical sciences in general, it has been argued that at any given time, even in the face of a fixed set of fossils and geological evidence (analogous to the closed form of the written text), the totality of the rest of science (theory, method, observations), which is constantly changing, enables an assessment of which of many possible *fabulae* are best supported (Jeffares 2010).

Under periodicity, which presented a narrative of recurrent, extraterrestrial perturbation of the biosphere, the search for evidence looked completely different. Planetary geologists and astronomers began to reread the record of impact structures (craters, astroblemes) for evidence of periodicity (Grieve et al. 1985). While this record is even more fragmentary and less well-dated then the fossil record, it eventually did yield periodicity (Rampino and Stothers 1984a; 1984b), and the hypothesis of impact periodicity continues to be pursued (Rampino, Caldeira and Prokoph 2019; Rampino, Caldeira and Zhu 2020). At stratigraphic boundaries marking extinction events, iridium

---

[17] This is not to say that the historical traces are all that is used in reconstructing the past. As Adrian Currie (2018) has argued, physical and mathematical modelling themselves can provide evidence for or against reconstructions of the past by determining which interpretations are physically or mathematically possible or impossible. Also, I will leave open for present purposes the nature of truth for statements about the past that arise in debates about social constructivism and scientific realism (Turner 2007).

anomalies were sought and sometimes detected (although for certain events, such as the end-Permian extinction, iridium anomalies have so far turned out to be spurious; Erwin 2015 and personal communication). Where iridium anomalies proved wanting, other markers of impact were sought: shocked quartz (with a distinctive crystalline lattice), microtektites (bits of molten rock associated with the high heat of impact), buckminsterfullerenes, osmium isotopes and soot (Raup 1986: 75–87). Markers of one type or another proved adequate to justify continued pursuit of the hypothesis. Meanwhile, astrophysicists, chiefly Berkeley astrophysicist Richard A. Muller, continued to scan the heavens searching for Nemesis, which as of 2007 was still an ongoing search. The pursuit of narrative closure does not always end in achieving it.

To summarize, emplotting all mass extinctions of the past 250 million years in the narrative of a cause that recurs with clocklike regularity enabled Raup and Sepkoski to resurrect the nomothetism of the 1970s in which they had been integrally involved by fitting a periodic model to the record of mass extinctions, yet at the same time to create a narrative, a narrative of recurrence which drove scientists from a number of different fields – astronomy, planetary geology, isotope geochemistry, mineralogy and palaeontology – to embark on a quest for narrative closure on the basis of a periodic pattern or cause.

In so doing, Raup and Sepkoski's research on extinction resolved an ongoing tension in the history of the earth sciences between uniformitarianism and catastrophism by putting forward an exemplar of a catastrophe (asteroid impact) that behaved according to a uniform periodicity rooted in the regularity of astronomical orbits. The Nemesis hypothesis was thus idiographic and nomothetic, catastrophist and uniformitarian, and it was a narrative explanation.

3.6   Rereading the Book of Nature through Diagrams

One step along the way to constructing a narrative of extinction is to 'read' and reread the stratigraphic record. In order to test whether patterns in the fossil record are consistent with a given causal narrative, such as sudden, catastrophic extinction, it is helpful to be able to investigate historical counterfactuals, which are narratives of events that could have happened, but did not. In the study of mass extinction, one of the templates for the formulation and articulation of counterfactual narratives has been the stratigraphic diagram. Stratigraphic diagrams do not operate alone to produce these counterfactual narratives, but in the context of tacit knowledge and 'ways of seeing' that are an extension of the practices of palaeontological and geological fieldwork. A common visual language and sets of practices makes possible the diagrammatic narratives that have been central to studies of mass extinction.

The stratigraphic diagram thus becomes a template for framing narratives of extinction and even for experimenting with alternative, counterfactual narratives; from reading through different configurations of *syuzhet*, scientists gain a sense of which *fabulae* are consistent with it, answering questions thrown up by the Nemesis hypothesis.[18]

For example, in his 1989 paper, 'The Case for Extraterrestrial Causes of Extinction,' David Raup presents a diagram, plotting the distribution of fossil occurrences of different ammonite species in a stratigraphic section of late Cretaceous age in Zumaya, Spain, based on the fieldwork of Peter Ward (Figure 3.3). Ammonites, cephalopods with a coiled morphology, are one of the taxa that became extinct at the K-Pg boundary. The question Raup sets out to answer is whether this extinction was gradual, stepwise or sudden. Here one must distinguish between apparent and actual patterns: the apparent pattern of last known fossils and the actual pattern of last surviving members of the species. If the actual pattern of ammonite extinction (and, by extension, the end-Cretaceous extinction of other species) was gradual leading up to the K-Pg boundary, then a sudden cause such as a bolide impact is not tenable. If the actual pattern of extinction was stepwise, then a multi-phase event such as a comet shower is not ruled out. And if the actual pattern of extinction was sudden, then an impact-caused extinction becomes viable.

The methodological problem palaeontologists face is that of stratigraphic range truncation: due to gaps in preservation or failure to find or identify species, there is often elapsed time between the last appearance datum (LAD) for any given species in the fossil record and the time that the species actually went extinct, a mismatch between apparent and actual patterns of extinction. This is a missing data problem. The consequence is that the fossil record of sudden, simultaneous extinction of many species can look as if the event were smeared out over geologic time: a sudden extinction event in the *fabula* will appear in the *syuzhet* as gradual, a phenomenon known as the Signor-Lipps effect (Raup 1986). Conversely, if there is a large hiatus in preservation or sampling, then a gradual extinction in which species became extinct one after another over an extended period of time will leave a record that looks as if species all became extinct simultaneously: a gradual extinction on the level of *fabula* will be read as sudden in the *syuzhet*. Alternatively, smaller hiatuses in preservation or sampling can mean an extinction is read as if it happened in a series of bursts – stepwise extinction – even if the extinction was gradual or sudden.

---

[18] This is a classic case of empirical underdetermination, such as is discussed by Miyake (Chapter 5) in the case of seismic data and underlying causal mechanism in the case of earthquakes.

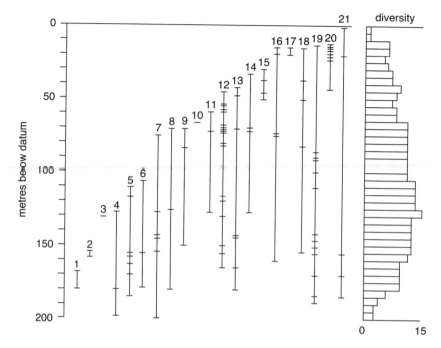

Figure 3.3 **Stratigraphic ranges of 21 lineages (i.e., species genus *Linnaeus*) of ammonites found at Zumaya, Spain**
Vertical scale marks distance in metres below the Cretaceous-Tertiary (today called the Cretaceous-Paleogene) boundary. Numbered vertical lines refer to ammonite lineages. Each horizontal tick mark designates a horizon at which a specimen of the lineage was found and identified. Note the 'gappiness' of the fossil records of the various lineages. For example, specimens of lineage 4 (*Pachydictus epiplectus*) were found and identified at 3 horizons: 200 m, 180 m, and 135 m below the Cretaceous–Tertiary boundary). The histogram on the right plots the number of lineages (inferred from first and last occurrences of specimens) in each 5 m interval (e.g., the 15 lineages who range through the 130 m to 125 m interval). Based on field data of Peter Ward.
Source: Raup (1989).

Raup (1989) points out a paradox in how palaeontologists have tended to read the fossil record. On one hand, palaeontologists know that the fossil record is gappy: absence of evidence does not (generally) constitute evidence of absence; the *syuzhet* requires interpretation in order to reconstruct the *fabula*. On the other hand, there is a tendency to read the last appearance datum as the time of extinction for a species. Raup believes this to be fundamentally a methodological problem, ultimately to yield to a quantitative treatment, but chooses to illustrate the point using an experiment – a thought experiment –

which happens to take the form of a visual, counterfactual narrative. Suppose, he asks, that all fossil occurrences of ammonites were eliminated beginning at a stratigraphic horizon 100 m below the K-T boundary: what would the fossil record of this sudden mass extinction look like? As can be seen from Figure 3.3, he argues, it would look gradual (with a spurious step introduced at the 125 m mark).

As has been pointed out elsewhere, palaeontology has a distinctive visual culture that places a premium on being able to show visually that which might also be demonstrated analytically or mathematically (Huss 2009). For example, when a palaeontologist looks at a stratigraphic diagram, he or she can visualize it as an idealized, synoptic representation of a rock outcrop embedded with specimens of fossil species, as well as the fruit of a great deal of integrative inference. It will be second nature for any geologist or palaeontologist to read this diagram from bottom (oldest) to top (youngest). Field skills and geologic training allow the interpreter to give the diagram a spatiotemporal reality that may not be perspicuous to others (Huss 2017). Embedded in such a diagram as that depicted in Figure 3.3 is a 'research narrative', as well as one of nature. Palaeontological field workers sought, found and identified fossils at certain horizons in the stratigraphic record. Tectonic forces may have distorted, tilted or completely inverted the sequence as found in the field. All is righted in the diagram. Laterally dispersed localities needed to be correlated using principles of stratigraphic inference to determine whether specimens of different species were found at the 'same' horizon. There are many such sketches of the reconstructive aspect of palaeontology that are encoded in a scientific diagram. While they need not be fleshed out each time, and the identities of those making the scientific contribution would itself need to be reconstructed from other sources, when palaeontologists look at a stratigraphic diagram they see encoded in it a community's research narrative.[19]

Yet Figure 3.3 also encodes a 'narrative of nature'. Beds of sediment were laid down, organisms lived and died and left fossilizable hard parts. Periods of erosion or depositional hiatus, along with dissolution of shells, create gaps in the rock and fossil records. Narratives of morphological change and differentiation – microevolution and macroevolution – leave their traces in the patterns

[19] David Sepkoski (2017) has written thoughtfully about the earth as an archive that stands in relation to other archives (synoptic databases among them). A more complete reconstruction of the field work that gave rise to a stratigraphic diagram of fossil occurrences could be achieved by tracking down individual museum specimens, field notes and metadata, but in many contexts of inquiry this level of detail is not needed to glean the temporal biodiversity patterns, the rise and fall of the number of species, represented in the diagram. Decisions always need to be made on how thick or thin research narratives need to be – for example, whether to foreground or background the work of individual scientists. On 'thick and thin description: thickening' research narratives, see Paskins (Chapter 13).

of fossil occurrences. Broad temporal trends in species gain and species loss, ultimately culminating in extinction, can be inferred from the patterns of diversity that are depicted in the running histogram jutting out from the right-hand side of the diagram. Tacit knowledge would enable most palae-ontologists to provide a narrative sketch of what they see in Figure 3.3. Experts on ammonites may be able to venture a richer narrative, but some elements of the causal story remain outstanding. While scientists broadly understand some of the processes that gave rise to the patterns of spatiotem-poral distribution of fossils in this diagram, ultimately, the causal analysis of evolution and extinction will need to be found elsewhere. The patterns in Figure 3.3 are the explanandum. Specific causal hypotheses are the explanans.

Figure 3.4 enables a visual reading of a counterfactual narrative: given the same evolutionary history and gappy stratigraphic distribution of fossils, what would the pattern of last appearances look like if extinction occurred suddenly at the 100 metre datum?[20] Because the temporal sequence of geological and evolutionary events leaves a spatial record – a vertical array of fossil occurrences organized into geologic strata con-sisting of depositional, erosional and quiescent horizons – the resulting visual chronology lends itself to a narrative treatment, including the formulation of alternative narratives to help assess the plausibility of the proposed narrative explanation under consideration. In the same fashion as Figure 3.3, the thought experiment depicted in Figure 3.4 draws upon the knowledge and interpretive habits of palaeontologists, who are now in a position to see that even cases of sudden, simultaneous extinction can leave a misleadingly gradual trace in the fossil record.

In the historical sciences, one often wishes to reconstruct what happened – to produce a historical narrative – based on physical traces, background theory and other assumptions.[21] One way to assess a pattern of physical traces as evidence for or against a proposed narrative is to ask whether a similar pattern would have been expected under an alternative narrative scenario. In these diagrams, the focal question is not what caused the extinction, but how to read the fossil record – what combination of species extinction and spotty preserva-tion does it reflect? Understanding their relative contributions can give rise to

---

[20] Raup (1989) uses this visual thought experiment to motivate the development of a non-parametric statistical technique to assess the effect of gaps on the pattern of fossil occurrences. He imagines repeatedly sampling imagined fossil records from the distribution of fossils and gaps found at Zumaya. Yet, strikingly, in presenting this technique, rather than presenting simply the numerical results, he translates those statistical trials into a diagram, another nod to palaeontology's visual culture.

[21] See Beatty (Chapter 20), for the need for plausible 'back stories' in evolutionary biology.

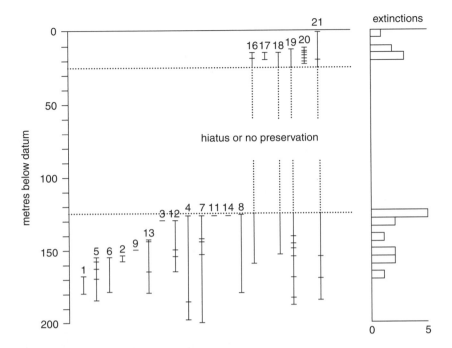

Figure 3.4 **Thought experiment on causes of extinction**
Here a thought experiment is posed: what if all lineages had suddenly become extinct at a datum 100 m below the Cretaceous-Tertiary boundary? Would the pattern of last appearances look sudden or gradual? Note that despite the instantaneousness of this hypothetical extinction event, the apparent pattern of die-off is gradual, with a spurious 'step' appearing at around the 125 m mark. The conclusion may be drawn that an extinction event that was in fact sudden and simultaneous may look gradual when filtered through the 'gappiness' of the fossil record. From data plotted in Figure 3.3. Source: Raup (1989).

a corrected pattern of species extinction, which is what, *qua* historians of life, palaeontologists seek to explain.

Ultimately, however, the narrative that explains the fossil record as we find it, that gives an account of the patterns therein, is relevant to the grander, causal narratives of mass extinction: extraterrestrial, climatological, ecological, volcanogenic, etc. At a minimum, the fossil patterns must be consistent with the proposed mechanism of extinction, but the search for additional evidence – of impact, climate change, trophic shift or volcanism – has taken scientists beyond the fossil patterns themselves to competing narratives of extinction and the evidence relevant to adjudicating among them.

3.7   Narrative Closure in Philosophical Context

Philosophers have disagreed about the epistemic underpinnings of narrative closure in the historical sciences. For starters, there remains the very real possibility that, depending on the question at issue, narrative reconstructions of the past are always one data point or a few data points away from being reopened such that scientists should always be open to the temporariness of narrative closure (Turner 2007). To this, I should add that narrative explanation is remarkably flexible and resilient in the way that components can be retained as well-established (e.g., suddenness of extinction, periodicity), even as evidence for other components of the narrative is found lacking, inconclusive or is even overturned (e.g., evidence for the existence of Nemesis). Second, there is an ongoing debate about what epistemically grounds narrative closure (Cleland 2002; Turner 2007; Forber and Griffith 2011). Cleland has argued that narrative closure is achieved when a 'smoking gun' is found: a piece of evidence that is consistent with one narrative but inconsistent with its rivals. In this view, the Chicxulub crater that has been dated to the end of the Cretaceous period played this role in establishing an asteroid impact as the cause of the K-Pg extinction. Yet Forber and Griffith point out that any given datum only has evidentiary value against a background of auxiliary assumptions, which in the historical sciences can be difficult to test. Hence, data that appear to rule in one hypothesis and rule out its rivals may prove to be indecisive, because their doing so is too sensitive to weak auxiliary assumptions: there is no one-to-one mapping between *fabula* and *syuzhet*. As we saw earlier, in the discussion of Raup's (1989) rereading of the stratigraphic record at Zumaya, evidence that the K-Pg extinction was gradual, based on a petering out of certain species as the K-Pg boundary is approached from below, can easily be shown to be consistent with sudden mass extinction if different assumptions are made about how preservation is expected to result in the observed fossil record. It is easy to 'explain away' inconsistencies in this way: one can 'save the narrative' by deflecting inconsistencies to auxiliary assumptions. Thus, Forber and Griffith (2011) have argued that a more promising and robust way to achieve closure that is likely to be less ephemeral is to ground historical inferences by a consilience of inductions (Whewell 1858), namely by finding lines of evidence that each depend on independent sets of auxiliary assumptions. They give the example of several different sets of evidence that were used to predict the size of the asteroid impact at the end of the Cretaceous and the degree to which they did or did not share auxiliary assumptions as crucial factors in assessing the strength of evidence, both in probabilistic terms and in reception by scientists (Forber and Griffith 2011). For my purposes here, I merely wish to note that historical science as the pursuit of narrative closure is consistent with both of these models.

3.8   Conclusions

A recurrent narrative such as the Nemesis hypothesis challenges some distinctions that have been used to set up oppositions between approaches in palaeontology. Simply put, a narrative of lawlike recurrence has both nomothetic and idiographic components consisting of the mathematical laws governing the periodic forcing agent as well as the overall causal narrative explaining which taxa became extinct, which survived and why. It also challenges the distinction between uniformitarianism and catastrophism, in a sense rendering bolide impact uniformitarian – a periodic catastrophe, as it were (Sepkoski Jr 1994).

The narrative of recurrent extinction known as the Nemesis hypothesis set in motion a search for narrative closure, and for communities of scientists, a quest for evidence that each mass extinction had been caused by an extraterrestrial impact. In the case of the K-Pg extinction, in which the dinosaurs, ammonites and a number of other groups perished, narrative closure was achieved with the discovery of an impact crater of approximately the size predicted on the basis of the iridium anomalies found around the world (Forber and Griffith 2011). This effectively closed off debate about alternative narrative explanations for that particular extinction.

The legacy of the Nemesis affair is far more complicated. For starters, the periodic pattern in mass extinction appears to be too stable to be compatible with the instability of the calculated orbit of the supposed companion star Nemesis (Melott and Bambach 2010)! Still, the pursuit of closure in the impact narrative is ongoing, especially on the part of Michael Rampino and colleagues (Rampino, Caldeira and Prokoph 2019; Rampino, Caldeira and Zhu 2020), but in general there is greater pluralism. Volcanism and deep ocean anoxia are among the proposed causal agents at horizons where evidence of impact is lacking (Rampino, Caldeira and Prokoph 2019), and three episodes of large-scale igneous province (LIP) eruptions are dated at times that coincide with the inferred ages of the three largest known impact craters, all of them falling at or near extinction peaks now computed as having a 27.5 million year periodicity (Rampino, Caldeira and Zhu 2020). This periodicity has been found to be statistically significant over the past 500 million years, extending it twice as far back in time as had been found in Raup and Sepkoski's original analyses (Bambach 2017; see also Erlykin et al. 2017). It is close to the half-period of passes of the solar system through the plane of the Milky Way galaxy – conjuring the image of a 'Galactic Carousel' (Rampino and Haggerty 1996). In a bit of brand differentiation, the hypothesis that the concomitant mass-extinction periodicity is due to the resultant galactically governed influx of asteroids, comets or even dark matter has been dubbed the 'Shiva hypothesis' (Gould 1984; Rampino and Haggerty 1996). Statistical searches for periodicity in the timing of mass extinctions, asteroid crater ages and oscillations through the galactic plane have been ongoing (Rampino and Stothers 1984a; Melott and Bambach 2014; Rampino, Caldeira and Prokoph 2019; Rampino, Caldeira and Zhu 2020). These analyses have also turned up an approximately 60-million-year periodicity in an isotopic signature in marine sediments that has given rise to a variety of alternative narratives involving internal drivers of plate tectonic activity, galactically driven increases in the influx of cosmic rays with effects on upper atmospheric ionization and climate and possible coupling between astronomical cycles and internal geodynamical cycles (Melott et al. 2012).

In other words, despite the pursuit of narrative closure, the science does not seem to be approaching it. Rather, narrative seems to be a rather flexible tool for adjusting to what scientists find as they 'read on'. So what does the Nemesis affair teach us about the pursuit of narrative closure in the case of periodicity of mass extinction? First, periodicity in the temporal pattern of the mass extinctions themselves has stood up to improved resolution of the data, revisions to the geological timescale (Melott and Bambach 2014) and the use of a range of different statistical methods (Rampino, Caldeira and Zhu 2020). Closure seems to have been achieved in the pattern in the timing of the mass extinctions themselves. Second, as might be expected when vastly different narratives compete, such as 'earth-bound' narratives of particular mass extinctions and astronomically driven recurrent causes of the periodic pattern, attempts to achieve narrative closure in one camp are met with attempts to keep the narrative open in another, sometimes by folding the objections in to produce a unifying narrative (Rampino, Caldeira and Zhu 2020). Third, the Nemesis narrative itself, while today finding few adherents, reoriented attitudes such that astronomical processes are deemed worthy candidates for driving biotic and geologic phenomena on planet Earth. Finally, in the case of periodic mass extinction, the search for narrative closure has been empirically and methodologically fruitful. Scientists really are pursuing narratives, seeking to assemble a causal story that can account for the apparent periodicity – we see this particularly in the attempt to connect galactic processes with oceanic, atmospheric and geological processes – drawing on that narrative to guide an empirical quest, and reading on in pursuit of evidence that can provide narrative closure, however elusive it may be.[22]

[22] For inviting me to a workshop on narrative science, I thank Mary Morgan. For discussion, I thank Chris Haufe and Joanna Huss. Audiences at the London School of Economics, Indiana University (especially Ana María Gómez López, Jordi Cat and Jutta Schickore) and The University of Chicago (especially Emma Kitchen) provided helpful feedback on a presentation I gave on earlier drafts of this chapter. *Narrative Science* book: This project has received funding from the European Research Council under the European Union's Horizon 2020 research and innovation programme (grant agreement No. 694732). www.narrative-science.org/.

Bibliography

Alter, S. G. (1999). *Darwinism and the Linguistic Image: Language, Race and Natural Theology in the Nineteenth Century.* Baltimore, MD: Johns Hopkins University Press.

Alvarez, L. W., W. Alvarez, F. Asaro and H. V. Michel (1979). 'Anomalous Iridium Levels at the Cretaceous/Tertiary Boundary at Gubbio, Italy: Negative Results of Tests for a Supernova Origin'. *Abstracts of the Geological Society of America* 11.7: 378.

(1980). 'Extraterrestrial Cause for the Cretaceous-Tertiary Extinction'. *Science* 208.4448: 1095–1108.

Bambach, R. K. (2017). 'Comments on: Periodicity in the Extinction Rate and Possible Astronomical Causes – Comment on Mass Extinctions over the Last 500 myr: An Astronomical Cause? (Erlykin et al.)'. *Palaeontology* 60.6: 911–920.

Claeys, P., J. G. Casier and S. V. Margolis (1992). 'Microtektites and Mass Extinctions: Evidence for a Late Devonian Asteroid Impact'. *Science* 257.5073: 1102–1104.

Cleland, C. E. (2002). 'Methodological and Epistemic Differences between Historical Science and Experimental Science'. *Philosophy of Science* 69.3: 447–451.

Currie, A. (2018). *Rock, Bone, and Ruin: An Optimist's Guide to the Historical Sciences.* Cambridge, MA: MIT Press.

Darwin, C. (1859). *On the Origin of Species by Means of Natural Selection, or the Preservation of Favoured Races in the Struggle for Life.* London: John Murray.

Davis, M., P. Hut and R. A. Muller (1984). 'Extinction of Species by Periodic Comet Showers'. *Nature* 308.5961: 715–717.

Eldredge, N., and S. J. Gould (1972). 'Punctuated Equilibrium: An Alternative to Phyletic Gradualism'. In T. J. M. Schopf, ed. *Models in Paleobiology.* San Francisco, CA: Freeman, Cooper.

Erlykin, A. D., D. A. T. Harper, T. Sloan and A. W. Wolfendale (2017). 'Mass Extinctions over the Last 500 myr: An Astronomical Cause?' *Palaeontology* 60.2: 159–167.

Erwin, D. H. (2015). *Extinction: How Life on Earth Nearly Ended 250 Million Years Ago.* Updated edn. Princeton, NJ: Princeton University Press.

Forber, P., and E. Griffith (2011). 'Historical Reconstruction: Gaining Epistemic Access to the Deep Past'. *Philosophy and Theory in Biology* 3.201306: 1–19.

Genette, G. (1980). *Narrative Discourse.* Trans. Jane E. Lewin. Oxford: Blackwell.

Glen, W., ed. (1994). *The Mass-Extinction Debates: How Science Works in a Crisis.* Stanford, CA: Stanford University Press.

(2002). 'A Triptych to Serendip: Prematurity and Resistance to Discovery in the Earth Sciences'. In E. B. Hook, ed. *Prematurity in Scientific Discovery: On Resistance and Neglect.* Berkeley, CA: University of California Press, 92–108.

Gould, S. J. (1984). 'The Cosmic Dance of Siva'. *Natural History* 93.8: 14–19.

Gould, S. J., and R. C. Lewontin (1979). 'The Spandrels of San Marco and the Panglossian Paradigm: A Critique of the Adaptationist Programme'. *Proceedings of the Royal Society of London. Series B* 205.1161: 581–598.

Grieve, R. A., V. L. Sharpton, A. K. Goodacre and J. B. Garvin (1985). 'A Perspective on the Evidence for Periodic Cometary Impacts on Earth'. *Earth and Planetary Science Letters* 76.1–2: 1–9.

Hoffman, A. (1989). 'Mass Extinctions: The View of a Sceptic'. *Journal of the Geological Society* 146.1: 21–35.

Huss, J. E. (2004). 'Experimental Reasoning in Non-Experimental Science: Case Studies from Paleobiology'. PhD dissertation, University of Chicago.

(2009). 'The Shape of Evolution: The MBL Model and Clade Shape'. In D. Sepkoski and M. Ruse, eds. *The Paleobiological Revolution: Essays on the Growth of Modern Paleontology*. Chicago: University of Chicago Press, 326–345.

(2017). 'Paleontology: Outrunning Time'. In C. Bouton and P. Huneman, eds. *Time of Nature and the Nature of Time*. Dordrecht: Springer, 211–235.

Jeffares, B. (2010). 'Guessing the Future of the Past'. *Biology and Philosophy* 25.1: 125–142. http://dx.doi.org/10.1007/s10539-009-9155-0.

Jevons, W. S. (1884). 'The Periodicity of Commercial Crises and Its Physical Explanation (1878), with Postscript (1882)', In *Investigations in Currency and Finance*. London: Macmillan, 206–220.

Lyell, Charles (1833). *Principles of Geology*. vol. 3. London: John Murray.

(1839). *Elements of Geology*. 1st American edn. Philadelphia: James Kay, Jr. & Brother.

McKinney, M. L. (1989). 'Periodic Mass Extinctions: Product of Biosphere Growth Dynamics?' *Historical Biology* 2.4: 273–287.

Melott, A. L., and R. K. Bambach (2010). 'Nemesis Reconsidered'. *Monthly Notices of the Royal Astronomical Society: Letters* 407.1: L99–L102.

(2014). 'Analysis of Periodicity of Extinction Using the 2012 Geological Timescale'. *Paleobiology* 40.2: 177–196.

Melott, A. L., R. K. Bambach, K. D. Petersen and J. M. McArthur (2012). 'An ~60-Million-Year Periodicity Is Common to Marine 87Sr/86Sr, Fossil Biodiversity, and Large-Scale Sedimentation: What Does the Periodicity Reflect?' *Journal of Geology* 120.2.

Muller, R. A. (1988). *Nemesis: The Death Star* (London: Weidenfeld & Nicholson).

Rampino, M. R., K. Caldeira and A. Prokoph (2019). 'What Causes Mass Extinctions? Large Asteroid/Comet Impacts, Flood-Basalt Volcanism, and Ocean Anoxia – Correlations and Cycles'. In C. Koeberl and D. M. Bice, eds. *250 Million Years of Earth History in Central Italy: Celebrating 25 Years of the Geological Observatory of Coldigioco*. Boulder, CO: The Geological Society of America, 271–302.

Rampino, M. R., K. Caldeira and Y. Zhu (2020). 'A 27.5-my Underlying Periodicity Detected in Extinction Episodes of Non-Marine Tetrapods'. *Historical Biology* 33.11: 3084–3090.

Rampino, M. R., and B. M. Haggerty. (1996). 'The "Shiva Hypothesis": Impacts, Mass Extinctions, and the Galaxy'. In H. Rickman and M. J. Valtonen, eds. *Worlds in Interaction: Small Bodies and Planets of the Solar System*. Dordrecht: Springer, 441–460.

Rampino, M. R., and R. B. Stothers (1984a). 'Terrestrial Mass Extinctions, Cometary Impacts and the Sun's Motion Perpendicular to the Galactic Plane'. *Nature* 308: 709–712.

(1984b). 'Geologic Rhythms and Cometary Impacts'. *Science* 226: 1427–1431.

Raup, D. M. (1986). *The Nemesis Affair: A Story of the Death of Dinosaurs and the Ways of Science*. New York: W. W. Norton.

(1989). 'The Case for Extraterrestrial Causes of Extinction'. *Philosophical Transactions of the Royal Society of London. Series B* 325.1228: 421–435.

Raup, D. M., S. J. Gould, T. J. Schopf and D. S. Simberloff (1973). 'Stochastic Models of Phylogeny and the Evolution of Diversity'. *Journal of Geology* 81.5: 525–542.

Raup, D. M., and J. J. Sepkoski (1982). 'Mass Extinctions in the Marine Fossil Record'. *Science* 215.4539: 1501–1503.

(1984). 'Periodicity of Extinctions in the Geologic Past'. *Proceedings of the National Academy of Sciences* 81.3: 801–805.

(1988). 'Testing for Periodicity of Extinction'. *Science* 241.4861: 94–96.

Schopf, T. (1972). *Models in Paleobiology*. San Francisco, CA: Freeman, Cooper.

Sepkoski, D. (2012). *Rereading the Fossil Record*. Chicago: University of Chicago Press.

(2017). 'The Earth as Archive: Contingency, Narrative, and the History of Life', in L. Daston, ed., *Science in the Archives: Pasts, Presents, Futures*. Chicago: University of Chicago Press, 53–84.

Sepkoski, J. J. Jr. (1982). *A Compendium of Fossil Marine Families*. Milwaukee Public Museum Contributions in Biology and Geology, 51. Milwaukee: Milwaukee Public Museum Press.

(1989). 'Periodicity in Extinction and the Problem of Catastrophism in the History of Life'. *Journal of the Geological Society* 146.1: 7–19.

(1994). 'What I Did with My Research Career: Or How Research on Biodiversity Yielded Data on Extinction'. In W. Glen, ed. *The Mass-Extinction Debates: How Science Works in a Crisis*. Stanford, CA: Stanford University Press, 132–144.

Stanley, S. M. (1987). *Extinction*. New York: Scientific American Library.

(1990). 'Delayed Recovery and the Spacing of Major Extinctions'. *Paleobiology* 16.4: 401–414.

Stigler, S. M., and M. J. Wagner (1987). 'A Substantial Bias in Nonparametric Tests for Periodicity in Geophysical Data'. *Science* 238.4829: 940–945.

(1988). 'Testing for Periodicity of Extinction: Response'. *Science* 241.4861: 96–99.

Turner, D. (2007). *Making Prehistory: Historical Science and the Scientific Realism Debate*. Cambridge: Cambridge University Press.

Walsh, R. (2001). 'Fabula and Fictionality in Narrative Theory'. *Style* 35.4: 592–606.

Wang, K., M. Attrep and C. J. Orth (1993). 'Global Iridium Anomaly, Mass Extinction, and Redox Change at the Devonian-Carboniferous Boundary'. *Geology* 21.12: 1071–1074.

West, R. (2001). 'All *sujet* and no *fabula*? *Tristram Shandy* and Russian Formalism'. In Jörg Helbig, ed., *Erzählen und Erzähltheorie im 20. Jahrhundert*, Festschrift für Wilhelm Füger. Heidelberg: Carl Winter, 283–302.

Whewell, W. (1858). *Novum Organon Renovatum*. London: John W. Parker.

Whitmire, D. P., and A. A. Jackson (1984). 'Are Periodic Mass Extinctions Driven by a Distant Solar Companion?' *Nature* 308.5961: 713–715.

4 The Narrative Nature of Geology and the Rewriting of the Stac Fada Story

Andrew Hopkins

Abstract

Geology can be characterized as 'earth history'. As such it relies on narrative forms in its explanations and interpretations. Unlike 'mere' stories, however, geological narratives are tightly constrained by physical laws, and they typically play an important role in geological reasoning. It is not uncommon for geological narratives to be rewritten when new evidence emerges or when theories change, as is illustrated here by the history of the changing interpretation of a particular stratigraphic layer in north-west Scotland, which had been regarded as unremarkable in the late nineteenth century. In the 1960s, it was re-evaluated as being the product of a violent volcanic eruption and was named the Stac Fada Member. A further reinterpretation in 2006 led to it being identified as the material ejected from a meteorite impact crater. This chapter examines the reasons behind the rewriting of these explanatory narratives and explores how narratives are used in geology.

4.1 Introduction

In an article entitled 'The Geologist as Historian', one of the twentieth century's foremost British geologists, H. H. Read (1889–1970), characterized geology as 'earth-history' (1952: 409). Read's point was that geology has a lot in common with human history, dealing as it does with the reconstruction of particular events that occurred in the past, albeit on a very different timescale (and minus a role for human agency). Read noted that in geology, as with its counterpart in the humanities, 'no event has ever been exactly repeated' (1952: 411). A possible point of contention emerges, however, when this analogy is extended to the mode of explanation in geology. It is uncontroversial to state that explanations in history have a narrative form, but suggesting that ultimately all explanations of events in time are narrative in structure (as asserted by Richards 1992: 22–23) may well cause unease among many geologists who do not wish their science to be associated with a term which may seem vague and unscientific, or which is suggestive of 'mere' storytelling.[1]

In this chapter, I will seek to emphasize and uphold the narrative nature of geology, a property which may not be widely understood or appreciated, even by its own practitioners, but which has a logic and rigour of its own (Frodeman 1995: 966). Unlike those in human history, narratives in geology are always strictly bounded by what is possible according to the general laws of physics and chemistry; they are also tightly constrained by what is geologically plausible, although there is always scope for daring or 'outrageous' hypotheses (Davis 1926), which are open to review and testing by the geological community.

Without being too prescriptive, the essence of a narrative statement can be usefully thought of as specifying two time-separated events, so that the prior event is understood to have given rise to, and thereby to explain, the later event;[2] it is typically expressed in the past tense (Danto 1962: 146). More complex narratives, involving multiple events and processes which have interconnected causal relationships, are built on this simple formulation. A narrative explanation therefore specifies the causal connections in a temporal sequence of events or processes. It is important to note that a narrative in a historical science such as geology is neither a chronicle (a chronological listing of disconnected events),[3] nor is it a merely descriptive exercise. It has been argued that a defining characteristic of historical sciences such as geology is that they rely on narrative sentences for understanding (Griesemer 1996: 66). As will become apparent in the course of this chapter, however, narratives in the geological literature are not always explicitly narrative.

4.1.1   Narrative Reasoning

While the most obvious function of narratives in geology is to communicate explanations or interpretations, narrative is also fundamental to geological reasoning. Mary Morgan (2017) describes the mental process of 'narrative ordering', in which what may initially appear to be disconnected events are able to be woven together into a coherent whole, thereby imparting meaning to them. Narrative ordering may take place in a variety of settings, such as in the act of individual reflection, in the course of writing or sketching, or in the process of discussion with others. The idea expounded by Morgan refers to common practice in the social sciences, but it describes well the route a geologist might follow in order to make sense of a collection of puzzling

---

[1] See Olmos (Chapter 21) on 'just so' stories.
[2] This is based on the characterizations of Arthur Danto (1962: 146) and Robert Richards (1992: 23).
[3] See Berry (Chapter 16) and Kranke (Chapter 10) for more-detailed discussions on chronicles.

observations. Interpreted events in geology also derive their meaning from being part of an overall story; that is, they only make sense when they contribute to and form a component of an overarching narrative. The theory of plate tectonics, involving the separation and collision of continents on a timescale of hundreds of millions of years, supplies the most obvious 'big picture' narrative in modern geology.[4] Robert Frodeman refers to this property in which 'details are made sense of in terms of the overall structure of a story' as 'narrative logic' (1995: 963). However, the term 'narrative logic' could usefully be extended to include the criteria employed in narrative ordering. Hence, narrative logic can be understood as having both an internal dimension, through the coherent ordering of related events, and an external dimension, via the relationship of those events to overarching ideas.

Counterfactual reasoning can serve as another powerful device in the geologist's mental toolbox, although it may not be explicit in many written accounts. Deliberately changing elements of a geological narrative or filling in gaps in the data to see what difference it makes to the overall picture can reveal flaws or strengths in a particular argument. Hence, counterfactual reasoning can help to expose narratives that do not make geological sense (i.e., do not display narrative logic), clearing the way for ones that do.[5]

4.1.2   The Impermanence of Geological Narratives

The evidence available to the geologist seeking to piece together 'the fantastic drama of the earth's crust' (Read 1952: 409) consists of traces[6] of events which ceased long ago but which have been left behind in rocks, fossils and landscapes. These traces, however, are prone to concealment, degradation, even complete destruction over the vast expanse of geological time: weathering and erosion act on rocks at the surface, while at depth, profound changes may be wrought by pressure, heat or geochemical reactions. It is also true that while some traces are particularly susceptible to elimination, certain geological processes leave no traces at all (e.g., Tipper 2015). As Kleinhans, Buskes and de Regt (2005: 290) conclude, a result of this incompleteness is that 'theories and hypotheses [in geology] usually are underdetermined by the available evidence'. Consequently, the word *interpretation* tends be used more frequently than *explanation* in geology.[7] The two terms are almost synonymous,

---

[4] Kleinhans, Buskes and de Regt (2005) point out, however, that the theory of plate tectonics often occupies an implicit, background role in local geological interpretations.

[5] See, for example, Beatty (2017) on narrative possibilities and this volume (Chapter 20), on counterfactuals.

[6] A trace may be defined as a 'downstream causal descendant' of a past event (Currie 2018: 56).

[7] For example, in the article by Branney and Brown (2011), which is discussed later in this chapter, *interpret* (or its derivatives) occurs 14 times, while there is only one instance of *explain* or its

although *interpretation* suggests something more provisional and hypothetical.[8] In the context in which it is generally used by geologists, an interpretation can be understood as a response to the question 'What caused these traces?' (see, for example, Faye 2010: 108–111).

In a discipline in which previously concealed or overlooked traces have a habit of eventually turning up, new ideas, analytic techniques and interpretative methods are constantly being developed. Old theories are regularly replaced, so narratives tend to come with a degree of implied uncertainty and provisionality and are often not assumed to be the last word. When the evidence changes or when theories are superseded it is not unusual for geological narratives to be modified or even to be completely rewritten, although intellectual inertia might retard the process of revision.

4.1.3   Central Subjects in Historical Narratives

The construction of a historical narrative requires the identification of a central subject (Hull 1975). Its purpose is to provide the coherence necessary for intelligibility (Frodeman 1995: 965–966) and to form 'the main strand around which the historical narrative is woven', a key requirement being continuity in space and time (Hull 1975: 262). In the rest of this chapter, the role of central subject will be occupied by the outcrop of a particular stratum of ancient rock situated in Scotland. The changing historiography of this layer will serve to illustrate both the narrative nature of geology and the impermanent character of many geological narratives.[9] The focus of the case study is on how geologists communicate through papers and articles within the community of fellow practitioners.

4.2   The Case of the Stac Fada Member

The cliffs along the Assynt coastline of Sutherland, north-west Scotland, are formed of some of the oldest sedimentary rocks in the British Isles. The reddish-brown outcrop has been known informally as the Torridonian since the late nineteenth century when it was recognized to be of Pre-Cambrian age.[10] Despite their great antiquity, the rocks are acknowledged to be remarkably well preserved. They were originally assumed to be unfossiliferous, although eukaryotic microfossils are now known to be present (Brasier et al. 2017).

---

derivatives. For a discussion of interpretation and explanation in historical sciences see Olmos (Chapter 21).

[8] See also Frodeman 1995 on the role of interpretation in geology.

[9] On narrative's functions in hypothesis testing, see Crasnow (Chapter 11).

[10] In formal stratigraphic nomenclature, Pre-Cambrian has now been replaced by (in order of increasing age) the Proterozoic, Archaean and Hadean Eons.

Writing in 1897, J. G. Goodchild of the Geological Survey of Great Britain drew a direct analogy between the ancient sediments that formed the Torridonian rocks and the modern sands currently being deposited in ephemeral rivers and lakes in the Sinai Desert, on the basis of their remarkably similar form and composition:

To my mind one of the most striking and significant illustrations of the principle upon which geologists interpret the records of the Past, by the study of the Present, is to be found in the Torridonian areas of the North-West of Scotland. If we review the conditions obtaining in the Sinaitic Peninsular [. . .] we find going on there to-day almost the exact counterpart of what must have taken place in Pre-Cambrian times in Sutherland and Ross. (Goodchild 1897: 220–221)

The Geological Survey studied and mapped the rocks of north-west Scotland in a major campaign that ran from 1883 to 1897. The results were published in a substantial Memoir several years later (Peach et al. 1907), in which descriptions of the Torridonian rocks occupied one of five subsections.

Following something of a hiatus during the early twentieth century, research on the Torridonian resumed in the 1960s. A research group was set up in the Geology Department of Reading University focused exclusively on furthering knowledge and understanding of the Torridonian (Stewart 2002: 3). Initial reconnaissance work revealed the rocks to be 'unexpectedly complex' (Gracie and Stewart 1967: 182), and it took several years of careful fieldwork to unravel the stratigraphic relationships. Among the unexpected complexities encountered by the Reading group was a particular layer which was noted to be generally between 10 m and 30 m thick. In the redefinition of Torridonian stratigraphy undertaken by the Reading group, this layer was named the Stac Fada Member (Stewart 2002: 5).[11] This rock unit, the outcrop of which stretches across more than 50 km of coastline (Figure 4.1), was regarded as unremarkable by the nineteenth-century Survey geologists,[12] but was noted by the Reading researchers to differ in a number of significant respects from the layers immediately above and below (Stewart 2002: 9–11).

D. E. Lawson's (1972) study of the Stac Fada Member described some of its constituents as angular shards of pumice, green particles of devitrified glass[13]

---

[11] Stac Fada is the 'type' location near the settlement of Stoer (Figure. 4.1) and *member* is a designation in the rock-stratigraphic classification hierarchy. The Stac Fada Member is part of the Bay of Stoer Formation, which itself is a sub-division of the Stoer Group (Stewart 2002: 5).

[12] Peach et al. (1907: 313) interpreted the layer as a sedimentary deposit which contained some fragments eroded from older igneous rocks.

[13] When volcanic magma cools rapidly it can form an amorphous glassy material which subsequently devitrifies into a crystalline silicate, and which commonly has the green colour seen in the Stac Fada Member.

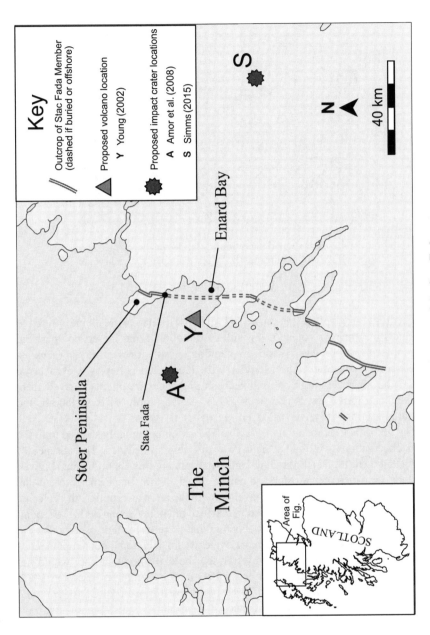

Figure 4.1 **Location map of the Stac Fada outcrop**

Figure 4.2 **Ball-shaped accretionary lapilli on the surface of a Stac Fada Member outcrop** The largest examples shown here are about 15 mm in diameter.
Source: Image courtesy of Renegade Pictures/Channel 4.

and accretionary lapilli[14] (as shown in Figure 4.2). These were all interpreted as products of a nearby volcano. As with Goodchild's Sinai Desert comparison, this interpretation was made by analogy with present-day processes. Accordingly, the Stac Fada Member was interpreted either as a pyroclastic flow – an airborne surge of fluidized ash and other fragments derived from a violent eruptive event (Lawson 1972) – or as more of a surface-bound volcanic mudflow (Stewart 2002).

The volcanic interpretation of the Stac Fada Member initially seemed to fit the field observations well and it held sway for several decades. The absence of a volcanic vent in the surrounding landscape was not seen as problematic, given the long-term effects of erosion and burial. Based on the distance that present-day accretionary lapilli are known to travel through the air in an eruption, the volcanic vent was suggested to have lain a short distance offshore (Young 2002: 7–8; point Y in Figure 4.1). However, it was apparent that there were some aspects that did not add up. For example, the lack of evidence for additional contemporaneous flows was regarded as 'curious' by Lawson: he explained that one 'would not really expect volcanic activity to cease after a single eruption' (Lawson 1972: 346, 360). Furthermore, several 'thorny problems' (Stewart 2002: 10–11) were identified in the geological evidence. For instance, indicators of transport

[14] Accretionary lapilli are distinctive pellets, generally pea-sized, with concentric internal structures. They are the volcanic equivalents of hailstones, and form by the successive build-up of thin layers of dust around nuclei as they are suspended by updrafts in plumes of hot gas and ash.

directions in the sediments showed that there had been an 'abrupt change' in the slope of the land from east to west 'immediately prior to deposition of the Stac Fada Member' (Stewart 2002: 10–11). This was difficult to explain in terms of the volcanic hypothesis given that the outcrops along the coast were estimated to have been located too far from the putative volcano to have been affected by any associated land movements. Although several modes of volcanic emplacement had been proposed, it was concluded that none of them satisfactorily explained all of the field observations (Stewart 2002: 10–11), and Stewart noted that, in 2002, the volcanic hypothesis remained 'controversial' (Stewart 2002: 65). Despite these unresolved issues, the phenomenon of Stac Fada volcanism was incorporated, albeit with a degree of incongruity, into the body of literature on Scottish geology. For example, a major synthesis of the tectonic and magmatic evolution of Scotland included a short section on Stac Fada volcanism, in which the authors referred to it as 'enigmatic' (Macdonald and Fettes 2006: 232–233).

4.2.1   Old Evidence, New Discovery

Since 2004, the Torridonian outcrop has formed part of the North West Highlands Geopark and has become a popular destination for geology undergraduate field trips. On an Oxford University field course in 2006, postgraduate geologist Ken Amor was serving as an assistant to the teaching staff. Amor had recently returned from Ries in Bavaria where he had been studying the rocks around one of Europe's few recognized meteorite craters. His attention was drawn to the distinctive green fragments of devitrified glass in the Stac Fada outcrop which had been highlighted by the Reading geologists. Although they were consistent with the prevailing volcanic interpretation, he had seen remarkably similar crystals close to the Ries crater, where they were interpreted to have been formed by the melting of the surface rocks in the impact event. There were no known instances of a major meteorite strike in the British Isles, but the green particles aroused Amor's curiosity. A microscopic examination of thin sections of the rock in question would provide a test of Amor's hunch, and on returning to Oxford he discovered that his department already held some thin sections that had been made from rocks collected on previous field trips.[15] However, he knew that the chances of finding anything new were not promising. 'How many countless eyes of undergraduates had looked at these very same thin sections over several decades and not spotted anything unusual'?[16]

---

[15] A thin section is a sliver of rock cut with a diamond saw and ground to a thickness of around 30 μm for mounting on a glass slide for analysis using a petrological microscope.

[16] Quoted in 'Walking Through Time: Scotland's Lost Asteroid . . . The Backstory', ToriHerridge. com: https://toriherridge.com/2016/09/23/walking-through-time-scotlands-lost-asteroid-the-backstory/.

What Amor hoped to see down his microscope were crystals of shocked quartz – a form of silica which bears the marks of the instantaneous application of stresses that are far higher than can occur in terrestrial processes, and which is regarded as an unequivocal indicator of a so-called 'hypervelocity impact'. Against his expectations, Amor did find grains of shocked quartz in the Stac Fada thin sections (Figure 4.3), and the implications of his discovery soon dawned on him. He later reflected: 'I remember thinking at the time that at that moment I was the only person [. . .] to realise that the UK had been struck by an asteroid [. . .] I didn't tell my supervisor for two days because I wanted to hold on to that discovery moment for a little longer'.[17]

Amor's moment of insight has led to the Stac Fada Member being reinterpreted as an ejecta blanket – the material violently thrown out of the crater

Figure 4.3 **Photomicrograph of a shocked quartz grain from the Stac Fada Member**
Showing two sets of intersecting lines (see inset sketch). These are planar deformation features (PDFs), which represent primary evidence for shock metamorphism. Image is approximately 0.35 mm across.
Source: Amor et al. (2008).

[17] See n. 16, above.

formed by the impact of a massive extraterrestrial body, overturning the previous interpretation of the layer as the product of a volcanic eruption and resolving many of the inconsistencies surrounding that explanation. For example, the sudden change in the slope of the land now made sense as a consequence of the impact. Amor published his findings in a co-authored paper two years after the breakthrough discovery (Amor et al. 2008). In addition to shocked quartz, the paper presented further evidence of an impact origin including the shocked form of another mineral (biotite) and key geo-chemical indicators such as anomalous chromium isotope values and elevated abundances of platinum group metals such as iridium. Subsequently, grains of shocked zircon were discovered in the Stac Fada deposit (Reddy et al. 2015), further confirming the impact ejecta interpretation. The melange of angular fragments and partially melted material in the Stac Fada Member is now routinely referred to as a *suevite*, the term for such a deposit created by an extraterrestrial impact.[18] The particles of devitrified glass and accretionary lapilli which had been assumed by the Reading geologists to be uniquely diagnostic of volcanic eruptions were evidently also capable of being formed in major meteorite impacts. This had been recognized from evidence at the Ries impact site for some time (Kölbl-Ebert 2015: 275), a fact of which Amor would have been aware. According to radiometric dating of the Stac Fada Member, the impact occurred approximately 1.2 billion years ago, placing it in the Mesoproterozoic Era (Parnell et al. 2011).[19]

There is no sign of the impact crater from which the Stac Fada Member was ejected. As with the now discarded volcanic interpretation and the absence of a volcano, this is not surprising given the burial or removal by erosion of much of the Torridonian outcrop in the last billion or more years. Different lines of reasoning by different geologists, but based essentially on the same evidence, have resulted in two different locations being posited for the crater (A and S in Figure 4.1). Amor et al. (2008; 2019) suggest that a point in the Minch Basin 15–20 km to the north north-west of Enard Bay, an area that would have been dry land at the time of impact, is the most likely location, while an alternative hypothesis by Simms has the crater deeply buried onshore about 50 km to the east of the outcrops along the coast (Simms 2015). Further data-gathering work in the form of expensive geophysical surveys or borehole drilling would be required to confirm or deny both proposals. However, Simms's location has been criticized as being inconsistent with the overarching narrative of the plate tectonic history of northern Scotland (Butler and Alsop 2019: 443), and, in an example of counterfactual reasoning, Amor et al. (2019) argue against Simms's

---

[18] The term was coined in 1901 for the Ries ejecta blanket, though at the time this was also assumed to have a volcanic origin (Stöffler and Grieve 2007: 25; Kölbl-Ebert 2015: 1).

[19] The Mesoproterozoic Era is a sub-division of the Proterozoic Eon.

location by pointing out that there would have been a topographic obstruction blocking the path of the ejecta blanket if it came from the east (Amor et al. 2019: 842).

Three years after the publication of the discovery paper (Amor et al. 2008), volcanologists Michael Branney and Richard Brown addressed the question of exactly how the Stac Fada Member might have been emplaced. Nothing remotely approaching the scale of the meteorite impact from which the deposit is believed to have originated has ever been observed in recorded history,[20] and terrestrial impact ejecta blankets are commonly not well preserved, limiting the availability of possible analogues. The example at Ries is a notable exception, however, and Branney and Brown (2011) highlight parallels between the Stac Fada deposit and the Ries suevite. The absence of a crater, however, means that there is no way of investigating the relationship between it and the ejecta.

The authors note that while there are 'important differences' between the ejecta from impacts and those from volcanoes – for example, the presence of shocked minerals and distinctive geochemistry – there are also 'striking similarities' (Branney and Brown 2011: 287–288). The distinctive components of the Stac Fada deposit, including the presence of devitrified glass and accretionary lapilli, as well as the order in which they were deposited, resemble those ejected from large explosive volcanic eruptions. These similarities have led them to deduce that emplacement mechanisms comparable to pyroclastic processes associated with volcanoes were at work and they have coined the analogous term, *impactoclastic* (Branney and Brown 2011: 276) to describe their model, which details how the Stac Fada impact ejecta blanket could have been deposited (as we can see in Figure 4.4).

4.3   Tracing the Narratives

4.3.1   *Geological Narratives, Explicit and Implicit*

At first glance, the writings of the geologists who have worked on the Torridonian outcrop since the late nineteenth century seem to contain few obvious instances of narrative statements of the kind discussed in the introduction, i.e., those which causally connect time-separated events. Notable

---

[20] A meteorite is defined as a fragment of an asteroid (a rocky, sub-planet-sized body) or comet (an amalgamation of rock, dust, ice and frozen gases) that has passed through the atmosphere and has collided with the surface of the Earth. However, all but the largest bodies tend to burn up and disintegrate or vaporize in the atmosphere, in which case they are referred to as meteors. The largest meteoric event in recorded history was an airburst which occurred over the remote region of Tunguska, Siberia, in 1908; a smaller meteor exploded above the Russian conurbation of Chelyabinsk in 2013 (Artemieva and Shuvalov 2016). Neither event resulted in a significant crater. Falls of much smaller meteorites, remnants of larger bodies which have broken apart, are not uncommon and are well attested in history (e.g., Marvin 1999).

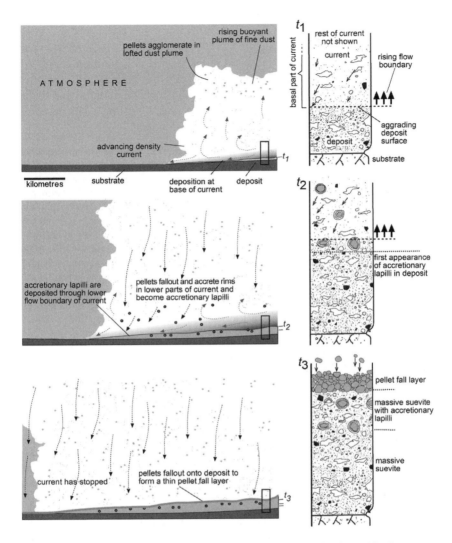

Figure 4.4 **The impactoclastic emplacement of the Stac Fada ejecta blanket**
For the benefit of the non-geologist reader, three pairs of images show the situation at
successive points in time immediately following the meteorite impact. Each pair
consists of a panel showing a cross-section through the dust plume thrown up by the
impact (on the left) and a column representing the vertical accumulation of different
types of debris deposited by the plume by that time (on the right). The time sequence, t_1
to t_3, runs from top to bottom. In the plume cross-sections, the crater lies out of frame to
the right and the plume moves from right to left through the time sequence. Along the
base of each of these cross-sections is the layer of debris deposited from the plume. This
increases in thickness with time as marked by the ticks labelled t_1, t_2 and t_3 at the bottom
right of each cross-section. The location of each column of debris is marked by
a rectangular outline in the bottom right of each corresponding plume cross-section. An
understanding of how the overall diagram is put together, along with some technical
(geological) knowledge, enables it to be read as a self-contained narrative.
Source: Branney and Brown (2011).

instances include parts of Goodchild's (1897) paper on the desert environment in which the Torridonian sediments were interpreted to have been deposited. For example, in the following passage he gives a straightforwardly narrative account of the weathering processes by which the Torridonian sands and shales would have been formed from the breakdown of pre-existing rocks based on observations made in the present-day Sinai environment:

[R]ain fell only occasionally, or practically never, and only on those occasions when thunderstorms happened to burst over the regions in question. At other times the arid conditions gave rise to great diurnal ranges of temperature. The rocks in consequence were heated soon after mid-day far above the temperature usual in more humid climates, and by early morning, owing to rapid radiation, had cooled down to the opposite extreme. In a rock composed of constituents of diverse mineral character differential expansion takes place, owing to their different coefficients of expansion. The felspars in the rocks [...] gave way under the strain set up by extreme expansion and contraction, due to the rapid changes of temperature. The ferro-magnesian minerals [...] in like manner splintered into fragments so small that they were easily blown away as dust by the wind. Little by little the rocks crumbled down, and of their wasted portions the larger part slid down the valley side as talus, to be eventually distributed and spread out in the bottoms of the wadies by the action of the occasional torrents arising during storms; the remainder, chiefly in the form of dust, was blown far and wide by the winds. (Goodchild 1897: 221)

Another noteworthy narrative passage occurs as part of the impactoclastic model of Branney and Brown (2011). The following excerpt accompanies an explanatory 'cartoon' (the main part of which is reproduced here as Figure 4.4):

Time frames (t_{1-3}) [depict] the generation and evolution of ash aggregates within an impactoclastic current. Turbulent entrainment of atmospheric air along the upper mixing zone of the current results in expansion and lofting, generating a buoyant dust plume. Within this [plume], ash pellets start to form (t_1). Once these pellets become too large to be supported by turbulence in the lofted plume, they drop to lower parts of the current, dry out, accrete concentric ash rims (t_2), and become deposited as fully formed accretionary lapilli, along with suevite from the base of the current. After cessation of the current, ash pellets fall out from the drifting buoyant dust plume and deposit directly on the top of the suevite (t_3). The absence of accretionary lapilli in the lower parts of deposits is due to the time lag between the onset of deposition from the base of the current and the formation of pellets, their descent into the current, their growth within the current into accretionary lapilli, and their subsequent deposition. (Branney and Brown 2011: 284–285)

The use of images to complement or clarify textual narratives is common in geology, and this text is designed to be read alongside the diagram. With appropriate geological knowledge and understanding of the context, however, the diagram itself could be read independently as a narrative, tracing as it does the temporal and spatial sequence of deposition caused by the transit of the waning dust plume.[21]

---

[21] A description to assist the non-geologist reader can be found below Figure 4.4.

In each of the passages by Goodchild (1897) and Branney and Brown (2011), the narrative structure of temporally arranged and causally connected sequences of events and processes is evident. Words and phrases denoting events and processes include 'differential expansion', 'splintered into fragments', 'turbulent entrainment' and 'deposition'; while causal connections are signalled by 'gave rise to', 'in consequence', 'results in' and 'generating', among others. Both passages also illustrate the role of physical laws in constraining geological narratives. For example, Goodchild refers to the differential expansion of minerals when heated, and the effects of turbulence, expansion and gravity in a hot dust plume form part of the account of Branney and Brown.

While these passages constitute examples of explicit geological narratives, in most of the other literature on the Torridonian and the Stac Fada Member considered here, narratives are generally more covert and implicit. Consider the following two sentences taken from A. D. Stewart's volume on the Torridonian:

Upward movement of the rift floor on the east arrested the growth of the alluvial wedge and formed a depression that trapped the Stac Fada mudflow and the lake sediments constituting the Poll a' Mhuilt Member that follows. (Stewart 2002: 21)

The palaeosol grades up through sandy claystone with corestones of gneiss (locally cut by the unconformity), into dusky red claystone. (Stewart 2002: 31)

The first sentence is manifestly narrative, linking as it does a chain of events causally connected by the 'Upward movement of the rift floor'. On the other hand, the second sentence appears at first to be a straightforward description. However, a geologist would also read this as a sequence of events. For example, the verb, 'grades up', while primarily a spatial expression, also serves as a proxy for temporal change due to the link between vertical succession and geological time in stratigraphy.[22] The change from palaeosol (fossil soil) to sandy claystone to red claystone indicates a series of environmental changes from humid to arid or semi-arid conditions; the corestones and the unconformity are also both the result of geological processes that have operated through time.[23] The sentence is therefore narrative when read in a certain way by a certain person (i.e., a geologist). The significance of phrases such as 'grading up'

---

[22] The phenomenon of diachronism, in which the age of a deposit may vary laterally, means that this relationship is not always entirely straightforward, however.

[23] 'Corestones' are the result of a certain type of chemical weathering and 'an unconformity' is a surface which represents a gap in time.

can be appreciated through the concept of *scripts* as described by David Herman in the field of cognitive narratology. Scripts allow the reader to 'build up complex (semantic) representations of stories on the basis of few textual or linguistic cues' (Herman 1997: 1051).[24] The ability to recognize cues entailed by geological terms derives from a geologist's specific training and experience, rather than from a general familiarity with routine life situations as in Herman's examples. Most of the narrative work in the geological papers referred to in this chapter is implicit and is performed by sentences which are nominally descriptive but which contain multiple geological cues. Unlike most (explicitly) narrative sentences, these tend to be written in the present tense.

Explicitly narrative sentences are particularly rare in some papers. Where they do occur, they tend to be restricted to the abstracts or the conclusion sections. For example, apart from one narrative sentence in the abstract of Amor et al. (2008), the entire paper is composed of dry geo-scientific prose in the form of (nominally) descriptive sentences laden with various cues which contain the implicit, underlying narrative of geological processes.[25] This form of presentation seems at odds with the cataclysmic drama of the discovery being reported. When he was interviewed for the BBC Radio 4 Today programme in 2019, Amor gave a very different style of account, which imagined the scene about 100 km from the point of impact:

The first thing you'd see would be this enormous fireball extending up from where the asteroid hit the surface. That would generate thermal radiation enough to ignite wood and paper. Shortly after that you would feel a seismic wave equivalent to a magnitude 8 earthquake. About 2.4 minutes later you would get the first debris – dust, hot bits of molten rock raining down on you. At 100 kilometres away it would be enough to cover about 6 inches depth. And then the final thing would be the 450 mph wind that would suddenly hit you as the air blast comes in.[26]

The tone of Amor's contribution was suggestive of the process of narrative ordering that he and his colleagues might have gone through when working out the causal sequence of events before the narrative got turned into the relatively bland text of a scientific paper.

---

[24] I am grateful to Kim Hajek for introducing me to this concept. See Andersen (Chapter 19) for a fuller discussion of Herman's use of scripts.

[25] The narrative sentence is: 'Field observations suggest that the deposit was emplaced as a single fluidized flow that formed as a result of an impact into water-saturated sedimentary strata' (Amor et al. 2008: 303).

[26] 'Scientists Close in on Hidden Scottish Meteorite Crater', BBC News (audio file): www.bbc.co.uk/news/science-environment-48560989.

4.3.2   Narrative Logic and Narratives Rewritten

The narrative sentences and statements discussed in the previous section may be thought of as narrative units or fragments,[27] each of which contributes to an extended narrative history of the Stac Fada Member. In little more than a century, the Stac Fada Member has been the subject of three of these narrative histories, each constituting a radical departure from its predecessor. The sequence might be summarized as follows:

1. The layer that came to be named the Stac Fada Member is an unremarkable part of the Torridonian outcrop, the sandstones and shales of which were deposited by rivers and lakes in a semi-arid environment (Geological Survey: late nineteenth century).
2. The Stac Fada Member was formed by a violent pyroclastic surge or volcanic mudflow derived from a nearby eruption (Reading Group: 1960s–2000s).
3. The Stac Fada Member represents the material violently ejected from the crater formed by a major meteorite impact (Oxford Group: since 2006).

The nineteenth-century geologists appear not to have recognized the distinctive nature or the significance of the Stac Fada Member, or perhaps they overlooked it altogether. This is not particularly surprising given the limited extent of the Stac Fada outcrop (Figure 4.1) and the extensive area and difficult terrain covered by the Survey geologists in mapping the north-west Highlands. The first change of narrative introduced the idea that the deposit was formed by volcanic activity, a familiar geological phenomenon. The second change, however, invoked a fundamentally different and novel causal explanation. The impact narrative was able to challenge the volcanic consensus largely on the basis of a piece of microscopic evidence which had been missed by all previous investigators. It also resolved some of the logical problems that had beset the volcanic narrative, such as the apparent occurrence of a solitary eruption and the evidence for the tilting of the land surface.

4.3.3   The Acceptance of the Impact Narrative: The Back Story

The volcanic interpretation of the Stac Fada Member was first expounded in the 1960s (Lawson 1965), a time when the possibility of extraterrestrial explanations was not even being considered by most geologists. Shocked quartz was not found in the deposit until 2006 because nobody had previously looked for it,

---

[27] Richards (1992: 25) coined the term 'narrites' to denote these smaller narrative units; it never seems to have caught on, however.

a deficit that can be explained at least partly by the fact that the potential significance of shock metamorphism and its distinctive petrology only began to be reported in the 1960s. The tendency for evidence to be overlooked when it does not form part of the observer's conceptual framework recalls an incident recorded by Charles Darwin (1809–82). Writing about a time before he knew of the theory of glaciation, Darwin recounted his experience of spending 'many hours' examining the rocks in a valley in North Wales 'with extreme care' while completely missing the abundant evidence for the glacial origin of the valley itself. He commented with hindsight that 'these phenomena are so conspicuous that [. . .] a house burnt down by fire did not tell its story more plainly than did this valley' (Darwin 1887: 57–58).[28] In the case of the Stac Fada Member, it should be noted that the failure to consider an impact origin is also mitigated to a significant extent by the absence of a crater, the interpretation of an impact ejecta blanket in the absence of a source crater being extremely rare. Even the Ries crater, which is well exposed, was interpreted as a volcanic edifice until shocked quartz was discovered there in the early 1960s.

The acceptance of the impact narrative should also be understood in the wider context of a disagreement that played out in the latter half of the twentieth century between the great majority of geologists who were only prepared to consider terrestrial explanations and those who were open to entertaining the possibility that solid bodies falling from space might act as geological agents (Marvin 1999). The dispute came to a head in 1980 with the publication of evidence that the impact of a major asteroid had caused the well-known mass extinction at the end of the Cretaceous, about 66 million years ago (Alvarez et al. 1980). The initial response was 'total uproar' (Marvin 1999: 105–109) and, by 1984, geologist Eugene Shoemaker (1928–97), who had worked on the Ries crater and had campaigned largely unsuccessfully for the acceptance of the evidence for extraterrestrial impacts since the early 1960s (Marvin 1999), felt obliged to lament the closed minds of many of his colleagues:

[M]ost geologists just don't like the idea of stones the size of hills or small mountains falling out of the sky. While they may concede, at an intellectual level, that such things might happen, at a visceral level, it still seems vaguely outrageous. In part this is due, I think, to an overdose of Lyellian uniformitarianism in their geological education, and in part, to their failure to view the Earth constantly as a member of the Solar system. (Shoemaker 1984: 1001)

Shoemaker's comment on 'Lyellian uniformitarianism' referred to the principle promoted by Charles Lyell (1797–1875), which, in its methodological

[28] I am grateful to Alok Srivastava for pointing me towards this historical example.

sense, holds that 'the observable present is a crucial resource in understanding the past' (Oreskes 2013: 595), and is thus indispensable to geological practice as exemplified by Goodchild's (1897) Torridonian-Sinai analogy (see section 4.3.1). Lyell, however, also implied that *only* explanations which invoked 'gradual change by processes intrinsic to the Earth' (Marvin 1999: 105) were admissible in geology, and he would not have been able to countenance the 'outrageous' possibility of a catastrophic meteorite impact – 'a process of random violence, originating outside the Earth' (Marvin 1999: 112) and capable of wreaking instantaneous devastation. The implication was that many of Shoemaker's geological contemporaries were still in thrall to Lyell on this matter. By the early 1990s, however, the Alvarez hypothesis had been greatly reinforced when the probable crater was located in Mexico (Marvin 1999: 109–112), and with the subsequent accumulation of evidence for many other crater-forming events in the geological record, the role of meteorite impacts in geology and in palaeontology[29] had become part of the mainstream by the early twenty-first century (e.g., French 2004).

Finally, should we consider the impact interpretation of the Stac Fada Member to be the last word, the final narrative? The history of science exclaims an emphatic 'No!' There seems to be no reason why more new evidence of as yet unknown significance might not turn up, or why new theories might not lead in a different direction. The consensus for the impact interpretation is quite strong at present, with most geologists with an interest in the region or in impact deposits coming down in favour. However, there are a few dissenting voices. Osinski et al. (2011) have cautioned that the Stac Fada deposit is 'not what it seems'; they point to several inconsistencies which they believe cast serious doubt on some of the details of the impact interpretation of Amor et al. (2008). The *Geological Excursion Guide to the North-West Highlands of Scotland*, published by the Edinburgh Geological Society (Goodenough and Krabbendam 2011), is also not convinced that the Stac Fada Member is an impact ejecta blanket.

Recently, evidence has emerged that shocked quartz can also be formed by lightning strikes, which threatens to remove its status as an unequivocal indicator of meteorite impacts ('Impact Geologists, Beware!' – Melosh 2017). This potentially replicates the situation that affected accretionary lapilli when they were relegated from their status as unambiguous evidence of volcanic eruptions upon their discovery at impact sites. It should be pointed out, however, that the impact interpretation is also supported by additional evidence such as anomalous geochemical markers which are not consistent with potential alternatives such as lightning strikes. Nevertheless, it remains to be seen whether factors which may emerge in the future will cause a further rewrite of the Stac Fada narrative.

---

[29] Huss (Chapter 3) discusses the role of meteorite impacts in mass-extinction events.

4.4   Conclusions

Robert Richards (1992: 23) is surely correct to claim that 'all explanations of events in time are ultimately narrative in character' (and it is evident that H. H. Read did not extend his analogy of geology with human history far enough explicitly to acknowledge this fact). Accordingly, the case of the interpretation and re-interpretation of the stratigraphic unit known as the Stac Fada Member demonstrates that geology is an inescapably narrative science that follows rigorous standards of internal and external narrative logic and is constrained by physical and chemical laws and by the norms of geology.

Narrative statements in the geological literature on the Stac Fada Member, and on the Torridonian of which it is a part, are commonly presented in the form of what appear at first to be descriptions of observations. However, these contain cues which the geologist automatically picks up, and by virtue of her training and experience she makes a range of default assumptions which are translated into causally connected temporal sequences. This phenomenon is particularly applicable to geology because of the relationship between space and time in the ways in which rocks accumulate, as well as in the clues to past environments which certain types of rock embody. More 'traditional' narrative passages that explicitly express the causal relationships between time-separated events also occur but are less common. Textual narratives may be accompanied by images that aim to add clarity but which may often be read as narratives themselves. Beyond the field of communication, a lot of narrative activity in geology is unseen, as it takes place in the minds and conversations of geologists as they try to make sense of observations that may initially be perplexing.

Geology is also an interpretive science, and narratives that attempt to answer the question, 'What caused these traces?', are inevitably accompanied by some degree of uncertainty. This is illustrated by the fact that key pieces of evidence, such as the presence of shocked quartz, may be overlooked for a variety of reasons, and by the observation that different conclusions may be drawn from the same field evidence – as illustrated by the disagreement over the most likely location of the missing crater. Geological narratives are therefore prone to be rewritten when new evidence or new ideas emerge. Whether the Stac Fada narrative will be rewritten again remains to be seen. In its latest version, the Stac Fada narrative provides a contribution to the body of knowledge relating to the susceptibility of the Earth to periodic meteorite impacts, a phenomenon which is now recognized as posing an existential threat to humanity.[30]

[30] I would like to thank Anne Teather, Dominic Berry, an anonymous reviewer and especially Mary Morgan for their helpful comments on earlier drafts. Special thanks are due to Jody Bourgeois for a particularly forensic critique. *Narrative Science* book: This project has received funding from the European Research Council under the European Union's Horizon 2020 research and innovation programme (grant agreement No. 694732). www.narrative-science.org/.

References

Alvarez, L. W., W. Alvarez, F. Asaro and H. V. Michel (1980). 'Extraterrestrial Cause for the Cretaceous-Tertiary Extinction'. *Science* 208.4448: 1095–1108.

Amor, K., S. P. Hesselbo, D. Porcelli, A. Price et al. (2019). 'The Mesoproterozoic Stac Fada Proximal Ejecta Blanket, NW Scotland: Constraints on Crater Location from Field Observations, Anisotropy of Magnetic Susceptibility, Petrography and Geochemistry'. *Journal of the Geological Society* 176.5: 830–846.

Amor, K., S. P. Hesselbo, D. Porcelli, S. Thackrey and J. Parnell (2008). 'A Precambrian Proximal Ejecta Blanket from Scotland'. *Geology* 36.4: 303–306.

Artemieva, N., and V. Shuvalov (2016). 'From Tunguska to Chelyabinsk via Jupiter'. *Annual Review of Earth and Planetary Sciences* 44: 37–56.

Beatty, J. (2017). 'Narrative Possibility and Narrative Explanation'. *Studies in History and Philosophy of Science Part A* 62: 31–41.

Branney, M., and R. Brown (2011). 'Impactoclastic Density Current Emplacement of Terrestrial Meteorite-Impact Ejecta and the Formation of Dust Pellets and Accretionary Lapilli: Evidence from Stac Fada, Scotland'. *Journal of Geology* 119.3: 275–292.

Brasier, A. T., T. Culwick, L. Battison, R. H. T. Callow and M. D. Brasier (2017). 'Evaluating Evidence from the Torridonian Supergroup (Scotland, UK) for Eukaryotic Life on Land in the Proterozoic', In A. T. Brasier, D. McIlroy and N. McLoughlin, eds. *Earth System Evolution and Early Life: A Celebration of the Work of Martin Brasier*. London: Geological Society, 121–144.

Butler, R. W. H., and G. I. Alsop (2019). 'Discussion on "A Reassessment of the Proposed 'Lairg Impact Structure' and Its Potential Implications for the Deep Structure of Northern Scotland" in *Journal of the Geological Society, London*, 176, 817–829'. *Journal of the Geological Society* 177.2: 443–446.

Currie, A. (2018). *Rock, Bone, and Ruin: An Optimist's Guide to the Historical Sciences*. Cambridge, MA: MIT Press.

Danto, A. C. (1962). 'Narrative Sentences'. *History and Theory* 2.2: 146–179.

Darwin, F., ed. (1887). *The Life and Letters of Charles Darwin, including an Autobiographical Chapter*. vol. 1. London: John Murray.

Davis, W. (1926). 'The Value of Outrageous Geological Hypotheses'. *Science* 63.1636: 463–468.

Faye, J. (2010). 'Interpretation in the Natural Sciences'. In M. Su árez, M. Dorato and M. Rédei, eds. *EPSA Epistemology and Methodology of Science*. Dordrecht: Springer, 107–118.

French, B. (2004). 'The Importance of Being Cratered: The New Role of Meteorite Impact as a Normal Geological Process'. *Meteoritics and Planetary Science* 39.2: 169–197.

Frodeman, R. (1995). 'Geological Reasoning: Geology as an Interpretive and Historical Science'. *Geological Society of America Bulletin* 107: 960–968.

Goodchild, J. (1897). 'Desert Conditions in Britain'. *Transactions of the Edinburgh Geological Society* 7: 203–222.

Goodenough, K., and M. Krabbendam, eds. (2011). *A Geological Excursion Guide to the North-West Highlands of Scotland*. Edinburgh: Edinburgh Geological Society.

Gracie, A., and A. Stewart (1967). 'Torridonian Sediments at Enard Bay, Ross-shire'. *Scottish Journal of Geology* 3: 181–194.

Griesemer, J. R. (1996). 'Some Concepts of Historical Science'. *Memorie della Societàitaliana di scienze naturali e del Museo civico di storia naturale di Milano* 27: 60–69.

Herman, D. (1997). 'Scripts, Sequences, and Stories: Elements of a Postclassical Narratology'. *PMLA* 112.5: 1046–1059.

Hull, D. L. (1975). 'Central Subjects and Historical Narratives'. *History and Theory* 14.3: 253–274.

Kleinhans, M., C. Buskes and H. de Regt (2005). 'Terra Incognita: Explanation and Reduction in Earth Science'. *International Studies in the Philosophy of Science* 19.3: 289–317.

Kölbl-Ebert, M. (2015). *From Local Patriotism to a Planetary Perspective: Impact Crater Research in Germany, 1930s–1970s*. Farnham. Ashgate.

Lawson, D. (1965). 'Lithofacies and Correlation within the Lower Torridonian'. *Nature* 207.4998: 706–708.

(1972). 'Torridonian Volcanic Sediments'. *Scottish Journal of Geology* 8: 345–362.

Macdonald, R., and D. Fettes (2006). 'The Tectonomagmatic Evolution of Scotland'. *Transactions of the Royal Society of Edinburgh: Earth Sciences* 97: 213–295.

Marvin, U. (1999). 'Impacts from Space: The Implications for Uniformitarian Geology'. In G. Y. Craig and J. H. Hull, eds. *James Hutton – Present and Future*. London: Geological Society, 89–117.

Melosh, H. (2017). 'Impact Geologists, Beware!' *Geophysical Research Letters* 44.17: 8873–8874.

Morgan, Mary S. (2017). 'Narrative Ordering and Explanation'. *Studies in History and Philosophy of Science Part A* 62: 86–97.

Oreskes, N. (2013). 'Why I Am a Presentist'. *Science in Context* 26.4: 595–609.

Osinski, G. R., L. Preston, L. Ferrière, L. Prave et al. (2011). 'The Stac Fada "Impact Ejecta" Layer: Not What It Seems'. *Meteoritics and Planetary Science* 46: A181.

Parnell, J., D. Mark, A. Fallick, A. Boyce and S. Thackrey (2011). 'The Age of the Mesoproterozoic Stoer Group Sedimentary and Impact Deposits, NW Scotland'. *Journal of the Geological Society* 168.2: 349–358.

Peach, B. N., J. Horne, W. Gunn, C. T. Clough et al. (1907). *The Geological Structure of the North-West Highlands of Scotland*. Memoirs of the Geological Survey of Great Britain. Glasgow: HMSO.

Read, H. H. (1952). 'The Geologist as Historian'. Reprinted in *Proceedings of the Geologists' Association* 81.3 (1970): 409–420.

Reddy, S., T. Johnson, S. Fischer, W. Rickard and R. Taylor (2015). 'Precambrian Reidite Discovered in Shocked Zircon from the Stac Fada Impactite, Scotland'. *Geology* 43.10: 899–902.

Richards, R. J. (1992). 'The Structure of Narrative Explanation in History and Biology'. In M. H. Nitecki and D. V. Nitecki, eds. *History and Evolution*. Albany: State University of New York Press, 19–54.

Shoemaker, E. M. (1984). 'Presentation of the G. K. Gilbert Award to Eugene M. Shoemaker'. *Geological Society of America Bulletin* 95.8: 1000–1001.

Simms, M. (2015). 'The Stac Fada Impact Ejecta Deposit and the Lairg Gravity Low: Evidence for a Buried Precambrian Impact Crater in Scotland?' *Proceedings of the Geologists' Association* 126: 742–761.

Stewart, A. D. (2002). *The Later Proterozoic Torridonian Rocks of Scotland: Their Sedimentology, Geochemistry and Origin*. London: Geological Society.

Stöffler, D., and R. Grieve (2007). 'Impactites'. In D. Fettes and J. Desmons, eds. *Metamorphic Rocks: A Classification and Glossary of Terms, Recommendations of the IUGS*. Cambridge: Cambridge University Press.

Tipper, J. C. (2015). 'The Importance of Doing Nothing: Stasis in Sedimentation Systems and Its Stratigraphic Effects'. In D. G. Smith, R. J. Bailey, P. M. Burgess and A. J. Fraser, eds. *Strata and Time: Probing the Gaps in Our Understanding*. London: Geological Society, 105–122.

Young, G. (2002). 'Stratigraphy and Geochemistry of Volcanic Mass Flows in the Stac Fada Member of the Stoer Group, Torridonian, NW Scotland'. *Transactions of the Royal Society of Edinburgh: Earth Sciences* 93.1: 1–16.

5   Reasoning from Narratives and Models: Reconstructing the Tohoku Earthquake

Teru Miyake

Abstract

This chapter examines the role of three kinds of narratives in producing knowledge about the rupture process of the Tohoku earthquake
of 2011. I show that each of the three kinds of narratives appears in
one of three stages on the way from data recorded of the earthquake to
a reconstruction of the rupture process. In the first stage, rupture
narratives are produced by computational tools called source models.
In the second stage, a set of details that is taken accurately to represent
features of the actual rupture process is distilled out of these conflicting rupture narratives through the use of a 'research narrative'. In the
third stage, these distilled details are strung together into an integrating narrative. This integrating narrative is used as a research tool for
formulating questions, the pursuit of which has led to the production
of further evidence about the rupture process.

5.1   Introduction

The ground shaking that an earthquake produces is the result of a complex
sequence of events that occur at a fault. This sequence of events is often given
a narrative account by seismologists. Here is an example of such an account of
the 2011 Tohoku-Oki earthquake. This is the massive earthquake that gave rise
to the tsunami that devastated the north-east coast of Japan and caused the
nuclear disaster at Fukushima.

On 2011 March 11, rupture of a frictionally locked region in the central portion of the
220 km wide megathrust fault commenced innocuously, with a magnitude 4.9 earthquake, but the rupture failed to arrest, continuing to expand for 150 s, spreading over the
full width of the boundary and along its length for 400 km. The rupture expanded
relatively slowly in the up-dip direction, with fault slip of ~30 m near the hypocenter,
spanning a region that had not failed since a great event in 869 CE, increasing to about
50 m or more near the trench. The rupture expanded more rapidly and erratically down-
dip to below the Honshu coast with slip of 1–5 m extending southward along the Miyagi,

Fukushima and Ibaraki Prefectures. Multiple source regions of large earthquakes of the last century re-ruptured sequentially, with short-period seismic waves released by this down-dip rupture being enhanced relative to the up-dip rupture. (Lay 2018: 4–5)

The events that are recounted here (e.g., the rupture 'expanded relatively slowly in the up-dip direction, with fault-slip of ~30 m', and later 'expanded more rapidly and erratically down-dip to below the Honshu coast with slip of 1–5 m') took place along a fault, deep within the earth. I will refer to the sequence of events at the fault, which played out over several minutes in the case of the Tohoku earthquake, as the *rupture process*. For each earthquake that occurs, there is a particular way in which these events play out – each earthquake has a unique rupture process. Knowing these rupture processes in detail would yield precious information about the faults on which they occur and their history, which can be used to make better determinations of seismic hazard.

The rupture process of an earthquake cannot be observed directly, since it takes place deep within the earth, but its effects can be observed at the earth's surface. The rupturing of a fault generates seismic waves that travel outwards in all directions from the fault. These seismic waves can be recorded on seismographs at the earth's surface. An earthquake can also result in permanent ground motion at the earth's surface, which can be recorded using GPS technology. Data on other effects of an earthquake, such as tsunamis, can also be recorded.

Reconstructing the rupture process of an earthquake from this recorded data is a particularly difficult problem, for several reasons. First, rupture processes are very complex, and highly contingent.[1] The way a rupture process unfolds is highly dependent on contingent features of the fault. Second, as I have already mentioned, seismologists generally do not have direct access to faults. This means that the contingent features of the fault are typically not known prior to the earthquake. Third, the data recorded from a major earthquake such as the Tohoku earthquake can come from observations of a number of different phenomena, such as seismic waves, permanent ground motion and tsunamis. This diverse data must be integrated in some manageable and principled way. In short, seismic reconstruction involves inferring from a wide variety of downstream effects a complex, highly contingent process that occurs on a fault that is not directly accessible .

An important tool for seismic reconstruction, *slip inversion*, produces models (called *source models*) that capture the rupture process . As we will see, a source model provides a narrative about a possible way the rupture process may have occurred. This narrative cannot, however, be straightforwardly regarded as an accurate account of the events at the fault as they actually

---

[1] On the importance of contingency in detailed evolutionary back stories, see Beatty, Chapter 20.

occurred. When a large number of different source models of the same earthquake are generated, they will generally conflict with each other, due to differences in the sets of data they utilize, the specific mathematical techniques used and the assumptions that go into these models. A problem that seismologists have faced when attempting to reconstruct the Tohoku and other earthquakes, then, is how to take such conflicting models and reconstruct the actual rupture process.

This chapter examines how seismologists have obtained increasingly detailed knowledge about the rupture process of the Tohoku earthquake in the face of this problem. I will give an account of the growth of this knowledge that is slightly unorthodox, but it exemplifies how thinking about narrative might help us to understand the growth of scientific knowledge.[2] I will focus in particular on three stages on the path from recorded data to increasingly detailed knowledge about the rupture process, and the role of narrative in each of those steps.

Here is an initial sketch of these three stages.[3] In the first stage, source models are used to produce, from recorded data, narratives that recount the rupture process in detail, which I call *rupture narratives*. As I have mentioned, these narratives generally conflict with each other due to differences in the data, techniques and assumptions that go into the source models. In the second stage, a set of details that is taken accurately to represent features of the actual rupture process is distilled out of these conflicting rupture narratives. This set of details is arrived at through the use of a *research narrative* that examines the evolution of source models. In the third stage, these distilled details are strung together into a model-independent rupture narrative, which I call an *integrating narrative*. This integrating narrative is used as a research tool for formulating questions, the pursuit of which has led to the production of further evidence about the rupture process.[4]

This chapter will proceed as follows. In section 5.2, I will lay down some basics about how earthquakes occur, the rupture process of an earthquake and the Tohoku fault. In section 5.3, I examine the construction of source models from data and present an example of a rupture narrative. In section 5.4, I show how details are distilled from source models through the use of a research narrative. In section 5.5, I present an example of an integrating narrative, and show how the pursuit of questions about this narrative results in further

[2] This chapter is thus complementary to the chapters by Andrew Hopkins (Chapter 4) and John Huss (Chapter 3), who also explore the nature of scientific knowledge in the earth sciences through the lens of narrative.

[3] My use of the word 'stage' here is intended to reflect not a temporal order, but an epistemic order, where one starts with data and there is a process of further and further refinement, ultimately resulting in detailed knowledge about the rupture process.

[4] Previous accounts of the relations between models and narratives are given in Morgan (2012) for economics, and Wise (2017) for chemistry.

evidence about the rupture process. In the concluding section 5.6, I briefly consider the functions of the three types of narratives just mentioned in the growth of knowledge about the rupture process of the Tohoku earthquake.

5.2   Earthquakes and the Tohoku Fault

Most earthquakes are generated at a fault, which may be thought of as a roughly planar surface within the earth where the ground on the two sides of the surface are slowly being pulled in opposite directions. A well-known example is the San Andreas fault, the two sides of which are moving a few centimetres a year relative to each other. If a fault were completely smooth and frictionless, the two sides would simply move very slowly past each other, and we would have no earthquakes. But faults are not frictionless. The two sides are rough, and there are portions, called *asperities*, where the two sides are locked together.

What happens when the forces on each side of the fault continue to act in opposite directions, while the two sides are locked together? Because rock is elastic, the rock around the fault will slowly bend due to the imposed forces, and it will store up elastic strain energy, much like a wooden ruler would store up elastic energy if you slowly flexed it. Points far away from the fault will tend to move slowly relative to each other, while the fault remains locked together. This will result in strain slowly accumulating in the material surrounding the fault as it gets pushed further and further out of equilibrium. The strain will continue to build until it is sufficient to overcome the friction that keeps the sides locked together. The two sides of the fault will then *rupture*, suddenly snapping back towards a position of equilibrium. The pent-up elastic energy is released, generating seismic waves.

The largest earthquakes occur on faults that can be hundreds of kilometres long. Several features of large earthquakes are particularly important for understanding this chapter. First, large faults do not rupture along their entire length all at once. The rupture initiates at a particular point on the fault. This rupture will then propagate to other parts of the fault. If the fault is hundreds of kilometres long, the rupture can take several minutes to propagate the entire length of the fault. This series of events at the fault is called the *rupture process*. Second, the state of friction on a large fault is generally heterogeneous. That is, there can be patches of the fault that are strongly stuck together (the asperities), while there can be other patches that are only weakly coupled. The patches that are weakly coupled rupture easily, while the asperities are resistant to rupture. When the asperities do rupture, however, they typically have built up a lot of elastic energy, so they tend to rupture much more forcefully than the weak patches. Thus, the particular way the rupture propagates will depend on contingent features such as the state of friction at various points of the fault.

These features are typically not directly accessible to seismologists, since the fault is buried deep within the earth.

I will now move to specific details about the Tohoku earthquake and the fault on which it occurred. The Tohoku earthquake occurred on a subduction zone off the north-east coast of Japan. There, tectonic forces are driving the Pacific plate underneath Japan and into the mantle, at a rate of roughly 8 centimetres per year. Figure 5.1 is a cutaway diagram showing the subduction zone, as viewed facing roughly northward. Northern Japan sits on top of the Okhotsk plate, towards the left side of the diagram. The fault on which the earthquake occurred is on the border between the Pacific and the Okhotsk plate. In the cutaway view, the fault is represented as a line at 12 degrees to the horizontal, with arrows indicating the relative motion of the two sides of the fault. The direction along the fault, at 12 degrees to the horizontal, is called the *dip direction*. In actuality, the dip angle of the fault is not known so accurately, and it may vary by a few degrees. Because the fault slopes downwards to the west, the western part of the fault that eventually goes underneath Japan is referred to as the *down-dip* part of the fault, while the eastern, *up-dip* part eventually reaches the ocean bottom at the Japan Trench, an extremely deep area of the Pacific Ocean off the coast of Japan.

Now it is fairly easy to visualize how large earthquakes occur on this fault. The Pacific plate is slowly getting pushed under the Okhotsk plate, but there are places where the two sides are locked together. The strain accumulates until it is enough to overcome the friction, and the two surfaces at the fault suddenly unlock. The upper surface jolts eastward and upward, releasing elastic energy in the form of seismic waves. The Tohoku earthquake ruptured an area of around 200 kilometres by 500 kilometres, and the entire rupture process took

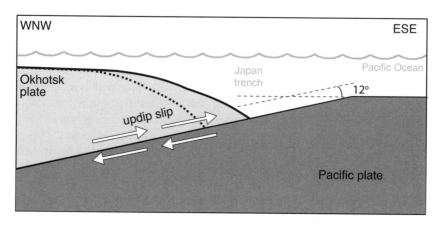

Figure 5.1 **Cutaway view of Tohoku fault**
Source: Figure kindly provided by Dr Jeroen Ritsema.

around 150 seconds. This motion also gave rise to a powerful tsunami that inundated the north-east coast of Japan. A detailed understanding of the rupture process, its connection to past and possible future earthquakes in the area, and the way in which it generated the tsunami, is of obvious importance for seismology, as well as the determination of seismic hazard along the coast of Japan.

The Tohoku earthquake was recorded on an unprecedented variety of instruments. Broadly, the data that were recorded for this earthquake can be categorized by the kind of phenomenon that was recorded. *Seismic data* are recordings of seismic waves. Strong motion seismic data is recorded at stations nearby an earthquake. These kind of data are recorded on several networks of different types of seismographs throughout Japan, including KiK-net, a network of over 600 strong-motion seismographs that are situated in boreholes; and K-NET, a network of over 1,000 strong-motion seismographs at the surface. *Geodetic data* are recordings of the deformation of the earth's surface. Such data are typically recorded using GPS technology. Most of the geodetic data for the Tohoku earthquake was recorded on a network of over 1,200 GPS stations distributed throughout Japan, called GEONET. *Tsunami data* are recordings of the tsunami caused by the earthquake. This type of data was typically recorded by offshore wave and tide gauges. These three categories do not exhaust all the types of data that were recorded for this earthquake. In addition, there were important data recorded of the motion of the ocean bottom at seafloor geodetic sites, data from deep drilling into the fault zone after the earthquake and even gravimetric data recorded by satellites.

5.3   Rupture Narratives: From Data to Details

The data collected from the Tohoku earthquake are rich and diverse, but they consist of recordings of the downstream effects of the earthquake, such as ground motions that occurred far away from the fault. Such data do not immediately reveal any details about the rupture process. An initial step towards a reconstruction of the rupture process is the use of *source models*,[5] which, as we will see, take this downstream data and provide a detailed account – albeit an unreliable one – of the rupture process.

Most source models for the Tohoku earthquake have been constructed using a method called *slip inversion*.[6] A good example can be seen in Figure 5.2,

---

[5] I use the term 'source model' throughout this chapter. Confusingly, seismologists use several different words – 'finite fault model', 'slip model', 'rupture model' – for roughly the same thing. There are some slight differences, but they may be treated as synonymous for the purpose of this chapter. The excerpts from Lay (2018) use some of these other words. Please read 'source model/s' whenever you encounter them.

[6] 'Slip inversion' is sometimes called 'finite fault inversion'. See Ide (2015) for a full description of how slip inversion works, including a brief history.

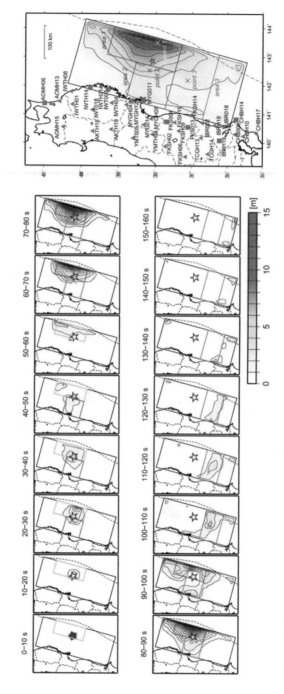

Figure 5.2 **Representation of the time progression of the rupture for the 2011 Tohoku earthquake**
On the left is a representation of the time progression of the rupture given in intervals of 10 seconds. On the right is a representation of the total slip distribution of the Tohoku earthquake.
From Suzuki et al. (2011: 3–4).

taken from Suzuki et al. (2011). This is an early source model that was produced entirely from seismic data recorded at 36 stations located throughout northern Japan.[7] Let us first examine the large figure on the right. An outline of the northern part of the Japanese island of Honshu can be seen towards the left. Just off the coast is a rectangle, which has a length of 510 km and a width of 210 km, oriented at a small angle in the north–south direction. This rectangle is a representation of the fault. We are here viewing the fault from directly above (in contrast to Figure 5.1, which is a cutaway view). The dip of the fault cannot be seen in this view, but in this model the dip angle was set at 13 degrees to the horizontal.

Slip is a measure of how much the two sides of a fault moved relative to each other during an earthquake. The contours and shading on the figure to the right are an indication of how much various parts of the fault slipped over the course of the Tohoku earthquake. The darker the shading, the more slip occurred. According to this source model, there was an area of very large slip of around 48 m near the Japan Trench (towards the right edge of the fault). The series of 16 small figures on the left are miniature versions of the figure to the right. Each of these small figures represents the amount of slip on the fault in each ten-second slice of time from the beginning of the earthquake to the end (reading from top left to right, and then bottom left to right). As I described earlier, when an earthquake occurs, various parts of the fault rupture in succession. We can think of these as a series of snapshots of this rupture process as it propagates. If we allow for a wide definition of 'narrative' that includes visual objects such as diagrams,[8] we can view this series as a visual narrative of the rupture process, indicating spatial changes of the fault over time during the Tohoku earthquake.

Source model studies also provide more straightforward textual narratives of the rupture process, along with such diagrams. For example, Suzuki et al. (2011) provides the following:

The total moment rate indicates that first remarkable moment release started 20 s after the initial break, when the rupture occurred around the hypocenter. Then, at approximately 40 s, the rupture proceeded northward along the trench axis and towards the down-dip direction. Somewhat later, the rupture also extends southward along the trench axis. The largest slip event occurred from 60 s to 100 s, with the rupture expanding towards the down-dip direction from the area along the trench axis. In this

---

[7] I chose Suzuki et al. (2011) as an example because it contains a particularly simple and clean visual representation of the rupture process. Many visual representations of source models are much more complex and include several layers of information. For those who are interested in these visual representations of rupture models and would like to see more examples, Lay (2018) contains a large variety of them.

[8] This volume presents many other examples of visual narratives in various fields. The uses of such narratives, and ways of reading them, are diverse. See, for example, the chapters by Teather (Chapter 6), Engelmann (Chapter 14), Kranke (Chapter 10), Hopkins (Chapter 4), Griffiths (Chapter 7), Bhattacharya (Chapter 8), and Paskins (Chapter 13).

stage, large slip occurred continuously far offshore of southern Iwate, Miyagi, and northern Fukushima prefectures. The last stage starts at around 100 s, where the rupture propagated southward in the area off Fukushima and Ibaraki prefectures. The entire rupture almost ceased within 150 s. (Suzuki et al. 2011: 4)

We can view slip inversion as taking seismic or other data as input, and outputting what I call *rupture narratives*, which are visual and textual narratives of a rupture process that includes quantitative details. Such details can include temporal details such as the timing of various sub-events within the rupture process. They can also include details about the rupture process as a whole, such as the total amount of slip that occurred at a particular part of the fault. Borrowing a term from Robert Meunier (Chapter 12), the rupture narratives produced by source models present themselves as 'narratives of nature' – narratives that recount a process as occurring in nature, independently of any human observers. As we will see, however, they are highly *model-dependent* – that is, many of the details within these narratives are artefacts of the data, techniques and assumptions that go into the source models.

An indication of this model-dependence is a wide variability among rupture narratives produced by source models of the Tohoku earthquake. Figure 5.3 is a comparison of 45 different source models of the Tohoku earthquake. Each of the lines represents the amount of total slip indicated by each source model. For ease of comparison, only the amount of slip along the corridor off the north-east coast of Japan indicated in the inset map is shown, extending from just underneath the coast to the Japan Trench. There is particularly wide variability in the up-dip regions, near the trench. Some models show slip of 50 m or more here, while other models indicate slip of 10 m or less.

How could there be such discordance between rupture narratives produced by various source models of the same earthquake? Broadly, there are two reasons. The first has to do with differences in the type of input data. I have mentioned that the data that were recorded for the Tohoku earthquake can be categorized into seismic data, geodetic data and tsunami data. Source models have been constructed using all of these types of data. Different types of data are sensitive to different features of the rupture, and thus models that rely on different types of data tend to emphasize different features. The second reason has to do with differences in the methods used to construct source models. This can include differences in the parameterizations used, differences in the idealizations and assumptions that go into the models, and differences in the mathematical and computational techniques that are used.

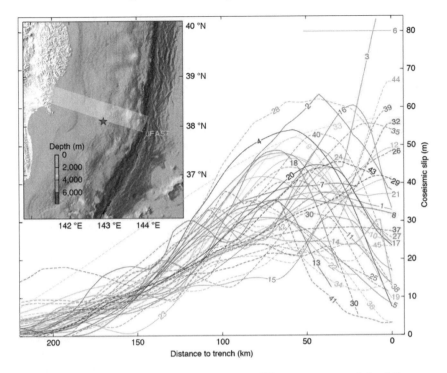

Figure 5.3 **Comparison of slip according to 45 different source models of the Tohoku earthquake**
Source: Lay (2018: 26), modified from Sun et al. (2017).

5.4   Research Narratives: Distilling Details from Source Models

The first source models of the Tohoku earthquake were produced and published in 2011, within months after the earthquake. A large number of source models of the earthquake have been produced since then – by 2017 there were at least 45 of them (Sun et al. 2017). From the start, there have been pronounced discordances between various source models of the earthquake. An important question for the reconstruction of the rupture process of the Tohoku earthquake has thus been: exactly what is one to conclude about the actual rupture process given the discordance between the source models? Is there a way of distilling out from these conflicting source models some set of rupture details that can be regarded as accurately representing the actual rupture process? One reasonable thought is that later source models are generally more accurate than earlier ones, since they presumably have more knowledge about the earthquake to draw upon. A more rigorous approach would examine in detail the evolution of source models since 2011 to determine whether indeed later models improve on earlier ones. This is the approach taken in Lay (2018), a review article of the Tohoku earthquake. Lay (2018) contains a very long and complex narrative that traces out the evolution of source models, with the aim of distilling out rupture details to which some degree of confidence can be attached. Borrowing again from Robert Meunier (Chapter 12), this is a *research narrative* – a narrative that provides an account of the activities of researchers.

Let us now take a closer look at the research narrative in Lay (2018). The general thrust of the narrative is to show how source models of the Tohoku earthquake have gone through an evolution, the result of which is that details in certain later source models have claim to being relatively accurate representations of details of the actual rupture process. For example, in a section of the narrative, Lay examines early source models based purely on geodetic observations made at onshore GPS sites. He notes that such source models 'can provide good resolution of the spatial distribution of slip if the observation configuration is favorable' (Lay 2018: 11). Unfortunately, it turns out that the observation configuration for the Tohoku earthquake is unfavourable – all of the GPS sites are on the Japanese mainland, which is on the down-dip side of the fault. This means that source models based purely on onshore geodetic data have poor sensitivity to slip that happens on the up-dip side of the fault, near the Japan Trench. This is significant, for, as Lay points out, although source models based purely on onshore geodetic data are largely consistent with each other, they are inconsistent with source models based purely on seismic data. Source models based purely on seismic data tend to show the largest slip happening up-dip, near the Japan Trench, as with the source model depicted in Figure 5.2, while source models based purely on onshore geodetic data tend to put the largest slip near the hypocentre, more towards the centre of the fault. One might surmise that the reason for this inconsistency is the unfavourable observation configuration for source models based on onshore geodetic data.

Recognizing this as a limitation, seismologists have attempted to address this problem in later source models utilizing onshore geodetic data by incorporating other types of data that are complementary to onshore geodetic data. Particularly important is a set of geodetic data taken by GPS/Acoustic stations located offshore, on the ocean bottom, which, according to Lay, has 'proved transformative for geodetic models of the 2011 Tohoku earthquake slip distribution' (Lay 2018: 12). Another important kind of additional data is time series data taken at GPS stations, called hr-GPS. Regarding the evolution of source models based on geodetic data, Lay states:

[T]here has been significant evolution of slip models inferred from geodesy, from the long smooth models with ~30 m peak slip near the hypocenter [. . .] to much more spatially concentrated and up-dip slip models with peak slip of 50 to 60 m at shallow depth when using hr-GPS time series [. . .] or from inclusion of up to 7 offshore GPS/ Acoustic measurements in static inversions. (Lay 2018: 13)

Significantly, these later source models are much more consistent with source models based on seismic data (note, for example, that the source model depicted in Figure 5.2 has a peak slip of 48 m in the up-dip, shallow part of the fault). In other words, the rupture narratives of these later source models look much more like the rupture narratives of source models based on seismic data.

Lay does similar analyses of the evolution of models based on other types of data, showing how later models have improved upon earlier models. Not only are the rupture narratives of later models more consistent with each other, but they are getting more detailed. He takes later source models that incorporate multiple types of data – called *joint inversions* – to be the most reliable. One reason is because data of different types can be complementary – they are sensitive to different aspects of the rupture process. Another reason is because the later source models generally address the shortcomings of earlier models. Lay summarizes the evolution of source models as follows:

The foregoing review of rupture models for the 2011 Tohoku earthquake shows progressive convergence of slip models, with slip being increasingly localized along strike and concentrated up-dip, extending all the way to the trench with slip ~50 m near 38.2°N. Over time the rupture models have progressed from quite smooth representations to more detailed slip distributions, especially for the geodetic and tsunami models. [. . .] Some of the differences among current models may represent the different parameterizations, but the similarity of the majority of joint inversion models [. . .] suggests that different parameterizations are at least not overwhelming the source information. (Lay 2018: 19)

We can think of the research narrative provided by Lay about the evolution of source models as providing a justification that the details that appear in the rupture narratives of the later models are relatively accurate. Greater confidence is placed on the later joint inversion models, but no one model is taken to be best, and the amount of confidence one can place in a particular detail is ultimately based on a judgement that takes into account the commonalities between particular source models, limitations due to the datasets and methods of construction and the overall evolution of source models.

5.5   Integrating Narratives: Pursuing Further Evidence

Typically, in review articles, the details that are distilled from source models are strung together into a new rupture narrative that is independent of any particular model. This is a 'narrative of nature' – one that is taken to represent the best current estimate of the actual rupture process. There is such a narrative in Lay

(2018), which he calls a 'strawman reference model that distills the features that appear most stable and/or plausible' (Lay 2018: 29). Although Lay calls it a model, it is not a model in the sense of the source models discussed earlier – it comes in the form of a textual narrative. As source models have evolved from the early source models based on single data sets to more detailed source models based on joint inversions, the details that were taken to be established about the Tohoku earthquake have also evolved. Thus, the model-independent rupture narratives that would be constructed by seismologists at any given time after the Tohoku earthquake would also evolve.[9] I will refer to rupture narratives of this type as *integrating narratives* because they are used as tools for integrating details of the rupture process with other seismological results.

In this section, I will show that integrating narratives play an important role in the production of new evidence about the rupture process of the Tohoku earthquake. First, let me provide, as an example, the 'strawman reference model'[10] of Lay (2018):

In terms of the primary slip zone, the joint models including tsunami information [. . .] provide good characterization of the rupture, with ~50 m of slip near or at the trench about 38° to 38.3°N. Shallow slip in the upper 10 km of the megathrust (from 8 to 15 km below the ocean floor) extends along strike from at least 37°N to 39.5°N, diminishing north and south of the central peak, which is near the site of the JFAST [Japan Trench Fast Drilling Project] drill hole. This is the Domain A zone of tsunami earthquake-like behavior discussed by [T. Lay, H. Kanamori, C. J. Ammon et al. 'Depth-Varying Rupture Properties of Subduction Zone Megathrust Faults', *Journal of Geophysical Research* 117(B04311): 1–21]. From 10 to 35 km depth, the large-slip region, with > 20 m of slip narrows to about ~150 km along strike, with the hypocenter within this zone, in what is called Domain B. Modest slip of 5 to 10 m is spread along strike, with down-dip Domain C concentrations of < 5 m offshore of Miyagi and offshore of Fukushima. These regions of prior M ~7.5 events during the past century appear to have re-ruptured with more high frequency radiation than the shallower regions. (Lay 2018: 29–30)

This is just the beginning of the first paragraph of the narrative. The thing to note about this narrative is that it does not just string together well-established details from rupture narratives. It also makes reference to 'domains' of the rupture process, the 'JFAST drill hole', and past earthquakes.[11] The later parts of this narrative, not shown here, continue on to discuss studies of afterslip (ground motions that occurred after the earthquake), seafloor deformation observations and specific earthquakes of the past. Integrating narratives can

---

[9] Such narratives can be found in earlier review articles of the Tohoku earthquake, such as Lay and Kanamori (2011), Tajima, Mori and Kennett (2013) and Hino (2015).

[10] Towards the beginning of Lay (2018) is another, simpler, version of the 'strawman reference model', from which I extracted the short narrative account given at the beginning of this chapter.

[11] JFAST, or the Japan Trench Fast Drilling Project, was a project that took place soon after the Tohoku earthquake to drill a borehole directly through the fault zone near the Japan Trench.

also highlight loose ends and open questions. Let me note again that this integrating narrative is from a comparatively late stage of analysis of the Tohoku earthquake. Earlier integrating narratives tend to be less detailed, they draw fewer connections to other studies and their references to past earthquakes are framed in a more speculative mode.[12]

Integrating narratives have provided a sort of framework[13] upon which certain questions about the Tohoku earthquake could be pursued. Some questions have to do with the connection between the rupture process of the Tohoku earthquake and its spatial and temporal context. More specifically, these questions can be about connections to past earthquakes, to seismic events immediately preceding the Tohoku earthquake, seismic events that came after it, and various features of the subduction zone on which it occurred, such as the locations of asperities, the distribution of accumulated strain, the composition of the rock in the subduction zone, and so on. Other questions have to do with anomalies or inconsistencies in the rupture process as laid out in the narrative. The pursuit of such questions has been a driving force for uncovering further evidence about the earthquake.

For example, a major open question has to do with the frequency characteristics of the rupture process. Just a few years before the Tohoku earthquake, a new technique for producing source models from seismic data, called *back-projection*, had been developed (Kiser and Ishii 2017). Back-projection is sensitive to high-frequency seismic waves, and the source model it produces is a kinematic image, not of slip, but of seismic radiation energy release over time. Since most of the energy being radiated at any given time during an earthquake originates from the rupture front, the back-projection image can be taken to show a kinematic image of the rupture front during the earthquake. Early back-projection studies of the Tohoku earthquake were systematically discordant with early slip inversion studies. The back-projection studies indicated that most of the seismic radiation was concentrated in the down-dip part of the fault. On the other hand, slip inversion studies indicated that the maximum slip was in the up-dip part of the fault, and the area of very large slip possibly extended all the way to the trench (Lay et al. 2011: 687). An early question about the rupture process was thus: why does back-projection appear to show that rupture occurred mainly down-dip, while slip inversion appears to show that the area of maximum slip was up-dip and close to the trench?

---

[12] See, for example, the narratives in Lay and Kanamori (2011: 37), or Ritsema, Lay and Kanamori (2012: 186–187).

[13] They are the sort of thing that Currie and Sterelny (2017) call 'scaffolds' in their work on historical reconstruction. See also Teather on scaffolding in archaeology (Chapter 6). They also appear to have much in common with narratives in Crasnow's (2017) account of process tracing. See also Chapter 11.

A possible answer is that different parts of the fault produced seismic radiation at different frequencies – the down-dip part producing more high-frequency radiation than the up-dip part. If this answer is right, then it gives rise to another question: is the difference in the frequency characteristics in different parts of the fault just a special feature of this particular earthquake, or is it a feature of the fault – in which case it ought to hold for other earthquakes as well? Koper et al. (2011: 602) suggested that the latter is the case – that the difference is due to depth-varying frictional properties of the fault.

The idea fits with the known history of the fault. The down-dip region corresponds to an area where large earthquakes of up to Mw 7.9 had repeatedly occurred over the past century, and these earthquakes would have had similar frequency characteristics as the down-dip part of the Tohoku earthquake. The up-dip region corresponds to an area that had not ruptured since 869 CE, but it also partially overlapped an area that is taken to have ruptured in 1896 during what is known as a 'tsunami earthquake'. Tsunami earthquakes have characteristics like those exhibited by this part of the fault during the Tohoku earthquake – with large slip but slow rupture velocities, leading to relatively more seismic energy being radiated at lower frequencies.

In this view, then, each part of the fault has its own rupture characteristics that are constant across earthquakes (these are roughly the 'domains' that Lay refers to in the extract above). The unusual feature of the Tohoku earthquake was that it ruptured both regions at the same time, so it combined the characteristics of both types of earthquakes. Given this view, the next question to ask would then be why there are regions with different rupture characteristics within the fault – does it have to do, for example, with the composition of materials in different areas of the fault? This has been probed by the use of seismic wave tomography (Tajima, Mori and Kennett 2013: 27) and studies (such as JFAST) where holes are drilled directly into the sea floor in the fault area (Lay 2018: 28).

The pursuit of questions such as these has improved the picture of how the Tohoku earthquake fits into its spatial and temporal context – whether it is, in some sense, a repeat of particular earthquakes in the past, for example. It has also opened up new lines of research that have contributed new evidence about the Tohoku earthquake. In some cases, this new information has been utilized to improve source models – thus contributing to the evolution of source models, and indirectly to the evolution of the integrating narratives themselves. Thus, there is a sort of mutual evolution of source models and integrating narratives, resulting in a more highly resolved, and more ramified, picture of the rupture process of the Tohoku earthquake. That Lay refers to the most recent version of an integrating narrative as a 'strawman reference model' is an indication that this is very much an ongoing process.

5.6   Conclusion

In this chapter, I have examined the growth of knowledge about the rupture process of the Tohoku earthquake, with a focus on the role of narrative. I have described three kinds of narratives in this chapter: rupture narratives, research narratives and integrating narratives. I would like to end with some considerations about the functions of these narratives in contributing to the growth of knowledge about the Tohoku earthquake.

5.6.1   Narratives as Filters

Let me begin with rupture narratives. It is not entirely correct to say that source model narratives are the outputs of source models, for the direct outputs of source models are simply large sets of parameters. But these sets of parameters must be put into a cognitively useful form: the textual and visual narratives that I call source model narratives. These narratives are the result of filtering out some of the needless complexity in source models. They allow seismologists to focus in on significant details. They also allow seismologists to readily make comparisons between source models in order to look for commonalities and differences. Side-by-side comparisons of visual representations are particularly powerful – Lay (2018) contains page after page of diagrams where a half-dozen source models are compared side by side.

5.6.2   Narratives as Arguments

The research narrative provided a justification for distilling certain details from source models. Rupture narratives formed an important ingredient for the research narrative, because the latter required a comparison between source models, and analyses of the assumptions and methods that were used in their production. Another important element of the research narrative given in Lay (2018) was a story about the evolution of source models that attempted to make a case that later source models are more accurate. The research narrative pulled together and organized these elements into a prolonged argument that certain details in the source models can be pulled out and regarded as well established, independently of any particular source model.

5.6.3   Narratives as Unifying Instruments

Integrating narratives of the Tohoku earthquake have strung together well-established details that are distilled from source models, with the help of research narratives. They locate the rupture process within a spatial and temporal context, and they provide a framework for the pursuit of further questions that may open up new lines of research into the Tohoku earthquake and other past and future earthquakes. We might regard integrating narratives as instruments for unification – bridging various empirical avenues and strengthening connections between them, perhaps with the aim of achieving Whewellian consilience.

Thus, the three types of narratives I have considered, all, in different ways, have made contributions to the growth of knowledge about the Tohoku earthquake. The fact that several different kinds of narratives are utilized by seismologists is perhaps not that surprising. The work that narratives do in enabling the growth of knowledge in seismology and other physical sciences, however, still needs to be better understood.[14]

References

Crasnow, S. (2017). 'Process-Tracing in Political Science: What's the Story?' *Studies in History and Philosophy of Science Part A* 62: 6–13.

Currie, A., and K. Sterelny (2017). 'In Defence of Story-Telling'. *Studies in History and Philosophy of Science Part A* 62: 14–21.

Hino, R. (2015). 'An Overview of the Mw 9, 11 March 2011, Tohoku Earthquake'. *Summary of the Bulletin of the International Seismological Centre* 48.1–6: 100–132.

Ide, S. (2015). 'Slip Inversion'. In H. Kanamori, ed. *Treatise on Geophysics*. vol. 4. *Earthquake Seismology*. 2nd edn. San Diego, CA: Elsevier, 215–241.

Kiser, E., and M. Ishii (2017). 'Back-Projection Imaging of Earthquakes'. *Annual Review of Earth and Planetary Sciences* 45: 271–299.

Koper, K. D., A. R. Hutko, T. Lay, C. J. Ammon and H. Kanamori (2011). 'Frequency-Dependent Rupture Process of the 2011 M_w 9.0 Tohoku Earthquake: Comparison of Short-Period P Wave Backprojection Images and Broadband Seismic Rupture Models'. *Earth, Planets and Space* 63.16: 599–602.

Lay, T. (2018). 'A Review of the Rupture Characteristics of the 2011 Tohoku-Oki Mw 9.1 Earthquake'. *Tectonophysics* 733: 4–36.

Lay, T., C. J. Ammon, H. Kanamori, L. Xue and M. J. Kim (2011). 'Possible Large Near-Trench Slip during the 2011 M_w 9.0 off the Pacific Coast of Tohoku Earthquake'. *Earth, Planets and Space* 63.32: 687–692.

Lay, T., and H. Kanamori (2011). 'Insights from the Great 2011 Japan Earthquake'. *Physics Today* 64.12: 33–39.

Morgan, Mary S. (2012). *The World in the Model: How Economists Work and Think.* Cambridge: Cambridge University Press.

[14] I would like to thank Mary Morgan for inviting me to present at the London workshop in 2019 and the subsequent online workshop in 2020. Thanks also to John Huss, Andrew Hopkins, Dominic Berry, Kim Hajek and other members of the Narrative Science Project for very helpful comments and discussion. This research was supported by the Singaporean Ministry of Education under its Academic Research Fund Tier 1 Grant (No. RG156/18-NS). *Narrative Science* book: This project has received funding from the European Research Council under the European Union's Horizon 2020 research and innovation programme (grant agreement No. 694732). www.narrative-science.org/.

Ritsema, R., T. Lay and H. Kanamori (2012). 'The 2011 Tohoku Earthquake'. *Elements* 8.3: 183–188.

Sun, T., K. Wang, T. Fujiwara, S. Kodaira and J. He (2017). 'Large Fault Slip Peaking at Trench in the 2011 Tohoku-Oki Earthquake'. *Nature Communications* 8: 14044.

Suzuki, W., S. Aoi, H. Sekiguchi and T. Kunugi (2011). 'Rupture Process of the 2011 Tohoku-Oki Mega-Thrust Earthquake (M9.0) Inverted from Strong-Motion Data'. *Geophysical Research Letters* 38.7: 1–6. https://agupubs.onlinelibrary.wiley.com/doi/full/10.1029/2011GL049136.

Tajima, F., J. Mori and B. L. N. Kennett (2013). 'A Review of the 2011 Tohoku-Oki Earthquake (M_w 9.0): Large-Scale Rupture across Heterogeneous Plate Coupling'. *Tectonophysics* 586: 15–34.

Wise, M. Norton (2017). 'On the Narrative Form of Simulations'. *Studies in History and Philosophy of Science Part A* 62: 74–85.

6   Stored and Storied Time in Archaeology

Anne Teather

Abstract

One of the primary goals of archaeology is to construct narratives of past human societies through the material evidence of their activities. Such narratives address how people led their lives and how they viewed and interacted with their world at different times in the past. However, the way archaeologists look at time is becoming increasingly disparate, fragmented and sometimes contradictory. While we now have more exact ways of dating past remains and deposits, and more sophisticated ways of examining how past humans may have engaged with their physical and social environments, there is some internal confusion as to the relative merits of alternative interpretations and evidence. In the research drive to determine a greater precision of dating and chronology, the effect that increased dating effort has on the accuracy of archaeological narratives has rarely been discussed. This chapter discusses the problems and opportunities for archaeological narratives in approaches to time.

6.1   Introduction

If you were thousands of years old, and in your youth had passed by the ancient city of Ur, Babylon,[1] at around 2000 BCE, you might have been lucky enough to visit Simat-Enlil, King Shulgi's daughter. Over a cold drink, she may have shown you a gift from her father: a bowl already over a hundred years old that he had inscribed as a gift to her (Thomason 2005: 74). This is the first documented case of people reusing and reappropriating already old materials in historical archaeology. One and a half thousand years later,[2] Nebuchadnezzar and Nabonidus, successive kings of Babylon, excavated and restored earlier structures at Ur. While doing so, they re-incorporated into their architecture inscriptions made by different kings, thousands of years earlier. Due to written records, we also know that Nabonidus's daughter, the princess En-nigaldi-Nanna, dug at the temple of Agade and had a room in her palace

---

[1] Now in modern Iraq.   [2] Around 550 BCE.

dedicated to items of antiquity, making her possibly the first known antiquarian (Daniel 1981: 14; Oates 1979: 162). These instances of the curation of already old artefacts are usually understood by archaeologists to be efforts to reinforce authority through emphasizing a connection to past rulers or ancestors.

In prehistory (i.e., without the benefit of written information), our understandings are built through narrative explanations based on the suite of physical evidence we encounter. The amount and nature of this evidence vary wildly – sometimes preservation is excellent and at other times very poor – but in general terms there is more evidence from time periods closer to the present day than from the more distant past. Already old materials incorporated into later deposits have been noted by archaeologists at prehistoric sites (Teather 2018; Knight, Boughton and Wilkinson 2019), and recent applications of absolute dating methods in prehistory have led to more instances of out-of-time artefacts being uncovered. I will argue in this chapter that this greater focus on attempting to ascribe more precision to an archaeological event has forced archaeologists to confront their own assumptions of what kinds of time they are trying to measure, and use, to frame accounts of past lives.

Archaeological information can be an important source of identity for human societies, providing an alternative narrative of life in the past and its connection to the present that can sit alongside origin myths, religion and history. As a uniquely integrated discipline of history and science, archaeology has made increasingly sophisticated use of narrative tools in the last 30 years. While it has always used narration to report and discuss evidence retrieved from past people's lives, this has been problematic. Narrative approaches have sometimes been seen as unduly historical and not scientific or objective enough; but scientific results are equally understood to be subjective and selective (or, at least, 'theory laden') and require explanation. In undertaking archaeological analysis, we are inextricably tied to both disciplinary approaches. This chapter traces the structure of knowledge creation in archaeology, how this applies to measuring time and how these are brought into coherent narrative form by archaeologists. In conclusion, archaeological information both contains stories and suggests stories: some that are ours and some that belong to the past.

6.2   Archaeological Knowledge and Narrative

6.2.1   What Is an Archaeological Narrative?

The craft of building an archaeological narrative is referenced in the profession as a type of interpretation, and in archaeology the study of types of interpretation, while often epistemological, is referred to as 'archaeological theory'. Archaeological interpretation is therefore the creation of narratives about the past, based on the evaluation of different kinds of facts.[3] For archaeologists, facts encompass a wide range of different categories of evidence. These might be historic, artefactual or architectural; comprise chemical or biological information; and, for both sets of these types of facts, be comparative or analogous. Wylie (2011: 302) has referred to these as facts of the record/mediating facts; and historical facts, which themselves comprise two types of facts – facts of the past and narrative facts. It is useful for the purposes of this chapter to think of facts as defined by Haycock (2011: 424): 'a fact is not of necessity something that is true; it is rather something that is taken to be true on the basis of current evidence in the context of a particular scaffolding of knowledge, ideas and beliefs that supports it'.[1]

Archaeology has different, and separate, types of scaffolding[5] that each constitute a body of interrelated facts, that themselves are composed of combinations of knowledge, ideas and beliefs of different kinds. For half a century, archaeologists have been familiar with visualizing this process, with much less sophistication than Chapman and Wylie's (2016) work, as a ladder of inference[6] where the further one travels up the ladder of knowledge, the further one is removed from the archaeological facts (or facts of the record, as Wylie (2011) might say). If we continue for the moment with a ladder analogy for scaffolding in archaeology, we can begin to see the types of scaffolding as separate self-supporting pillars of knowledge, or chronicles, that create genealogies[7] by a process of colligation (Morgan 2017: 88–89).[8] For this chapter it might be easier to visualize them as subject-based – for example, one genealogy might be that of pottery production (following a particular technical and historical trajectory and incorporating chronicles encompassing the types of clay, temper and firing times, experimental work and ethnographic analogy); another genealogy might be an account of animal husbandry (following animal domestication, genetics, behaviour, meat or dairy yield and comparative ethnographic information of the composition of different kinds of herds for different economic purposes). For example, in northern Europe the remains of sheep, cattle and pig in domestic species form a consistent contribution to past economies from 4000 BCE to the modern period, but their

---

[3] For the remainder of this chapter and for the sake of clarity I will refer to archaeological 'narratives' rather than 'interpretations'.

[4] This is referred to in some disciplines as an 'axiom'.

[5] Defined in the index to Chapman and Wylie (2016: 252–253) as conceptual, inferential, institutional, provisional, reconfiguration, reification and technical.

[6] Initially proposed by Hawkes (1954) to illustrate that religious and ritual beliefs are further away from other types of knowledge such as economic and have to be reconstructed through inference.

[7] On chronicles and genealogies, I follow the approach taken by Berry (Chapter 16).

[8] Morgan (2017: 89) explains the use of the term colligation as 'to capture the way a scientist both brings together, and assembles, a set of similar elements framed under some overall guiding conception, or categorization schema'.

proportion in deposits changes depending on the subsistence focus.[9] The chronicles might be broadly the same in many instances, but branch into different genealogies and narratives.

6.2.2   Construction of Archaeological Narratives

Narratives in archaeology weave between these chronicles and genealogies, as if with ribbon, creating individual cat's cradles by encompassing different facts from different chronicles and genealogies. For example, I conducted a synthesis of prehistoric human burials with strike-a-light kits[10] that determined that the overwhelming majority occurred with male burials between 2200 and 2000 BCE (Teather and Chamberlain 2016). While already considered to be a gendered practice, this research showed it was more common, very strongly male-related, often seen in higher status burials and, as a product of new radiocarbon dating, the duration and peak occurrence of the practice could be ascertained. In terms of scaffolding, this paper relied on many different chronicles of knowledge: experimental work; chemical work on the degradation of iron pyrites in soils over time; a genealogy of situating the practice within European prehistoric analyses of similar types of burials; and finally, the metaphorical work of Lakoff and Turner (1989) (a genealogy based on the use of textual analysis and material culture in archaeology) to suggest that death may have been seen as a type of journey for the dead men during this time period and requiring a portable source of light and/or heat. Each of these elements as brought to that paper have their own histories and scaffolds of knowledge in archaeology and cognate disciplines of anthropology and ethnology, but it was the authors' preference and choice to bring them together in this particular narrative. Other authors could use the same starting point of evidence and produce a different cat's cradle of narrative.[11] The success of this particular approach was that it has stimulated more attention during excavation to record and identify these otherwise quite functional and unremarkable objects; the thorough synthesis accompanied with a compelling narrative proposing a rich metaphorical significance has affected field practices. In Berry's terms (Chapter 16), the authors are present in the archaeological narrative through this process. Yet, the motive of the original research question or puzzle that stimulated that work is not actually mentioned in that paper.[12] As

---

[9] We can ascertain that some economies might be cattle-based (Neolithic) compared to ones that might be sheep-based (Iron Age); or a high proportion of older female cattle may suggest a dairying economy etc.

[10] A combination of a flint tool and iron-rich stone used for fire-lighting.

[11] Archaeologists refer to this as 'interpretation'.

[12] Morgan (2017: 90) writes that 'Stephen Turner [in *Sociological Explanation as Translation* (New York: Cambridge University Press, 1980] argues that sociological explanations are "translations" – they arise from comparisons which raise puzzles'. Puzzle here refers to both

a specialist in artefacts made from chalk in the Neolithic, I have proposed that most chalk artefacts mimic artefacts made from different substances, such as stone or wood (Teather 2017; Teather, Chamberlain and Parker Pearson 2019). I was puzzled[13] by chalk 'charms'[14] found in a small number of burials of predominantly women and children, and their visual resemblance to iron pyrite strike-stones that were in a few adult male burials (Figure 6.1).

No recent research had been conducted on strike-a-light burials so I had to complete it myself and having done so can argue (and will do so further in a monograph in preparation) that strike-a-light burials may have been male-dominated in this period, but that there was a metaphorical past connection in a different material within the burials of women and children. Therefore, the strike-a-light burials are male-gendered, but a similar practice included women and children in a different, and potentially socially subversive, way. In order to argue that position and create a convincing narrative, the research had to be completed in this sequence.

These examples show that archaeological narratives appear to map well onto a narrative science framework. I will now turn to focus on archaeological dating methods and how these fit into this proposal.

Figure 6.1 **Iron pyrites (left) and chalk charms (right) from the burial of a female, dated to 3600 BCE**
Cissbury, West Sussex, Shaft 27.

the query that emerged through comparison (as described by Turner), but also its narrative implications.

[13] Morgan (2017: 94) suggests that 'puzzles are generally solved within the existing community norms – that is, they provide narrative explanations considered satisfactory to those scientific communities for sound epistemic reasons'.

[14] These are small, rounded pieces of chalk that are decorated with short wavy incised lines to suggest a rough surface and are visually similar to natural nodules of iron pyrites.

6.3   Archaeological Dating

Chronologies in archaeology are manifold and can refer to temporal
changes in certain types of artefact or in modes of an economy or social
system. In effect, they are types of chronicle that seek to order selected
events by the inclusion and exclusion of information. Relative and
absolute chronologies (Figure 6.2) sit side by side in archaeology and
can include many different facts of the record, but are constructed in
different ways.

6.3.1   Relative Dating

In the history of archaeology, an interest in relative chronology began in
earnest with typological studies of antiquities that initially made

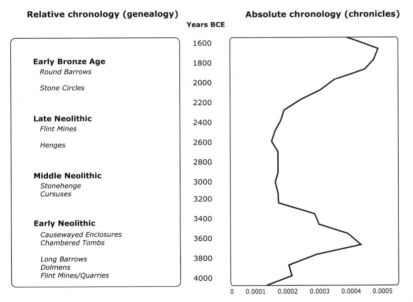

Relative chronologies as genealogies
are largely unaffected by the density
of dates as they broadly cover cultural
trends defined by monument types.

Post-excavation projects aimed at dating
archival material create a varying density
of radiocarbon dates as data points
(data from Woodbridge *et al.* 2014:221).

Figure 6.2 **Schematic representation of narrative reasoning in archaeological
chronologies for British prehistory**
This shows the end phases of the Stone Age and the Beginning of the Bronze Age,
relative chronologies (left) and absolute chronologies (right).

a categorical separation of geological objects and human-made artefacts. In the 1880s, Oscar Montelius created extensive relative typologies of archaeological artefacts across Europe with the goal being to devise a chronology for the broad cultural sequence (see Trigger 1989: 155– 156). Yet it was Christian Jürgensen Thomsen who advocated an approach to determining chronology through the typology of 'closed finds', repre- senting objects found together in burial, hoard or other groupings, which suggested they were deposited at the same time (Trigger 1989: 76). As a result, he was able to organize prehistoric material culture into a Three Age System that defined a narrative progression from the Stone Age to the Bronze Age and the Iron Age (Marila 2019: 94–95). The left-hand side of Figure 6.2 shows the end phases of the Stone Age into the Bronze Age.

Therefore, it was not only the individual artefact that was important but what it was found with – i.e., its systematic co-occurrence with other artefacts. For example, it was not simply that a bronze axe was discovered with a particularly distinctive pottery vessel together with a human burial in a mound, but rather that bronze axes were repeatedly found with that type of pottery with burials in those types of earthen mound. Thus, the first chron- ologies in archaeology were the product of a scaffolding of facts – such as the typologies of artefacts commonly found together and through stratigraphic sequences,[15] where deposits are discovered above or below other deposits. The terminologies for this are specified as the *terminus post quem* and the *terminus ante quem*: something has to be later (*post*) or earlier (*ante*) than something else either because of where it appears in the archaeological sequence or what it contains, but a firmer date cannot be specified. For example, a coin hoard might contain currency minted in AD 83, AD 104 and AD 200, and therefore it would have a *terminus post* of AD 200 in that it would have had to have been deposited after AD 200, even though it contained earlier coins. If this hoard were buried under a sixth-century Anglo-Saxon brooch, the coin hoard would have been deposited between AD 200 and 600, as the brooch provides a *terminus ante* of AD 599.

Relative dating in archaeology is therefore predicated on two distinc- tions: first, that combinations of portable material culture,[16] as assemblages,[17] are distinctive to a particular culture and indicative of

---

[15] Archaeological stratigraphy records the physical and spatial properties of cultural deposits examined through excavation. The relationships physically expressed are recorded through written descriptions and scaled drawings, where each event is recorded as a different numbered context, occurring before or after another, and so sequenced into relative time.

[16] Portable material culture describes artefacts such as pottery, stone tools and human remains, as opposed to structural remains such as architecture or foundations.

[17] Assemblages here refer to a grouping of artefacts commonly found in a society at a particular time period that would include portable material culture and fixed cultural information, such as architecture.

the subsistence, economics and social relations of that past population; and second, that each archaeological context such as a burial is physically separated from another context as a time capsule (making it a closed context) that we can discern through stratigraphic excavation. Therefore, typologies (systems of classification of evidence into distinct categories) and seriation (the sorting of evidence into a temporal sequence) were first established alongside chronology and relative dating, and all continue to be core elements of archaeological research practice. Assemblages, formed of typologies, closed contexts and seriation, can be seen as the scaffolding for chronicles, that form the basis of narratives.

Types of comparative assessments are also key to relative dating but might be seen more effectively as genealogies rather than as chronicles (Berry, Chapter 16). For example, aerial surveys using light detection and radar (LiDAR), or other remote sensing methods, produce images that can be compared with known excavated sites to produce an initial identification and assessment. The kinds of question that are approached here primarily rely on size and form: a 30 m circular banked and ditched enclosure may suggest an earthen henge monument like Avebury in Wiltshire (refer to the left-hand side of Figure 6.2), whereas a 10 m × 10 m square enclosure suggestive of stone-built foundations may indicate a Romano-Celtic temple.

Broader-scale narrative sequences are therefore built from aggregations of evidence from many excavated sites. Alongside typological considerations of materials present in archaeological deposits, they enable temporal chronologies to be brought together. Only limited comparisons may be possible between different archaeological sites unless cultural expressions such as pottery, buildings or monument styles are widespread and very similar. Where these are present, it allows for broad regional- and continental-scale syntheses, a particular kind of narrative. Relational dating is therefore reliant on a series of comparative interpretations of artefact and site typological data that are assessed and applied to each example, and are effectively genealogies. These are nested within a wider geographical and temporal understanding of comparative data as chronicles. Narratives can emerge at these different scales: the artefact, the period, the site, the region, the continent. While relative chronologies still have their uses for broadly determining an outline chronology, they can fail where we cannot easily determine chronologies from artefacts (e.g., through typologies where artefacts are degraded/when material culture changes little through time) or when we attempt to compare sites that are geographically separated.

6.3.2 Absolute Dating

The advent of absolute scientific dating in the mid-twentieth century had a significant effect on archaeological chronologies. It meant that materials and deposits could be placed within a chronicle of a measured time-scale of calendar years (refer to the right-hand side of Figure 6.2) rather than being reliant on relative factors alone. Initial methods included dendrochronology (tree-ring dating) and radiometric (commonly, radiocarbon) dating. Radiocarbon dating measures the decay of an unstable isotope of carbon from the time the living organism ceased respiration, and as such provides a date of 'death' as a probabilistic range.[18] Dendrochronology can identify the year at which the living tree was felled, and as such is potentially much more accurate than any other form of dating but is only applicable to suitable surviving timbers that preserve a sufficient number of tree rings to allow them to be matched to a reference chronology. Direct dating methods such as these are preferred particularly for human and animal remains, and due to the prevalence of these organic materials in deposits, radiocarbon dating has proved to be a key tool in discerning chronologies.

However, these have now been joined by further physico-chemical dating methods. These include uranium series dating of cave minerals, optically stimulated luminescence (OSL) to date sediments to when they were last exposed to sunlight, and palaeomagnetic dating that can be used to date deposits such as hearths that have been subject to an episode of burning. Amino acid racemization measures time-dependent changes in protein molecules (Demarchi and Collins 2014) and ceramic rehydroxylation (RHX) is able directly to date pottery through the chemical reaction on firing the pottery.

Indirect methods are also increasingly important. For example, molecular clock (or coalescent analysis dating) analyses molecular diversity to determine the sequence and timing of past demographic and evolutionary events that have left traces in both modern and ancient genomes. This means that we can assess the diversity of ancient populations in particular areas, and pinpoint times that they began to diverge from each other. These different methods speak to the different aspects and scales of the archaeological record; whether the research aim concerns the activities of living people such as firing pottery, or a much wider scale such as questions of genetics and evolution. They all result in establishing archaeological chronologies for the methods and assist in creating genealogies and narratives that refine archaeological sequences and constrain the duration of temporal events.

Absolute dating is seen as a preferred and more robust, scientific way of determining age[19] as it can order events and material culture within a fixed

---

[18] The decay of radiocarbon is measured, but the natural decay of atmospheric radiocarbon fluctuates between years. Therefore, radiocarbon dates are provided numerically 4350 ± 30 BP (before present) with an error range to encompass issues there may have been in the laboratory or with the material. These numbers then require calibration and produce a further probability of dates, and are then presented as, for example, 2900–2750 cal BC.

[19] Particularly for prehistoric archaeology.

temporal chronology and make it possible to compare within and between sites in a greater range of archaeological settings. While it might be considered better and more objective science to discard the application of anything but the broadest use of relative chronology, there are some instances where relative dating techniques are the preferred method. Relative chronologies are still very informative for styles of prehistoric rock art, which by their nature are often impossible to date using absolute methods. For historic periods, relative chronologies such as seriation are more often employed as many of the dating methods (apart from dendrochronology) are not more accurate. Therefore, clay pipes, for example, can be attributed to periods of manufacture and sometimes individual makers (e.g., Williamson 2006), and as they are commonly discarded when broken their period of use is reflected in deposits.

In summary, our archaeological chronologies have been subject to a shift and expansion in reasoning with the advent of the direct dating of materials. From a relative-based chronological approach, which is inherently genealogical, we now include an absolute chronology, or a chronicle, that does not rely on the storied or narrative aspect of the archaeological record but simply ranks the dates in sequential order (Figure 6.2, right-hand side). Absolute dating chronicles can be used to generate different genealogies and narratives without attention to the existing genealogy of relative chronology. This has resulted in a suite of new approaches to examining archaeological time.

6.4   From Dates to Narratives: Impacts of Recent Studies

As detailed above, relative and absolute dating provide different ways to build narratives. For relative chronology, narratives are built piecemeal from the contributions of seriation, typologies and closed contexts – often by individual scholars working on sets of material with continual reference to the existing genealogies and chronicles. When syntheses are produced, they are agreed in chronicles by the process of accepting individual facts within the scaffolding of each chronicle. Therefore, these have been subjected to continuous revision and refinement over time (Chapman and Wylie 2016), as new discoveries have been made or inferences or assumptions have been successfully challenged. This process produces rich and subtle narratives that often incorporate previously overlooked evidence and challenge earlier interpretations or narratives, such as the Glastonbury Iron Age village where the excavators' results were revisited and reworked five times between the 1960s and 1990s (Chapman and Wylie 2016: 108–136).

On a wider scale and using data from the British and European Neolithic period, the span of data derived from relative dating originates from the accretion and consensus of primary analyses of archaeological finds. These initially consist of collections of multi-authored specialist reports of pottery, flint and other materials, that when taken together lead to a consensus of date or date range that subsequently fits into the broader relational chronological framework. At this stage, it is usual for a few 'range-finder' radiocarbon dates to be obtained to confirm the proposed artefact-based chronology. Therefore, at an initial and primary assessment stage, three to five radiocarbon dates may be taken per site to confirm and establish, or conversely challenge, the range of activity and sequence proposed through the relative dating methods. When information from many sites is aggregated, this results in a fairly even spread of absolute dates throughout the chronological sequence, as seen with the numbered dates on the right-hand side of Figure 6.2.

In the last decade and a half, advanced computational approaches have permitted new analyses of absolute dating evidence undertaken on archives of previously excavated materials that have challenged this traditional means of narrative building in archaeology. Intensive radiometric dating studies on archival material that have often focused on a particular category of site or research question have increasingly been undertaken. These dates supplement the range-finder dates typically gained through excavation and have the effect of creating blocks of data in the temporal framework and consequently an uneven distribution of information (please refer to the tightly spaced lines on the right-hand side of Figure 6.2). Two approaches have been particularly prominent: new absolute dating on sequential stratigraphic layers within archaeological deposits to produce Bayesian statistical analyses of site chronologies; and large-scale geo-spatial analysis of aggregated radiocarbon dates. These projects produce narratives and chronologies that are based on either inclusive or reductive reasoning. Inclusive narratives use absolute dating to establish how events are chronologically and spatially related. They are inclusive, as they seek to address all the findings uncovered, and they are narrative in that they provide an explanation for those findings through attention to the type of artefactual material under study and the sequence revealed. Reductive chronologies use absolute dating to provide a more precise date for an event. They are chronologies in that precise time measurement is their goal; and they are reductive in that they exclude dating evidence that does not easily confirm their chronological goal. This is because they assess any type of material culture as solely contributing in terms of chronological (and not cultural) evidence.

6.4.1   *Case Studies Producing Inclusive Narratives*

Absolute dating can help archaeologists establish how events are chronologically and spatially related. When projects consider all dates obtained and attempt to explain them by attending to questions of artefact type and sequencing, they produce inclusive narratives. Two examples of inclusive research-led projects that involved gaining new radiocarbon dates follow.

Parker Pearson et al. (2019) presented the results of a large-scale project on dating the Early Bronze Age phenomenon of Beaker burials[20] in Britain that date from 2600 to 1700 BCE. The aim[21] of this project was to discover and assess the level of mobility of those buried people during their lives. The skeletal remains from 370 individuals were investigated, with 17 found to date earlier than the period under study and 19 later, together with a number of examples that could not be dated by absolute methods (Parker Pearson et al. 2019: 426). Constraining the dates of the 'Beaker phenomenon' is an important archaeological question but one that has become more current due to ancient DNA analyses that have suggested that Beaker people arrived in the UK as migrants from continental Europe (Olade et al. 2018).

The second project directly dated archival material from mining and quarrying sites that were thought to date from the start of the Neolithic in Britain and continental north-west Europe (6500–1500 cal BC: Schauer et al. 2019b; 2019a; Edinborough et al. 2020). The goal here was to determine if stone axes (the primary products of mines and quarries) were in increased demand for clearing forest during the economic change to agriculture at this time, and, if so, precisely when this occurred.[22]

For both projects, the radiocarbon dates as new 'facts of the record' are treated as fully correct, even if they do not answer the research question. The impact of the small proportion of dates that fall outside the timespan of the research question is minimal, but the anomalous dates are nonetheless recorded and discussed informatively. In the first example on Beaker burials, the absolute dating related to one phenomenon (the death and subsequent burial), the one dating material type (the body) and the accompanying analyses on mobility (the body) have created a different genealogy of Beaker practices through new radiocarbon dates. Each burial is contextualized in the project publication within its geographical location at death, location/s of the life of the past person, the character of the grave (mound/cist/cut grave) and its accompanying grave goods. In terms of evidence, it could be argued that this research dating programme has enriched the primary archive and allows further reconsideration of that primary archive in future through informing different chronicles, even when the information is unexpected. The second example of the summed probability dating has enhanced the suite of absolute dates available for mining and quarrying and so enhanced the primary archive. In both cases, the computer modelling for activity trends takes place with the achieved raw dates to detach

[20] Human burials in mounds with a distinctive pottery style and often other grave goods.
[21] Or puzzle.
[22] Detailing the premises for the mathematical models that this project relies on is out of the scope of this chapter (but see Schauer et al. 2019a).

them from their immediate context within the archaeological site[23] and to re-contextualize the dates as proxies for population.[24]

At these different scales of analysis, the recontextualized narrative permits the puzzle to be convincingly answered while the non-compliant results are explicable either by producing new knowledge, or by reassessing the facts of the record. For flint mining, radiocarbon dates from Church Hill, West Sussex, demonstrated that there was an early Bronze Age phase that had been previously suspected but not substantiated (Edinborough et al. 2020). In the case of Beaker human remains from Linch Hill, Oxfordshire, that surprisingly produced a Neolithic radiocarbon date, it was suggested that the human remains from this site were mislabelled at some point post-excavation. These remains now have the correct attribution for future researchers (Parker Pearson et al. 2019: 426); the known fact has been corrected. Therefore, inclusive narratives seek to use the chronicles created to produce as many genealogies and narratives as are required, even if they were not the primary goal.

6.4.2   Case Studies Producing Reductive Chronologies

Some projects use absolute dating with the unique aim of precise time measurement. In the process, they evaluate material culture only in chronological (not cultural) terms, while excluding dating evidence that doesn't 'fit' their chronological goal. Such studies tend to produce reductive chronologies.

Bayesian analysis has been an innovation in archaeology in recent years that uses a statistical process to constrain the probability range that radiocarbon dates provide, thereby increasing dating precision. This type of methodology is increasingly used on large datasets to examine trends,[25] but is also used by some researchers for analyses within a small physical scale of archaeological remains in a stratigraphy-based Bayesian analysis. Here, the stratigraphic record is used to enable the research team to restrict the radiocarbon dates by considering whether a datable deposit is higher or lower in the depositional sequence than another datable deposit (*terminus post quem* or *terminus ante quem*).

'Gathering Time', a project led by Alasdair Whittle early this century, focused on dating causewayed enclosures, an early Neolithic type of

---

[23] Miyake (Chapter 5) refers to this kind of process as a 'source model' and a 'rupture narrative'. See also Wise (2017).

[24] The premise is that the number of radiocarbon dates is directly proportional to the amount of remains and so the amount of activity, and so directly reflects the population size (for a larger population you would expect more sites/monuments/burials than for a smaller population).

[25] It was incorporated into the Beaker people project to examine which areas had Beaker burials before or after others by aggregating the radiocarbon dates of individual sets of remains.

monument.[26] Deposits from ditches were multiple dated to ascertain an accurate construction date in order to map the temporal distribution of UK enclosures. For example, at Whitehawk enclosure, in Sussex, an additional 38 radiocarbon dates were gained from deposits in the 4 concentric ditches in addition to the two dates already obtained from Ditches III and IV (Whittle, Healy and Bayliss 2011: 214–226). The original two post-excavation radiocarbon dates suggested that the enclosure was built and used within a broad range of time between 3710 and 3090 cal BC.[27] New dates were taken assuming that the digging of the original ditches and the accumulation of chalk rubble within them constituted one phase and that silting above this was a secondary phase. Using their analyses,[28] the conclusion is that Whitehawk was constructed between the mid-37th and 36th centuries BC, and 'in primary use for 70–260 years (*95% probability*), probably for 100–115 years (*4% probability*), or 155–230 years (*64% probability*)' (Whittle, Healy and Bayliss 2011: 226). This project concluded with assessing the radiocarbon dates from enclosures, alongside other modelled regional dates, to propose that, as a phenomenon, causewayed enclosures were constructed at slightly different times in different regions in southern Britain, although all between 3710 cal BC and 3630 cal BC, with some later in Wales (to 3550 cal BC: Whittle, Healy and Bayliss 2011: 694). Apart from these new chronicles, the further narrative conclusions of this project are based on the estimation that the Neolithic transition[29] began in each of the causewayed enclosure areas two hundred years prior to the actual construction of causewayed enclosures, thus the chronicles have been combined with other models in order to create a plausible narrative for the Neolithic transition (q.v. Whittle, Healy and Bayliss 2011: 727–729).

In order to be successful, this method either completely excludes – or assigns very low prior probability weightings – to radiocarbon dates that are regarded as erroneous or that simply lie too far outside the acceptable range of possible dates. This rationale creates new chronicles by detachment, permitting the separation of dates into acceptable and unacceptable categories that lead to inclusion or rejection. The expectation is for conformity. By filtering the dates, two chronicles are created: one normative and used as a source model and one

---

[26] These are monuments constructed of 2–4 concentric circles of sausage-shaped ditches with causeways in between them.

[27] Ditch III, sample I-11846, produced a date of 4700 ± 130 BP; at Ditch IV, sample I-11847 produced a date of 3690–3090 cal BC, 4645 ± 95 BP.

[28] Ditch I dates are proposed of 3635–3560 cal BC (*95% probability*), Ditch II, 3675–3630 cal BC (*72% probability*), Ditch III, 3660–3560 cal BC (*95% probability*, or 3650–3600, *68% probability*), Ditch IV, 3650–3505 cal BC (*95% probability*) but refine this to 'probably 3635–3610 cal BC (*18% probability*) or 3600–3530 cal BC (*50% probability*)' (Whittle, Healy and Bayliss 2011: 225), but suggest that for both Ditch II and Ditch IV these later dates may be from later deposits placed into these ditches.

[29] The transition from hunter-gathering to domesticated lifestyles including monument building, pastoralism and farming.

that remains peripheral.[30] The new normative chronicle is based on this filtered data, and these chronicles are aggregated to include multiple archaeological sites that have been assessed in a similar way (another chronicle) to lead to an overarching narrative of the beginning and end dates of particular types of site. A genealogy is not necessarily produced (as chronology is the goal, not the identification of causation), but rather new chronicles for each archaeological site – that are then aggregated again into another larger chronicle. The narrative is created through reference to existing archaeological genealogies.

These stratigraphy-based Bayesian methods have been responsible for a greater number of radiocarbon dates as facts, but the density of the information is intra-site rather than inter-site. The multiple radiocarbon dates produced are separated in the archaeological record not by archaeological site but by archaeological layer.[31] They are therefore not representative of different activities in different places, but rather episodes of similar activity at the same location in a broadly temporally similar or adjacent time.

6.4.3   Summary: From Dates to Narratives

The absolute dating of archaeological material as a separate and secondary procedure that takes place on already excavated material held in museum archives is often completed with a particular research question in mind. From this initial stage, a new chronicle begins to be constructed that will ultimately assess certain categories of archaeological material, and not others. Monuments or burials that were excavated by different people, decades apart, become one synthesized category for the purposes of the project, producing new chronicles. These may lead to new genealogies and narratives at different scales, although that is not always certain. If the research question is primarily temporal, the narrative implications may be largely to construct new chronicles. Whatever their primary aim may have been, their enduring influence is in enhancing the archaeological record with more absolute chronological data as facts of the record.

Alongside these new methods, new ways of accessing radiocarbon dates have also been developed. For example, the University of Kiel in Germany has created RADON,[32] a free online database of radiocarbon dates across Europe. It is possible to compare the radiocarbon dates gained on archaeological material from different countries and their regions through this database. Interestingly, the UK has almost double the number of radiocarbon records of

---

[30] The process of eliminating the rejected dates is euphemistically termed 'chronometric hygiene'. Excluded dates gain a reason or attribute of rejection, commonly categorized as 'outliers' (Teather 2018).

[31] This might mean they are only centimetres apart.

[32] See Hinz, Furholt and Müller (2012). RADON can be found at https://radon.ufg.uni-kiel.de.

any other country,[33] perhaps reflecting an uneven approach to dating in the UK – or perhaps the inclusion of UK dates in the database has been higher for other reasons. Nevertheless, in the RADON chronicle of chronology, any dates from peripheral chronicles and excluded from a stratigraphically based Bayesian narrative have been reapplied to their site context: the distinction in the Bayesian narrative between 'right' and 'wrong' dates is removed. This reincorporation is important, as it allows all the primary data for each site to be held together as a single site chronicle.

6.5   Conclusion

This chapter began with a discussion of the use of already old material culture in later deposits in ancient Babylon. Between the third and first millenniums BCE, the connection between the curation, deposition in graves and/or powerful display of material culture initially had the purpose of legitimizing the power of individuals in their leading societal role, both locally and regionally. In the British Neolithic (4000–2000 BCE), it is possible that old bones were already incorporated into pits or burial chambers, which may have been for the purpose of integrating past material with those of that present (Teather 2018), and, later in prehistory, human bodies appear to have been deliberately mummified, curated and buried at a later date (Booth, Chamberlain and Parker Pearson 2015). For archaeologists, time and temporality are different faces of the same coin: time is simply a clock; temporality encompasses the human experience of time and is not easily measured.

I have sought to discuss how narratives in archaeology are created through chronicles and genealogies. Archaeologists are familiar with only achieving a temporary success with our narratives; research in our archives and in the field is a continual process, and new discoveries can quickly destabilize existing narratives. By separating our narratives into the use of facts, chronicles and genealogies, it allows us to comprehend the complex structure of archaeological knowledge and how we construct it. New information is readily incorporated into our existing genealogies and chronicles.

I have chosen to discuss chronology in archaeology, and how absolute dating methods have moved our primarily genealogical reasoning into establishing chronologies which, in the end, produce only chronicles. Further, the different methodologies within absolute dating projects have resulted in a diversity of composed narratives. In particular, the creation (or not) of filtered chronicles and use of detached narratives (narratives detached from context) have been

---

[33] As of 10 December 2020, 4,656 records; the next nearest figure is France, with 2,765.

seen to be pivotal to this process, and I have termed these inclusive narratives and reductive chronologies.

While more absolute dating allows us to create more chronicles of absolute chronologies, these are not equal. Some will provide us with more adjacent temporal moments that do not necessarily produce better understandings of the archaeology but answer questions of time. Normative approaches will produce a normative view of human behaviour. But the peripheral, inconvenient and subversive facts construct entirely different chronicles, genealogies and narratives. These are where the human stories lie.[34]

References

Booth, T., A. Chamberlain and M. Parker Pearson (2015). 'Mummification in Bronze Age Britain'. *Antiquity* 89.347: 1155–1173.

Chapman, R., and A. Wylie (2016). *Evidential Reasoning in Archaeology*. London: Bloomsbury.

Daniel, G. (1981). *A Short History of Archaeology*. London: Thames & Hudson.

Demarchi, B., and M. Collins (2014). 'Amino Acid Racemization Dating'. In W. Rink and J. Thompson, eds. *Encyclopedia of Scientific Dating Methods*. Dordrecht: Springer, 13.

Edinborough, K., S. Shennan, A. Teather, J. Baczkowski et al. (2020). 'New Radiocarbon Dates Show Early Neolithic Date of Flint-Mining and Stone Quarrying in Britain'. *Radiocarbon* 62.1: 75–105.

Hawkes, C. (1954). 'Archaeological Theory and Method: Some Suggestions from the Old World'. *American Anthropologist* 56: 155–168.

Haycock, D. B. (2011). 'The Facts of Life and Death: A Case of Exceptional Longevity'. In P. Howlett and Mary S. Morgan, eds. *How Well Do Facts Travel? The Dissemination of Reliable Knowledge*. Cambridge: Cambridge University Press, 403–428.

Hinz, M., M. Furholt, J. Müller, C. Rinne et al. (2012). 'RADON: Radiocarbon Dates Online 2012. Central European Database of 14C Dates for the Neolithic and Early Bronze Age'. *Journal of Neolithic Archaeology* 14: 1–5. https://doi.org/10.12766/jna.2012.65.

Knight, M. G., D. Boughton and R. E. Wilkinson, eds. (2019). *Objects of the Past in the Past: Investigating the Significance of Earlier Artefacts in Later Contexts*. Oxford: Archaeopress Publishing.

Lakoff, G., and M. Turner (1989). *More Cool than Reason: A Field Guide to Poetic Metaphor*. Chicago: University of Chicago Press.

[34] I am grateful to Mary Morgan for the invitation to contribute to her project and thank her, Kim Hajek, Dominic Berry and Andrew Hopkins for their warm welcome and continuing intellectual generosity. Julian Thomas, Catherine Frieman and Stephen Shennan were kind enough to comment on an earlier draft of this paper, and I am grateful to John Huss and an anonymous reviewer for their insights and comments. Any errors or omissions are my own. *Narrative Science* book: This project has received funding from the European Research Council under the European Union's Horizon 2020 research and innovation programme (grant agreement No. 694732). www.narrative-science.org/.

Marila, M. M. (2019). 'Slow Science for Fast Archaeology'. *Current Swedish Archaeology* 27.1: 93–114.

Morgan, Mary S. (2017). 'Narrative Ordering and Explanation'. *Studies in History and Philosophy of Science Part A* 62: 86–97.

Oates, J. (1979). *Babylon*. London: Thames & Hudson.

Olalde, I., S. Brace, M.E. Allentoft, I. Armit et al. (2018). 'The Beaker Phenomenon and the Genomic Transformation of Northwest Europe'. *Nature* 555.7695: 190–196.

Parker Pearson, M., A. Sheridan, M. Jay, A. Chamberlain et al., eds. (2019). *The Beaker People: Isotopes, Mobility and Diet in Prehistoric Britain*. Oxford: Oxbow Books.

Schauer, P., A. Bevan, S. Shennan, K. Edinborough et al. (2019a). 'British Neolithic Axehead Distributions and Their Implications'. *Journal of Archaeological Method and Theory* 27: 836–859.

Schauer, P., S. Shennan, A. Bevan, G. Cook et al. (2019b). 'Supply and Demand in Prehistory? Economics of Neolithic Mining in Northwest Europe'. *Journal of Anthropological Archaeology* 54: 149–160.

Teather, A. (2017). 'More than "Other Stone": New Methods to Analyse Prehistoric Chalk Artefacts'. In R. Shaffrey, ed. *Written in Stone: Function, Form, and Provenancing of a Range of Prehistoric Stone Objects*. Southampton: Highfield Press, 303–321.

(2018). 'Revealing a Prehistoric Past: Evidence for the Deliberate Construction of a Historic Narrative in the British Neolithic'. *Journal of Social Archaeology* 18.2: 193–211.

Teather, A., and A. T. Chamberlain (2016). 'Dying Embers: Fire-Lighting Technology and Mortuary Practice in Early Bronze Age Britain'. *Archaeological Journal* 173.2: 188–205.

Teather, A., A. T. Chamberlain and M. Parker Pearson (2019). 'The Chalk Drums from Folkton and Lavant: Measuring Devices from the Time of Stonehenge'. *Journal of the British Society for the History of Mathematics* 34.1: 1–11.

Thomason, A. K. (2005). *Luxury and Legitimation: Royal Collecting in Ancient Mesopotamia*. Aldershot: Ashgate.

Trigger, B. G. (1989). *A History of Archaeological Thought*. Cambridge: Cambridge University Press.

Whittle A., F. Healy and A. Bayliss, eds. (2011). *Gathering Time: Dating the Early Neolithic Enclosures of Southern Britain and Ireland*. Oxford: Oxbow Books.

Williamson, C. (2006). 'Dating the Domestic Ceramics and Pipe-Smoking-Related Artifacts from Casselden Place, Melbourne, Australia'. *International Journal of Historical Archaeology* 10.4: 323–335.

Wise, M. Norton (2017). 'On the Narrative Form of Simulations'. *Studies in History and Philosophy of Science Part A* 62: 74–85.

Woodbridge, J., Fyfe, R. M., Roberts, N., Downey, S., Edinborough, K. and Shennan, S. (2014). 'The Impact of the Neolithic Agricultural Transition in Britain: A Comparison of Pollen-Based Land-Cover and Archaeological 14C Date-Inferred Population Change'. *Journal of Archaeological Science* 51: 216–224.

Wylie, A. (2011). 'Archaeological Facts in Transit: The "Eminent Mounds" of Central North America'. In P. Howlett and Mary S. Morgan, eds. *How Well Do Facts Travel? The Dissemination of Reliable Knowledge*. Cambridge: Cambridge University Press, 301–324.

III

Accessing Nature's Narratives

When nature is seen as narrating itself, narrative becomes a constituent feature of scientific accounts

7   Great Exaptations: On Reading Darwin's Plant Narratives

Devin Griffiths

Abstract

Drawing on narrative theory, performance studies and the history and philosophy of science, this chapter explores the distinct kinds and functions of what we might call *plant narratives* – the stories we tell about botanical life, but also the stories that plants tell us. Charles Darwin's botanical studies developed various techniques to study plant behaviour and record their movements in time. These methods drew scientific observers into an experimental 'dance' that aligned human and plant actions in order narratively to reconstruct evolutionary histories, especially histories of *exaptation*. These culminated in his last study, *The Power of Movement in Plants* (1880), which uses extensive illustrations to record and then reconfigure these individual micro-histories as what Darwin termed the 'life history of a plant'. Ultimately, its holistic account integrates these individual narratives and evolutionary history through a unified narrative, a conclusive *Bildungsroman* detailing a generic plant's experiences over the course of its life.

7.1   Can Plants Tell Stories?

This is the question Darwin set out to answer in his final book, *The Power of Movement in Plants* (Darwin and Darwin 1880), published shortly before he died. This idiosyncratic question summed up Darwin's life-long attempt to understand the common history of all life, and to devise strategies for telling it. Using a variety of innovative techniques, Darwin eventually figured out how to record what he termed 'the life-history of the plant', putting special emphasis on the way plants interacted with the world around them, sensing changes in environment, reacting to stimuli, deciding their fate (Darwin and Darwin 1880: 548). Darwin's argument for the ability of plants to feel and react, even to think, was controversial in his time, but opened up entirely new avenues of research into plant physiology, from plant signalling (the relay of information), to chemotaxis and tropism (respectively, movement and growth in response to stimuli). Today, research into these phenomena is commonplace. But I wish to take up perhaps the most radical implication of Darwin's plant studies: plants *do* have stories to tell, and if we listen closely, they can tell them to us.

These stories bear little comparison to a Jane Austen novel; in the stories Darwin recorded, we do not find plant-based Elizabeth Bennets, waiting to see whether Mr. Darcy will deign to join the dance. But these narratives do catch a perhaps more delicate interaction in which, as literary historian Gillian Beer has put it, 'observer and observed are in a dance of accord' (Beer 2017: 31). Drawing on the history and philosophy of science, performance studies and narrative theory, I will explore the implications of Darwin's plant studies for the place of narrative in science.

It is generally recognized that Darwin's scientific accounts were organized by narratives – various stories that attempted to explain how specific relationships, structures and behaviours evolved in the past (Levine 1988; Beer 2000). The key role that narratives play in Darwin's accounts underlines the importance of narratives to science in general, but also the importance of considering how scientific narratives are structured by wider practices of storytelling. In my earlier studies of Darwin's science, I have emphasized the necessarily *fictive* quality of the stories produced by Darwin's studies, insofar as they retroject a persuasive narrative on the basis of incomplete evidence (Griffiths 2016). As Greg Priest points out, these 'conjectural historical narratives' were sometimes organized by Darwin into diagrams, as in the famous tree of life from the *Origin* (Priest 2018). Darwin described the stories he imagined as 'castles in the air', retrospective fictions tethered to empirical grounds through the meticulous but necessarily partial assembly of historical data and present observations. In this way, we might take Darwin's castles as proof of the claim that new scientific narratives, and new scientific theories in general, are produced by the scientific imagination; they are, as Alistair Crombie put this, 'designed in the mind' (Crombie 1988). Similarly, Erin James has suggested that narratives about plants tend to stage a kind of ventriloquist act, in which plants serve as a vehicle for the expression of human opinions and perspectives (James 2017).

The present chapter departs from this line of human-centred thinking, by asking: to what degree were Darwin's narratives recordings of narratives from nature itself? And how did the objects of Darwin's studies intervene in telling their own stories? Darwin's narratives often operate on at least two levels. On the one hand, he hypothesized long-term stories of adaptive evolution to bridge the evidentiary gaps in the distribution of traits within current and past species, explaining how complex behaviours and traits might evolve from simpler precursors. But in his later works he also placed increasingly heavy emphasis on the contingent and idiosyncratic way that specific traits were adapted to new purposes. And he sought to document this contingency, which operated at the level of species history, through smaller-scale narratives that described the lives of individual specimens, in detailed micro-histories of their growth and contingent change. Darwin's last major work, *The Power of Movement in Plants*, marks the culmination of these efforts. In allowing plants to draw their contingent behaviour on the page, he enlisted them in his efforts to narrate (from the perspective of individual plants) how they grow, subsequently articulating these accounts to reconstruct (from the perspective of the species) how they once evolved.

Central to that approach was a set of techniques that allowed plants to inscribe their growth on various media, writing their lives into Darwin's science. Figure 7.1 is a late example – a graphic reproduction of the trail a plant root left on a glass slide as it grew. What would it mean to read these graphic traces as scientific narratives? Narrative theorist Mieke Bal defines narrative as 'a text in which an agent relates ("tells") a story in a particular medium'. Bal further defines 'story' as

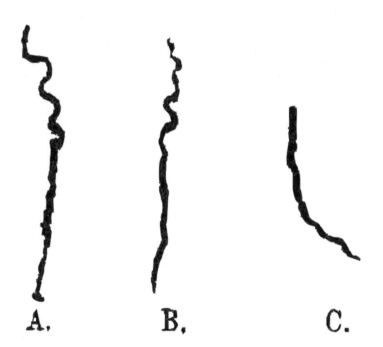

A.   B.   C.

Figure 7.1 *Phaseolus multiflorus*'Tracks left on inclined smoked glass-plates by tips of radicles in growing downwards. A and C, plates inclined at 60°, B inclined at 68° with the horizon'.
Source: Darwin and Darwin 1880: 29. Reproduced, with permission, from John van Wyhe, ed., *The Complete Work of Charles Darwin Online* (http://darwin-online.org.uk/converted/pdf/1880_Movement_F1325.pdf).

a set of 'chronologically related events that are caused or experienced by actors', emphasizing that agents are 'not necessarily human' and that the 'medium' may be visual as much as textual (Bal 1997: 4). Darwin's plant illustrations certainly *seem* to meet a minimal description of narrative as a linear account of events experienced by some agent. Here, we see in the various squiggles, and in the alteration of thicker and thinner strokes, the varying pressure and direction of the root of a plant as it senses and attempts to grow around the slide. And yet, to take these illustrations seriously as narratives, as pieces of the 'life-history of the plant', we must rethink our basic intuitions about where scientific narratives come from. More than the narratives *about* science that Robert Meunier discusses (Chapter 12), or the 'narratives *of* nature' that John Huss studies (Chapter 3), they are narratives *from* nature. Such scientific narratives are not simply produced by the scientists and projected on the world, but rather are generated through interaction with that world, elaborated through what Andrew Pickering has termed a 'dance of agencies' – both natural and human (Pickering 1995).

The following chapter has two movements. In the first, I'll explore the methods of Darwin's plant science, attending to the performative intricacy of scientific experimentation as a collaborative dance that highlights the contingent, narrative aspects of plant development – akin to the 'reticulate approach' identified by Elizabeth Haines (Chapter 9). In the section that follows, I'll examine how these contingent histories, co-elaborated by Darwin and his plant subjects, are articulated together at the close of the work as a 'life-history' – in effect, a *Bildungsroman* that reconfigures individual events in relation to an overarching evolutionary thesis. On the basis of this account, I propose a way of understanding scientific narratives as moving between entanglement with the world and reconceptualization, and between the narration of contingent events and their reconfiguration into higher-order narrative genres, by means of a process that draws multiple actors, human and non-human, into alignment, cultivating multi-level, multiply-scaled stories about how the world works.

7.2   Plant Narrative and the 'Dance of Agencies'

At the core of Darwin's botanical research program was the question of how plants and other forms of life interact. Over the course of several decades, especially after his move to Down House, some fifteen miles from London, Darwin let his imagination run riot, exploring an astonishing range of methods to entice plants into cooperating with his studies. He tickled them with horse hair and pencil; he gave them a spin in rotating boxes; he played them music; he fed them sweetmeats, and sometimes, just meat. These experiments, simple, elegant, intimate, produced results that often astonished the botanical world, and the greenhouse at Down became an object of fascination for visitors. The results were published in a series of pathbreaking works of botany (Darwin 1862b; 1865; 1875a; 1875b; 1876; 1877; Darwin and Darwin 1880).

All of Darwin's plant studies demonstrate a fascination with the relation between plant and animal life, and Darwin's insistent assertions that plants possessed sensitivity, an ability to move, digest and even think, much like animals. Darwin's grandfather, Erasmus Darwin, was famous for drawing analogies between plant and animal life, between their modes of reproduction and growth, and was eventually notorious for insisting that these analogies indicated a common nature and a more basic, shared history of evolution (Priestman 2000). In adopting his grandfather's commitment to the shared nature of plant and animal life, Charles took an unusual perspective on the possibilities of plant agency, and a unique interest in documenting their sensitive and responsive engagement with the world – sensing and capturing insects for nourishment, grasping and then climbing neighbouring trees and structures.

As Darwin exposed increasingly complex plant behaviours and adaptations, others argued that these elaborate behaviours defied explanation by the gradual means of natural selection. In 1871, St. George Jackson Mivart summed up these objections, arguing that, even if natural selection might operate on such adaptations after they evolved, it could not explain how they first developed. The elaborate adaptive structures of orchids, and the power of twining plants to climb trees, illustrated the 'incompetency of "natural selection" to account for the incipient stages of useful structures' (Mivart 1871: 35).[1] It was the most succinct statement of what Stephen Jay Gould would later term the '5 percent of a wing principle': variations in the wing structure of flying birds might experience selective pressure, but an incipient wing would seem to be useless for flight and therefore non-adaptive (Gould 2002: 1220).

Darwin recognized this as a serious challenge to the comprehensiveness of the theory of natural selection and immediately set out to answer it. The following year, he added an entirely new chapter to the sixth edition of *On the Origin of Species*, responding at length to Mivart's critiques. In the only new chapter ever added to that work, Darwin placed heavy emphasis on the sensitive actions of plants as examples in which the 'incipient stages of useful structures' might have developed 'incidentally' from other adaptive traits (Darwin 1872: 198). All plants, he noted, seemed to have some capacity to move, and this movement is often coordinated with a basic sensitivity to specific influences, like sunlight and gravity. This innate sensitivity gave them an 'incidental' sensitivity to touch, much as 'the nerves and muscles of an animal are excited by galvanism' or electrical stimulus, despite such sensitivity being non-adaptive (Darwin 1872: 198). These incidental abilities,

---

[1] For further discussion of Mivart's critique, and Gould's discussion, see Beatty (Chapter 20).

Darwin argued, could be the building blocks of much more complex adaptive behaviours, like the behaviour of climbing and insect-eating plants.

As Gould explains, this marked a significant shift in the emphasis Darwin placed on such 'exaptations' – a term coined by Gould and Elisabeth S. Vrba to describe cases of functional repurposing (Gould and Vrba 1982). As Gould later explained, Darwin gave exaptation a 'vital role in establishing the contingency and unpredictability of evolutionary change', with the consequence that 'historical [i.e., *narrative*] explanation' became central to his evolutionary histories (Gould 2002: 1224–1225). Exaptations effectively differentiate the historical origin of a trait and its current function, locating the contingency of evolutionary development in selective events that repurpose a given trait. Mivart's critique helped Darwin recognize the importance of such examples of exaptation both as a way to explain 'incipient' adaptive structures and as a way to underline the essential contingency of natural selection. In light of Mivart's critique, the exaptive development of plant behaviour took on a further significance, not recognized in Gould's analysis. If exaptation allowed plants to develop animal-like behaviour, moving as well as reacting to their environment, this showed that complex behaviours could emerge contingently by repurposing traits to serve new functions. Yet the convergence of plant and animal behaviour *also* demonstrated that complex adaptations could be achieved by radically different exaptive pathways. Darwin's long-standing interest in the analogy between plant and animal life took on enhanced importance in demonstrating the *unexceptional* as well as contingent evolution of animal behaviour. Demonstrating the agency of plant life was the linchpin of this analysis because it drew attention to both the complexity of plant behaviour and its analogy with animal action. For the rest of his career, Charles Darwin would doggedly pursue this strategy, working to prove, first, that plants exhibited forms of agency, second, that these behaviours could be explained as the exaptation of traits that did not originally serve their present purpose, and, finally, that these mechanisms were distinct from the (equally contingent) adaptations undergirding animal behaviour.

After making these revisions to *On the Origin of Species*, Charles launched a series of studies to solidify his argument that plants not only moved, but that this movement was a purposeful behaviour exapted from previously existing traits. This required the development of a variety of new experimental techniques for registering both the contingency of plant behaviour and the contingency of their evolutionary history. Working with his children, Francis, George and Horace, he produced a considerably revised, second edition of *Climbing Plants* that nearly doubled its length (Darwin 1875b) as well as an in-depth study of the carnivorous behaviours of *Insectivorous Plants* (1875a). These works stunned botanists by showing that virtually all plants exhibited some movement, not just growth, in response to their environment. One outcome of his study of insectivorous plants demonstrated, via various chemical and electrical experiments, that plants possess what he termed 'nervous matter', distinct from the nerve tissue of animals (Darwin 1862a).

At the time, studies of plant physiology were growing considerably more sophisticated. In part, this was due to rigorous new techniques developed by Julius Sachs in his lab at the University of Würzburg. Sachs's 'auxanometer', which mechanically registered plant growth, is one example (Figure 7.2a). Although the auxanometer provided a precise way to study plant growth, it did so with a significant limitation: it could only register growth monotonically along one dimension. Sachs believed that all of the processes that controlled plant movement and growth were rooted in the direct impact of external factors like light, humidity and temperature on the physics and chemistry of growth. This 'mechanics of growth', he argued, would eventually explain apparently 'discontinuous variations of growth' as the interaction between different continuous processes (Sachs 1887: 552). The auxanometer expresses this understanding of plant growth, carefully measuring the vertical growth, normalized as monotonic movement along a single axis, in order to disentangle the influence of various factors. When plotted alongside controlled changes in temperature, humidity or illumination, Sachs believed that the auxanometer would reveal that apparent changes of behaviour were not contingent, irregular events, but rather the unfolding of basic physio-chemical processes.

Walter Bryce Gallie and various literary theorists have argued that events are significant to a narrative if they are both non-deterministic and have consequences for later events, affecting the outcome of the narrative (Gallie 1964; Barthes 1975; Chatman 1978). As Beatty summarizes the distinction, meaningful narratives have 'turning points' that are defined both by their temporal and causal relation to later events (the way later events are *contingent upon* their outcome) and because turning points are *contingent* per se (they are not necessary, and might have gone some other way) (Beatty 2016: 36–37). In such cases, as Mary S. Morgan explains, time serves as a 'material in which we see the dependency of relations or the unfolding of events' (Morgan 2017: 87). Insisting, by contrast, that plants react in a strictly deterministic fashion, Sachs insisted that plant movement was not contingent per se. As a parallel example of a non-eventful, and so non-narrative, sequence of events, Gary Saul Morson (2003: 61) imagines a description of the movements of Mars that records only where the planet was in each subsequent month, ad infinitum. Such accounts, as Morgan points out, are merely 'chronicles': they 'order events through time', but do not seek to explain 'the relations between them' (2017: 86). In a similar fashion, Sachs's interpretation of plant recordings translated the seeming eventfulness of plant growth into the continuous action of physical processes, demoting the narrative of plant behaviour into a chronicle of plant response not

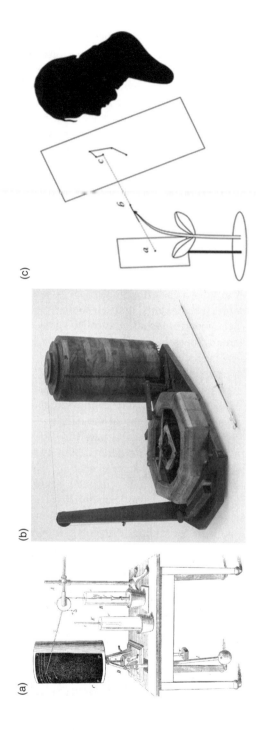

(a)

(b)

(c)

Figure 7.2a **Auxanometer**
Source: Sachs (1874). The Rare Book and Manuscript Library, University of Illinois at Urbana-Champaign.
Figure 7.2b **Horace Darwin's self-recording auxanometer**
Source: Nall, Taub and Willmoth (2019: 12).
Figure 7.2c **Experimental design for Charles Darwin and Francis Darwin's plant nutation observations**

essentially different, if seemingly more complicated, from the way planets respond to the interplay of gravitational forces.

Darwin's studies of exaptation were designed to underline the *narrative* rather than simply *chronological* character of evolutionary explanations by making room for the agency of the plants – helping them to function as narrators of their own story by allowing them to record their active response to their surroundings. Impressed with Sachs's technique, Charles initially asked his son Horace to make a replica of Sachs's instrument (Figure 7.2b; Horace was an accomplished instrument-maker), and he helped his son Francis secure an invitation to study with Sachs in his lab. Yet they soon abandoned the auxanometer, developing alternative techniques that gave the plants greater freedom of movement. Charles had first begun to try and record their movements in 'On the Movements and Habits of Climbing Plants' (Darwin 1865). Placing a hemispherical glass over the tendril and plotting its revolving movement over the course of one workday using a pencil, he confirmed Henri Dutrochet's earlier studies of the 'circumnutation', or revolving movement, of pea tendrils and demonstrated that this rotation sometimes reversed (Darwin 1865: 65). But he lamented that he could not affix the pencil to the plant itself, allowing it to draw its own movement more accurately. Fifteen years later, Charles and Francis announced a breakthrough: they finally devised a scheme to get plants to draw. After smoking glass plates to deposit a layer of carbon, they suspended them at an oblique angle beneath germinated seeds, allowing the small root stems or 'radicles' to trace a pattern as they moved across the plate, seeking soil (Figure 7.1).

Taking the pencil out of their own hands and so allowing the plants to trace their own course permitted the Darwins graphically to capture not only the waving path of the roots but variations in force as the tips bent towards and away from the inclined plates. The varying thickness of the line traced by the root tips marks fully contingent narrative events in which the actor (here, the root tips), confronted by an obstacle (the slide), attempts to overcome it. The eventful and non-monotonic nature of each track is underlined by an accompanying textual narrative, which emphasized that 'Their serpentine courses show that the tips moved regularly from side to side; they also pressed alternately with greater or less force on the plates, sometimes rising up and leaving them altogether for a very short distance' (Burkhardt et al. 2019: 29). As Francis privately noted, the fact that the tips of the plant roots only lightly touched the plates, rather than 'pressing hard' on them, suggested that plants sensed the obstacles and tried to move around them, like hands feeling in the dark (Burkhardt et al. 2019: 27). This thoroughly contingent action is what discriminates these root tracings from mindlessly deterministic behaviour, distinguishing the former as micro-narratives that, per Bal's definition, describe a series of 'related events that are [both] caused [and] experienced by actors'.

The various experiments performed by the Darwins on root tips showed a complex form of agency that actively responds to and discriminates between various kinds of stimulus, including light, moisture, physical pressure and the pull of gravity, in order to decide the course pursued by the plant 'in penetrating the ground' (Darwin and Darwin 1880: 573). In translating these graphic narratives into text, the Darwins rearticulated the sinuous narrative inscribed by the root tips upon the glass plates, characterizing them as a sequence of turning points, significant events in which the root tip, acted on 'simultaneously' by 'two, or perhaps more, of the exciting causes', effectively changed its mind, pursuing one course rather than another (Darwin and Darwin 1880: 574). For this reason, the radicle provided the primary evidence that plant behaviour was contingent per se. They concluded that such tracks demonstrate the agency of the plant root:

It is hardly an exaggeration to say that the tip of the radicle thus endowed, and having the power of directing the movements of the adjoining parts, acts like the brain of one of the lower animals; the brain being seated within the anterior end of the body, receiving impressions from the sense-organs, and directing the several movements. (Darwin and Darwin 1880: 573)

Sachs forcefully rejected the analogy between plant and animal cognition, complaining that 'Charles Darwin and his son Francis [. . .] on the basis of experiments which were unskilfully ['*ungeschicht*', or clumsily] made and improperly explained, came to the conclusion, as wonderful as it was sensational, that the growing point of the root, like the brain of an animal, dominates the various movements in the root' (Sachs 1882: 843; 1887: 689). The debate between Sachs and the Darwins over the status of plant behaviour – whether plants have the capacity to 'direct' their movements – turned on this question: do plants have the capacity to make meaningful changes in the course of their lives; in other words, are they narrative agents? Sach's auxanometer provides a signal example of the nineteenth-century turn towards 'mechanical objectivity', which asserted that increasingly sophisticated instrumental recordings would allow nature to speak for itself (Daston and Galison 2007). And yet, any scientific apparatus makes assumptions about a phenomenon under study. Even as such self-recording instruments were designed to produce a neutral or 'universal' language of nature, as Soraya de Chadarevian explains, 'they exerted a normative power on "nature" itself [. . .] forc[ing] the phenomena to inscribe their movements on paper' in the restricted terms furnished by the apparatus (de Chadarevian 1993: 290). Conversely, the relative looseness of the Darwins' unmechanized experimental set-up is precisely what allowed plants to demonstrate their wider degree of agency, narrating their own stories.

If Sachs wanted his plants to *behave*, marching with regular, lawlike action, Darwin wanted his plants to *dance to their own tune*. Insisting on the analogy between plant and animal agency, the Darwins narrowed the distance between the agency of the scientists and the agency of their object of study. Their smoked glass experiments provide a robust example of the methodological adjustments that Pickering has described as a *dance of agencies*, in which the human agency of the scientists continually adapts the experimental protocol in order effectively to frame the 'material agency' of the phenomena being studied (Pickering 1995: 103). As used by Pickering, 'dance' furnishes a metaphor that captures how scientists observe and then actively adjust the experimental protocol when physical phenomena fail to behave in the expected manner (the scientist leads).

The Power of Movement in Plants is striking, in part, because it insists that plants have agency, too – that within the dance of agencies, *plants can lead the scientist*. This dance is evident in its teeming illustrations of plant movement. The movement of really large plant structures had long been clear. But the movement of tiny shoots, stems and roots was generally so minute it had gone unnoticed. The problem was to connect the human scale to the plant scale. While they remained ignorant of the mechanisms underpinning plant sensation, the Darwins had considerable success recording the mechanisms that underpinned the various forms of plant movement itself. To each structure of interest, they glued a small glass filament, with a bead of wax at the end. Behind the filament, they staked a card with a black dot as an index. And on the other side, they placed a pane of glass perpendicular to the filament, measuring the distance between all three. As the filament moved, they used ink to trace the alignment between bead and index on the pane of glass, taking note of the time (Figure 7.2c). The whole movement was magnified up to thirty times by the differential ratio between bead, index and pane of glass (AB, AC). The result was nearly two hundred illustrations of plant movement that ranged across the gamut of vegetal life.[2] Take the illustration given in Figure 7.3a, an observation of a fava bean leaf, which captures two days in the life of this plant in the Darwin household. To make each observation, the Darwins had to move with the plant; aligning plant structure, environmental index, glass, pencil, hand and body, at specific moments in time. Individual dots mark observations, moments at which one of the Darwins hovered in alignment with the filament and index card, and marked their line on the glass. Solid lines connect sequential observations; dotted lines indicate periods overnight when the Darwins slept.

In attending to the embodied situation of these experimental events, I take a note from the field of performance studies, which, as Barbara Kirshenblatt-

---

[2] Jonathan Smith has given extensive attention to the various genres that Darwin drew upon in developing these illustrations, noting that this movement away from more idealized representations of plant movement, to more accurate inscriptions of plant movement in *The Power of Movement in Plants*, marks a turn to a messier aesthetic (Smith 2006: 150).

Figure 7.3a *Vicia faba*

'Circumnutation of leaf, traced from 7.15 P.M. July 2nd to 10.15 A.M. 4th' (woodcut)

Source: Darwin and Darwin 1880: 234. Reproduced, with permission, from John van Wyhe, ed., *The Complete Work of Charles Darwin Online* (http://darwin-online.org.uk/converted/pdf/1880_Movement_F1325.pdf).

Figure 7.3b *Brassica oleracea*

'Conjoint circumnutation of the hypocotyl and cotyledons during 10 hours 45 minutes' (woodcut)

Source: Darwin and Darwin 1880: 16. Reproduced, with permission, from John van Wyhe, ed., *The Complete Work of Charles Darwin Online* (http://darwin-online.org.uk/converted/pdf/1880_Movement_F1325.pdf).

Figure 7.3c *Brassica oleracea*

'Heliotropic movement and circumnutation of a hypocotyl towards a very dim lateral light, traced during 11 hours, on a horizontal glass in the morning, and on a vertical glass in the evening' (woodcut)

Source: Darwin and Darwin 1880: 426. Reproduced, with permission, from John van Wyhe, ed., *The Complete Work of Charles Darwin Online* (http://darwin-online.org.uk/converted/pdf/1880_Movement_F1325.pdf).

Gimblett explains, emphasizes 'practice and event [as] a recurring point of reference', focusing attention on questions of 'presence, liveliness, agency, [and] embodiment'.[3] Performance studies furnishes a strategy for reading experimental histories, in the spirit of John Dupré and Daniel H. Nicholson's 'Manifesto for a Processual Philosophy of Biology', as embodied engagements with nature and its 'hierarchy of processes, stabilized and actively maintained at different timescales' (Nicholson and Dupré 2018: 3). Note the synchronies of the interaction between the Darwins and their plants. To make this alignment work, a series of different temporalities have to come into each other's sway; from stable processes of the physical apparatus (the relative stability of the environment, index card and glass slide, the quick-drying varnish that secured the filament to leaf, the mutability of ink); to the different rhythms of the living agents drawn together by that apparatus. Each slide traces this drama of bodies in motion. Far from clumsy, each mark, each plate, captures another step in an extended attempt – stretching over multiple decades – to learn how to dance with plant life, how to follow its lead.

7.3   Genre and the Reconfiguration of Narrative Levels

The simplicity and sensitivity of the experimental design proved to be its virtue, allowing the Darwins to show that virtually all plant structures moved, and allowing plants to expose their quivering, wakeful life to human view. *The Power of Movement in Plants* demonstrated the near universality of circumnutation (circular plant movement) across plant species, and across the parts of the plant, from roots and shoots, to leaves and petals, to branches and trunks. Using careful microscopic work, the Darwins also verified that plant movement was produced by the combined action of two traits – variations in the growth of cells on opposite sides of the supporting structure, and more specialized plant structures called 'pulvini', in which cells on one side or the other could periodically expand or contract (Darwin and Darwin 1880: 113–116). This, in turn, allowed them to track how circumnutation had been exapted to serve various new functions. They traced myriad examples of heliotropism and apheliotropism (bending towards or away from light sources), geotropism (growing towards the earth) and reactions to temperature and other stimuli.

An important example of this strategy is given in their exploration of *Brassica oleracea*, or cabbage plant. When we first encounter cabbages in *The Power of Movement in Plants*, the Darwins document how the seedling, despite lacking a pulvinus, rotates clearly from its base early in its growth, providing a fine example of circumnutation (Figure 7.3b). Later in the study,

---

[3] Barbara Kirshenblatt-Gimblett, 'Performance Studies', a report written for the Rockefeller Foundation (1999) (quoted in Schechner and Brady 2013: 3).

they return to these seedlings to demonstrate how that behaviour is bent towards different purposes. Moving a similar cabbage seedling near a partially veiled window in the morning they observe how its rotation is deflected and elongated in the direction of the light, only returning to its more circulate rotational behaviour after sunset, at 5.15 p.m. (Figure 7.3c). The strong linear movement from the bottom right to upper left corner of the pane in Figure 7.3c provides a 'striking' contrast, they note, to the orderly rotation of Figure 7.3b. Various comparisons like these demonstrate how circumnutation has been adapted to serve a variety of functions, moving towards and away from light, towards and away from gravity, and responding to touch.

In this way, each vacillation in plant movement is tied to a swerve in that plant's evolutionary history. The extraordinary number of illustrations – five times as many illustrations in a single volume than included in any of Darwin's other works – demonstrate the variety of events that constitute a plant's life, and the variety of ways different plants might respond to them. In each, the periodicity of circumnutation provides a background pattern, an elliptical expectation of behaviour that casts any deviation into sharp relief. The Darwins chart deviations in the size, direction and periodicity of these ellipses through the study – the term 'ellipse' is itself used nearly two hundred times. Against this elliptical expectation, any sharp deviation of plant movement stands out as a clear fork in the road, the marked reaction of the plant to some stimulus. To put this differently, the ellipse functions in these images as a kind of *narrative scaffold*, a generic pattern that highlights concrete and consequential events in the narrative.[4]

In essence, each illustration, with its swerves and turns, magnifies a micro-narrative, or better, a micro-history, co-written by the Darwins and their plant subjects. Yet these events do not only mark consequential happenings in the life of the plant; they also index turning points in the evolution of plant life, past moments when circumnutation was exapted to serve a new function. The larger argument set out in *The Power of Movement in Plants* depends on a multipart analogy between these micro-histories of individual plants, detailed through both their self-inscription and accompanying textual account, and the evolutionary narratives of species history, an analogy that reads differences in the behaviour of individual species as distinct histories of exaptation and adaptive refinement.

For most of the study, over the hundreds of accounts of the growth and movement of individual plants, this analogy is implicit; the authors generally seem to rely on their audience's knowledge of the wider evolutionary argument

---

[4] I am suggesting that such conventions or generic models underwrite narrative *scaffolding* – the process, discussed by Anne Teather in her chapter (Chapter 6), through which data and empirical objects are assembled into narratives. Line Andersen, in her contribution to this collection (Chapter 19), similarly describes such conventions as narrative 'scripts'.

of all of Charles Darwin's studies. This coy positioning of evolutionary argument is abandoned in the conclusion, which gathers all of the observations into a single, unified story. In the final chapter, the study's scientific narrator draws the various plant micro-histories together, asking that 'we will in imagination take a germinating seed, and consider the part which the various movements play in the life-history of the plant' (Darwin and Darwin 1880: 548). The speculation that follows traces a generic seedling from germination, through various events, to its ultimate flowering as a tree – summing up the life events typical for plants in general. The result is a 27-page novella (or mini-novel) that gathers the various experiments explored over the course of the study and organizes them into a unitary narrative that strings together various 'adaptive movements' (Darwin and Darwin 1880: 551). Throughout the passage, this narrator slips into present-tense, active formulations that emphasize the plant's agency, as when 'our seedling now throws up a stem bearing leaves' (Darwin and Darwin 1880: 558). When we look at a tree, we see a solid object tossed by the wind, but, in fact, 'each petiole, sub-petiole, and leaflet' quivers with activity, activity that marks its continuous reaction to the light, moisture, gravity and other stimuli of its surroundings. Reviewing all the actions that constitute a tree's life, the narrator comments, 'All this astonishing amount of movement has been going on year after year since the time when, as a seedling, the tree first emerged from the ground' (Darwin and Darwin 1880: 558). All of the illustrations of the book, from the elaborate dance steps of the inked filament tracks to the tracings of rootlets on smoked glass, are organized through this single tree's story, which takes seed and blooms in the mind's eye.

The 'life-history' serves a key function in mediating between the micro-histories of individual plant growth and evolutionary history – a relation set out in Table 7.1. Like *David Copperfield* or *Great Expectations*, the 'life-history of a plant' gathers various micro-histories and observations into an unfolding story within which an individual actor confronts various challenges and successfully overcomes them. In essence, the Darwins repurpose the *Bildungsroman* (the contemporary 'novel of development' that dominated early to mid-nineteenth-century fiction) as a scaffold capable of interpolating these micro-histories into a single compelling narrative. The accession of this new generic model is marked by specific shifts of narrative point of view (or 'focalization'), as well as tense, character of action and agent.[5] All underline a shift from contemporary conventions of scientific monographs. Over the last several centuries, scientific prose has come to rely on passive constructions that minimize the focalization of the scientist–narrator.[6] And for much of *The Power of*

---

[5] See Wittenberg (2018: 35–37) and Hajek (Chapter 2) for a discussion of 'focalization'.
[6] This shift has been minutely traced by Gross, Harmon and Reidy (2011) and is also discussed in Meunier (Chapter 12).

Table 7.1 *Narrative levels in Charles Darwin and Francis Darwin, The Power of Movement in Plants (1880)*

Narrative Level	Genre	Agent	Event Type	Narrative Tense	Focalization
specimen/ illustrative plates	micro-narrative/ micro-history	plant specimen (+ scientist via passive construction)	specific action	simple past (+ past perfect actions of scientist)	third person limited
'life-history of the plant'	*Bildungsroman*/ novella	generic plant	life event (germination, root growth, etc.)	present	first person plural
evolutionary history of plants	evolutionary narrative	natural selection	selection and adaptation, (especially exaptation)	past perfect + present perfect	third person speculative, omniscient

Movement in Plants, the Darwins similarly deploy passive constructions that describe the experimental design paired with simple-past descriptions of what the plants did. By contrast, the first-person plural 'we' that narrates the 'life-history' ripples with personality, even as it draws the reader into the act of imaginative engagement. It also facilitates a periodic shift into possessive constructions within the recounting of the 'life-history' (including 'our seedling'), which suffuse the narrative with a sense of familiar responsibility maintained between the narrator's description, the reader's implication and the seedling itself, who is now recognized not as an individual scientific specimen but as a more charismatic agent – effectively, our hero. A shift in tense, from the past-tense constructions of the micro-histories to a present-tense unfolding of life events similarly marks a shift in temporal relations and in the character of the events narrated. As many narrative theorists have pointed out, the time of telling and the original timing of the events described in a narrative can never precisely align in either speed or duration (Wittenberg 2018: 14–15). The 'life-history' exacerbates this contrast to powerful effect, dramatizing the distillation of entire life cycles, abstracted across various species, into a handful of crisply plotted pages. The accession of the *Bildungsroman* structures this turn towards holist integration, centring the account on a unitary actor, sequence and perspective, reconfiguring the events studied throughout the treatise as a series of developing chapters in the life of an individual plant.

The 'life-history' dramatizes the status of plants as active agents within their own narratives. This dramatization, in turn, calls attention to the links between the demonstrated actions of plants, with their various powers, and the antecedent action of natural selection, through which adaptations (especially exaptations) shaped these behaviours and the development of each species. The intermediate status of the 'life-history', as the middle stratum of three narrative levels, is made explicit at its close, which situates its story between the micro-histories traced by individual plants and an overarching narrative of species evolution. Listing the various forms of movement studied over the course of the preceding pages, the Darwins assert that 'it has now been shown' that these 'important classes of movement all arise from modified circumnutation', and they finally, and explicitly, identify the power which has given this plant its various abilities: these uses of the 'power to bend [. . .] might gradually be acquired through natural selection' (Darwin and Darwin 1880: 569–570).

If, as Ian Duncan argues (2019), the *Bildungsroman* emerged alongside the birth of modern anthropology as a formal model through which an individual life could register the larger story of the human species, the narrator of *The Power of Movement in Plants* uses their own 'life-history' to set out an analogous story about the longer evolution of all plant species. Morgan describes narrative 'configuration' as a process that 'make[s] things cohere – a process of [. . .] making an account that is consistent with all the evidence, that offers a coherency within that account, and that has some explanatory credibility' (Morgan, 2017: 93). Narrative genres like the *Bildungsroman* are powerful aids, as they can provide shared standards for the consistency, coherence and credibility of an account. But the 'life-history' also demonstrates the role that genres play in *reconfiguring* previous narratives. Mediating the plant's relation to evolutionary time, the 'life-history' reshapes the micro-histories it draws upon. It is worth thinking more about the role literary genres play in structuring scientific narratives, insofar as they provide such patterns of structure, and establish a 'horizon of expectations', that, in Hans Robert Jauss's influential account, conditions how readers interpret narratives (Jauss 1970). Reviewed with the evolutionary implications of the 'life-history' in mind, and so, tacitly reconfigured and placed in relation to a wider story, each individual illustration does not simply present the tale of a few hours in the life of an individual plant, but also a concretized retelling of a longer evolutionary history. In light of Darwin's science, each trait, each action, is pregnant with a history of evolutionary change.

7.4   Conclusion

The dynamic, interactive narratives of *The Power of Movement in Plants* suggest that we should pay less attention to discrete narratives than to the *relation that obtains between various narratives and their world* and especially *to narration as a process that mediates between a description of events and events in a world, setting them into relation.* If narrative, including graphic narratives, do not simply capture, but coordinate events, this means their effect depends on the specific way they coordinate wider patterns, as well as the various purposes to which they are later applied. Perhaps most scientific narratives work this way. Certainly, narratives help scientists to identify and respond to – to sync up with – temporal patterns in the world, and so to coordinate their inscription into the scientific record. Narratives have a peculiarly powerful ability to draw us into an alignment with the world, to train our attention on patterns of action and exceptional events. If we usually think of scientific narratives as perspicuous fictions, 'designed in the mind' to model aspects of the world, the unusually active role that plants play in telling their stories within *The Power of Movement in Plants* suggests that scientific narratives are in part produced by, rather than simply applied to, the world they describe – that the dance of agencies is also a dance of authorships.

Richard Bellon has recently observed the profound influence that Darwin's plant narratives had on a generation of plant ecologists to come, encouraging them to attend to the sociability of plants and their intertwined evolutionary histories (Bellon 2009). Michael Marder, taking stock of the wealth of recent studies that have built on the pioneering work of the Darwins to explore the ability of plants to interact with their environment, and even communicate with each other, underlines their core implication: we need to understand plants as 'not only a *what* but also a *who*, an agent in its milieu' (Marder 2016: 42). Plant narratives mediate, pulling our attention, entraining our thoughts, bringing us into contact with nature. In this way, Darwin's studies, which draw naturalist, specimen and world into a delicate movement, continue to test the conformations of interspecies relations, the anthropology of the inhuman.[7]

Bibliography

Bal, M. (1997). *Narratology: Introduction to the Theory of Narrative*. 2nd edn. Toronto: University of Toronto Press.

Barthes, R. (1975). 'An Introduction to the Structural Analysis of Narrative'. Trans. L. Duisit. *New Literary History* 6.2: 237–272.

Beatty, J. (2016). 'What Are Narratives Good For?' *Studies in History and Philosophy of Science Part C* 58: 33–40.

Beer, G. (2000). *Darwin's Plots: Evolutionary Narrative in Darwin, George Eliot and Nineteenth-Century Fiction*. 2nd edn. Cambridge: Cambridge University Press.

  (2017). 'Plants, Analogy, and Perfection: Loose and Strict Analogies', in J. Faflak, ed. *Marking Time: Romanticism and Evolution*. Toronto: University of Toronto Press, 29–44.

Bellon, R. (2009). 'Charles Darwin Solves the "Riddle of the Flower"; or, Why Don't Historians of Biology Know about the Birds and the Bees?' *History of Science* 47.4: 373–406.

Burkhardt, F. et al., eds. (2019). *The Correspondence of Charles Darwin*. Cambridge: Cambridge University Press.

Butterfield, H. (1997). *The Origins of Modern Science*. New York: Free Press.

Chatman, S. (1978). *Story and Discourse*. Ithaca, NY: Cornell University Press.

Crombie, A. C. (1988). 'Designed in the Mind: Western Visions of Science, Nature and Humankind'. *History of Science* 26: 1–12.

Darwin, C. (1862a). Letter to J. D. Hooker, 26 September [1862]. Darwin Correspondence Project. University of Cambridge. www.darwinproject.ac.uk/letter/DCP-LETT-3738.xml.

  (1862b). *On the Various Contrivances by Which British and Foreign Orchids Are Fertilised by Insects, and on the Good Effects of Intercrossing*. London: John Murray. Darwin Online. http://darwin-online.org.uk/converted/pdf/1862_Orchids_F800.pdf.

[7] I am very grateful for all of the assistance that I received in thinking about this essay, including two workshops hosted by the Narrative Science Project at the London School of Economics, organized by Mary S. Morgan, Dominic J. Berry and Kim M. Hajek; a workshop with the Vcologies workgroup, hosted by Deanna Kreisel at the University of Mississippi; participation in the Victorian Conference at the CUNY Graduate Center, organized by Thalia Schaffer; and discussions in the Nineteenth-Century Seminar at Cambridge University, at the invitation of Ewan Jones. *Narrative Science* book: This project has received funding from the European Research Council under the European Union's Horizon 2020 research and innovation programme (grant agreement No. 694732). www.narrative-science.org/.

(1865). 'On the Movements and Habits of Climbing Plants'. *Journal of the Proceedings of the Linnean Society of London* 9.33–34: 1–128. Darwin Online. http://darwin-online.org.uk/content/frameset?itemID=F1733&viewtype=text&pageseq=1.

(1872). *The Origin of Species by Means of Natural Selection.* 6th edn. London: John Murray. Darwin Online. http://darwin-online.org.uk/content/frameset?pageseq=1&itemID=F391&viewtype=text.

(1875a). *Insectivorous Plants.* London: John Murray. Darwin Online. http://darwin-online.org.uk/converted/pdf/1875_Insectivorous_F1217.pdf.

(1875b). *The Movements and Habits of Climbing Plants.* 2nd edn. London: John Murray. Darwin Online. http://darwin-online.org.uk/converted/pdf/1875_Plants_F836.pdf.

(1876). *The Effects of Cross and Self Fertilisation in the Vegetable Kingdom.* London: John Murray. Darwin Online. http://darwin-online.org.uk/converted/published/1876_CrossandSelfFertilisation_F1249/1876_CrossandSelfFertilisation_F1249.html.

(1877). *The Different Forms of Flowers on Plants of the Same Species.* London: John Murray. Darwin Online. http://darwin-online.org.uk/content/frameset?itemID=F1277&viewtype=text&pageseq=1.

Darwin, C., and F. Darwin (1880). *The Power of Movement in Plants.* London: John Murray. Darwin Online. http://darwin-online.org.uk/converted/pdf/1880_Movement_F1325.pdf.

Daston, L., and P. Galison (2007). *Objectivity.* Princeton, NJ: Princeton University Press.

de Chadarevian, S. (1993). 'Graphical Method and Discipline: Self-Recording Instruments in Nineteenth-Century Physiology'. *Studies in History and Philosophy of Science* 24.2: 267–291.

Duncan, I. (2019). *Human Forms: The Novel in the Age of Evolution.* Princeton, NJ: Princeton University Press.

Gallie, W. B. (1964). *Philosophy and Historical Understanding.* London: Chatto & Windus.

Gould, S. J. (2002). *The Structure of Evolutionary Theory.* Cambridge, MA: Harvard University Press.

Gould, S. J., and E. S. Vrba (1982). 'Exaptation: A Missing Term in the Science of Form'. *Paleobiology* 8.1: 4–15.

Griffiths, D. (2016). *The Age of Analogy: Science and Literature between the Darwins.* Baltimore, MD: Johns Hopkins University Press.

Grimaldi, D. A., and M. S. Engel (2007). 'Why Descriptive Science Still Matters'. *BioScience* 57.8: 646.

Gross, A. G., J. E. Harmon and M. S. Reidy (2011). *Communicating Science: The Scientific Article from the 17th Century to the Present.* West Lafayette, IN: Parlor Press.

James, E. (2017). 'What the Plant Says: Plant Narrators and the Ecosocial Imaginary'. In M. Gagliano, J. C. Ryan and P. Vieira, eds. *The Language of Plants: Science, Philosophy, Literature.* Minneapolis: University of Minnesota Press, 253–272.

Jauss, H. R. (1970). 'Literary History as a Challenge to Literary Theory'. Trans. E. Benzinger. *New Literary History* 2.1: 7–37.

Levine, G. L. (1988). *Darwin and the Novelists: Patterns of Science in Victorian Fiction.* Chicago: University of Chicago Press.

Marder, M. (2016). 'Ethics and a Pea (Or If Peas Can Talk, Should We Eat Them?)'. In M. Marder, ed. *Grafts: Writings on Plants*. Minneapolis, MN: Univocal Publishing, 41–43.

Mivart, St. George Jackson (1871). *On the Genesis of Species*. London: Macmillan.

Morgan, Mary S. (2017). 'Narrative Ordering and Explanation'. *Studies in History and Philosophy of Science Part A* 62: 86–97.

Morson, G. S. (2003). 'Narrativeness'. *New Literary History* 34.1: 59–73.

Nall, J., L. Taub and F. Willmoth, eds. (2019). *The Whipple Museum of the History of Science*. Cambridge: Cambridge University Press.

Nicholson, D. J., and J. Dupré (2018). 'A Manifesto for a Processual Philosophy of Biology'. In D. J. Nicholson and J. Dupré, eds. *Everything Flows: Towards a Processual Philosophy of Biology*. Oxford: Oxford University Press, 3–48.

Pickering, A. (1995). *The Mangle of Practice: Time, Agency, and Science*. Chicago: University of Chicago Press.

Priest, G. (2018). 'Diagramming Evolution: The Case of Darwin's Trees'. *Endeavour* 42.2–3: 157–171.

Priestman, M. (2000). *Romantic Atheism: Poetry and Freethought, 1780–1830*. Cambridge: Cambridge University Press.

Sachs, J. (1874). *Arbeiten des botanischen Instituts in Würzburg*. Liepzig: W. Engelmann.

    (1882). *Vorlesungen über Pflanzen-Physiologie*. Leipzig: W. Engelmann.

    (1887). *Lectures on the Physiology of Plants*. Trans. H. M. Ward. Oxford: Clarendon Press.

Schechner, R., and S. Brady, eds. (2013). *Performance Studies: An Introduction*. London: Routledge.

Smith, J. (2006). *Charles Darwin and Victorian Visual Culture*. Cambridge: Cambridge University Press.

Tutt, J. W. (1896). *British Moths*. London: G. Routledge & Sons.

Wilberforce, S. (1860). 'On the Origin of Species, by Means of Natural Selection; or the Preservation of Favoured Races in the Struggle for Life (Review)'. *Quarterly Review* 108: 225–264.

Wittenberg, D. (2018). 'Time'. In M. Garrett, ed. *The Cambridge Companion to Narrative Theory*. Cambridge: Cambridge University Press, 120–131.

Debjani Bhattacharyya

Abstract

Throughout the nineteenth century, shipwrecks during tropical cyclones in the Indian Ocean resulted in extended legal battles in the
Marine Court of Enquiry in Calcutta. This chapter explores how
cyclones became an object of scientific curiosity at the intersection
of the imperial legal world and marine insurance. It explores the
court records, consisting of legal depositions about the wrecks by
mariners and insurance agents, ships' logs with barometric readings,
and diaries kept by the captain and pilots, which formed a significant
archive for the colonial scientist Henry Piddington (1797–1858),
made famous for coining the term 'cyclone'. Piddington narrativized
storm observations by condensing accounts from multiple sources
and created a 'storm card' to finally develop a theory of tropical
cyclones. His storm narratives and the accompanying visualization
through the storm card shaped the very object – the cyclone – as
a scientific category of investigation, transforming storm memories
into a narrative science of forecasting.

8.1   Introduction

The English language does not contain a native word to express the *more
violent forms of wind*. We have borrowed a great many since we became the
great merchants of the East, but hurricane and tornado are Spanish,
typhoon, we believe, Chinese, though dictionaries derive it from the
Greek, simoom Arabic, and cyclone pure Greek, with a conventional
meaning imposed upon it by science. [. . .] Storm is the only native word
of any force, and an Englishman's idea of a storm does not tempt him to
sympathize greatly with the sufferers from its violence. Accustomed only
to the winds of the north, which bring catarrh and consumption, but leave
wooden houses standing for years, which seldom last many hours, and are
never destructive except at sea, his power of imagining wind is limited, and
he reads a story like that of the catastrophe at Calcutta with a feeling of pity
in which there is just a trace of something like contempt. People out there
must be very weak or arrangements very bad for a mere wind to work all

that destruction, throw 'Lloyds into a panic, and impede the systole diastole of Her Majesty's foreign mails.'

Anon., The Spectator (12 November 1864) (my emphasis)

On 5 October 1864, as the monsoon winds were retreating from the littorals of the Bay of Bengal, a devastating cyclone struck, killing 80,000 people, drowning the city of Calcutta and washing away large swathes of coastal villages. Thirty-six ships were lost in the storm, and of the 195 ships docked at Calcutta point, 182 were damaged, with an estimated combined loss of approximately 1 million pounds sterling (Gastrell and Blanford 1866: 145). While the loss of life, cattle and property were staggering, the coasts of the Bay of Bengal were no stranger to the cyclonic battering. Moreover, meteorology as a public science had also gained a solid footing in England and its colonies (Golinski 2007; Anderson 2010; Carson 2019). Yet, surprisingly, *The Spectator* wrote that the English language did not have the capacity to narrate what happened on that fateful October day. How do we then understand and historicize the semiotic confusion expressed in the opening epigraph by the anonymous writer of *The Spectator*? The focus of the question should not be the English language, but perhaps the narrative and representational possibilities and crisis produced by the storm under consideration itself.[1] While one may ascribe some of the writer's confusion to the 'blinkered' vision of colonial writings about colonized environments and climate, a deeper engagement with the writer's lament that science has merely imposed a 'conventional meaning' upon the fury of the wind is necessary.[2]

By the sixteenth century, we can witness the emergence of a scientific curiosity about storms by Iberian theologians and lawyers investigating hurricanes in the Caribbean. One of the most noteworthy among them was López Medel, who was a high court judge and served in the appellate courts in Santo Domingo, Guatemala and New Granada from 1540 to 1550, overseeing shipping and trading disputes (Schwartz 2015: 17). He wrote about *buracanes*, which he defined as a 'meeting and dispute of varied and contrary winds', later recognized as circular winds and defined as cyclones by the president of the Marine Court of Enquiry in Calcutta, Henry Piddington (1797–1858), almost three centuries later. What kinds of science did these men of law in the colonies produce? How did the legal search for plausible narratives influence a particular narrative science of storm forecasting?

Tropical storms in the Bay of Bengal emerged as a problem of knowledge as the East India Company was expanding its trade in Britain's eastern colonies.

---

[1] Here I draw upon the works of Arnold (2014) and Huang (2013) on the invention of 'tropics' as a shorthand for both environmental othering and the quest for empirical difference in the colony through writing, cartography and painting, among others.
[2] I draw upon Guha's usage (2002) of 'blinkered' to describe colonial knowledge.

Turning to Piddington's cyclone research allows us to historicize his new science of 'cyclonology', which was a product of Victorian science, but also of the colonial legal and trading world in which he found himself. This science, as he wrote, was not meant to be conducted 'in the state room of science, but in the cabin-table' of ships and docks (Piddington 1848: xiv). In the process, he narrativized historical storms in his works. This chapter argues that through his storm narratives and the accompanying visualization of the storm card, he shaped the very object – the cyclone – as a scientific category of investigation. His science used conversational language to conceptualize, for the sailors, the phenomenal world of the storms as wind movement in which one can discern patterns and tendencies through rigorous training of the eye and use of the storm card itself.[3] He began writing storm memoirs in 1839 and continued to write them until he died in 1858. His storm writings were geared towards achieving a discernible order in the stormy skies with the purpose of predicting the direction of the storm and plot its track. This was meant to help both mariners and jurors. In Mary Morgan's definition, what Piddington's narratives did, was 'create a productive order amongst materials with the purpose to answer why and how questions' (Morgan 2017: 86).[4] By organizing the patterns of historical cyclones from ships' logs, reports and court depositions, he wanted to understand why and how cyclones formed. A narrative science of cyclones, replete with storm memoirs and a diagrammatic representation in the storm card, ordered the interpretation of winds to make cyclones both trackable and predictable in the service of the marine insurance market. Storms became a problem of knowledge precisely because these 'violent forms of winds' created panic among underwriters in the colony and metropole, and as the epigraph wonderfully captures, they 'impede[d] the systole diastole of Her Majesty's foreign mails'. If fire, capture and piracy were known risks associated with maritime routes, tropical storms became the 'unknowns' of the expanding insurance markets.[5] As we saw, the 1864 cyclone devastated the very sinews of global trade and credit that, by the nineteenth century, tightly stitched together far-flung geographies from the Caribbean, Coromandel, Malacca and Bengal to the ports of England.[6]

Through the eighteenth century, the process of interpreting the skies and understanding the causes of storms navigated a terrain between providential

[3] The storm cards that emerged as a technological tool can be compared with scientific articles and notebooks discussed by Robert Meunier (Chapter 12).

[4] Piddington's storm narratives may be thought in relation to the thick narratives that Mat Paskins's chapter deals with (Chapter 13).

[5] Guerrero (2010: 240–241) argues that unknowns and uncertainties always fetch a very high premium in insurance. In medical cases, underwriters assess uncertainty and unknown very differently (Parson 2015).

[6] Kingsbury (2018) gives a detailed account of how the 1876 cyclone laid the groundwork for early experiments in austerity.

design, folk traditions and emerging science about geological, chemical and meteorological phenomena. Scholars have documented this as a historical transition from Aristotelian astrometeorology, through the ascendance of Renaissance observational sciences and what was known as 'rustic' weather knowledge, to nineteenth-century dynamic climatology (Golinski 2007; Anderson 2010; Coen 2018). Yet there was a parallel tradition of knowledge production that sometimes intersected with Victorian science, and other times remained firmly locked within the worlds of trade, insurance and legal spheres. Oftentimes this parallel world could be found along the ports, docks and observatories spread across the globe: Barbados, Mauritius, Bengal, Madras and Manila. Indeed, it was narrative storm memoirs written by colonists engaged in a range of professions from planters to shipmasters or legal actors who would shape both the form and content of weather science as well as frame the diagrammatic representation of the storm as a circular image. This parallel body of knowledge followed the routes of imperial capital and was sustained by a nautical marketplace.

The search for scientific cyclone forecasting emerged from narratively ordering accounts of historical storms, which were converted into a diagrammatic tool to depict, plot and track tropical winds. This, in turn, created laws of predictable wind patterns, which would allow one to read cyclonic motions that deviated from wind tracks. Indeed, for Britain's expanding empire in the east, the problem of estimating risks of trade and administering compensation following shipwrecks created Piddington's new science of cyclonology. Faced with the exigencies of global trade, the Bay of Bengal became a laboratory for nineteenth-century weather science. Turning to Piddington's writing and the curious scientific tool – the storm card – allows us to document how a narrative science of cyclone forecasting emerged from the interstices of imperial trade. It shows how in the process of narrativizing memories of tropical storms, the cyclone as an object of knowledge came into being in the texts and the diagrams. The meaning-making and meaning-conveying process of narrating the science of storms was shaped by the traffic in language, imaging and metaphors between weather observers and shipmasters' logbooks as they brushed with the colonial marine and admiralty courts through the nineteenth century.

8.2   The Nautical Marketplace

Piddington's legal and scientific world was embedded in the nautical marketplace. Throughout the eighteenth century, the East India Company lost nearly one-quarter of its ships sent to Asia.[7] For instance, between 1760 and 1796, it

---

[7] Papers on Marine Subjects, IOR/L/MAR/C/325, British Library, London.

lost 20 per cent of its ships to shipwreck on their way to Asia (Bowen, McAleer and Blyth 2011: 118). A Select Committee on Shipwrecks reported to the House of Commons in 1836 that England was losing nearly 3 million pounds sterling per annum (£ 2,836,666) and had lost 894 lives to shipwrecks.[8] This report was prepared with the help of the accounting books of Lloyds and so only reflects cases of ships insured by Lloyds. The report gives details of the reasons the ships were wrecked or floundered and crew drowned. Among the many causes for wrecks, two bore the highest responsibility. First, the committee wrote that often instruments of navigation (namely depth recorders, barometers and chronometers) were either faulty or absent, or the crew was not sufficiently trained to use them (Jennings 1843). Second, they pointed out that the widespread use of premium-based marine insurance might mean that shipmasters and merchants were indulging in risky voyages in stormy seas, and as a result there was a higher incidence of shipwrecks. While there is no existing data that links the use of premium-based marine insurance to increased numbers of shipwrecks, the report indexes some of the assumptions prevalent within the expanding nautical marketplace of the early nineteenth century. The specific concern for this Committee, widely reflected in the world of nautical writing too, was that the expansion of marine insurance had allowed shipmasters to transfer the risk of shipwreck to the underwriters, which ultimately transferred the risk to the British public (*Nautical Magazine* (1836): 593). The result was fierce battles in the imperial admiralty courts adjudicating liability over wrecked ships and ultimately flinging blame for the wrecks onto 'the plainest sailor', to use one of Piddington's oft-used descriptors, who routinely failed to navigate the cyclonic and turbulent waters of the Indian Ocean.

The British Indian Navy and their hydrographers had been charting the oceanic currents and coastal tides in the Indian Ocean since the 1760s. In order to make long-distance shipping safer, Piddington furthered the project by developing a usable science of storm forecasting for sailors. Piddington, who grew up in south-east England, worked his way up to command ships to India. He settled in Bengal in 1824 and remained there till his death, serving as the foreign secretary to the Agricultural Society of India, a secretary to the Asiatic Society of Bengal, curator to the Museum of Economic Geology (a first of its kind in the world) and, more importantly for this chapter, as the president of the Marine Court of Enquiry in Calcutta from 1830 to 1858. Following his death, he became famous for his meteorological pursuits and was known for coining the term 'cyclone'. He described the storms which he saw in the Bay of Bengal as 'coiled snakes', for their circular motions, and came up with the

---

[8] 'Report of Select Committee on Shipwrecks', *The Nautical Magazine* 5 (1836): 588–600. https://archive.org/details/nautical-magazine-1836/page/587/.

name cyclone to distinguish them from trade winds, which blew in straight lines (Markham 2015: 35–37).

Piddington's scientific pursuits into storms emerged out of his life as a shipmaster, but the scaffold of his storm narratives was shaped by his work in the Marine Court. In the 1830s, the Marine Court was a simple affair. The Calcutta court was housed in a small room that served as a court once a week or less, depending on the availability of a mariner's jury (which, prior to the coming of steam, depended on the monsoon winds), and this same room served as the meeting room or exchequer on other days. It was only in 1836 that a special court of enquiry was set up in England and its imperial ports, dedicated to establishing the 'fact of the wreck' and to creating mechanisms to 'censure owners or commanders of vessels' or acquit them honourably from charges of having caused the wreck. It was also tasked with suspension of certificates or licences should shipmasters be found to be incompetent. These courts were to be funded from the ship registration fees (*Nautical Magazine* 1836: 596–597). Piddington's presidency over the Marine Court of Enquiry in Calcutta was during this moment of transition, when government oversight was increasing and standardization of practices and the pedagogy of mariners were being discussed within both the East India Company in India and the House of Commons in London.[9] He was familiar with the legal arguments and counter arguments made to establish the fact and narrative of the wreck during the onset of a cyclone. He not only heard mariners, pilots and witnesses narrate the sighting of storms, but he collected their barometric readings and read their logs documenting disputes about how to steer the ship in a cyclone. Apart from this, he was simultaneously poring over the archive of prior cases as he sought to lend structure to the procedure of settling disputes. What emerges is the way he used these multiple different narrative accounts in order to distinguish the contingent wind patterns from their predictable movements, thereby developing scientific taxonomies of various kinds of winds and a law of storms in the Indian Ocean.

Michael Reidy's work on the development of British marine science has already documented how the imperial imperatives to sail unencumbered and safely through the littorals of England drove tidal science in the nineteenth century. The Admiralty, he shows, turned to science to advance its overseas empire (Reidy 2008: 5–7). Along with the Royal Navy, the rise of marine insurance conglomerates like Lloyds of London from the latter half of the eighteenth century was coterminous with British imperial expansion in the east and the rise of scientific storm forecasting. That some colonial legal actors, who

---

[9] Instruction had long been an interest for Piddington. While most of his writing is dedicated to training deck-hands and shipmasters, it is also present in his writings about the act of curation, when he was president of the Museum of Economic Geology, where he first articulated his idea of instruction and industry as a joint venture (Sarkar 2013–14: 162).

were busy adjudicating on trading issues, were also obsessed with reading patterns in storms should elicit further investigation. Storms were documented and narratively ordered in ships' logs and shipmasters' diaries. The physical sites that enabled such documentation – often understood as the field and laboratory of weather science – have been documented by historians of science as floating observatories (ships) and weather stations spread across the empire (Reidy 2008; Naylor 2015). If ships were floating observatories, they were also carrying bundles of letters, queries and texts across the empire, which were exchanging, plotting and tracking information about the very winds that carried them on the vast oceanic expanse. Knowledge of the atmospheric world was stabilized at multiple sites and through various genres — memoirs, barometric tables, diagrams, legal writings. Along with the scientific work from ships and observatories, a curious scribal culture emerged from the late eighteenth century in the British Empire, similar to the epidemiological narrative cultures documented by Engelmann in this volume (Chapter 14). This curious narrative outpouring saw planters and lawyers write storm narratives and cyclone memoirs. Apart from navigators who needed to understand the wind infrastructure that fuelled their trade, underwriters, lawyers and financiers took an active interest in those elemental phenomena that had the ability to disrupt the imperial financial machinery. The amateur scientific writings about storms and legal petitions and court decisions about wrecks formed a polyphonous world that laid the groundwork for nineteenth-century storm science.

Many of the wrecks occurred in the Bay of Bengal, especially in the last stretch of the journey as the ships navigated from the tip of the Bay to the port in Calcutta, the then capital of the British India. It was a rain-fed, tidal and changeable landscape. Mariners' and hydrographers' early attempts at control began with sketching the coasts of the Bay of Bengal.[10] From 1753, the East India Company began employing an official hydrographer, Alexander Dalrymple. Under Dalrymple's oversight the official process of systematizing coastal charts began. Navigating into the port of Calcutta, which was situated almost 100 miles from the Sagar Islands in the Bay of Bengal, was difficult as the ships would have to sail through a network of mangrove islands, tidal sand flats and seasonal salt marshes, which annually changed shape, disappeared or sometimes suddenly reappeared especially during the summer months of tropical cyclones. Logs of ships warned that when storms and 'hurricanes' occurred at the mouth of the river Hooghly (or Hughli), sailing can become disastrous because the sea inundates the low-lying alluvial lands and ships often founder (Reid 1838: 284). Rudyard Kipling, who considered this stretch

---

[10] See the contrasting tide charts and maps in the following collections from the seventeenth century: Private Papers of Barlow, IOR/X/9128, British Library; Papers Concerning New Harbour in Bengal, IOR/H/Misc/396:1765–1809, British Library; and Dalrymple (1785).

among the most dangerous as far as navigability was concerned, wrote about the River Hughli thus: 'Men have fought the Hugli for two hundred years, till now the river owns a huge building, with drawing, survey, and telegraph departments, devoted to its private service, as well as a body of wardens, who are called the Port Commissioners' (Kipling 1923: 28).

8.3   Storm Science in the Courts

Two representative shipwreck cases debated in the Marine Court in Calcutta reveal how the legal 'fact of wreck' was established and show the legal imperatives that drove Piddington's science. The first case was debated following the founding of the Marine Court and almost half a century prior to Piddington's term. A sloop, *Betsey Galley*, wrecked at the mouth of Bay of Bengal as it was reaching the port of Calcutta on a stormy evening on 25 April 1778. The *Betsey* was wrecked upon the Long Sand in the Bay of Bengal at the mouth of the delta, with 13 members and its cargoes going under water before reaching the port of Calcutta.[11] *Betsey*'s wreck was fiercely debated in the Marine Court in Calcutta over four months. The petitioners were Capt. John Raitt and Mr. Weller (the merchant invested in the sloop), who claimed to the Court that Thomas Broad, the master attendant in charge of the pilot schooner to the *Betsey Galley*, did not offer any assistance and must be held responsible for the wreck. The Committee of Insurance deposed in the Marine Court and supported the claim against Thomas Broad, deeming him the negligent master of the pilot schooner, responsible for the wreck and seeking to debar him from future navigational duties. As the petitioners pointed out, it was a dark summer's night and the ship was sailing fast through the waters of the Hughli, and Broad's pilot boat failed to keep ahead of the *Betsey Galley*.[12] Moreover, Broad also rendered no assistance after the wreck, although it was no more than a few leagues ahead. However, the one-sided incriminations of a shipmaster against his attendant should hardly surprise anyone or be enough to establish the reason for the wreck.

Broad's deposition, on the other hand, pointed out that the storm during the months of April can wreak havoc in these areas. April is the nor'wester season and is marked by sudden storms and coastal surges which can make riverine travel and navigation tricky in the Bengal delta. Caught in the turbulent waters of the Hughli, Broad pointed out that he steered the boat based on the direction of the incoming gales, which he had successfully done many times, yet the winds changed course and *Betsey* foundered before Broad could do anything. Following the adversarial interrogation of the admiralty courts, Broad was

---

[11] *Betsy Galley* Case, Home Public No. 6–12, National Archives of India (NAI), New Delhi.
[12] By 1801, ships were debarred from navigating without pilots at night (Phipps: 1832: 36).

called for questioning, which consisted of questions about the usual role of the pilot schooner during storms and about whether he felt that he performed his duties. Like a well-honed defendant, he answered questions about the usual duties and responsibilities mostly thus: 'It is sometimes usual and sometimes not'. And for questions where they tried to assess his opinion, he offered stock answers. For instance, to the question: 'How come the ship [was] lost?' Broad's answer was: 'If you put any particular questions to me I shall answer them'. Thereafter he demurred and the interrogation remained inconclusive.

Yet, the mariner's jury and the judge concluded that Broad's 'obstinacy and misconduct' were to blame. How did they reach this conclusion? The Committee of Insurance and the merchant's jury turned to another source to ascertain the truth about the wreck, namely Broad's prior mistakes of navigation. The committee in whose interest it was to locate blame on the negligence of the master attendant or the pilot navigator offered depositions in the court documenting prior instances when Thomas Broad failed in his duties while attending other ships.[13] Turning to precedence made the wreck appear to be caused not by the cyclone, but instead due to Broad's habitual navigational misconduct. As legal historians have pointed out, reputation and credibility were deeply entangled in courtroom decisions through the eighteenth and nineteenth centuries, especially prior to the arrival of expert evidence and forensic criminology (Golan 2009: 5–51). Even then, and to an extent now, credibility performs a critical role in establishing the plausibility of the narratives offered.

Upon hearing all the testimonies, the judge decided that the total loss of the vessel was owing to an error in judgement on Broad's part, and was not due to the nor'wester that suddenly set in. Such legal decisions were often based not on the availability of the evidence such as the ships' logs, charts of depth sounding and barometric pressure, a studied understanding of wind direction or testimonies about the unnavigability of the channel, but rather character assessments of those steering the ship or pilot boats. Indeed, in multiple cases, the moment of wreck is often reconstructed by turning to other instances of failure of the shipmaster's or pilot's duty, including character assessments – such as 'wanting in attention' or 'given to liquor'.[14] These character deficits also defined the ability to develop a studied understanding of the laws of storms. What bothered Piddington's scientific temper was the excessive role the personal character, social standing, or networks of credibility, and the ability of the defendant to draw upon powerful witnesses, played in establishing the depth and nature of human error. Within the space of the Marine Court,

---

[13] *Betsy Galley*'s wreck was followed by the wreck of *Snow Mars* where Captain French was held responsible. This was followed by a letter from the insurance company suggesting measures for the careful observance of duties by pilots. Original consultation, 9 November 1778, No. 9, NAI.

[14] 'Report on the Wrecks in Indian Waters, 1865'. British Library.

trying to separate human miscalculation from unavoidable natural disaster was complicated.

By the time Piddington began presiding over the Marine Court, the ability to forecast natural disaster remained mostly poor and the nature of adjudication of wrecks navigated a terrain not very different from the one we witnessed in the case of the *Betsey Galley*. The *Barge Amherst* was partially wrecked in October 1838, mid-way on its voyage from Myanmar (Burma) to Calcutta.[15] Dalrymple's work as the Company's official hydrographer had transformed the landscape of navigation prints, with official charts in circulation by the last decade of the eighteenth century. He was followed soon after by James Horsburgh, who served the Company from 1810 to 1836, keeping extensive records of the tides of the Bay of Bengal coasts. Horsburgh also introduced the need to take extensive depth soundings to detect shoals and shifts in the coastline, while regularly updating those surveys.[16] By 1832, the Royal Navy recognized that the tidal charts for India were more complete and detailed than the ones pertaining to the English coasts.[17] The arrival of Horsburgh and his diligent publication of official nautical charts introduced a new standard of judgement. In cases of accidents, ships which were found to be in possession of non-official charts could be penalized. However, given that the route from Burma to Calcutta was so treacherous, Horsburgh's directions were considered insufficient. A mariner under the pseudonym 'Nautics' suggested that 'Should ships frequenting Rangoon, attend only to Mr. Horsburgh's directions, without waiting for a pilot (which at times they may be compelled to do from stress of weather) they will surely run aground and suffer considerable damages' (Phipps: 1832: 145).

The *Amherst* was supposed to set sail from Kyaukphu one early October morning in 1838. However, the ship was delayed due to low winds. When the ship finally set sail, it reached a rock face then known to sailors as the Terribles. Unable to stay on course, the *Amherst* hit those rocks on the night of 22 October and was damaged, but managed to reach Calcutta, half-damaged, with its logbooks intact. In this instance, the logbook, the detailed notes of arguments and conversations kept by both Captain Bedford and attendant Captain Jump, would have allowed the Marine Court to establish that the swinging barometric pressure and winds veered the ship off its course. The notes, the witness depositions and the log show that Jump disagreed with Capt. Bedford's directions, who insisted that the ship should have continued to sail in the direction it was headed. Had he followed Jump's chart, the ship might have been saved from hitting the rocks.

[15] *Marine Index* 2, 9–11 (9 January 1839). West Bengal State Archives (WBSA), Kolkata.
[16] Papers of James Horsburgh, MSS Eur F305, British Library.
[17] Beaufort to Captain Horsburgh, 1 November 1832, PRO, ADM.1.3478, National Archives, Kew.

There is a twist in this case. The day after *Amherst* dropped anchor in Calcutta following this fateful journey, Capt. Jump deposited his papers with the port authorities as Piddington had required all sailors to do. Thereafter, Jump quietly slipped out of Calcutta that very afternoon, boarding a ship to Bombay and then London and in the process forfeiting part of his pay. The court spent a considerable time deciphering Jump's sudden disappearance and gathering evidence of his prior conduct in their attempt to piece together his character. The court ultimately decided his fate in absentia. It ruled that Jump could not man another Company ship or ship in his Majesty's service as he was deemed too incompetent. His incompetence, the court declared, was not his ability to decipher winds, but in his inability to be judicious enough, first, to disregard his master's misreading and veer the ship in the right direction, and, second, not to stay back in Calcutta to offer witness in the court of law. The archival trail breaks off here, and we do not know if, along with barring Jump from Company duty, the merchants invested in *Amherst* were duly compensated for their partial loss.

What these court minutes reveal is how the multiple iterations and reconstructions of the wreck in the courtroom were embedded within the socio-political hierarchies of the world outside. According to the court's decisions, ships sank or foundered more often because of human error stemming from altercations between master and pilot, inexperienced pilots or drinking and 'rottenness of native crafts' than because of the turbulence of the seaboard. Legal decisions, as we know, are a product of 'social, political, epistemic struggle', and these struggles set the background for discerning the nature of wind patterns and the causes of wrecks (Raman, Balachandran and Pant 2018: 2). This narrative reconstruction of the moment of wreck, which made human character central, was crucial to adjudicating damage claims throughout the first half of the nineteenth century. These resources left Piddington, with a vast set of storm narratives, to construct his science in the service of the mariners. He wanted his science to act as a protection not just from cyclones but also wanted to protect sailors and pilots like Broad and Jump, who were being fleeced by the insurance agents and the mariner's jury who shifted the liability for wrecks during cyclones onto them.

The legal disputes in the Marine Court were geared towards the search for plausible narratives about a shipwreck. One may divide Piddington's legal archive into two sets of evidence: one was recalling the memory of the onset of the storms and the other was an observational set of evidence. The testimonies of shipmasters, pilots and sailors constituted the memory evidence, which would often include not just notes about the storm but also character judgements about the people involved. Observational evidence comprised that which was written down in the ships' logs, like wind direction, daily logs, temperature and barometric pressure charts. They were descriptions of observed phenomena rather than recalled memory and were either verbal testimony or written petitions. Court decisions often emerged by pitting various storm and character narratives against one another to arrive at a plausible description of the facts of the wreck. If the legal enquiry was geared towards establishing a plausible argument about shipwrecks in order to locate the liabilities incurred in the damages, Piddington's cyclonology attempted to standardize the narratives of storms through his scientific writings and the storm card.

8.4   From Memories to Prediction: The Making of the Storm Card

In the twenty odd years following his entry into the Marine Court in Calcutta, Piddington consulted on multiple cases and analysed 250 ships' logs from mariners plying in the Bay of Bengal and collected storm observations from port masters in various ports in India. In 1839, Piddington published his first storm observations in the *Journal of the Asiatic Society of Bengal*. Between 1839 and 1851, he published 23 memoirs of cyclones, each one taking up between 11 and 100 pages. These memoirs were like working notes, where he collated logs from ships that were caught in the gales, along with observatory notes, reports in newspapers and notes from port masters to plot the movements of cyclones in the Bay of Bengal and to develop his hypothesis. Following the publication of his first cyclone memoir in 1839, he began to receive multiple logs and extracts that were then preserved at India House (which furnished him with accounts of storms from 1780 to 1841) and built his own 'storm library' (Piddington 1848: 7). The accumulation of storm writings in the form of logs, observations, reports and his own collection of memoirs comprised his attempt to understand how winds in their interaction with the world around them – reacting to atmospheric heat, thermodynamics, oceanic currents – developed into cyclones. In his writings, storms, much like a narrative plot, had a beginning, a middle and an end. Akin to Darwin's plants' 'life-histories', which are considered by Devin Griffiths elsewhere in this volume (Chapter 7), the cyclone emerges as a scientific object through a 'two-way traffic' between representation and scientific discovery.

Unlike Darwin's visual narratives, Piddington's were primarily textual and tabular, tracing the transformation of regular winds into circular storms. This allowed him, among other things, to complete the puzzle that Medel ascribed to the indistinctive directions of the *buracanes* winds, laying the groundwork for the development of a rotational theory of winds.[18] He standardized the

---

[18] German geographer Bernhardus Varenius had understood the whirlwind nature of hurricanes as early as 1650, and by the nineteenth century the idea of circular winds had taken hold among the mariner–scientists who were studying oceanic winds. Colonel William Reid's *Law of Storms* (1838), which was a direct influence for Piddington, lays out most of the features of circular storms, but stops short of naming them cyclones (Sen Sarma 1997).

definition of a cyclone – which was far from the scientific imposition of a conventional meaning upon a strong gust of wind, as *The Spectator* claimed. In order to come up with a name for this wind, Piddington moved away from terminology expressing strength to those expressing direction. He clarified that 'cycloidal' was a known word expressing 'a relation to a defined geometrical curve, and one not sufficiently approaching our usual views, which are those of something nearly though not perfectly circular'. He then proposed to use a single word 'cyclone', which would be used to express 'the same thing in all cases; and this without any relation to the strength of the wind' (Piddington 1848: 11). This laid the foundation for his practical new science of cyclonology, which he developed over three books: *The Horn-Book of Storms for the India and China Seas* (1844), an expanded version, published as *The Sailor's Horn-Book for the Law of Storms* (1848) and a textbook entitled *Conversations about Hurricanes: For the Use of Plain Sailors* (1852).

His science was geared 'to enable the plainest ship master, then, clearly to comprehend this science in all its bearings and uses' (Piddington 1848: i). Piddington's goal was to ease adjudication and at the same time to instruct the seamen by developing a science of storms through his 'thick narratives' (Paskins, Chapter 13). Piddington wanted storm science to act as a form of insurance and protection against wreckage. If mariners were preparing their logs with an eye towards the centrality of the logbook in documenting navigational knowledge and for adjudicating potential settlement cases, then Piddington was prospectively archiving the same logs with an eye towards creating a database from which to develop a systematic way to discern the law of storms.

His practices for assembling an archive for the law of storms involved a process of acquiring and retrieving material, reconfiguring that material and then transcribing this body of information into a narrative interpretive framework. For Piddington's new science of cyclonology it was the process of reconfiguration that drove the interpretive framework. Each storm that Piddington adjudicated upon, observed in situ, read about in logbooks and heard during deposition was situated in deep historicity.[19] The monsoon, the capability of the navigator, the observer, the reputability of the pilot all shaped his archive of storm writing. Piddington's life narratives of winds with the storm as denouement can be understood as exemplifying colligation (Morgan 2017). Piddington wanted to produce 'law like knowledge' of storms that was based on cases but utilizing modes of inquiry and methods of organizing vast amounts of data that would be systematic enough to mimic the natural world and thereby produce knowledge that could become universal (Creager, Lunbeck and Wise 2007). Such a method would result in producing usable

---

[19] On the historicization of other natural events, like earthquakes by seismologists, see Miyake (Chapter 5).

evidence within both the scientific and legal domains. In his attempt scientifically to order the storm he devised the storm card, a tool that would make storm
tracks discernible and protect mariners against wrecks with the hope that it
would also help in the adjudication of cases.

Piddington first introduced his storm card in the *Sailor's Horn-Book*. It was
meant to serve as a card of practical utility that he produced for the use of 'plain
sailors'. The storm cards were developed as a diagrammatic representation of
wind pattern and direction during a storm that was circular, with the basic
assumption that there were certain laws that governed the wind movement
within this circularity. Thus, they were highly schematic visualizations of wind
movements, which taught: 'how to *avoid Storms*; how best to *manage in Storms*
when they cannot be avoided; and how to *profit by Storms*!' (Piddington 1848:
xiii). As can be seen in Figure 8.1, Piddington's storm cards were translucent
sheets which the sailor would place upon a map to understand the track of the
storm and determine the direction to steer the ship. There were two separate
cards for the two hemispheres, with the eye of the cyclone visualized vis-à-vis
the wind direction. The sailor could plot the eye on the map and avoid it. The
storm card was a perfect representation of the wind directions as the cyclone
gathers strength, , and as such it sought to highlight the 'sensory character of
much natural language' (Wise, Chapter 22).

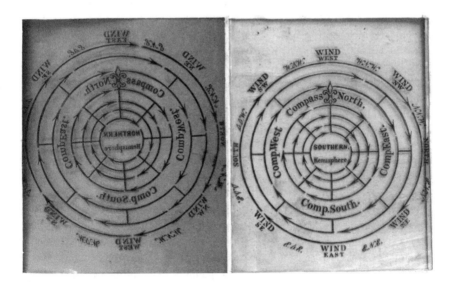

Figure 8.1 **Piddington's storm card, 1848**
Source: British Library, London, digitized as part of the Google Books project.

Through his work in the courts Piddington had deduced that there were three kinds of dangers to a vessel in a cyclone: 'the veering of the wind; the excessive violence of it near the centre; and the sudden calms and shifts and awful sea at the centre' (Piddington 1848: 103). The biggest challenge was that while most seamen knew not to be in the centre of what mariners often called the 'water-spouts', there was no scientific ordering of their tacit knowledge. The lack of any science to govern their observations had to do with the fact that seamen were 'not accustomed to consider the winds as tangent lines to a circle, and the bearing of the centre perpendicular to them, the consideration of "how the centre bears," even with the aid of the Storm Card, may hence sometimes be found puzzling' (Piddington 1848: 105). Piddington's storms cards were accompanied by a tabulated explanation of the wind depicted in the cards. Moreover, directions for using the storm cards – i.e., avoiding the centre, heaving with the direction of the wind, or profiting from it – were illustrated by ships' logs elucidating how other ships managed or failed in cases of storms in the multiple oceans and coasts. His reasoning was that a sailor had more felicity with reading tables and logs than diagrams, and the accompanying narratives and tables will teach them how to use the storm card more successfully. Moreover, juxtaposing the logs which he had accumulated with the storm cards allowed him to reconstruct possible cyclone scenarios and devise ways to improve upon managing in these cyclones. The storm cards were widely used and reprinted in many sailing manuals and laid the groundwork for a prescriptive science of storms. Figure 8.2 shows a further development of Piddington's storm cards, reproduced in a textbook for sailing, in 1891. The narrative directions on how to manage in a storm have been condensed and moved into the centre of the card. This recipe-narrative condenses the various scenarios for managing in cyclones. As these directives became part of the storm visualization, the storm card is transformed from a navigational into a pedagogical tool.

Such a schematic visualization of the storm in advance of aerial and satellite photography should not be taken as a given.[20] For Piddington to plot this diagram of the storm, the science of cyclones had to move away from an understanding of storms as a meeting of disputed and contrary winds. This was no mean feat given that eighteenth- and nineteenth-century storm observers would have seen a storm from a single vantage point (for Piddington, who had worked as a sailor, this would have been the deck of a ship), so that tornadoes, waterspouts, hurricanes and tropical storms were often strong winds that violently changed directions and were accompanied by thick clouds (Walker 1989: 483). A particular kind of 'epistemic switch' (Brian Hurwitz, Chapter 17) was necessary to move from visualizing and narrating tropical storms as contrary wind patterns to the bird's-eye view of this neat and cycloidal representation.

---

[20] On narrative-making through aerial photography, see Haines (Chapter 9).

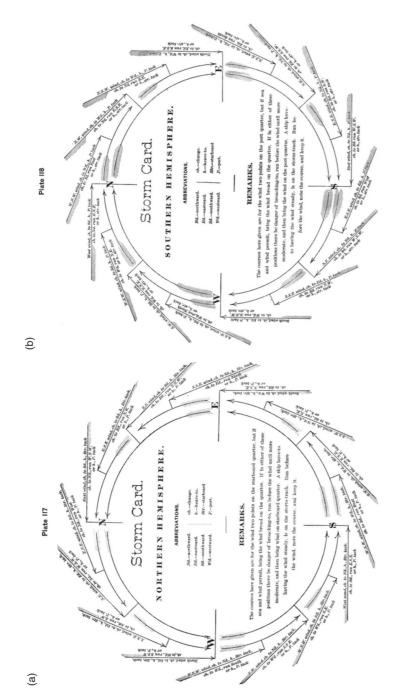

Figure 8.2 **S. B. Luce's recreation of the storm card, from *The Textbook of Seamanship* (1891)**
Source: Made available by US National Archives.

Piddington's work built upon accounts of hurricanes in the Caribbean Seas, ships' logs and court cases involving coastal landfalls of cyclones in the Indian ocean. Gilbert Blane's account of the 1780 hurricane that struck Barbados had confirmed for Piddington that there was a need for developing terminological specificity to distinguish between straight and rotatory winds, and that with some observation, tracking wind direction and training one's eyes, one would be able to discern patterns in these rotatory winds well enough to predict the direction of the tropical cyclone. Apart from Blane, Piddington had access to accounts of storms in the Coromandel from the south-eastern coasts of the Indian peninsula given to him by the Master Attendant of Madras Port, Capt. Christopher Biden. He was simultaneously reading American meteorologist William Charles Redfield's work, which had already described the storms of the north Atlantic Ocean as 'progressive whirlwinds', i.e., that they were always rotatory and that they moved in a plottable track (Piddington 1848: 4). In 1838, William Reid, who was stationed as the governor of the Bermudas, published *An Attempt to Develop the Law of Storms by Means of Facts*, where he documented that the storms that struck the Caribbean coasts were storms that rotated clockwise in the southern hemisphere and anticlockwise in the northern (Piddington 1848: 5).

Following on from these writings, Piddington announced both the reason for developing a law of storms and the two principles that made storms discernible and plottable. He declared that storms would gradually become understood as a trackable wind movement, which any good sailor could navigate in. The first principle laid down the wind motion and direction, and Piddington showed that winds circulate in two motions on two sides of the equator and that it was both a straight and a curved motion, which made the winds systems circulate as they were 'rolling forward at the same time' (Piddington 1848: 8). The second principle proved that in the northern hemisphere wind moved from east to west, 'or contrary to the hands of a watch', while in the southern hemisphere the wind motion lay with the hands of the watch. These two central principles of Piddington's 'new science of cyclonology' rendered the sky with discernible wind patterns. His storm science, visualized through the card, would allow sailors to train their eye to recognize deviations from the pattern and therefore cyclonology would ultimately act as a form of insurance and protection against wreckage: 'to enable the plainest ship master, then, clearly to comprehend this science in all its bearings and use' (Piddington 1848: i). As someone presiding over the Marine Court in Calcutta, he worked with a very specific definition of law:

Theory and *Law*. The seaman may best understand these two words by his quadrant. As long as people who paid attention to these things *supposed* that light when reflected from a mirror was always so at a certain angle depending somehow on the direction in which the original light fell upon it, this was a theory. When it was *proved* by experiment that the angle of reflection was equal to the angle of incidence this became the *Law* of reflection, and when Hadley applied it to obtain correct altitudes, and to double the angle by the two reflections of the quadrant, he used it for a nautical object of the first importance and of daily practical utility. These are the three great steps of human knowledge and progress. The theory, or supposition that a thing always occurs according to certain rules, the proof or Law that it does and will always so occur, and the application of that Law to the business of common life. (Piddington 1848: 7–8)

For Piddington, the storm card is a distilled version of the law of nature applied to the business of common life – his science that should be conducted in the cabin tables of a ship. Piddington's storm science was geared towards teaching sailors to recognize the centre of the cyclone and to devise methods to avoid it. According to him, the safest way of managing a vessel in a storm is by following the wind direction and sailing on its rotatory or circular course rather than straight through it. In order to do that a sailor had to see a particular kind of storm – not one where strong winds blew in multiple directions, but one where there was a circular pattern to it with a centre that one must, at all cost, avoid. However, he was quick to point out that what the sailor is discerning with the storm card are not tracks of storms, but the 'tendency of the paths of the usual Cyclones' (Piddington 1848: 42). For this reason, his directives to use the storm cards were accompanied by excerpts of shipmasters' logs which he meticulously collected from ships that docked at Calcutta and Madras.

Storm cards not only order the moments before the storm, but also make historical wind movements legible and transform them into a set of universal signs to be read and deciphered in order to avert a wreck. And given his role in the Marine Court, he also hoped that they would ease adjudication about wrecks. The storm card was a technical tool that helped the shipmaster verify the wind direction. By standardizing storm science, Piddington had also hoped to develop plausible narratives about the moment of the wreck were they to occur, and plot when and where mistakes were made. He was also fully aware of the difficulties of rendering the volatile tropical skies into a set of laws and diagrams. Therefore, Piddington recommended that mariners follow the storm card, but cautioned against 'the mischievous and ignorant notion that there is any fixed law for the tracks of these terrific meteors, especially in narrow seas with volcanic islands or continents within, or near to, or limiting them' (Piddington 1848: 62). Moreover, Piddington saw his storm card as an evolving tool and he requested the sailors to offer feedback for improving upon the tool. Indeed, the storm card made the sailor's tacit knowledge into a discernible evidence of his ability to read wind direction reflecting his capability as an experienced sailor. Thus, the storm card performed two functions: it was a critical tool of pedagogy for sailors and it sought to standardize the narrative science of cyclones.

8.5   Conclusion: Narrating Imperial Cyclonology

In the Bay of Bengal, the line between what was knowable in the 'blooming, buzzing' (Daston 2016: 60) world of storms and gales shaped the material practices of rowing, towing and navigating the seaboard and in the process was translated into empirical knowledge through storm narratives. As mentioned above, Piddington was not the first weather observer, nor was he the first to write about winds and hurricanes. What makes Piddington's work stand out is the legal and imperial imperatives that drove his cyclonology. He was driven by a desire to bring order to the process of administering justice, protecting the plainest shipmaster against storms and equally from the wreckages of the inequitable justice system of the Marine and Admiralty courts.[21] Piddington's cyclonology emerged out of what Morgan and Wise called a backward understanding of the event, whose narratological cognition and reconstruction happens after the fact, i.e., after he had listened to multiple accounts of the storm that wrecked ships. In that, he was very much the 'confused and reflective participant' who, 'when confusion is resolved, [becomes] the narrator throwing explanatory light on the situation' (Morgan and Wise 2017). For example, in *Conversations about Hurricanes* (Piddington 1852), meant to be a book of dialogic pedagogy between three sailors, Capt. Wrongham, one of the fictive sailors, tries to understand if the storm card is a form of 'prognostication'. He comments, 'our knowledge then would all be fore-knowledge, both as to what happened and what in all probability was going to happen' (Piddington 1852: 93). With this form of foreknowledge acting as insurance against wreckage, the jury's ability to judge and place liability for the storm would be resolved efficiently. The storm card, a product of his new science of cyclonology, was also a product born of an encounter with the legal world of the Marine Court.[22]

[21] As a bid to reform the court, he submitted multiple petitions between 1848 and 1853 in attempts to change the nature of the jury and the process of judicial inquiry. See 'Paper on Defect of Marine Courts of Enquiry, by Mr. Piddington', 394–395, IOR/E/4/822, British Library.

[22] *Narrative Science* book: This project has received funding from the European Research Council under the European Union's Horizon 2020 research and innovation programme (grant agreement No. 694732). www.narrative-science.org/. This chapter was drafted during my fellowship at the Shelby Cullom Davis Seminar, Princeton University, in 2019–20, and it has benefited immensely from comments from the three editors of this book and also from Angela Creager, Rohit De, Mary Mitchell, Gyan Prakash, Anupama Rao, Judith Surkis, Francesca Trivellato and all my co-fellows at the Davis Seminar. I am incredibly grateful to be affiliated to CASI, University of Pennsylvania, which made it possible for me to access primary and secondary sources necessary to finish the chapter.

Bibliography

Anderson, K. (2010). *Predicting the Weather: Victorians and the Science of Meteorology*. Chicago: University of Chicago Press.

Arnold, D. J. (2014). *The Tropics and the Traveling Gaze: India, Landscape, and Science, 1800–1856*. Seattle: University of Washington Press.

Bowen, H. V., J. J. McAleer and R. J. Blyth (2011). *Monsoon Traders: The Maritime World of the East India Company*. London: Scala Publishers.

Bowrey, T. (1805). *A Geographical Account of Countries Round the Bay of Bengal, 1669 to 1679*. Cambridge: Haklyut Society.

Carson, S. (2019). 'Ungovernable Winds: The Weather Sciences in South Asia, 1864–1945'. PhD thesis, Princeton University.

Coen, D. R. (2018). *Climate in Motion: Science, Empire, and the Problem of Scale*. Chicago: University of Chicago Press.

Creager, A. N. H., E. Lunbeck and M. Norton Wise, eds. (2007). *Science without Laws: Model Systems, Cases, Exemplary Narratives*. Durham, NC: Duke University Press.

Dalrymple, A. (1785). *Collection of Nautical Papers concerning the Bay of Bengal: Published at the Charge of the East India Company from the MSS by A. Dalrymple*. London: George Bigg.

Daston, L. (2016). 'Cloud Physiognomy'. *Representations* 135.1: 45–71.

Defoe, D. (2005). *The Storm*. Ed. R. Hamblyn. London: Penguin.

Gastrell, J., and H. Blanford (1866). *Report on the Calcutta Cyclone of the 5th October 1864*. Calcutta: Government of Bengal.

Golan, T. (2009). *Laws of Men and Laws of Nature*. Cambridge, MA: Harvard University Press.

Golinski, J. (2007). *British Weather and the Climate of Enlightenment*. Chicago: University of Chicago Press.

Guerrero, L. L. (2010). 'Insurance, Climate Change and the Creation of Geographies of Uncertainty in the Indian Ocean Region'. *Journal of the Indian Ocean Region* 6.2: 239–251.

Guha, R. (2002). *History at the Limit of World-History*. New York: Columbia University Press.

Huang, H. (2013). 'Inventing Tropicality: Writing Fever, Writing Trauma in Leslie Marmon Silko's Almanac of the Dead and Gardens in the Dunes'. In L. Wylie, O. Robinson, P. Hulme and M. C. Fumagalli, eds. *Surveying the American Tropics: A Literary Geography from New York to Rio*. Liverpool: Liverpool University Press, 75–100.

Jennings, E. (1843). *Hints on Sea-Risks: Containing Some Practical Suggestion for Diminishing Maritime Losses Both of Life and Property Addressed to Merchants, Ship-Owners and Mariners*. London: R. B. Bate.

Kingsbury, B. (2018). *An Imperial Disaster: The Bengal Cyclone of 1876*. New York: Oxford University Press.

Kipling, R. (1923). 'An Unqualified Pilot'. In *Land and Sea Tales for Scouts and Guides*. London: Macmillan, 28–33.

Markham, C. R. (2015). *A Memoir on the Indian Surveys*. Cambridge: Cambridge University Press.

Morgan, Mary S. (2017). 'Narrative Ordering and Explanation'. *Studies in History and Philosophy of Science Part A* 62: 86–97.

Morgan, Mary S., and M. Norton Wise (2017). 'Narrative Science and Narrative Knowing: Introduction to Special Issue on Narrative Science'. *Studies in History and Philosophy of Science Part A* 62: 1–5.

Naylor, S. (2015). 'Log Books and the Law of Storms: Maritime Meteorology and the British Admiralty in the Nineteenth Century'. *Isis* 106.4: 771–797.

Parson, C. (2015). 'Insuring the Unknown'. *Human and Experimental Toxicology* 34.12: 1238–1244.

Phipps, J. (1832). *Guide to the Commerce of Bengal, for the Use of Merchants, Ship Owners, Commanders, Officers, Pursers and Others Resorting to the East Indies; But Particularly of Those Connected with the Shipping and Commerce of Calcutta.* Calcutta: Master Attendant's Office.

Piddington, H. (1839). 'Researches on the Gale and Hurricane in the Bay of Bengal on the 3rd, 4th, and 5th of June, 1839 … with reference to the Theory of the Law of Storms in India', Part I and Part II, *Journal of the Asiatic Society of Bengal* 8: 559–590; 631–650.

(1844). *The Hornbook of Storms for the India and China Seas.* Calcutta: Bishop's College Press.

(1848). *The Sailor's Horn-Book for the Law of Storms: Being a Practical Exposition of the Theory of the Law of Storms, and Its Uses to Mariners of All Classes in All Parts of the World, Shewn by Transparent Storm Cards and Useful Lessons.* London: Smith, Elder.

(1852). *Conversations about Hurricanes: For the Use of Plain Sailors.* London: Smith, Elder & Co.

(1855). 'A Twenty-Fourth Memoir on the Law of Storms, being the Calcutta and Sunderbund Cyclone of 14th and 15th May, 1852'. *Journal of Asiatic Society of Bengal* 24: 397–461.

Raman, B., A. Balachandran and R. Pant (2018). *Iterations of Law, Legal Histories from India.* Delhi: Oxford University Press.

Reid, W. (1838). *An Attempt to Develop the Law of Storms by Means of Facts, Arranged According to Place and Time; and Hence to Point out a Cause for the Variable Winds, with the View to Practical Use in Navigation, Illustrated by Charts and Wood Cuts.* London: J. Weale.

Reidy, M. (2008). *Tides in History: Ocean Science and Her Majesty's Navy,* Chicago: Chicago University Press.

Sarkar, S. (2013–14). 'The Museum of Economic Geology'. *Vidyasagar University Journal of History* 2: 161–186.

Sen Sarma, A. K. (1997). 'Henry Piddington (1797–1858): A Bicentennial Tribute'. *Weather* 52.6: 187–193.

Schwartz, S. B. (2015). *Sea of Storms: A History of Hurricanes in the Greater Caribbean from Columbus to Katrina.* Princeton, NJ: Princeton University Press.

Walker, J. M. (1989). 'Pre-1850 Notions of Whirlwinds and Tropical Cyclones'. *Weather* 44.12: 480–487.

9   Visual Evidence and Narrative in Botany and War: Two Domains, One Practice

Elizabeth Haines

Abstract

This chapter compares work done by Hugh Hamshaw Thomas (1885–1962) in two domains. First, in palaeobotany; second, in military intelligence in the First and Second World Wars. In each, Thomas investigated landscape processes using fragmentary visual evidence: plant evolution from fossils, enemy behaviour from aerial photographs. I propose we understand the connection between those domains by drawing together two, largely separate, scholarly discussions: (i) on the construction and evidential use of photographic archives; (ii) on evidence and causal explanations in the historical sciences. Through analysis of Thomas's palaeobotanical and military work I situate narrative as the *central and unifying principle* of a practice in which neither evidence collection nor explanatory accounts were prior. This unifying 'narrative practice' was reticulate, multi-scalar and dynamic, as revealed by contemporary figures of speech that sought to describe it (working 'like Sherlock Holmes', 'reading the book of nature', thinking 'like a river').

9.1   Introduction

This chapter explores connections between the intellectual work done by Hugh Hamshaw Thomas (1885–1962) in two separate domains: first, as an academic palaeobotanist; and second, as a military intelligence officer during the First and Second World Wars. In both domains, Thomas relied on the use of fragmentary visual evidence (and photography in particular) to attempt to understand landscapes. In his palaeontological work he was looking at fossil plants and past environments; in his military work he was piecing together landscapes of enemy activity. Rather than considering the visual fragments as elements in a process of 'mapping' those landscapes, I emphasize the way in which, in both domains, they were part of narratives.

In what follows, I review the wide range of material and intellectual resources drawn upon by Thomas to undertake this narrative work under the umbrella of what I call a 'narrative practice'. I argue that this 'narrative practice' included particular techniques for handling and analysing visual material, the accretion of visual evidence into archival architectures and the inculcation of epistemic virtues, with and alongside the construction of conjectural accounts about historical processes. In other words, the nature and the usefulness of the archive was predicated on narrative techniques and outcomes. An exploration of the figurative terms used by Thomas's peers to characterize the kind of work he was engaged in allows us to see 'narrative practice' as a unified whole.

Thomas pursued his career in palaeobotany almost exclusively at the University of Cambridge, where he was Fellow at Downing College, and later university lecturer in Botany. He was awarded the prestigious Darwin Centenary medal in 1958 (Harris 1963). That academic career was punctuated by war and work with aerial photointerpretation. In the First World War, he was Photographic Officer for the 5th Wing of the Royal Flying Corps. It has been claimed that the success of the British campaign in Palestine and Egypt (in which T. E. Lawrence famously also played a part) was to a great extent attributable to Thomas's contribution. In the Second World War, Thomas was responsible for producing the Manual of Photointerpretation used by the Allied Central Interpretation Unit, and, as Chief of Third Phase Interpretation, initiated the allied investigation into rocket development at Peenemunde (Smith 1985: 189).

Both of Thomas's domains of work relied heavily on visual records. Figure 9.1 is an image from a First World War manual on the use of aerial photography: 'The Interpretation of Aeroplane Photographs in Mesopotamia'.[1] It is one of a series of sample images that offered military officers an introduction to the different physical and social features of the terrain as viewed from the sky. 'The study of photographs', the manual explains, 'is only of value in so far as the results may be turned to practical account, either in the way of assisting tactical operations, or of obtaining information regarding the Enemy's intentions and dispositions'.[2] In other words, the photographs were to be used to generate narrative conjectures about what the enemy would do next.

Figure 9.2 is visual evidence from Thomas's most important paper, published in 1925. It depicts a fossil that he had collected in Yorkshire. In the paper in which the figure was published, Thomas was presenting a new fossil species that he had named 'the Caytoniales'. Thomas was proposing that the Caytoniales were an entirely new order of plants that corrected a 'missing' link between ferns and flowering plants. He was offering an important narrative conjecture about the evolution of plant life.

---

[1] Royal Air Force GHQ Mesopotamia (1918), 'Notes on Aerial Photography Part II: The Interpretation of Aeroplane Photographs in Mesopotamia'. Baghdad. AIR10/1001, National Archives, Kew.
[2] 'The Interpretation of Aeroplane Photographs in Mesopotamia', 4.

Town of Kulawund, partly ruined, near Kifri.   Height taken from : 10,000 ft.
A.   Town.   B.   Kifri road.   C.   Cultivation.   D.   Irrigation cuttings.   E.   Tracks (some
across cultivation, implying passage of troops).   F.   Qarah Tappah road.

Figure 9.1 **The Town of Kulawund, partly ruined, near Kifri**
From Royal Air Force GHQ, Mesopotamia (1918).
Source: Royal Air Force GHQ Mesopotamia (1918). 'Notes on Aerial Photography Part
II: The Interpretation of Aeroplane Photographs in Mesopotamia', 46. AIR10/1001,
National Archives, Kew.

Each of these forms of visual evidence was being woven into narrative conjectures in a similar way. To explore the connections between his use of images in each context I make two propositions. First, that we should frame Figures 9.1 and 9.2 as constitutive of a 'narrative practice' that Thomas deployed in each domain.[3] Second, that we should understand that practice as a hybrid association of multiple forms of activity.

Historical studies of the different activities that I consider to lie within Thomas's narrative practice have often been pursued in separate fields. From the historical

---

[3] While this chapter focuses on Thomas's work in palaeobotany and military intelligence, the nature of his personal networks (archaeology, forensic science, geology) and the multiplicity of his other contributions (in broader discussions about botany and evolution, in ecology and in the history of science) and outputs (as a curator of geological, botanical and instrument collections as well as an author of reports and publications) suggest that the sets of skills involved in his 'narrative practice' were of use in many contexts (Harris 1963; Cleal and Thomas 2019).

Figure 9.2 **Photograph of a fossil collected by Thomas in Yorkshire**
'Part of an infructescence showing its attachment to a larger branch, also isolated fruits in which the outlines of seeds can be made out. No perianth scars can be found on the axis or on the branch' (original caption).
Source: Thomas (1925: plate 12), fig. 16 (× 2.5).

epistemology of early twentieth-century photography we know that the status of each of these photographic reproductions as evidence was, as John Tagg puts it, a 'complex historical outcome [. . .] [of] certain institutional practices and historical relations' (Tagg 1988: 4–5). Both the aerial photograph and the depicted fossil fell within institutional practices and historical relations that could be described as 'a colonial habitus' (Pinney 2008). Both items were implicated within British efforts to know in order to dominate, motivations that through the nineteenth century had prioritized data collection on a massive scale in attempts to map physical, biological and social processes of increasing scale and complexity (Driver 2000; Coen 2018). That effort included innovation in modes of producing, refining, labelling and categorizing visual evidence within archives (Rudwick 1976; Kelsey 2007; Tucker 2013). This scholarship provides us with a strong basis from which to consider how the role of visual evidence in each of Thomas's domains was being developed and negotiated.

We can also draw on a rich and flourishing scholarship on the relationship between evidence and causal explanations in the historical sciences, particularly biology, palaeontology and archaeology. A scholarly resurgence of interest in this field has offered several further important strategies for thinking about Thomas's narrative practice, particularly with respect to visual evidential materials, many of which are discussed in this volume.[4] Such scholarship has emphasized that narrating the past can entail a variety of epistemic techniques for drawing together and assessing evidential elements (Richards 1992; Turner 2000; Morgan 2017). Scholars, in particular Alison Wylie, have shown that these approaches require certain epistemic skills that might include opportunism, flexibility, a respect for ambiguity and epistemic humility (Chapman and Wylie 2016), and that those modes offer different scopes for developing and extending historical claims (Cleland 2011). Adrian Currie's *Rock, Bone, and Ruin* (2018) situates many of these alongside each other.

This chapter connects and extends these two largely separate discussions in several ways. First, it situates narrative as the *central and unifying principle* of an epistemic practice that encompassed multiple simultaneous activities, in which neither evidence collection nor explanatory accounts were prior. Second, this chapter contextualizes that practice within early twentieth-century figures of speech: working 'like Sherlock Holmes', 'reading the book of nature', and thought as being 'like a river'. The interplay of these figurative terms allows us to characterize the unifying narrative practice as reticulate (networked), multi-scalar and dynamic; qualities that have resonance with more recent descriptions of explanatory practices in the historical sciences. This characterization is reinforced when we follow the translation of Thomas's narrative practice in palaeobotany into the arena of military intelligence.

9.2   A Narrative Practice? Finding Traces?

9.2.1   Fossils

To begin, we need to ask how the items in Figures 9.1 and 9.2 were being used as evidence. The dark spots on the surface of the rock fragment depicted in Figure 9.2 were identified by Thomas in his 1925 paper as seeds from a newly identified prehistoric plant species. However, at the moment he was formulating that proposition, the function of the plant organ he had identified, the nature of the organism it came from and the nature of the prehistoric environment which hosted the plant were all unknown. For Thomas to arrive at his conclusion, several different and interlinked questions need to be resolved simultaneously.

---

[4] See, for example, Engelmann (Chapter 14), Griffiths (Chapter 7) and Hopkins (Chapter 4).

- How might I recognize or define a prehistoric plant organ?
- How is this organ related, developmentally, to others, such as leaves, in an individual plant? Does its tissue share common properties with other organs?
- What is the role of this structure in the development of a species or evolutionary branch? Is this an early flowering plant (angiosperm)?
- What does the presence of a fossil with these characteristics tell us about the environment? For example, how was pollen transmitted?
- What inadvertent effects might have been produced in the preservation and preparation of the fossil that falsely suggest botanical features?

In Thomas's (1925) paper we can follow how he asked and answered these parallel questions of the fossil (Figure 9.2). He observes to the reader that some of the seed-membranes were not cleanly extracted and therefore appear 'broken up' (Thomas 1925: 359). They could nonetheless be interpreted. The proto carpel and fruit-bearing structures appeared to share some characteristics with their equivalents in modern flowering plants: they had a foliar-type tissue, i.e., shared some qualities with leaf tissue (Thomas 1925: 306). They didn't have the scars that would indicate they were originally part of hermaphrodite flowers (Thomas 1925: 315). Crucially, however, for Thomas, the structure depicted in Figure 9.2 seemed to *function* more like carpel than the reproductive organs of a seed-bearing fern. The important difference being whether the ovule was open to the air (ferns) or closed with a stigma and pollen-tube leading to it (flowering plants) (Thomas 1925: 325). Thomas made the assertion that he could identify a proto stigma.

In Thomas's account, all these questions are deeply interrelated. For example, in recounting the difficulty that he had in extracting the remains of seeds and seed membranes from the rock matrix Thomas offers some insight into the potential for representational artefacts to weaken his interpretation – that the effects of his process might misguide our 'reading'. However, that account also informed his suggestion that the seeds were stuck in a fleshy fruit, adding weight to an account of the plant as a proto angiosperm (flowering plant) rather than a seed-bearing fern. We see how interconnected questions, therefore produced interconnected narratives (of the process of extraction and of the function of the plant organ) from which Thomas arrived at an overarching narrative in which the fossil represented a transitional stage between a fern and the first flowering plants.[5]

9.2.2   Aerial Photographs

In considering how the aerial photograph in Figure 9.1 was interpreted, the guide from which it was drawn identifies a similar set of interconnected

---

[5] On narrative and evidentiary coherence, see Miyake (Chapter 5).

questions. The overarching question addressed to the image in Figure 9.1 would be: what can we understand about enemy strategy from this landscape feature? Here, as with the fossil fragments, there was a scarcity of material to interpret, and multiple unknowns. The network of questions would include:

- How can a particular track in use by the enemy be identified?
- What is the role of the track in relation to other elements in the image? (How is it networked with other resources?)
- What is the role of this track in the context of an immediate goal of the enemy?
- What does the nature of the track tell us about the broader capabilities and intentions of the enemy?

The section in the guide on interpreting tracks is particularly revealing.

They disclose dumps; battery positions; headquarters; wire which is otherwise invisible, and gaps through it; patrol paths; observation posts; in villages, those houses which are important centres; advanced listening posts [. . .] the evidence of numerous tracks and shortcuts leading across irrigated and cultivated fields may be taken to denote the presence in the neighbourhood of bodies of troops. The local population would not make sufficient use of these foot-paths to cause destruction to growing crops.[6]

As with the fossil evidence, a *set* of interconnected questions at different scales were all being developed as the basis for narratives that functioned in dialogue with each other. We see much more than a mapping of landscape features. Each question tested an emergent overarching narrative that would be a conjectural account of the enemy's activities. Importantly, however, some of these questions were not all posed or answered once the photograph was in hand. Some were determined in advance, some emerged as by-products of the processing or preparation of the photographs (gridding, annotating), while others emerged from direct analysis of a single image. To understand the resulting narratives requires expanding our field of analysis to a narrative practice that encompasses all of these stages. Figurative descriptions of this kind of work that were contemporary to Thomas offer us ways to take such a holistic approach.

9.2.3   From Questions to Narratives?

Sherlock Holmes made his first public appearance just two years after Thomas's birth and within a short space of time, was being used as a methodological role model in a variety of contexts (see 'New Habits of Media Use', section 9.4.1, below).[7] Carlo Ginzburg has placed Holmes in

---

[6] 'The Interpretation of Aeroplane Photographs in Mesopotamia', 61.
[7] On the use of Sherlock Holmes as a role model in narrative exploration, see Crasnow (Chapter 11).

a late nineteenth-century zeitgeist that connected the *modus operandi* of the consulting detective with an emergent episteme that touched art history, psychology, and medicine (Ginzburg 1989). Picking up Ginzburg's perspective, it is sensible to propose that to think 'like Sherlock Holmes' meant to formulate whodunnit and howdunnit narratives using a network of traces of an event. We could take, for example, an episode from Holmes's debut in *A Study in Scarlet* in which he lends his attention to minute aspects of the appearance of a watch. After an examination of a series of scratches and pawnbrokers' marks, Holmes connects them into a conjectured biography of the watch which is intertwined with a conjectural biography of its former owner.

The second trope that was applied to visual practices such as Thomas's by his contemporaries was that of reading evidence 'like a book'. Critical histories of the use of aerial photography in the social sciences such as Hauser (2007) and Haffner (2013) have drawn attention to early twentieth-century use of the 'book' trope to describe the interpretation of aerial views. In one of Haffner's examples, an early advocate of the value of the aerial view in human geography argued that air photographs let you 'read the land as one reads a great open book' (Haffner 2013: 27).[8] The study by Lorraine Daston and Peter Galison of visual epistemic authority (2007) has drawn attention to the use of this expression in other modes of scientific practice as indicative of authority through 'trained judgement'.

The act of interpretation is characterized slightly differently in these two cases. Haffner suggests in *The View from Above* that reading a landscape process from an aerial view 'was simply a matter of noting what had existed before versus what had appeared' (Haffner 2013: 13). In Hauser's account, the longer (past) human histories that are evidenced by aerial views are described as being stored below the ground, a history simply waiting to be revealed by new angles of flight and light (Hauser 2007). Daston and Galison go somewhat further in describing what visual interpretation requires. They cite examples of introductory texts in early twentieth-century scientific atlases that exhorted their readers to consider image interpretation as the 'skills required to read a new language using an unfamiliar alphabet and a different script' (Daston and Galison 2007: 328). A common trope for this mode of analysing of visual evidence, Daston and Galison argue, was the intuitive scanning of facial physiognomy – the rapid and possibly subconscious recognition and comparison of complex and minute differences within facial appearances. Attention to those differences could be learned.

Yet both of the above retrospective accounts using the figure of the book have flattened the role of narrative that it implies. We can reinvigorate the narrative component of the book trope if we reconnect it to Holmes tracing the

---

[8] Haffner here cites Brunhes, *Leçons de géographie: cours moyen* (1926), 1.

contours of a human life from the incidental features of a watch. Conan Doyle's emphasis on the *unique* possession of these skills in the consulting detective, encourages us to put aside the idea that the interpretation of the phenomena under scrutiny is determined by the phenomena itself (is read). Instead, we can investigate how interpretations were 'written' or constructed.

A third figurative expression, which sheds further light on the first two, is provided by the philosophical writing of Agnes Arber, one of Thomas's botanist colleagues at Cambridge.[9] For our purposes, her most useful expression was one in which she compared biological thought to a river. Through this expression, Arber was rejecting the idea that thought was strictly linear and proposing that it was better imagined as a reticulum (a network) moving through three dimensions in one direction. She saw the flow of thought that would eventually produce a narrative account as including eddies and currents, and therefore, by consequence, as dynamic. In Arber's words (drawing from yet another three-dimensional analogy), a biological explanation does not 'grow by accretion of ready-made parts, as a building', but rather 'in passing from phase to phase [. . .] suffers transformation from within' (Arber 1954: 69).

Bringing these three expressions together, we have: (i) the gathering of evidence to narrate a crime as offered by the figure of the consulting detective; (ii) attention to the relationship of phenomena within and across images as offered by the analogies to learning to read; (iii) the conceptualization of biographical thought as a river, encompassing data collection, analysis and the production of an explanatory narrative. All these reinforce the value of exploring a narrative *practice* that encompasses multiple modes of work. Understanding how a practitioner might seek to develop such a practice is best understood by returning to the context in which Thomas received his botanical training and then to the reworking of that practice in a military context.

9.3   Linking Vision and Narrative in Thomas's Scientific Work

9.3.1   Botanical Visual Cultures

Scholars have demonstrated that training in botany and biology in the late nineteenth century had a very strong emphasis on the visual. Botanical and biological knowledge were primarily transmitted as a visual practice. Posters, chalkboard drawings and field outings were the primary access to understanding plant life in schoolchildren. However, they remained the key tools for teaching at undergraduate level, as Thomas's own mentor, Albert Charles

---

[9] In the development of this framework, Arber explicitly engaged with the philosophy of biology, as discussed in the recent revival of interest in her work (Flannery 2003; Feola 2019).

Seward, testified in a paper for the *New Phytologist* (Seward 1902). By the turn of the century, British botanists had become actively interested in the possibilities that were offered by photographs as a research resource. In 1901, the British Association for the Advancement of Science (BAAS), Section K (Botany), joined other BAAS groups who had begun their own systematic image collections.

Kelley Wilder (2008) and others have identified the changing uses of photography in science in this period as a tool for both classification and measurement. Scholars have also explored how image technologies and analytical techniques were co-developed (Hentschel 2002; Kelsey 2007). Photography operated with and alongside gridding, labelling and diagramming practices that highlighted or abstracted particular aspects of a phenomenon, as well as indexing practices, which placed each image within ordered and signifying relationships to places, people, specimens and to other images. The epistemic architectures for botanical photography calibrated scientific visions of the vegetative world and formed the basis of a disciplinary visual practice (Hughes 2016). In other words, visual record systems were set up that would allow comparisons *between photographs* as epistemic objects and *identify particularity within* individual visual records (Rheinberger 2015). They trained a botanical or palaeobotanical observer's attention to groupings of related features within a visual record, allowed them to find new or unknown features to explore, and cultivated in them an 'exceptionally seeing eye' (Meinig 1979: 199).[10]

There is more at work here than what has been called expert training in 'pattern recognition' (Daston and Galison 2007: 329). The background to the publication of Figure 9.2 offers excellent insight into how several layers of material and epistemic work built the photograph into a disciplinary visual archive through reticulate *narratives*. The fossils that came to be known as Caytoniales were found in Cayton Bay in Yorkshire. Their geographical origin was recorded in their new names, in the archival records alongside the original specimens in the Sedgwick Museum, Cambridge, and noted in the scientific paper in which they were presented (Thomas 1925: 302). Figure 9.2 shows a rock from Cayton Bay, on the surface of which the fruits of *Caytonia Sewardii* are identified (magnified to 2.5 × their original size) to allow inspection of the form of the plant organs at the relevant level of visual detail. Once logged, such rocks were then subject to extensive further preparation including boiling for several weeks in specific chemicals, passing through alcohols and slicing with a microtome. This process revealed the chemical composition of different parts of the fossilized plant matter creating further sets of microscopic images at a cellular level that were identified as different plant organs.

---

[10] With thanks to Mat Paskins for highlighting this aspect.

We start to see how these processes created epistemic objects that had a networked relationship to each other, crossing multiple scales of place (from the cliff-face to the laboratory) and of plant function (from prehistoric landscape ecologies to organ to cell). These relational features of the image are accounted for through questions that produced various smaller narratives, smaller narratives which are held in unity by the overarching narrative of Caytoniales as a missing link between ferns and flowering plants. That overarching narrative includes the relationships between site, process, archive and conjecture, linking place and plant function in a reticulate and partly predetermined and partly emergent manner.

9.3.2   Writing Botanical Relationships

The role of narrative in the use of visual archives has received less critical attention than the social and political processes that shape those archives' production. Yet the narrative techniques involved were just as much subject to culturally specific processes, to disciplining and to epistemic virtues. Here, I propose an analysis of two particular narrative 'challenges' that demonstrate this. The first of these challenges was the deceptively simple task of translating the profoundly visual experiences of research of plant lifeforms into verbal accounts.[11] Another was the entanglement of botany within arguments about the nature and destiny of life forms.

The relationship between categorizing and narrating plant physiology is revealed by a banal but profoundly perplexing problem that was offered to readers of the *New Phytologist* by the botanist Leonard Alfred Boodle in 1903. Boodle's worries about botanical accounts of the vascular structure in leaves give us an insight into the connection between botanical visual and narrative practices. He argued that an individual vascular structure needed to be traced *from* the stem towards the leaf in order to observe the increasing complexity of the structure. Boodle is specifically arguing against a proposition by Heinrich Anton de Bary (1884) that the course of vascular structures could be most easily understood by tracing them down from the base of the leaf *into* the stem. Important to both Boodle and Bary was that the order in which you approached the description might unwittingly commit you to a different perspective on the evolutionary or developmental precedence in the plant's physiological features. A narrative artefact might impede your interpretation.

As Boodle pointed out, 'In many cases according as one describes the vascular and other tissues as traced upwards or downwards, one is easily led to use phrases which commit one to a different opinion as to their morphological nature in the two cases' (Boodle 1903: 108). While you could choose

---

[11] See Miyake (Chapter 5), on translating seismic data registrations into earthquake records.

a direction for descriptive purposes, 'the topographical statement must be reworded according to the view arrived at of the first origin of these tissues' (Boodle 1903: 109). We see a focusing of attention on the epistemic pitfalls that could occur as a result of clumsy expression. One might suggest that Boodle is simply arguing for a clarity of thought that distinguishes between the spatial relation and physiological origin of an organ in a scientific description. However, the case material itself (the relationship between the leaf and the stem) demonstrates that he is also arguing for a particular kind of 'openness' required to solve problems in a field of knowledge with so few parameters. To return to the terms set up in this chapter, he is arguing for a dynamic narrative practice.

A second challenge in constructing botanical narratives lay in the question not just of process but of the destiny of plant forms, a question that interested Thomas himself. This was the question of emergent differences in plant tissue (stem, leaf, bud) and whether there was an essential or original nature of plant tissue. This question had been under debate for decades and was the focus of Goethe's essay 'On the Metamorphosis of Plants' (1790) that had been seminal in the discipline. Although Thomas rejected the thesis that Goethe put forward in that essay (that the leaf was the urform of plant matter), he celebrated the epistemic openness of Goethe's investigation. In a manifesto for plant morphology from 1933, Thomas directly quoted Goethe to underline a point on epistemic method: 'The thing now to be aimed at is to keep habitually in view the two contrary directions in which variations are developed' (Thomas 1933: 47). For Thomas, narratives about process and progress had to be dynamic, open to radical reshaping.

In Thomas's case, as those of many of his colleagues, that openness included scepticism about oversimplifying evolutionary processes. In his advocacy for a '*new* morphology', in 1933 Thomas affirmed the value of using the forms of both living and fossilized plants to construct narratives about evolutionary history. However, Thomas argued, these narratives should be assembled with caution. In particular, he emphasized, one should not rely on the stability of concepts such as the leaf, petal or stamen (Thomas 1933: 48; 1934). One might also be cautious about over-reliance on simplistic models for the mechanisms of inheritance when similar characteristics might emerge across different, and widely separated, groups (Thomas 1934: 176; Winsor 1995).

Thomas's manifesto bears the echoes of one made by his predecessors as President of Section K (Botany) of the BAAS, William Henry Lang. Lang had also argued for the value of morphological study some years earlier, in 1915, suggesting that it offered the means to untangle an interrelationship of plant species that looked 'more like a bundle of sticks than a tree' (Lang 1915: 242). When we understand the opacity of the biological mechanisms at the heart of Thomas's narrative accounts of plant life, it becomes clear why it was important to hold so many questions open simultaneously. We can see why the narratives about fossil plants needed to be dynamic. Thomas was bringing together multiple narratives that were all under formulation at the same time, and which, borrowing Arber's expression, when passing from 'phase to phase' suffered 'transformation from within' (Arber 1954: 69).[12]

Casting our minds back to the analogy of the 'book of nature', the relationship of narrative to traces evidenced in plant tissue reveals how much more like writing than reading this process was. We see that the archive was bound into narrative relations and we see the vital importance of narrative skills in Thomas's work as a palaeobotanist. That training included attention to narrative artefacts that might inadvertently be introduced into verbal description. It included the measured use of existing explanatory mechanisms. It also included the capacity simultaneously to explore multiple networks of significance at multiple scales in order to produce a reticulate, multi-scalar architecture of narratives that constrained and supported each other and that were flexible enough to accommodate instability in the identity of the narrated objects. Leaves, stems or proto stamens might be called upon to play entirely new roles in narratives that shifted with and around them.

9.4 Thomas's Narrative Practice and Military Intelligence: The 'New Morphology' of War

9.4.1 New Habits of Media Use

If Thomas's narrative practice was a valued technique for botanical enquiry, the capacity to narrate enemy behaviour might seem an even more obvious and essential part of warfare. Retrospectively, it also seems common sense that the military would use aerial photography to construct such narratives. The scholarship of military historians has often suggested a continuous enthusiasm for photographic and aerial surveillance technologies since the mid-nineteenth century. Yet in fact there was dissent and difficulties in their uptake into intelligence practices.

It is often stated that the use of photography by British soldiers and military engineers was already widespread in the nineteenth century, particularly in imperial endeavours (Mattison 2008). The Royal Engineers School at Chatham began to teach photography in 1856. Photography was in use in reconnaissance during the 1867–68 campaign in Abyssinia. The Royal Engineers also began to develop balloon technology to generate photography from the air during the nineteenth century. By 1878, there were four balloon sections with men also

---

[12] See Kranke (Chapter 10), for further discussion of the relationship between phylogenetic representations and narrative.

trained in photography (Mead 1983: 19). Balloons and unmanned kites were used for reconnaissance by the British in the South African conflicts, particularly from 1898 to 1902 (Mead 1983: 25). In this reading, nineteenth-century institutional uptake paved the way for a spectacular deployment of photography in the First World War, when millions of photographs were produced, and the merits of the aerial view were proven.

These accounts of the use of remote sensing in war overplay its roots. In fact, to paraphrase Elizabeth Edwards, while photographs offered the armed forces 'a whole new different class of knowledge', it had to be 'recognized, contained and utilized within [. . .] existing habits of media use' in the early twentieth century (Edwards 2014: 175).[13] That process was not a straightforward one. In the early twentieth century, the British armed forces' habits of use for photography were not analytically oriented. While in previous decades botanists, archaeologist and geologists had enthusiastically begun to assemble visual records for analytical purposes, in military contexts, photography appears to have been used largely to other ends. In the late nineteenth century, the British military's primary official use of photography was in fact as 'the readiest and most accurate mode of copying' officers' hand sketches of the terrain, maps and charts (Holland and Hozier 1870: 360). Where original images were being made in an official capacity, they were not being used for field intelligence, but rather to document scenes 'after the fact'. Some images were made of peoples and places in an ethnographic vein. Some images were made of military manoeuvres in progress or completed military engineering projects. Some images were made to document the effects of bombardment or sieges (Bolloch 2004; Sampson 2008). In sum, photography was being used to record seen events rather than to make conjectural narratives about unseen events. Even the military value of the panoptic view from the air was not a given. The potential of balloons and kites for military reconnaissance was tested in South Africa; however, this had been due to strong advocacy by enthusiasts, and in the face of some reluctance by officials.

During this period of prevarication about photography, the whole field of military intelligence was, however, in flux. At the end of the nineteenth century, concerns about the nature of combat in any future European wars were prompting institutional change. It was recognized that British field intelligence expertise was minimal and ill-adapted either to fighting across large fronts in wars that involved civilians, industry and distributed resources or to meeting 'guerrilla' forces in relatively unknown terrain. Previously, field intelligence activities had been mustered in response to each crisis; there was no ongoing training. Nonetheless, British failures in South Africa had caused some anxiety, and in 1907 the first attempts were made to provide a permanent core of experts

---

[13] Here Edwards cites Gitelman and Pingree (2003: xii).

in the form of a field-intelligence training course (Siegel 2005). The impact of this course was doomed to be limited, as there were only around eight attendees each year. Its content was also not very rigorous. Attendees were given little instruction in observation; instead, it was recommended that they closely examine 'the adventures and methodology of Sherlock Holmes' (Siegel 2005: 136).

Within the first year of combat in the First World War, attitudes to photography began to change. In February 1915, the Royal Flying Corps began to innovate cameras for use from aeroplanes, and the aerial photographs that were produced were taken up eagerly in the field. The technologies and techniques for photographic capture, organization and analysis were innovated rapidly. British military success in Palestine and Sinai was attributed to aerial photography (and, Air Chief Marshall Salmond said, to Thomas's work in particular, as Photographic Officer, RFC 5th Wing) (Harris 1963).

In both the Western and the Eastern conflicts in the First World War, aerial photography was eventually integrated into forms of hybrid media produced from cartography, photography, annotation and diagramming. However, the *narrative* power of the visual material in each arena was not the same. Contemporaries who observed the enthusiastic uptake of aerial photography in the East attributed it, at least in part, to the difference in the amount of pre-existing knowledge that the British had about the environment and societies in that region (Dowson 1921). Although the changing scale and pattern of warfare was devastating and unprecedented in the West, the campaigns in France and Belgium were being fought in terrains that had familiar social and physical geographies, and of which the British had detailed topographical maps. It was relatively easy to identify changes in the landscape that were due to enemy action. In the East, the aerial photographs that were taken carried the additional epistemic burden of providing more general understandings of the physical terrain, as well as the socio-cultural habits and material dispositions of enemy forces.[14] Narrative conjectures about the enemy were riskier, and there was a greater need to link questions and visual evidence in a reticulate, dynamic, multi-scalar fashion.

Despite the importance of this work during the First World War, in the years that followed the conflict the British armed forces did not continue to develop further techniques to analyse aerial photographs. The military seemed relatively indifferent to the powerful potential of photointerpretation for intelligence purposes, despite some effort by Thomas and other veterans from the war in Palestine and Mesopotamia, including T. E. Lawrence.[15] When the British

[14] 'The Interpretation of Aeroplane Photographs in Mesopotamia', iv.
[15] Thomas and his Eastern Front colleagues communicated the epistemic potential of aerial photography in lectures and publications aimed at geographers, photographers, scientists and politicians. See, for example, Thomas (1920).

entered into conflict in 1939, that lack was fairly rapidly felt, and Thomas himself, along with other photo-interpretation veterans, were re-recruited in military intelligence.

When Thomas was transferred back to strategic photointerpretation work his first role was to review the RAF manual of photointerpretation. He subsequently became leader of the 'third phase' of the Joint Forces photointerpretation work, in which specialist groups produced long-term studies of changes on the ground (Rose 2019). His interests led to the founding of two new photo-interpretation sections in the third phase dedicated to topography and to industry. Thomas eventually led his section in one of the most celebrated moments in the history of military intelligence, the identification and destruction of German V-2 rocket capabilities at Peenemünde.

9.4.2   A Military Narrative Practice

The epistemology of military photointerpretation has been subjected to far less academic analysis than the epistemological role of photography in academic science. However, we know that it was not only the 1908 British military intelligence officers who invoked Sherlock Holmes for methodological instruction. As Paul K. Saint-Amour notes, instruction manuals for photointerpretation regularly called on Holmes as a role model (Saint-Amour 2003: 359). We are learning from our association of Holmes, books and rivers in Thomas's botanical narrative practice that the apparently simple evocation of 'Holmesian' practice conceals a highly complex process.

First, we see the emergence of visual infrastructures in military intelligence that were intended to produce 'exceptionally seeing eyes'. Very early use of photography in First World War military intelligence often involved unwieldy photomontage panoramas of enemy encampments. These were not easy to circulate or report upon. One of Thomas's key contributions in the East was organizing systematic, gridded, aerial photography of the unknown landscapes in which war was unfolding. That photography was abstracted to produce topographic maps of the terrain, *from* which significant changes could be more easily observed and *upon* which they could be more easily notated. We see the first steps in practice similar to the organizing principle for visual materials in botany, creating systems from which similarity and difference across sets of features could be discerned (Rheinberger 2015).[16] That system-atizing of photography also allowed the production of 'atlas-like' documents such as the one from which Figure 9.1 is drawn: 'The Interpretation of Aeroplane Photographs in Mesopotamia (Part 2)'. These guides taught

---

[16] See also Hughes (2016) for a deeper exploration of how similar practices in ecology allowed the identification of vegetative/landscape relationships as ecological 'objects'.

observers how to find visual indicators of enemy activity in and against non-signifying landscape features.

Ultimately, such documents allowed for reticulate and multi-scalar narrative accounts of enemy activity. We see this in the gradual extension of the role of aerial photography in the East during the years of the First World War. Initially it was used for identifying targets and calculating the effect of bombing campaigns. Later it was deployed to more obviously narrative ends: creating track maps that showed human movement in the landscape or estimating troop deployments based on types of tents and shelters (Sheffy 2014: 189). During the Second World War, the value of narrative practices such as Thomas's for that photo-interpretative work was more substantially recognized. Via his scientific networks, Thomas was responsible for the recruitment of 'many men and women accustomed in their professional lives to examining subjects in depth', recognizing potentially signifying features of an image, 'and pursuing a "lead" until its nature and purpose were established' (Halsall 2005).

As this new military narrative practice emerged, we see that an emphasis was placed on working in a dynamic way. In the context of rapidly and regularly providing reports, the need to keep an open mind, as a technique for disciplining the imagination, was highly valued. André-H. Carlier (a French photo-interpretation expert from the First World War) noted the activity required someone who would, 'not give in to his imagination, and be willing to surround himself with all sources of information, ignoring none' (quoted in Haffner 2013: 13). Art historian Ernst Gombrich recounted the story of a photographic interpreter who had been crucial to the success at Peenemünde whose imagination was later insufficiently disciplined. In looking at a photographic trace, Gombrich warns:

There can be no professional stocking of minds with an infinite variety of possibilities. All the professional should learn, and obviously never learns, is the possibility of being mistaken. Without this awareness, without this flexibility, interpretation will easily get stuck on the wrong track. (Gombrich 1969)

Here we have advocacy for flexibility in the construction of narratives that is similar to that which we encountered in the context of studying plants. We see that the objects whose traces were being both sought in and narrated from the aerial photographs had conjectural definitions that might be highly unstable. Was this site an ordinary factory, or a laboratory for a secret Nazi weapon? Dynamism in narrative accounts, while managing the possibility of being mistaken, were qualities common to each of Thomas's domains.

Overall in military narrative practice based on aerial photography we can identify epistemic techniques similar to those deployed in palaeobotany. We see the collection and organization of fragments of traces into visual evidence, into archives and atlases. We see the need to find ways to train attention on small details of landscape change, of similarity and of difference that would be accounted for by mini narratives that would provide 'leads'. We then see those accumulate in overarching narratives. We see the need to work flexibly, considering multiple narratives simultaneously and at multiple scales so that parallel accounts of visual traces might test each other's coherence. We can see that this emergent military narrative practice paralleled skills that Thomas had honed in his scientific work.

9.5   Conclusion

Bringing together the two domains of Thomas's work offers us a clearer view of each. Existing scholarship on the architectures of visual knowledge in botany and ecology allow us to identify the emergence of similar practices when the armed forces developed new habits of media use around aerial photography. Juxtaposing Thomas's two domains serves to reinforce that such narratives were not read *from* the visual record but mutually developed *with* the accretion of visual evidence. The co-construction of archival architectures *with* narrative required working in a way that was reticulate (networked), multi-scalar and dynamic, characteristics that are reflected in the various figurative expressions used by Thomas's peers. We can recognize this practice as particularly well adapted to investigating complex processes for which evidence was scarce. In addition to its contributions to understanding the role of narrative in scientific practice, this chapter provides new perspectives on the visual cultures of twentieth-century biology, and on the legacy of military aerial photography in civilian spatial sciences.[17]

Bibliography

Arber, A. (1954). *The Mind and the Eye: A Study of the Biologist's Standpoint.* Cambridge: Cambridge University Press.

Bary, A. de (1884). *Comparative Anatomy of the Vegetative Organs of the Phanerogams and Ferns.* Oxford: Clarendon Press.

Bolloch, J. (2004). *War Photography.* Paris: Musée d'Orsay.

Boodle, L. A. (1903). 'On Descriptions of Vascular Structures'. *New Phytologist* 2.4–5: 107–112.

Chapman, R., and A. Wylie (2016). *Evidential Reasoning in Archaeology.* London: Bloomsbury.

[17] I'd like to offer my sincere thanks to Robert Meunier, and an anonymous reviewer, as well as to the Narrative Science Project team for their feedback, suggestions and encouragement during the writing of this essay. *Narrative Science* book: This project has received funding from the European Research Council under the European Union's Horizon 2020 research and innovation programme (grant agreement No. 694732). www.narrative-science.org/.

Cleal, C. J., and B. A. Thomas (2019). *Introduction to Plant Fossils*. Cambridge: Cambridge University Press.

Cleland, C. E. (2011). 'Prediction and Explanation in Historical Natural Science'. *British Journal for the Philosophy of Science* 62.3: 551–582.

Coen, D. R. (2018). *Climate in Motion: Science, Empire, and the Problem of Scale*. Chicago: University of Chicago Press.

Currie, A. (2018). *Rock, Bone, and Ruin: An Optimist's Guide to the Historical Sciences*. Cambridge, MA: MIT Press.

Currie, A., and K. Sterelny (2017). 'In Defence of Story-Telling'. *Studies in History and Philosophy of Science Part A* 62: 14–21.

Daston, L., and P. Galison (2007). *Objectivity*. New York: Zone Books.

Dowson, E. M. (1921). 'Further Notes on Aeroplane Photography in the Near East'. *Geographical Journal* 58.5: 359–370.

Driver, F. (2000). *Geography Militant: Cultures of Exploration and Empire*. Oxford: Wiley.

Edwards, E. (2014). 'Photographic Uncertainties: Between Evidence and Reassurance'. *History and Anthropology* 25.2: 171–188.

Feola, V. (2019). 'Agnes Arber, Historian of Botany and Darwinian Sceptic'. *British Journal for the History of Science* 52.3: 515–523.

Flannery, M. C. (2003). 'Agnes Arber: Form in the Mind and the Eye'. *International Studies in the Philosophy of Science* 17.3: 281–300.

Ginzburg, C. (1989). *Clues, Myths, and the Historical Method*. Baltimore, MD: Johns Hopkins University Press.

Gitelman, L., and G. B. Pingree, eds. (2003). *New Media, 1740–1915*. Cambridge, MA: MIT Press.

Goethe, Johann Wolfgang von (1790). *Versuch die Metamorphose der Pflanzen zu erklären*. Gotha: Carl Wilhelm Ettinger.

Gombrich, E. H. (1969). 'The Evidence of Images: I The Variability of Vision'. In C. S. Singleton, ed. *Interpretation: Theory and Practice*. Baltimore, MD: Johns Hopkins University Press, 35–68.

Haffner, J. (2013). *The View from Above: The Science of Social Space*. Cambridge, MA: MIT Press.

Halsall, C. (2005). 'Dr Hugh Hamshaw Thomas'. BIOG Hamshaw Thomas, Medmenham Collection, Cambridgeshire. www.medmenhamcollection.org/collection201510/collection.php.

Harris, T. M. (1963). 'Hugh Hamshaw Thomas. 1885–1962'. *Biographical Memoirs of Fellows of the Royal Society* 9: 287–299.

Hauser, K. (2007). *Shadow Sites: Photography, Archaeology, and the British Landscape, 1927–1955*. Oxford: Oxford University Press.

Hentschel, K. (2002). *Mapping the Spectrum: Techniques of Visual Representation in Research and Teaching*. Oxford: Oxford University Press.

Holland, M. T. J., and C. H. Hozier (1870). *Record of the Expedition to Abyssinia*. London: HMSO.

Hughes, D. (2016). 'Natural Visions: Photography and Ecological Knowledge, 1895–1939'. PhD thesis, De Montfort University, Leicester. http://hdl.handle.net/2086/13307.

Kelsey, R. (2007). *Archive Style: Photographs and Illustrations for U.S. Surveys, 1850–1890*. Berkeley: University of California Press.

Lang, W. H. (1915). 'The British Association. Section K Botany'. *Nature* 96.2400: 242–248.

Mattison, D. (2008). 'Aerial Photography'. In J. Hannavy, ed. *Encyclopedia of Nineteenth-Century Photography*. London: Routledge.

Mead, P. (1983) *The Eye in the Air: History of Air Observation and Reconnaissance for the Army, 1785–1945*. London: HMSO.

Meinig, D. W. (1979). 'Reading the Landscape: An Appreciation of W. G. Hoskins and J. B. Jackson'. In D. W. Meinig, ed. *The Interpretation of Ordinary Landscapes*. Oxford: Oxford University Press, 195–244.

Morgan, Mary S. (2017). 'Narrative Ordering and Explanation'. *Studies in History and Philosophy of Science Part A* 62: 86–97.

Pinney, C. (2008). 'The Prosthetic Eye: Photography as Cure and Poison'. *Journal of the Royal Anthropological Institute* 14.S1: S33–S46.

Rheinberger, H.-J. (2015). 'Preparations, Models, and Simulations'. *History and Philosophy of the Life Sciences* 36.3: 321–334.

Richards, R. J. (1992). 'The Structure of Narrative Explanation in History and Biology'. In M. H. Nitecki and D. V. Nitecki, eds. *History and Evolution*. Albany: State University of New York Press, 19–54.

Rose, E. P. F. (2019). 'Aerial Photographic Intelligence during World War II: Contributions by Some Distinguished British Geologists'. In E. P. F. Rose, J. Ehlen and U. L. Lawrence, eds. *Military Aspects of Geology: Fortification, Excavation and Terrain Evaluation*, Geological Society of London Special Publications 473. London: Geological Society, 275–296.

Rudwick, M. J. S. (1976). 'The Emergence of a Visual Language for Geological Science 1760–1840'. *History of Science* 14.3: 149–195.

Saint-Amour, P. K. (2003). 'Modernist Reconnaissance'. *Modernism/modernity* 10.2: 349–380.

Sampson, G. D. (2008). 'Military Photography'. In J. Hannavy, ed. *The Encyclopedia of Nineteenth-Century Photography*. London: Routledge, 929–931.

Seward, A. C. (1902). 'Note on Botanical Teaching in University Classes'. *New Phytologist* 1.1: 14–17.

Sheffy, Y. (2014) *British Military Intelligence in the Palestine Campaign, 1914–1918*. London: Routledge.

Siegel, J. L. (2005). 'Training Thieves: The Instruction of "Efficient Intelligence Officers" in Pre-1914 Britain'. In P. J. Jackson and J. L. Siegel, eds. *Intelligence and Statecraft: The Use and Limits of Intelligence in International Society*. Westport, CT: Praeger Publishers, 127–138.

Smith, C. B. (1985). *Air Spy: The Story of Photo Intelligence in World War II*. Falls Church, VA: American Society for Photogrammetry Foundation.

Tagg, J. (1988). *The Burden of Representation: Essays on Photographies and Histories*. London: Macmillan.

Thomas, H. H. (1920). Aircraft Photography in the Service of Science. *Nature* 105.2641: 457–459.

(1925). 'The Caytoniales, a New Group of Angiospermous Plants from the Jurassic Rocks of Yorkshire'. *Philosophical Transactions of the Royal Society of London. Series B* 213.402–410: 299–363.

(1933). 'The Old Morphology and the New'. *Nature* 131.3298: 47–49.

(1934). 'The Nature and Origin of the Stigma'. *New Phytologist* 33.3: 173–198.

Tucker, J. (2013). *Nature Exposed: Photography as Eyewitness in Victorian Science.* Baltimore, MD: Johns Hopkins University Press.

Turner, D. (2000). 'The Functions of Fossils: Inference and Explanation in Functional Morphology'. *Studies in History and Philosophy of Science Part C: Studies in History and Philosophy of Biological and Biomedical Sciences* 31.1: 193–212.

Wilder, K. (2008). *Photography and Science.* London: Reaktion Books.

Winsor, M. P. (1995). 'The English Debate on Taxonomy and Phylogeny, 1937–1940'. *History and Philosophy of the Life Sciences* 17.2: 227–252.

10 The Trees' Tale: Filigreed Phylogenetic Trees and Integrated Narratives

Nina Kranke

Abstract

In this chapter, phylogenetic trees are discussed in the context of narrative science and historical explanation. Phylogenetic trees are predominantly bifurcating dendrograms that biologists use to represent evolutionary trajectories and patterns of shared ancestry. By means of a case study I explain how these diagrams are constructed, show that specialists read them as narratives and argue that they represent narrative explanations. Some phylogenetic trees not only consist of a branching structure and taxa names but include additional visual and textual elements. These filigreed trees are used in different contexts to represent integrated narratives (e.g., narratives of species migration, political and pedagogical narratives) that extend beyond evolutionary narratives of origin and divergence or narratives of shared ancestry. My chapter shows that diagrams as visual representations can be the central element of a scientific narrative and that narrative is used to create coherence between heterogeneous materials.

10.1   Introduction

Phylogenetic trees are predominantly bifurcating tree diagrams that biologists use to represent evolutionary trajectories and patterns of shared ancestry. In the past two decades, phylogenetic trees have become more flashy, colourful and visually sophisticated compared with tree diagrams of the 1980s and 1990s. In addition to the basic structure of connected lines with taxa names, these filigreed trees contain other graphic and textual elements like images of animals or plants, coloured areas or lines and symbols. Several authors have discussed phylogenies from the perspectives of narrative science and historical explanation (e.g., Cleland 2011; Griesemer 1996; O'Hara 1988; 1992). It has been argued that phylogenies are somewhat prior to evolutionary histories in the sense that they are more descriptive and therefore more objective or that they provide a scaffold for evolutionary histories. My chapter will show that, even if this is the case in some respects, phylogenetic trees are all the more interesting for it.

206

Robert O'Hara (1988), for example, claims that the relationship of phylogeny to evolutionary history is a relationship of chronicle to history.[1] According to him, a chronicle is a 'description of a series of events, arranged in chronological order but *not* accompanied by any causal statements, explanations, or interpretations' (O'Hara 1988: 144; emphasis original). Following Arthur Danto, he argues that histories, on the other hand, contain 'a class of statements called *narrative sentences*' (O'Hara 1988: 144; emphasis original). In addition to his view of phylogenies as chronicles, O'Hara advocates a scaffold view[2] of phylogeny, meaning that he understands phylogenies as the basis for evolutionary histories. James Griesemer (1996) disagrees with O'Hara with respect to his view of phylogenies as interpretation-free chronicles. He argues that phylogenies are the result of several methodological decisions (e.g., the choice of phylogeny construction method, choice of outgroup) and claims that phylogenetic analysis 'produces something more theoretically charged than chronicle' (Griesemer 1996: 67). However, like O'Hara, Griesemer subscribes to the chronicle–history dichotomy and to the scaffold view of phylogeny and evolutionary history. He writes: 'I agree that cladistic analysis aims at something prior to evolutionary narrative in the way that chronicle precedes history' (Griesemer 1996: 67). Neither O'Hara nor Griesemer seems to believe that phylogenies do much, if any, explanatory work.

My discussion of a study conducted by Maria Nilsson and her collaborators (2010) shows that phylogenies and phylogenetic trees are much more interesting than interpretation-free chronicles. In fact, the phylogeny construction process requires several decisions (e.g., which taxa to include, which characters to use) that potentially affect the outcome of the analysis and is based on fundamental assumptions about molecular evolution. I show that the chronicle–history dichotomy is misleading in the case of phylogenetic trees and evolutionary histories because tree diagrams as the central and only comprehensive representation of phylogenies are read as evolutionary narratives, provided that the reader is familiar with the specialist conventions.[3] I argue that all phylogenetic trees, even plain ones, represent narrative explanations, and the informed reader can derive narrative sentences from them. My discussion of the filigreed marsupial tree constructed by Nilsson et al. (2010), and other examples of filigreed phylogenetic trees, shows that, by adding graphic and textual elements to the basic tree structure, narratives can extend beyond phylogenetic narratives of origin and divergence, including narratives of species migration and political and

---

[1] See Berry (Chapter 16) for a more detailed discussion of the chronicle–history distinction.
[2] See Teather (Chapter 6) on the role of scaffolding in archaeology.
[3] See Hopkins (Chapter 4) and Andersen (Chapter 19) for discussions of the role of expert knowledge in reading scientific narratives. See Hajek (Chapter 2) for a detailed discussion of the relation between discourse/narration and reader.

pedagogical narratives. I conclude that filigreed phylogenetic trees are used to represent integrated narratives, and so contain more epistemic features than have been recognized thus far.

10.2   Phylogenetic Analysis: Reconstructing the Past

As a historical science, evolutionary biology shares characteristics with other natural sciences, such as physics and chemistry, as well as other historical sciences, such as anthropology and archaeology (Harrison and Hesketh 2016; Kaiser and Plenge 2014; Tucker 2014). Just like human historians who study the origin and trajectory of events (e.g., wars, revolutions), evolutionary biologists are, among other things, concerned with accounting for unique, localized events that happened in the past – for example, the origin and evolutionary trajectory of species (see Beatty, Chapter 20; Currie 2014). Since the events of interest are not directly accessible or observable, both human historians and evolutionary biologists need to find other ways to gain knowledge of the past.

One way of reconstructing the past is to look for traces[4] (Cleland 2002) or clues (Gardiner 1961: 74; Ginzburg 1979) and infer past events from this evidence. This type of trace-based reasoning is frequently compared to detective work where the investigator tries to reconstruct the crime based on clues that they find at the crime scene (Cleland 2002: 490; Ginzburg 1979: 276; see also Haines, Chapter 9). The investigation usually starts with the discovery of a puzzling phenomenon and the question of how, when or why it came to be as it is (Roth 2017: 44). While historians visit archives to find records that can be used as clues, in molecular phylogenetics the traces are part of the organism itself, namely its genome which is seen as an archive containing the historical record of its lineage (Bromham 2016: 329). Since methods of phylogenetic analysis are comparative, scientists also need molecular data of closely related taxa to construct a phylogenetic tree. The main assumption is that the more similar the genomes of two populations are, the closer they are related to each other. If genomes of two populations are very similar to each other, researchers assume that they have diverged rather recently. In the following section, I will give a more detailed account of the different steps of a phylogenetic analysis by using Nilsson and her collaborators' (2010) study as an exemplary case.

10.2.1   Constructing the Marsupial Tree

One of the first steps to construct any phylogenetic tree is to choose which organisms should be included in the analysis. In this case, Nilsson and her collaborators (2010) decided to include representative species of all seven

---

[4] See Crasnow (Chapter 11), for a discussion of traces in narratives.

marsupial orders. In total, the researchers ran their analysis with representative specimens of twenty species plus one outgroup (a reference species that is only distantly related with the group of interest). Another important step that needs to be made in the beginning of a phylogenetic analysis is the choice of characters, the traits or features of organisms taken to matter. While morphological characters are still used by some researchers, most phylogenetic trees that have been published in the last two decades are at least partially based on molecular characters (e.g., DNA or rRNA sequences). As phylogenetic markers, Nilsson et al. (2010) used retroposons, also called *jumping genes*, because these DNA fragments are transcribed into RNA, then 'jump' to a different place in the genome where they are inserted through reverse transcription. Once a retroposon has been inserted in the ancestral germline, it can become fixated in the ancestral population and is inherited by all descents. One can thus conclude that if a certain number of retroposons is present in two or more marsupial species, they are more closely related to each other than to species that do not share these retroposons. According to the researchers, retroposons exhibit low insertion site preferences, which makes it highly unlikely that the same retroposon was inserted twice in the same place in the genome of two different species (Nilsson et al. 2010). The scientists thus assume that when two marsupial species share a retroposon at a certain place in the genome, it was inherited from a common ancestor. On these grounds, they claim that 'the shared presence of retroposed elements at identical orthologous genomic locations of different species, families, or orders is a virtually homoplasy-free indication of their relatedness' (Nilsson et al. 2010).

The researchers used sequence data from databases but also received marsupial DNA samples from collaborators in Australia. The data were then used to preselect potential phylogenetically informative retroposon loci. Altogether, the group found 53 phylogenetically informative markers (Nilsson et al. 2010). These 53 characters were plotted in a presence–absence table, and analysed. Nilsson and her collaborators used parsimony analysis to find the most parsimonious tree (of all possible tree patterns, the tree diagram that minimizes the total number of character state changes is to be preferred). They used a program referred to as 'PAUP* (Phylogenetic Analysis Using Parsimony *and other methods)' to analyse the data and generate the tree topology. The procedure shows that phylogenetic analysis is a comparative approach, with similarity as the ordering principle. These similarities, however, were inherited from a common ancestor and can thus be used to reconstruct phylogenetic relationships.

10.2.2   Reading Tree Diagrams as Visual Narratives

If the researchers had decided to publish their results in a systematics journal that is devoted to phylogenetic theory and practice, they could have stopped here and published the plain tree diagram as a depiction of the phylogenetic narrative (Figure 10.1). But since the scientists published the article in *PLoS Biology*, they integrated the phylogenetic narrative with narratives from other fields to create an appealing story that is more likely to get published in journals with a broader thematic scope. To depict the integrated narrative, the researchers turned the plain tree that consists of connected horizontal lines and species names into an attractive image that contains additional visual elements (Figure 10.2). I use the labels *basic structure* or *plain tree* to refer to phylogenetic trees that only consist of connected lines and names of biological taxa, and *filigreed tree* to refer to tree diagrams that include the basic structure and additional visual and textual elements. While this is a type of scaffolding where the plain tree is used as the basis for the filigreed tree, I do not use the labels *plain* and *filigreed* to distinguish between chronicles and histories or to imply that phylogenies are prior to evolutionary histories. Instead, I argue that both plain and filigreed trees are read as narratives and depict evolutionary histories.

The plain tree (Figure 10.1) is not part of the main paper by Nilsson et al. (2010) but can be found in the supplementary material. Given that phylogenetic trees are used within the framework of evolutionary science, the temporal aspect of these diagrams seems obvious. However, there are several misunderstandings about how to interpret the internal nodes, the relationship among taxa and the time axis (Gregory 2008). In the mid-twentieth century, most phylogenetic trees contained actual ancestors and depicted ancestor-descendant relationships (of extant or extinct species).[5] Today, however, phylogenetic analysis is focused on sister group relationships and it is assumed that contemporary species cannot be each other's ancestors. Although the branching diagrams that do not contain any specified ancestors could be interpreted as cladograms that merely depict patterns of character distribution (Wiley 1981: 98; Eldredge and Cracraft 1980: 10), most contemporary scientists who practise phylogenetic analysis understand the branching diagrams that they produce as phylogenetic trees, implying a process of change over time, and commonly refer to them as *phylogenies*. The internal and unnamed nodes of phylogenetic trees are interpreted as actual (but unknown) or hypothetical common ancestors. However, they also represent speciation events (the divergence of one cohesive population into two descendent populations), and/or the emergence of unique characters (Gregory 2008).[6] In any case, the internal nodes represent an event (speciation event) or species (extinct ancestor) that happened or existed at an earlier point in time. The tips of the branches represent the present and the rest of the tree represents the past. Regardless of the interpretation of the internal nodes, the connected lines of the tree diagram represent the

---

[5] For examples, see Mayr (1942: 285); Simpson (1951: 148).
[6] See Maddison and Maddison (2000: 37 ff.) and Podani (2013) for discussions of different interpretations of phylogenetic trees.

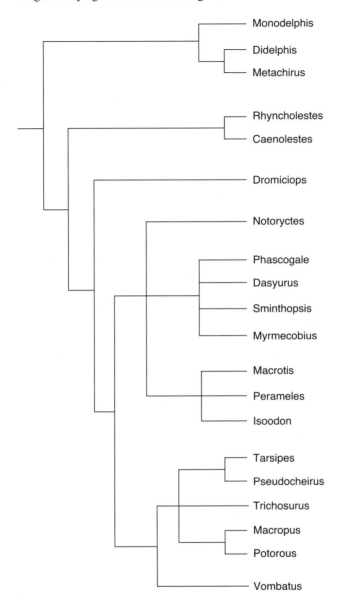

Figure 10.1 **Plain marsupial tree**
Source: Nilsson et al. (2010). Please see Figure 10.2 for further
source information.

pathways that eventually led to the currently existing species. In a phylogenetic tree that represents all living beings, one could trace all lines back to the so-called last universal common ancestor (LUCA). Since the root of the plain tree is on the left and the tips of the branches on the right, its timeline runs from left to right.[7]

The basic structure of the marsupial tree (Figure 10.1) depicts the following phylogenetic narrative of origin and divergence that is similar for all phylogenetic trees: the marsupial clade originated, and over time the ancestral population underwent character changes. Then, the ancestral population diverged into separate populations that again underwent character changes. One of these populations eventually evolved into Didelphimorphia, with three extant species, and the other population underwent further speciation events. Further character changes and the next speciation event occurred and separated the population that eventually evolved into Paucituberculata, from the population that evolved into Microbiotheria and the Euaustralidelphian orders.[8]

The filigreed tree (Figure 10.2) is the central element of the paper by Nilsson et al. (2010) and was created by Jürgen Schmitz, the project's principal investigator. To create the filigreed tree, he added images of seven marsupials as representatives of each of the orders to the basic structure (e.g., the order Diprotodontia is represented by a kangaroo). The names of the marsupial orders were added in grey (red in the original figure); the phylogenetically informative retroposon insertions are shown as white dots and different shading was used for the South American and Australasian lineages. The grey lines represent South American and the black lines represent Australasian marsupial lineages, which is made clear by additional images of the continents South America and Australia. The names Australidelphia and Euastralidelphia were also added to the plain tree.[9]

With the main narrative and target audience in mind, Schmitz first created the diagram and then constructed the text to provide more detailed information and explanation (Schmitz, personal communication, 11 April 2018). While the main function of the diagram is to depict a 'narrative of nature' (the evolution and spread of the marsupial clade), some visual elements have a dual function and also represent the researchers' narrative of science (what the scientists did to get the results).[10] Representations of the retroposons, for example, show how many retroposons are shared by members of a clade but also tell the reader that

---

[7] This example shows only one of many ways of arranging phylogenetic trees. There are also vertical phylogenetic trees with the root at the bottom or at the top (like Darwin's famous tree diagram in *On the Origin of Species* (1859); see section 10.4 for other examples) and circle trees with the root at the centre and the tips at the outer edge (Gregory 2008: 126; Baum and Smith 2012: 48).

[8] See Morgan (2017) for detailed discussions of narrative ordering.

[9] See Morgan (Chapter 1) and Hajek (Chapter 2) for discussions of time and time ordering in the historical sciences'.

[10] See Meunier (Chapter 12) on the distinction between a 'research narrative' and a 'narrative of nature'.

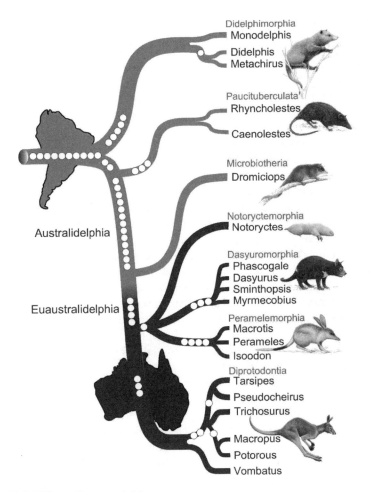

Figure 10.2 **Filigreed marsupial tree**
The original caption for Figure 10.2 is: 'Phylogenetic tree of marsupials derived from
retroposon data. The tree topology is based on a presence/absence retroposon matrix
(Table 1 https://journals.plos.org/plosbiology/article/figure/image?download&size=ori
ginal&id=info:doi/10.1371/journal.pbio.1000436.t001) implemented in a heuristic
parsimony analysis (Figure S3 https://doi.org/10.1371/journal.pbio.1000436.s007). The
names of the seven marsupial orders are shown in red, and the icons are representative of
each of the orders: Didelphimorphia, Virginia opossum; Paucituberculata, shrew opossum;
Microbiotheria, monito del monte; Notoryctemorphia, marsupial mole; Dasyuromorphia,
Tasmanian devil; Peramelemorphia, bilby; Diprotodontia, kangaroo. Phylogenetically
informative retroposon insertions are shown as circles. Gray lines denote South American
species distribution, and black lines Australasian marsupials. The cohort Australidelphia is
indicated as well as the new name proposed for the four 'true' Australasian orders
(Euaustralidelphia)' (Nilsson et al. 2010: 4).
Source: https://journals.plos.org/plosbiology/article/figure/image?download&size=ori
ginal&id=info:doi/10.1371/journal.pbio.1000436.g002

retroposons were used as characters for the phylogenetic analysis. Some of the visual elements were added to make the diagram look more appealing and raise the readers' interest. For this purpose, Schmitz hired a professional artist to draw pictures of marsupials. The main function of the additional elements, however, is to create an image that 'speaks for itself', meaning that the informed reader understands the central argument of the paper just by looking at the diagram (Schmitz, personal communication, 11 April 2018).[11]

Schmitz created a diagram that emphasizes the most important findings of the analysis, namely that there is a clear divergence between Australasian and South American marsupials, that Microbiotheria is more closely related to South American marsupials than to Australasian marsupials and that the four Australasian orders share a single origin with Microbiotheria suggesting one single migration event from South America to Australia (Nilsson et al. 2010). Nilsson et al. take the finding that all Australasian marsupials share four retroposons that are not present in *Dromiciops gliroides* (the only extant species of Microbiotheria) as evidence that Microbiotheria is more closely related to South American marsupials than to Australasian marsupials. In the filigreed tree, these retroposons are represented as four white dots located at the transition area from grey to black. Schmitz emphasized the divergence between Australasian and South American marsupials by using grey lines for South American lineages and black lines for Australasian lineages. The analysis by Nilsson et al. suggests that the species *Dromiciops gliroides*, the only survivor of the order *Microbiotheria*, is not nested within the Australasian orders. Based on these findings, the researchers suggest nomenclatural changes and 'propose the new name Euaustralidelphia ("true Australidelphia") for the monophyletic grouping of the four Australasian orders Notoryctemorphia, Dasyuromorphia, Peramelemorphia and Diprotodontia' (Nilsson et al. 2010: 4–5). The way the tree diagram was arranged horizontally instead of vertically, with South America to the left of Australia, visually represents the migration event from South America to Australia. Since the filigreed tree (Figure 10.2) represents both time and geographical information, it is read from top-left to bottom-right. Interestingly, the continents are represented in their current state as separate land masses, although the migration event supposedly occurred when South America, Antarctica and Australia were still connected by land bridges (Schmitz 2010).

By analysing the text of Nilsson et al. (2010), it becomes clear that the migration narrative is not only created on the basis of the phylogeny but through integration with narratives from other fields such as palaeontology and geology. The following excerpt illustrates that the group incorporated the fossil record and biogeographical evidence into the phylogenetic narrative.

---

[11] The diagram's caption contains the detailed information on what is represented and how.

The fossil Australian marsupial *Djarthia murgonensis* is the oldest, well-accepted member of Australidelphia. Thus, combined with the lack of old Australidelphian fossils from South America, the most parsimonious explanation of the biogeography of Australidelphia is of an Australian origin. However, the poor fossil record from South America, Antarctica, and Australia does not exclude that *Djarthia*, like *Dromiciops*, could be of South American origin and had a pan-Gondwanan distribution. (Nilsson et al. 2010: 3)[12]

An integration of the phylogenetic narrative with narratives from other historical sciences like palaeontology and geology is facilitated by similar narrative conventions of a central subject (protagonist) that changes over time (Hopkins, Chapter 4; Huss, Chapter 3; see also section 10.3, below). The fact that researchers in other fields follow the same narrative conventions makes it easy to integrate heterogenous elements to form one coherent narrative. Broadening a narrative by integrating it with narratives from other fields is one way of creating a thicker scientific narrative (see Paskins, Chapter 13).

The integrated narrative that is represented by the filigreed marsupial tree can be phrased like this: the marsupial clade originated and over time the ancestral population underwent character changes. Then, the ancestral population diverged into separate populations that again underwent character changes. One of these populations eventually evolved into Didelphimorphia, with three extant species, and the other population underwent further speciation events. Further character changes and the next speciation event occurred and separated the population that eventually evolved into Paucituberculata from the other population, that again underwent character changes over time. Then the next speciation event occurred and one of the descendent populations eventually evolved into Microbiotheria. Members of the other descendent population migrated from South America to Australia, which constituted the origin of the superorder Euaustralidelphia.[13]

While specialists can read these narratives directly off the diagrams, the untrained reader needs additional information to understand the trees' narratives. To be sure, the filigreed tree's caption provides information on how to interpret the added visual elements, but the authors assume that the reader understands the basic structure without further information. To be able to read the diagram as

---

[12] Here, the scientists refer to an extinct species that is not represented by the tree diagram because they used only extant organisms for their analysis. The fossil *Djarthia murgonensis* is part of a palaeontological narrative that Nilsson et al. (2010) use to extend their phylogenetic narrative.

[13] Neither the text nor the diagram by Nilsson et al. (2010) provides details of the migration narrative. However, Schmitz published a more comprehensive narrative of marsupial migration elsewhere (Schmitz 2010). From this publication we learn that he indeed believes that speciation has occurred through migration from the 'South American' part of Gondwana to the 'Australian' part of Gondwana via land bridges instead of through geographical separation when the supercontinent Gondwana split up. This view is also illustrated by the text excerpt where he refers to the competing hypothesis of a pan-Gondwanan distribution of marsupials (Schmitz 2010: 7).

a narrative, the reader thus relies on background knowledge and needs to be familiar with the specialist conventions (see Andersen, Chapter 19; Merz 2011; Vorms 2011). The resemblance of the basic structure of phylogenetic trees with human family pedigrees and the cultural practice of representing kinship and genealogy with tree images and branching diagrams might facilitate the understanding of phylogenetic trees as representations of shared ancestry (Gregory 2008; Hellström 2011; Russell 1979). However, there are common misunderstandings in the interpretation of phylogenetic trees that show how difficult it is for non-specialists properly to understand phylogenetic trees (Meir et al. 2007).[14]

10.3   How Phylogenetic Trees Represent Narrative Explanations

So far, I have established that specialists read plain and filigreed phylogenetic trees as narratives. In this section, I argue that the informed reader can also derive narrative explanations from them. Or, from the perspective of the author, phylogenetic trees are used to represent narrative explanations.

Arguably, not every narrative is explanatory. However, when they offer solutions to puzzles, narratives qualify as explanations (Morgan 2017; Roth 1989). As Mary Morgan puts it, 'what narratives do above all else is create a productive order amongst materials with the purpose to answer why and how questions' (2017: 86). In the case of the phylogenetic analysis of marsupials, the material at hand (molecular sequences) was ordered in terms of similarity to answer the question of how the seven marsupial orders are related to each other (Nilsson et al. 2010). The phylogenetic tree of the marsupial clade represents an answer to this question. The scientists were particularly interested in the phylogenetic position of Microbiotheria. This relationship, however, is only one of the many evolutionary relationships that are represented in the tree diagram. In this sense, the diagram stands for itself because it is more detailed than the text and includes relationships that are not mentioned in the text. Thus, the visual narrative is more comprehensive than the written one that focuses only on the most disputed phylogenetic relationships.

In addition to being answers to puzzles, narrative explanations show 'what happened at a particular time and place and in what particular circumstances' (Gardiner 1961: 82). Thus, they are mostly concerned with token events – for example, a particular war or revolution – not with finding regularities of how wars or revolutions come about. They don't merely explain an occurrence but show how things came to be as they are by referring to events that happened at an earlier point in time (Beatty, Chapter 20).[15] To be sure, the marsupial tree

---

[14]   See also Gregory (2008); O'Hara (1992); Omland, Cook and Crisp (2008); and section 10.4.1, below.

[15]   See also Ereshefsky and Turner (2020); Little (2010: 29); Martin (1986: 72–73); Roth (2017: 44).

represents the origin and evolution of a particular clade and its exact branching pattern is probably unique to the marsupial clade. However, the tree diagram also represents type phenomena like speciation and emergence of traits. Moreover, it is 'exemplary as a concrete problem solution that can be extended to give an explanation to similar phenomena elsewhere' (Morgan 2017: 94). Phylogenetic trees not only represent explanations of the origin and evolution of biological taxa but are also used in other disciplines such as linguistics to represent the origin and diversification of languages (Atkinson and Gray 2005).

The events that are included in a narrative explanation are events that made a difference to the outcome (Beatty 2016; 2017). In the temporal series, the outcome B is contingent upon at least one previous event A in the sense that B could not have happened if A had not happened in the past (Beatty 2016). To be more precise, B is contingent upon the pathway that connects B with previous events (Desjardins 2011). In the tree diagram, the difference-making events are represented as a temporal series of internal nodes (speciation events) and lines that connect the nodes (emergence of traits). The tree diagram by Nilsson et al. (2010) thus represents an explanation of how recent marsupial species came to be as they are by referring to speciation events and divergence that happened earlier in time. The existence of recent marsupial species is contingent upon the existence of their ancestors and the evolutionary pathway that eventually led to their occurrence. However, the tree diagram is rather thin on detail because it contains no exact information on ancestors or difference-making events such as speciation events (except for the migration event from South America to Australia) and loss or acquisition of traits.

Narrative explanations also include narrative sentences that 'give descriptions of events under which the events could not have been witnessed, since they make essential reference to events later in time than the events they are about' (Danto 1985: xii; see also Roth 2017). An example is: 'The Thirty Years War began in 1618' (Danto 1985: xii). Thus, only in hindsight, when we know how the narrative ends, are we able to identify its beginning and unfolding (Martin 1986: 74). Narrative sentences can be derived directly from the tree diagram by Nilsson et al. (2010), and phylogenetic trees in general. For example, 'Microbiotheria originated before Notoryctemorphia, Dasyuromorphia and Peramelemorphia' or 'The first divergence within the marsupial clade gave rise to Didelphimorphia'.

Narrative explanations are characterized as 'connected account[s] of [an] entity's development in time' (White 1963: 4), or, as Roth puts it (2017: 45), a narrative is 'unified by showing the development of a subject over time'. These statements express a notion of coherence that is captured by the concept of *central subject* (White 1963; Hull 1975; Ereshefsky and Turner 2020) and corresponds to the concept of *protagonist* in narratology (see Hajek, Chapter 2). The role of central subjects is 'to form the main strand around which the historical narrative is woven' (Hull 1975: 255). Examples for central subjects are Napoleon (Hull 1975: 262) and the Hawaiian Island archipelago (Ereshefsky and Turner 2020). The central subject in Nilsson et al.'s (2010) narrative is the marsupial clade because this entity forms the main strand of the evolutionary narrative. The migration event from South America to Australia is singled out as a particularly important event in the life of the clade because it led to the formation of a new superorder.

To be sure, the scientists present their explanation of the origin and evolution of marsupial orders in the text of the research paper; however, narrative explanations are also represented by phylogenetic trees in a more immediate manner. I have shown that they represent answers to a puzzle, temporal series with difference-making events, token phenomena, and revolve around a central subject. I have also argued that an informed reader can derive narrative sentences directly from the diagram. The basic structure depicts all elements of a narrative explanation discussed in this section and thus already represents a narrative explanation (phylogenetic narrative). The filigreed tree with additional elements (e.g., images of continents), however, represents a broader narrative explanation about migration. In the following section, I discuss examples that show further ways of modifying phylogenetic trees to represent narrative explanations.

10.4   Use of Phylogenetic Trees in Different Contexts

The use of phylogenetic trees extends beyond biological systematics. In this section I will give two examples of the use of phylogenetic trees in other fields to show their functions in different contexts. Like the marsupial tree (Figure 10.2), the diagrams discussed here are filigreed trees that include different types of additional textual and graphic elements. These examples illustrate two things. First, phylogenetic narratives are not always represented in the same form. Even though the diagrams discussed in this section are based on a branching structure, they are arranged and read in different ways, particularly with respect to the time axis. Second, filigreed trees are modified to represent narratives that extend beyond evolutionary histories or common ancestry and differ in terms of narrative content.

10.4.1   *Phylogenetic Trees in Museums*

Phylogenetic trees can be found in many museums, science centres, zoos, aquariums and botanical gardens. The phylogenetic tree entitled 'vertebrate diversity' (Figure 10.3) is part of a permanent exhibit at the University of Kansas Natural History Museum. The diagram contains a vertical tree diagram (root at the bottom), with schematic images of species at the eight tips (extant

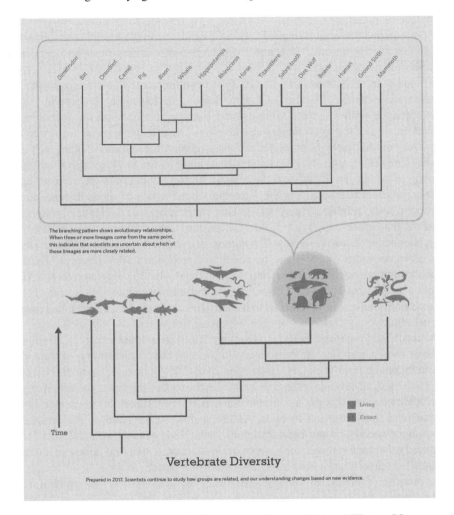

Figure 10.3 **Vertebrate tree at the University of Kansas Natural History Museum**
Reproduced, with permission, from the Kansas Natural History Museum.

and extinct species represented by different shades).[16] The tips represent fishes, birds plus reptiles (in one group), mammals and amphibians. Unlike phylogenetic trees in scientific papers, this tree diagram includes an arrow indicating temporal directionality from bottom to top. The top of the diagram shows an

---

[16] See Morgan (Chapter 1) for a discussion of the relationship between narratives and other forms of scientific representation.

extended mammal branch with seventeen tips. The tree designers included both common and scientific names of species. In addition to the legend with the two colours that represent living and extinct species, the diagram also contains a short explanatory text.

One of the functions of Figure 10.3 is to communicate scientific research to a broad audience. The tree diagram depicts phylogenetic relationships in accordance with scientific findings, and the explanation, that the branching pattern represents evolutionary relationships, enables people who are completely unfamiliar with phylogenetic trees to get a basic understanding of the diagram. The explanatory text states that some of the phylogenetic relationships are unresolved: 'When three or more lineages come from the same point, this indicates that scientists are uncertain about which of those lineages are more closely related' (Figure 10.3). This either means that scientists disagree about the respective phylogenetic relationships or that phylogenetic analyses produced inconclusive results. The authors also mention that new evidence can lead to revisions of phylogenetic relationships (Figure 10.3, bottom). These additional remarks help the audience understand what the diagram represents, but also informs about the character of scientific research and its results. The schematic images of vertebrates can easily be understood by a broad audience including young children. Another important function of the vertebrate tree is to teach 'phylogenetic literacy' (Gregory 2008) to a broad audience. Studies have shown that there are misconceptions about the representation of time in phylogenetic trees (Gregory 2008; Meir et al. 2007; Omland, Cook and Crisp 2008). Instead of reading the time axis from the root of the tree to the tips, many students believe that the location of the tips is a representation of temporality and read time from left to right, assuming older species are on the left and younger species on the right (Gregory 2008: 134; Meir et al. 2007: 72). To avoid misinterpretations, the authors of the vertebrate tree thus added an arrow labelled 'time' that indicates the time axis from bottom to top.

Another function of phylogenetic trees in museums is 'to make links between specific exhibits and the broader tree of life' (MacDonald 2014). When scientists refer to the *tree of life*, they usually mean a phylogeny of all living beings, but also the concepts of common ancestry and biodiversity (MacDonald and Wiley 2012: 14). The bottom part of the vertebrate tree does not contain any details like species names because its main function is to show that all vertebrates are related to each other. The schematic images of different vertebrates depict the diversity within this group. Like many other phylogenetic trees in museums, the extended mammal branch of the vertebrate tree also includes humans. In contrast to other phylogenetic trees in museums or zoos, however, the human branch is not emphasized in any way and does not have a central position (see MacDonald and Wiley 2012 for examples). This way of representing humans in a phylogenetic tree allows visitors to see who our closest relatives are and at the same time communicates that humans are one species among many with no special position on the tree of life. In general, the arrangement of the branches might help to correct the common misconception of 'ladder thinking' with higher and lower species (Gregory 2008: 127–128; Kummer, Clinton and Jensen 2016: 393; O'Hara 1992).

Compared to the marsupial trees (section 10.2), the vertebrate tree shows an alternative way of representing phylogenetic narratives. The trees differ with respect to the direction of time and the taxonomic level of the central subject. The marsupial trees are arranged horizontally, with the root on the left, but the vertebrate tree's branching structure is arranged vertically, with the root at the bottom. Thus, time on the vertebrate tree is not read from (top-) left to (bottom-)right, but from bottom to top. While the marsupial trees show the evolutionary history of marsupial orders, the vertebrate tree represents the evolutionary history of four groups of vertebrates and a more fine-grained evolutionary history of mammals. This shows that phylogenetic narratives can be developed on different taxonomic levels. The trees also differ with respect to narrative content. The filigreed marsupial tree emphasizes an important turning point in the evolutionary history of the marsupial cade (migration event), but the vertebrate tree's narrative was developed to include a narrative of connectedness thorough common ancestry. Thus, the vertebrate tree is read both as a narrative of evolutionary history (emphasis on time) and as an ancestor narrative (emphasis on shared ancestry). Elements of self-reference in the diagram (arrow, explanatory text) enable readers to interpret and understand not only this particular phylogenetic tree but phylogenetic trees in general.

10.4.2   Phylogenetic Trees in Animal Rights Debates

In a flyer entitled 'Brother Chimp, Sister Bonobo: Rights for Great Apes!' published by the Giordano Bruno Foundation, the authors included a phylogenetic tree of great apes (Giordano Bruno Foundation 2011: 5; Figure 10.4). In contrast to most phylogenetic trees in scientific papers the tree in the flyer contains information on taxonomic ranks (e.g., superfamily, family, genus/species) at the nodes. The prevailing phylogenetic classification identifies chimpanzees (*Pan troglodytes*) and bonobos (*Pan paniscus*) as the closest extant relatives of humans (*Homo sapiens*), with chimpanzees and bonobos as members of the genus *Pan* and humans as members of the genus *Homo*. Given the close phylogenetic relatedness of humans, chimpanzees and bonobos, the authors argue that the latter two should be placed into the genus *Homo* and renamed *Homo troglodytes* and *Homo paniscus*, respectively (Giordano Bruno Foundation 2011: 4–5). Interestingly, this demand is already implemented in their great ape tree. The ultimate demand of the Giordano

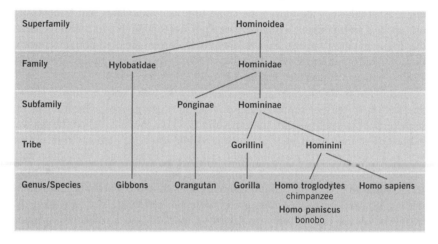

Figure 10.4 **Great ape tree**
Source: Giordano Bruno Foundation (2011). Reproduced, with permission, from
Volker Sommer original author and image maker.

Bruno Foundation, however, is not the renaming of chimpanzees and bonobos,
but the recognition of fundamental rights for great apes (Giordano Bruno
Foundation 2011: 6). According to the authors their updated classification
'would not only be scientifically consistent, it would also have psychological
knock-on effects – as it would deflate our exaggerated sense of importance and
motivate us to grant our closest relatives the respect they deserve' (Giordano
Bruno Foundation 2011: 5).

Like the vertebrate tree (section 10.4.1), the great ape tree is arranged
vertically, but with the root at the top. Similar to the marsupial trees (section
10.2), the vertebrate tree is also a phylogeny of a mammalian clade, but it is not
used to represent a narrative of the origin and evolution of great apes. Instead,
the great ape tree is read as a narrative of common ancestry of humans,
chimpanzees and bonobos. Thus, the emphasis of the great ape narrative is
not on the temporal aspect of evolution but on the genealogy[17] of great apes.
Like phylogenetic trees in scientific papers, the great ape tree does not include
ancestors, but the placement of taxonomic ranks at the nodes makes the
diagram look more like a human family pedigree that includes ancestor
names at the nodes. The narrative of common ancestry of humans, chimpanzees
and bonobos is also expressed in the title of the flyer that refers to chimpanzees
and bonobos as our brothers and sisters, implying that we share the same
'parents'. The authors also refer to common ancestry when they argue that

---

[17] See Berry (Chapter 16) for a discussion of genealogies.

humans, chimpanzees and bonobos should be placed in the same genus: 'Today it is an undisputed fact that humans are the closest living relatives of chimpanzees and bonobos. The genome of these three species differs only by a fraction – between 6.4 per cent and 0.6 per cent, depending on the methods of measurement. Some scientists would therefore like to unite them in a single genus, *Homo*' (Giordano Bruno Foundation 2011: 5).

The phylogenetic tree represents the scientifically recognized phylogenetic relationships (chimpanzees and bonobos as the sister species of humans) but not the prevailing scientific nomenclature. The renaming of chimpanzees and bonobos places them in the same genus as humans (Homo), thereby distorting the prevailing scientific classification that places chimpanzees and bonobos in the genus *Pan*. The tree diagram represents an explanation of why chimpanzees and bonobos should be renamed (Giordano Bruno Foundation 2011: 4–5). However, the main purpose of including the great ape tree in the flyer is not to represent scientific findings, but first and foremost to represent a political narrative that explains why fundamental human rights (e.g., the right to life, the right to individual liberty) should be extended to other great apes. The example of the great ape tree thus shows how political and scientific narratives are woven together and represented in a visual representation. To be sure, the common ancestry of humans, chimpanzees and bonobos is only part of the narrative that explains why the 'community of equals' should be extended beyond humans,[18] but the flyer discussed here focuses on this particular aspect of the argument (Giordano Bruno Foundation 2011: 6).

10.5   Filigreed Trees and Integrated Narratives

The construction process of the marsupial tree clearly shows that phylogenies are not descriptive chronicles. In fact, there are many decisions that potentially affect the outcome of the analysis such as the choice of characters, species, outgroup and method of data analysis. There are also fundamental assumptions about molecular evolution (e.g., retroposon insertions) that form the basis of phylogenetic analysis and the interpretation of the tree diagram. I have argued that a specialist audience reads phylogenetic trees, even plain ones, as evolutionary histories and that all phylogenetic trees represent narrative explanations. The scaffold view of phylogeny and evolutionary history as advocated by O'Hara and Griesemer is thus misleading because it implies that phylogeny is something prior to or separate from evolutionary histories. It is true that plain trees are scaffolds for more filigreed versions of trees, but not in the sense that

---

[18] Other reasons mentioned in the flyer are the 'complex mental landscape' of great apes that 'includes consciousness, emotions and sophisticated cognitive abilities' and the evolution of our moral sense (Giordano Bruno Foundation 2011: 6–7).

filigreed trees depict evolutionary histories while plain trees depict something prior to evolutionary histories. I have shown that the filigreed marsupial tree is read as a narrative that includes geographical aspects of the evolution of the marsupial clade, namely the divergence of South American and Australasian marsupials after a migration event. It is thus used to represent a coherent narrative that resulted from integration of a phylogenetic narrative with narratives from geology and palaeontology. The examples of the use of filigreed trees outside of academic evolutionary biology show that they are also used to represent narratives that extend beyond evolutionary histories of clades. These narratives are formed through integration of an ancestor narrative with political demands or integration of a phylogenetic narrative with a pedagogical narrative. All diagrams discussed in this chapter contain the basic branching structure of a phylogenetic tree but differ in narrative content and reading of the diagrams.[19]

Bibliography

Atkinson, Q. D., and R. D. Gray (2005). 'Curious Parallels and Curious Connections: Phylogenetic Thinking in Biology and Historical Linguistics'. *Systematic Biology* 54.4: 513–526.

Baum, D. A., and S. D. Smith (2012). *Tree Thinking: An Introduction to Phylogenetic Biology*. Greenwood Village, CO: Roberts & Company Publishers.

Beatty, J. (2016). 'What Are Narratives Good For?' *Studies in History and Philosophy of Science Part C* 58: 33–40.

(2017). 'Narrative Possibility and Narrative Explanation'. *Studies in History and Philosophy of Science Part A* 62: 31–41.

Bromham, L. (2016). *An Introduction to Molecular Evolution and Phylogenetics*. 2nd edn. Oxford: Oxford University Press.

Cleland, C. E. (2002). 'Methodological and Epistemic Differences between Historical Science and Experimental Science'. *Philosophy of Science* 69.3: 447–451.

(2011). 'Prediction and Explanation in Historical Natural Science'. *British Journal for the Philosophy of Science* 62.3: 551–582.

Currie, A. (2014). 'Narratives, Mechanisms and Progress in Historical Science'. *Synthese* 191: 1163–1183.

Danto, A. C. (1985). *Narration and Knowledge*. New York: Columbia University Press.

[19] My research was funded by the German Research Foundation (DFG) – 281125614/GRK2220. I would like to thank Jürgen Schmitz for his time, patience and insights into his work. I would also like to thank Mary Morgan, Jim Griesemer and an anonymous reviewer for valuable feedback and comments. Special thanks to the editors of this volume and the other contributors and participants of the writers' workshop for stimulating discussions. I also thank Oliver Höltker for his assistance. *Narrative Science* book: This project has received funding from the European Research Council under the European Union's Horizon 2020 research and innovation programme (grant agreement No. 694732). www.narrative-science.org/.

Darwin, C (1859). *On the Origin of Species by Means of Natural Selection, or the Preservation of Favoured Races in the Struggle for Life.* London: John Murray.

Desjardins, E. (2011). 'Historicity and Experimental Evolution'. *Biology and Philosophy* 26.3: 339–364.

Eldredge, N., and J. Cracraft (1980). *Phylogenetic Patterns and the Evolutionary Processes: Method and Theory in Comparative Biology.* New York: Columbia University Press.

Ereshefsky, M., and D. Turner (2020). 'Historicity and Explanation'. *Studies in History and Philosophy of Science* 80: 47–55.

Gardiner, P. L. (1961). *The Nature of Historical Explanation.* Oxford: Oxford University Press.

Ginzburg, C. (1979). 'Clues: Roots of a Scientific Paradigm'. *Theory and Society* 7.3: 273–288.

Giordano Bruno Foundation (2011). *Brother Chimp, Sister Bonobo: Rights for Great Apes!* www.giordano-bruno-stiftung.de/sites/default/files/download/greatapes2.pdf.

Gregory, T. R. (2008). 'Understanding Evolutionary Trees'. *Evolution: Education and Outreach* 1.2: 121–137.

Griesemer, J. R. (1996). 'Some Concepts of Historical Science'. *Memorie della Societàitaliana di scienze naturali e del Museo civico di storia naturale di Milano* 27: 60–69.

Harrison, P., and I. Hesketh (2016). 'Introduction: Evolution and Historical Explanation'. *Studies in History and Philosophy of Science: Biological and Biomedical Sciences* 58: 1–7.

Hellström, N. P. (2011). 'The Tree as Evolutionary Icon: TREE in the Natural History Museum, London'. *Archives of Natural History* 38.1: 1–17.

Hull, D. L. (1975). 'Central Subjects and Historical Narratives'. *History and Theory* 14.3: 253–274.

Kaiser, M., and D. Plenge (2014). 'Introduction: Points of Contact between Biology and History'. In M. Kaiser, O. R. Scholz, D. Plenge and A. Hüttemann, eds. *Explanation in the Special Sciences: The Case of Biology and History.* Dordrecht: Springer, 1–23.

Kummer, T. A., J. W. Clinton and J. L. Jensen (2016). 'Prevalence and Persistence of Misconceptions in Tree Thinking'. *Journal of Microbiology and Biology Education* 17.3: 389–398.

Little, D. (2010). *New Contributions to the Philosophy of History.* Dordrecht: Springer.

MacDonald, T. (2014). 'Horse Evolution: Updating Natural History Museum Exhibits with Trees'. Trees for Museums and Zoos. https://evolution.berkeley.edu/evoli brary/article/0_0_0/evotrees_zoos_07.

MacDonald, T., and E. O. Wiley (2012). 'Communicating Phylogeny: Evolutionary Tree Diagrams in Museums'. *Evolution: Education and Outreach* 5.1: 14–28.

Maddison, D. R., and W. P. Maddison (2000). *MacClade 4.* Sunderland, MA: Sinauer Associates, Inc. http://ib.berkeley.edu/courses/ib200/readings/MacClade%204%2 0Manual.pdf .

Martin, W. (1986). *Recent Theories of Narrative.* Ithaca, NY: Cornell University Press.

Mayr, E. (1942). *Systematics and the Origin of Species from the Viewpoint of a Zoologist*. New York: Columbia University Press.

Meir, E., J. Perry, J. C. Herron and J. Kingsolver (2007). 'College Students' Misconceptions about Evolutionary Trees'. *American Biology Teacher* 69: 71–76.

Merz, M. (2010). 'Designed for Travel: Communicating Facts through Images'. In P. Howlett and Mary S. Morgan, eds. *How Well Do Facts Travel? The Dissemination of Reliable Knowledge*. Cambridge: Cambridge University Press, 349–375.

Morgan, Mary S. (2017). 'Narrative Ordering and Explanation'. *Studies in History and Philosophy of Science Part A* 62: 86–97.

Nilsson, M. A., G. Churakov, M. Sommer, N. Van Tran et al. (2010). 'Tracking Marsupial Evolution Using Archaic Genomic Retroposon Insertions'. *PloS Biology* 8.7: 1–9. https://journals.plos.org/plosbiology/article/file?id=10.1371/journal.pbio.1000436&type=printable.

O'Hara, R. J. (1988). 'Homage to Clio, or, Toward an Historical Philosophy for Evolutionary Biology'. *Systematic Zoology* 37.2: 142–155.

    (1992). 'Telling the Tree: Narrative Representation and the Study of Evolutionary History'. *Biology and Philosophy* 7.2: 135–160.

Omland, K. E., L. G. Cook and M. D. Crisp (2008). 'Tree Thinking for All Biology: The Problem with Reading Phylogenies as Ladders of Progress'. *BioEssays* 30.9: 854–867.

Podani, J. (2013). 'Tree Thinking, Time and Topology: Comments on the Interpretation of Tree Diagrams in Evolutionary/Phylogenetic Systematics'. *Cladistics* 29.3: 315–327.

Roth, P. A. (1989). 'How Narratives Explain'. *Social Research* 56.2: 449–478.

    (2017). 'Essentially Narrative Explanations'. *Studies in History and Philosophy of Science Part A* 62: 42–50.

Russell, C. (1979). 'The Tree as a Kinship Symbol'. *Folklore* 90.2: 217–233.

Simpson, G. G. (1951). *Horses: The Story of the Horse Family in the Modern World and through Sixty Million Years of History*. New York: Anchor Books.

Schmitz, J. (2010). 'Die Wanderung der Beuteltiere'. *Biologie in unserer Zeit* 40.5: 294–295.

Tucker, A. (2014). 'Biology and Natural History: What Makes the Difference?' In M. Kaiser, O. R. Scholz, D. Plenge and A. Hüttemann, eds. *Explanation in the Special Sciences: The Case of Biology and History*. Dordrecht: Springer, 347–366.

Vorms, M. (2011). 'Formats of Representation in Scientific Theorizing'. In P. Humphreys and C. Imbert, eds. *Models, Simulations, and Representations*. New York: Routledge, 250–273.

White, M. (1963). 'The Logic of Historical Narration'. In S. Hook, ed. *Philosophy and History: A Symposium*. New York: New York University Press, 3–31.

Wiley, E. (1981). *Phylogenetics: The Theory and Practice of Phylogenetic Systematics*. New York: John Wiley & Sons.

IV

Interlude

11 Process Tracing and Narrative Science

Sharon L. Crasnow

Abstract

Process tracing is a familiar analytical tool in a number of sciences. Successful process tracing pulls together what is already known, believed or assumed and the various events, activities and entities in a case study in order to construct a narrative of the case. Several chapters in this volume offer accounts of narrative science that are explored through process tracing. These examples are analysed to reveal how various aspects of process tracing inform narrative and how narrative, in turn, aids process tracing in an iterative process of interpretation and reinterpretation of evidence, testing, development and revision of hypotheses, and the explanation of singular events.

11.1 Introduction

'Process tracing' is used in theorizing about, testing for and identifying causal mechanisms operating within a case. A number of chapters in this volume are concerned with such questions of causality, and deal with them through analysing the process through which some particular outcome occurred; and the accounts that they offer of the role of narrative are similar to the patterns of reasoning that are used in process tracing. For this reason, I have read them with process tracing in mind and I offer my reflections on that exercise.

My interest in how narrative and process tracing are related stems from exploring how case studies have been used in political science. Political science case studies are detailed investigations of some particular event, time period, region or country. A case study includes many different types of information – interviews, archives, journalistic accounts and other data (both quantitative and qualitative). Researchers typically organize and present this information in a narrative form. Some of the goals of case study research are explaining and understanding the occurrence of some particular event, testing a hypothesis about how an event came about and discovering similarities among this and other events that might allow for generalization. I have argued elsewhere that the narrative presentation of case studies is not coincidental to these goals, but rather aids in achieving them (Crasnow 2017). Narrative does this by giving the interconnections among the various entities, activities and events detailed in the case, indicating how some events are contingent on others. Narrative pulls together disparate evidence into a coherent whole. Thus the narrative through which a case is presented contributes to understanding the case, to seeing various aspects of it as evidence for causal claims, and our understanding of how the case is both different from and similar to other cases. In these ways, narrative contributes to the production of knowledge.

I begin this account of the relation of process tracing to narrative knowledge with an explanation of process tracing as practised in political science and then connect the method more directly to narrative through an analogy with the reasoning involved in solving a mystery. This is an analogy that appears frequently in the political science methodological literature. I next sketch an example of process tracing in political science, using it to illustrate the way that evidence is identified and used to support a claim that particular causal mechanisms are operating and how they shape the narrative of the case. I then turn to some examples chosen from the chapters in this volume. I begin with the historical sciences – specifically with geology – but I do not confine myself to those disciplines as a number of other chapters contain ideas about narrative science that are relevant to process tracing as well. I close with some thoughts on what might be learned from this exploration and a suggestion for further inquiry.

11.2   Process Tracing in Political Science

In political science, because a case study is primarily undertaken to investigate causality, the interconnections among the various aspects of the case are causal and the primary thread of the narrative is given through hypothesized causal mechanisms. Process tracing is tracing the workings of those mechanisms through identifying evidence that the mechanisms were operating. The mechanisms make sense of how the various aspects of the case fit together and so provide the overarching narrative ordering of the case.

Political methodologists Derek Beach and Rasmus Brun Pedersen offer the following account of process tracing: 'Process-tracing is a research method for tracing causal mechanisms using detailed, within-case empirical analysis of how a causal mechanism operated in real-world cases' (Beach and Pedersen 2019: 1). Beach and Pedersen clarify that a mechanism is not to be identified with the starting point or any of the intervening stages of the process. Mechanisms are not themselves causes, but rather are triggered by causes (Beach and Pedersen 2019: 3). In other words, they emphasize the dynamic nature of mechanisms together with the idea that process tracing is not a matter of picking out one or a few features that are connected causally, but rather of tracing a process from its inception to its conclusion, both showing how it operated in this case and identifying the evidence that it did so.

The question of how to think of mechanisms is not fully agreed upon among social scientists or philosophers of science, but there is enough overlap in accounts to support a characterization that captures features that are generally agreed on (Machamer, Darden and Craver 2000; Glennan 2002; Tabery 2004; Bogen 2005; Reiss 2008; Illari and Williamson 2012). Such an account of causal mechanism is as follows: mechanisms have parts – these parts may be identified as activities, events or entities, but are in some sense discrete (if only analytically); the parts are organized in some way – a mechanism has a structure; and there is an active element that constitutes the inter-relationship of the parts – all characterizations of mechanisms include the idea that an effect is brought about, produced, propagated or maintained through the inter-relationship of the parts (Crasnow 2017: 8).

The mechanisms relevant to political science accounts are psychological, social and political. Process tracing analytically unpacks mechanisms allowing the researcher to seek evidence that the component parts of any particular mechanism are present and operating in the case. The mechanism is the means through which these parts are understood as part of the same process. Acknowledging the integrative nature of causal mechanisms involves recognizing that process tracing is not just a matter of collecting evidence of the parts through the examination of the case but involves finding evidence that the various parts are connected to each other as well, that is, indicating how they are parts of the whole (the mechanism), which is more than the collection of its parts.

When process tracing is focused on examining the operation of a mechanism in a specific case, it may be probative or exploratory – does the proposed mechanism provide a plausible account of the case? If the answer is yes, then further investigation of how that mechanism operates in the case is warranted. Or it may be that the case is investigated as a means of testing the hypothesized mechanism for this case. In such circumstances, the success of the test is a way of garnering support for the hypothesized mechanism as the best (or better) explanation of the case. If the test fails, the failure motivates a search for alternative accounts – other mechanisms that are consistent with the observed traces. These may be other known mechanisms or may suggest new hypotheses. In this way process tracing informs theory development. The case might also be investigated to see why a mechanism that was expected to operate failed in this particular context – in which case, process tracing might suggest the limits of theory or motivate theory revision. Similarities of this case to other cases in which known mechanisms operate may suggest new candidates for consideration in this case. In addition, when researchers are confident of their account they may seek other similar cases in which the mechanism operates.

In order to seek traces of a process, some sense of what that process might be – for processes that are causal mechanisms, what mechanism might be operating – already has to be on the table. It might be only vaguely sketched out, to be filled in later should it prove promising, or it may already be well-developed theoretically and investigated elsewhere. The hypothesis serves to guide the researcher towards the appropriate sorts of intervening factors expected to be present if that mechanism was indeed operating. These are the entities, activities and events that are elements of the causal mechanism or indicators of its operation (recognized as indicators because of other things that are known or believed). Process tracing is undertaken to see if these elements are present. When theorizing (developing a hypothesis), the various elements of the case may be seen to resemble elements of other known mechanisms, suggesting that the same or a similar mechanism may be operating in this case. Alternatively, expectations produced by such similarities may be overthrown motivating a search for differences that suggest alternative causal hypotheses. Finally, which activities, entities or events are relevant to the case – what the boundaries of the case are and what the case should be understood as a case of – depend in part on how they fit the hypothesized causal mechanism and what the outcome of the mechanism is understood to be, which, in turn, depends on the research question that guides the investigation of the case.

11.3   Solving Mysteries

Political methodologists often use the analogy of a detective story to illustrate how process tracing works. The elements of the detective's case are the traces of the mechanism that process tracing is seeking. They are like clues of the sort that a detective might pick out in trying to solve a murder mystery. In classic murder mysteries, the murder itself has not been directly observed but there are clues (traces) that suggest what happened and who was responsible. As the detective finds out about the movements of the suspects, a timeline is constructed that suggests certain hypotheses about what might have happened and ultimately who among the suspects is guilty. The detective might start with standard hypotheses suggested by a body of knowledge about crimes of this type. For example, it is known that most victims are killed by those close to them or that often the person who discovers the body is the murderer. These hypotheses can be tested against what is discovered in the case as the detective follows various implications of the hypothesis together with what is known and makes predictions about what other traces might be found that would be consistent or inconsistent with that hypothesis.

Information about the psychology of the subjects – specifically about their relationships with each other and their possible motivations – is put together with the timeline. As these elements are pulled together, they further shape hypotheses and so the study of the case supports both the development and testing of those hypotheses. These hypotheses are hypotheses about who committed the crime, but that determination also depends on knowing how and why it was done. Identifying the culprit requires pulling together a variety of evidence on means, motive and opportunity.

Each of these lines of investigation might be pursued separately. If a suspect has an alibi for the time of the murder this is a clue that counts against her being the murderer – she did not have the opportunity. If a will is discovered that provides a motive, this may suggest a new suspect. If the murder weapon is found, this provides evidence of the means. Each of these aspects of the crime has its own traces (clues that are relevant to who, how and why), but the overarching account of the production of the outcome requires putting these all together.

I am belabouring this analogy for several reasons. First, it is an analogy that appears repeatedly in the political science literature on process tracing (e.g., 'The Adventure of Silver Blaze' is used by Beach and Pedersen in the appendix of their 2019 book). Second, and relatedly, it illustrates how process tracing supports the various different goals of case study research previously mentioned: explaining and understanding a particular event, testing a hypothesis about how an event came about and seeking similarities and difference with known cases (although the goal is to solve *this* murder, it is not unlikely that something more general about murders might emerge from the investigation). Third, it suggests that process tracing is an iterative process. Proposed hypotheses are discarded or revised in response to discovered clues (traces of the process through which the murder was accomplished). Reinterpretation of evidence may occur when new information comes to light (perhaps the alibi is found out to be a lie after more careful investigation). The mystery story analogy shows how a narrative works not only to suggest hypotheses and test them, but also how working back and forth between hypothesis and evidence to pull together a *backstory* (Beatty, Chapter 20) is often part of what is required to understand how and why something happened. When new information becomes available, old evidence may need reinterpretation in order to achieve a coherent whole – an account of who did the murder, in what way, and why.

Process tracing – the tracing of the causal mechanism that connects an initial causal factor with an outcome of interest – functions as an important analytical tool for identifying and organizing evidence in an iterative process and not only as a tool for testing hypotheses. Process tracing involves the interplay of what we already know, believe or assume, the hypotheses we are considering (suggested by what we know believe or assume) and the various traces (events, activities and entities) in the case that are seen as evidence when brought into relation with the hypothesis. Consequently, the traces only become evidence relative to some particular hypothesis and against the background of other knowledge, beliefs and assumptions. The integrative and iterative nature of process tracing provides one way of constructing a narrative of the case.

11.4   An Example from Political Science

Elisabeth Wood (2003) investigates the civil war in El Salvador in order to solve what she takes to be a puzzle about the case: why was there broad participation in the insurgency even though the costs of participation were so high and the rewards so minimal? Most alternative explanations postulate particular ways in which insurgents are motivated by expected gains (or avoiding higher costs). She argues that such accounts do not adequately track the pattern of participation in El Salvador. 'The relevant literature on revolutions, collective action, and social movements provide some guidance but not adequate answers to the puzzle of high-risk collective action in the Salvadoran context' (Wood 2003: 10). Participation, while broad, was not universal. Only about a third of the poor, rural class (*campesinos*) participated in the insurgency – a large enough rate of participation to have significant effect, but standard hypotheses, such as Marxist accounts of class struggle, do not square with the large number of 'free riders'. In addition, the level of participation is puzzling given the high level of risk and the minimal reward.

To answer her research question, Wood looks more closely at the difference between those who participated (approximately one-third of the rural poor) and those who did not. Her primary evidence is extensive interviews (more than 200) over multiple years both during and after the civil war. Using this information, together with details about the timing of particular events during the civil war, when they happened in relationship to other events, and the varying levels of participation, she concludes that three psychological mechanisms were responsible for collective action in El Salvador. She calls the first 'participation' a term that she defines more precisely than mere involvement in the insurgency. Participation in this sense is the desire to be involved in activities that reflect moral commitment. She identifies the influence of Liberation Theology on *campesinos* as fuelling this motivation. The second is 'defiance'. The government response to strikes was believed by many *campesinos* to be an overreaction to legitimate means of protest for unfair working conditions (strikes and peaceful demonstrations). This perception fuelled and justified defiance as a motivator of collective action. The third is what she calls 'pleasure in agency'. Participants reported a pride and sense of authorship in having been involved in making history.

Wood traces the operation of each of these mechanisms through patterns of responses to her interview questions, together with specific documented instances of collective action and government response that occurred through the decade of civil war. For example, the repressive government response to strikes increased the perceived risk of participation (and so further maximized the cost and for relatively small benefit) and yet resulted in increased participation in the insurgency. Interview responses indicate that the repressive efforts of the government were perceived as unjustified and met with outrage and defiance.[1]

Her book-length case study provides an account of the civil war through these and other mechanisms, incorporating particular events and shifts in strategy in response to such events (increased repression on the part of the regime resulting in a shift from political mobilization to armed insurgency). Fundamentally, the argument is that the narrative of the case told through the psychological (emotional and moral) mechanisms that Wood identifies provides a better account than alternatives.[2] The traces or clues to the importance of the emotional and moral factors are primarily in the results of her interviews.[3] These responses yielded recurring themes: injustice of pre-war land distribution; desire for land; the contempt with which they had been treated; brutality of the government responses to non-violent strikes and demonstrations; fear during the war; suffering of their families; post-war assertion of political and social equality; authorship of changes; and pride in participation (Wood 2003: 18).

While I only sketch Wood's argument here, it is worth noting several ways in which process tracing is operating. There is, first of all, the tracing of evidence that the psychological mechanisms that she postulates were operating. She notes these in both their appearance in interviews with those who were participants in the insurgency and their absence in those interviews with those who were not. Additionally, responses appear to appeal to these mechanisms as motivations for action, revealing causal connections.

While her goal is to explain the case through the operation of these mechanisms (to explain a singular case), she is also using the case to develop hypotheses. We can see this particularly well in her development of the notion of pleasure in agency which results from an interpretation of a number of themes in interviews. She is also engaging in theory revision or augmentation since she does not entirely reject material explanations in the literature on insurgency and collective action but does find them inadequate. Finally, she is also making use of background knowledge against which she identifies the puzzle that she wants to address but also through which she recognizes as evidence particular events that occur in her case – for example, the escalation of insurgent activity in the wake of violent

---

[1] And putting them together brings about 'narrative closure'. See Hajek's introduction, Chapter 2.

[2] Wood does not reject the idea that other mechanisms were operating but only that without making reference to the emotional and moral features that emerge out of her interviews the accounts are limited. Most importantly, they do not make sense of the differences between those who participated and those did not, whereas her account accommodates the two-thirds non-participation.

[3] For example, she had volunteers from one of her study areas participate in a map-drawing workshop where they produced maps of the farmlands in their area showing use of the land before and after the civil war.

government repression following strikes and peaceful protests in the early 1980s. Escalation of resistance is one of the known responses to repression. Finally, Wood offers a coherent narrative of the civil war in El Salvador that solves the puzzle that she originally saw in the case.

These are some of the ways that process tracing works in political science.[4] Do we see something like this elsewhere in the sciences? Are there ways that narrative functions elsewhere in the sciences that might help understand how narrative and process tracing function together better? I turn to the chapters of this volume with these questions in mind.

11.5   Tracing A Singular Event: The Rupture Process of the Tohoku Earthquake

I begin with Teru Miyake's analysis of research on the 2011 Tohoku earthquake, in Chapter 5, that I think bears the greatest similarity to the example of Wood's use of process tracing. He opens his chapter with what he describes as a narrative of the 'rupture process' of this particular earthquake. As he notes, each earthquake has its unique rupture process – much as each insurgency has its own trajectory.

The rupture process of this quake (as with all quakes) is a complex sequence of interconnected events. Calling it a process indicates that it is more than a chronology of events. We can see from the paragraph that begins Miyake's chapter that this sequence is one in which the events are causally connected. The origin of the Tohoku earthquake is a 'wide megathrust fault' that ruptures at 'a frictionally locked region in the central portion' of the fault. The description of the fault (wide megathrust fault) carries causal implications, as does the information that this region of the fault was frictionally locked. The descriptions are dynamic. This initial rupture 'failed to arrest', 'continuing to expand for 150 s, spreading over the full width of the boundary and along its length for 400 km' (Miyake, Chapter 5, quoting Lay 2018: 4–5). This description is not just what happens next but how it is connected to what went before and, because the expansion is described as continuing, what happens after the initial rupture. As the account continues, the events are sequenced and positioned in causal relation to each other through terms that would be familiar to geologists as having causal import. This sequencing is supported by a variety of evidence, which we can think of as traces of the earthquake. The various stages identified in the paragraph that summarizes the rupture process appear to be components of the specific mechanism that accounts for *this* earthquake.

These rupture processes cannot be directly observed but are reconstructed from 'traces' – among them seismographic data, permanent shifts in the earth's

---

[4] Another example from political science appears in Crasnow (2017).

surface and data from tsunamis. While Miyake notes that the lack of direct observation of the fault is one of the factors that makes it difficult to reconstruct the rupture process and to study earthquakes more generally, this sort of problem is far from unique to earthquakes, as we have seen in the previous discussion. Miyake's account of how these difficulties are tackled – how these traces are recovered and become evidence for the rupture process – involves narrative in two ways that he identifies.[5] The first involves determining how to treat the seismographic data as reliable evidence. One way this is accomplished is through a technique relied upon in the discipline (slip inversion) and its use to inform source models. Determining reliability is accomplished through what Miyake calls earthquake '*rupture narratives*'.

While I do not want to discuss these narratives in terms of process tracing, it is worth pointing out that this example illustrates how data does not directly speak as evidence but comes to be understood as evidence through its relationship to other background beliefs, assumptions and knowledge. There is a similarity here with Wood's account in that she takes one chapter of her book to discuss her methods – most particularly the use of interviews. These interviews involved the interviewees recalling events from the past and hence their use as evidence depends on knowledge, beliefs and assumptions about how memory functions. Wood argues, for example, that memories of particular types of events – 'those that rank as highly intense (in a variety of cognitive and biological measures) tend to be better remembered' (Wood 2003: 33).

The second type of narrative that Miyake discusses – 'integrating narrative' – displays a number of the characteristics of iteration and integration that I have focused on in the discussion of process tracing in section 11.4. In Miyake's account, integrating narratives pull together a variety of seismological information including evidence (data that appears to be most stable) from source models, but also what is known about the movement of the earth, data from tsunamis, information about the history of the fault (past earthquakes) and seismic events immediately preceding the earthquake. Integrating narratives of the 2011 Tohoku earthquake are revised and adjusted as new information becomes available or is reinterpreted.

Miyake works through examples of how the introduction of new techniques and data produced results that were in conflict with slip inversion results. These conflicts are tackled through understanding what each additional technique and data set reveals and reconciling the accounts by adjusting the integrating narrative to reflect the new understanding. As Miyake points out, this, in turn, raises new questions – specifically whether the particular process that

---

[5] Miyake also identified 'research narratives', which I do not discuss, but which are related to Meunier's distinction (Chapter 12) between a 'research narrative' and a 'narrative of nature', recounting what happened.

gave rise to this data is unique to this particular rupture or is due to features of the fault (and hence relevant to future ruptures). When traces do not fit current understanding (as in the case of Wood's consideration of available explanations for collective action and insurgency), this lack of fit motivates new research questions. The drive to integrate the data into a coherent whole – to show how the events are initiated and proceed through various stages to a particular outcome – is disrupted by what does not fit and calls for a new or revised account. Miyake notes that such events raise open questions to be addressed through the iterative process of creating an integrating narrative. It directs researchers in their search for further evidence – evidence that either the event is unique or that it is based in characteristics of the fault. What counts as evidence will depend on what else is known (about the fault, about the geology of the region more generally, as well as geological theory); some of what is learned from the Tohoku earthquake may alter what is believed to be known and thus change how other traces are interpreted.

In the typical political science case, the search for traces is guided by the hypothesized mechanism (although, as we have seen in the Wood example, that mechanism is often complex). In the case of the earthquake, the mechanism is also highly complex. While the data speak primarily to the component parts of this complex process, the goal is to put these parts together into a whole – to narrate the rupture process of this earthquake.

11.6   Testing a Hypothesis

In another example (Chapter 4), Andrew Hopkins gives an account of the search for an explanation of a rock formation in north-west Scotland – the Stac Fada Member. In section 11.5, I focused on process tracing as tracing the operation of causal mechanisms in a particular case, but process tracing also serves as a means of testing hypotheses. Often these two modes of inquiry are intertwined. We see that in this chapter, where several different hypotheses are considered as explanations for a singular event – the formation of the Stac Fada Member.

Hopkins's account discusses the understanding of the geology of a particular region of Scotland in the nineteenth century. At that time, the proposed causal mechanism through which the geology was explained was based on an analogy with the contemporary sedimentary formation on the Sinai Peninsula. This hypothesis is proposed because of the similarities between that present-day activity and the formation as it was known at the time. The source of this hypothesis in the observation of contemporary geological events highlights that what are considered relevant or 'live' hypotheses depends in part of the state of both our empirical and theoretical knowledge at the relevant point in time. Background knowledge, beliefs and assumptions shape the narrative ordering of events and the understanding of what counts as evidence for their explanation.

The discovery of the Stac Fada Member in the 1960s called for explanation. Its discovery disrupted expectations since the mechanism that accounted for the surrounding region was not consistent with this formation. Among the traces that did not fit were 'angular shards of pumice, green particles of devitrified glass and accretionary lapilli' (Hopkins, Chapter 4) observed in this newly discovered formation. The Stac Fada exhibited features that appeared to be consistent with a volcanic hypothesis (although there were some aspects that raised questions – for example, the lack of a volcanic vent and the unusual absence of other volcanic activity in the area). Also disconcerting was evidence in the formation of an 'abrupt change' shifting land from east to west. If we think of process tracing as fitting evidence to the hypothesis, anomalies like this that do not fit into the narrative told through the hypothesis are problematic. They call for explanation and so pose new research questions. Or, if the other things that were known, believed or assumed at the time indicate why they may be discounted as irrelevant, they may be put aside. In this case, for example, it is known that erosion and burial over the long period of geologic times can eliminate or hide relevant evidence like a volcanic vent. The 'abrupt shift' is harder to account for, however, and thus challenges expectations and opens space for consideration of an alternative hypothesis. Other factors in what hypotheses might be considered were, as Hopkins notes, the prevalence of uniformitarianism – the methodological assumption that the processes that should be appealed to when offering geological explanations should only include those that are currently observable – and the seeming 'outrageous' nature of proposing extraterrestrial causes. The proposed hypotheses – in the nineteenth century prior to the discovery of the Stac Fada and after its discovery in the 1960s – reflect this constraint in that they rely on similarities between processes in regions contemporaneously observed and the region under investigation.

It is not until the early 2000s that Ken Amor proposed an alternative based on his comparison of the traces in the Stac Fada Member with traces he was familiar with from his study of a meteor crater in Bavaria. He first notes the devitrified glass, which suggested a similar cause since he had observed such traces previously in Bavaria. While such similarities cannot establish any hypothesis, they offer the opportunity to seek other traces of the proposed causal mechanism. When Amor took a sample specifically to look for such traces, he found shocked quartz in the sample – also consistent with his hypothesis. The fit of these traces with his hypothesis guides further research and informs a sketch of an account of what happened.

In this example, process tracing is used as a means of hypothesis testing and development – the lack of fit throws doubt on the viability of the hypothesis and motivates a consideration of alternatives. When a hypothesis fits with traces, it continues to be a live possibility; if it does not fit, and anomalies cannot be accounted for, it may cease to be viable.

In addition, the comparison of cases suggests hypotheses through the identification of similarities between cases – first for the geology of the region prior to the discovery of the Stac Fada Member (the similarity to the Sinai Peninsula) and then after the discovery of the formation (similarity to meteor crater). Also of interest is the role of the uniformitarian framework, dominant during the early part of this period, which appears to have constrained the set of viable hypotheses, as Hopkins points out. The meteorite hypothesis only emerges as a real possibility when the traces of such an event had been observed elsewhere, uniformitarianism has loosened its grip and an extraterrestrial explanation becomes a live option, as Hopkins notes. The abrupt shift of land from east to west is no longer an anomaly but now relevant since it can be understood as a trace of the meteor impact but not of the volcanic hypothesis. Telling the history of the geology of the region through the meteor hypothesis thus offers a more coherent account than the volcanic hypothesis. Although the site of the impact crater cannot be identified, this is not thought to be problematic for the same reasons that the missing volcanic vent is not considered an issue – erosion and burial can make such evidence inaccessible over time.

Hopkins describes the consideration of alternative hypotheses in terms of reinterpretation of evidence, but what is noteworthy is that the reinterpretation results from what the various bits of evidence are understood to be traces of – that is, how they are made relevant through the hypothesis and background knowledge. The hypothesis and what else we know come together with the events, activities and entities found in the case to make a more coherent account of the geology of the region. If there are traces that do not fit, that may leave open the possible consideration of a new hypotheses *if* they are deemed relevant given what else is known. It could be, of course, that more than one hypothesis provides a causal mechanism consistent with the details of the case. In Hopkins's account, we find out that shocked quartz could also be caused by a lightning strike. But can the lightning strike hypothesis fit the abrupt change that shifted land from east to west into an integrated whole?

11.7   Narrative as a Tool for Process Tracing; Process Tracing as a Tool for Narrative

Both these examples illustrate the iterative and integrative nature of how process tracing informs the construction of a scientific narrative by working back and forth between theory, evidence and background knowledge, beliefs and assumptions. In fact, although I have treated them as illustrative of different modes of process tracing, they also indicate how these modes are not mutually exclusive but in fact complement each other in the research process. These various stages of process tracing both aid in the goal of constructing a narrative of the case (offering a response to the research question) and are informed by that goal.

A number of other chapters recount incidents in science that can be understood as engaging with aspects of reasoning involved in process tracing. For example, Elizabeth Haines (Chapter 9) describes Hugh Hamshaw Thomas as working in two disciplinary realms. His understanding in each of these realms is informed by what is taken to be true at the time in each and the questions that arise for those contexts. His understanding of the fossils that he finds – traces that he argues indicate a missing link species between ferns and flowering plants – is determined by the need to fit the narrative into an overarching evolutionary narrative. In the realm of aerial photography, how to interpret what is photographed requires some hypotheses about what features are relevant – which can be understood in relation to others given that particular hypothesized activities could take place there. Ideas about what such activities would involve are needed both to suggest and to limit the possible interpretations of the aerial photographs. For the former, the theoretical framework (evolutionary theory) aids in determining which traces are potentially significant. For the latter, consider Haines's recounting of Gombrich's comments about the photographic interpreter who is useless because he has too much imagination. He sees too many features as potentially significant. The theoretical framework is necessary to constrain what sorts of connections can be made.

Such constraints are not always positive, however. Englemann's examination of plague narratives (Chapter 14) offers an illustration of how the fixation on soil as a cause of plague limited understanding of the disease. In this case, false beliefs and assumptions lead to a focus on traces consistent with a hypothesized mechanism that turns out to be wrong. Relevant traces are not recognized as relevant because the search is based on flawed, probably racist, background beliefs. An incomplete examination of traces may not turn up evidence that challenges the hypothesis and alters the conception of how the case should be viewed and what counts as evidence. Noticing the importance of what may be anomalous traces – things that do not fit the narrative a particular hypothesis offers – can bring about what Hurwitz refers to as 'epistemic switches' in his chapter discussing anecdotes (Chapter 17).

For Hurwitz, in the context of medical knowledge, an important feature of anecdotes is how they allow for the reframing of information so that it can become evidence. In the example of the Bouvart anecdote about the Marquis, greetings are reframed as part of the diagnostic context and so become evidence, whereas the prevailing conceptions of evidence had previously deemed them irrelevant. Something like this also occurs in the case of Viagra, when the angina research project reaches a dead end and a new framework for research is suggested through anecdote. The expectation that ordinary interactions are irrelevant to diagnosis, the expectation that only what is relevant to the treatment of angina is of interest, are shifted. In each of these cases, the anecdote itself does not give us an overarching story – does not reveal the causal mechanism – but switches awareness of what is already in front of researchers, altering its significance. The research that follows is what establishes the medical claims that are later made, but Hurwitz makes the strong assertion that, without anecdote, treatments and cures might never be found. Without dislodging negatively constraining prior beliefs and assumptions new ones are not possible.

The way background knowledge, beliefs and assumptions constrain what hypotheses are open to consideration and its close connection to what counts as evidence is also illustrated in Bhattacharayya's chapter (Chapter 8). She notes that reconstructing shipwrecks plays a crucial role in establishing legal responsibility for the disasters. But, to fit the events together, some sense of what the plausible causal connections are through which the fitting can be done must already be at hand. Think, for example, of Beatty's discussion of the explanation of the location of the eyes on flatfishes (Chapter 20). Evolutionary theory both constrains and suggests the plausible connections shaping the narrative. Bhattacharayya's discussion of two legal cases offers some clue to what sorts of ways of fitting these events together were considered live possibilities. The two sorts of evidence thought relevant in these cases were evidence about the character of the actors (particularly the various shipmasters involved) and evidence related to the storm itself. In the case of character, past performance, behaviour after the shipwreck, history of drinking and other features thought related to character come into play. These are treated as evidence of failure to behave adequately under the specific circumstances of the storm. Piddington's work argues for preferencing a narrative of how the storm might be expected to unfold and the appropriateness of the response of the shipmaster given that expectation. The storm cards function as a way of shifting the standard of evidence to expectations about how shipmaster and crew ought to behave given how the course of the storm was thought likely to progress.

11.8   Conclusion

I began by noting how process tracing works to support the construction of narrative in political science case studies. I next explored some of the examples of narrative science in this volume with process tracing in mind. What stands out is the various ways that process tracing calls for the use of theory, background beliefs and assumptions to identify and make use of relevant elements of the case in order to construct a coherent and complete narrative. Process tracing thus functions as a tool for colligating all of these features into a narrative in the sense that Mary Morgan describes (2017; and Chapter 1). Part of process tracing involves identifying what it is that should count as traces – making the case that particular elements count as evidence, or, put another way, that they should be part of the narrative. Wood does this when she offers research about memory to support her use of interview data as evidence of the psychological mechanisms that she claims are operating. Miyake's source model narratives function in a similar way. Process tracing pulls together disparate sorts of evidence produced through a variety of methods and integrates them into a unified account. It does this through theorizing about those connections and working back and forth between theory and the elements of the case in an iterative fashion until a satisfactory account can be constructed. Given that process tracing must start from connections that we know, believe or assume, it carries with it the danger of falsely limiting our understanding so that evidence is not recognized as such. However, because the method is iterative, process tracing also allows for reinterpretation through epistemic shifts.

I close with a final thought about process tracing and narrative that suggests an area of further investigation. Process tracing connects the particular with the general through the use of theory and causal mechanism as means of structuring narrative. How features of a case or, more generally, data are to be interpreted as evidence depends on other things that we know, believe or assume, not only about this case but about others. And what we hope to take away from the case is knowledge that will be useful in other locations as well. The interplay of the particular and the general, something crucial for knowledge of the empirical world, strikes me as an important feature of narrative and why it has such fundamental appeal to human beings.[6]

Bibliography

Beach, D., and R. B. Pedersen (2019). *Process-Tracing Methods: Foundations and Guidelines*. 2nd edn. Ann Arbor: University of Michigan Press.

Bogen, J. (2005). 'Regularities and Causality; Generalizations and Causal Explanations'. *Studies in History and Philosophy of Biological and Biomedical Sciences* 36.2: 397–420.

Crasnow, S. (2017). 'Process Tracing in Political Science: What's the Story?' *Studies in History and Philosophy of Science Part A* 62: 6–13.

Glennan, S. (2002). 'Rethinking Mechanistic Explanation'. *Philosophy of Science* 69. S3: 42–53.

[6] *Narrative Science* book: This project has received funding from the European Research Council under the European Union's Horizon 2020 research and innovation programme (grant agreement No. 694732). www.narrative-science.org/.

Herman, D. (1997). 'Scripts, Sequences, and Stories: Elements of a Postclassical Narratology'. *PMLA* 112.5: 1046–1059.

Illari, P. M., and J. Williamson (2012). 'What Is a Mechanism? Thinking about Mechanisms across the Sciences'. *European Journal of Philosophy of Science* 2: 119–135.

Lay, T. (2018). 'A Review of the Rupture Characteristics of the 2011 Tohoku-Oki Mw 9.1 Earthquake'. *Tectonophysics* 733: 4–36.

Machamer, P., L. Darden and C. Craver (2000). 'Thinking about Mechanisms'. *Philosophy of Science* 67: 1–25.

Morgan, Mary S. (2017). 'Narrative Ordering and Explanation'. *Studies in History and Philosophy of Science Part A* 62: 86–97.

Ragin, C. (1992). '"Casing" and the Process of Social Inquiry'. In C. Ragin and H. S. Becker, eds. *What Is a Case? Exploring the Foundations of Social Inquiry*. New York: Cambridge University Press, 217–226.

Reiss, J. (2008). *Error in Economics: Towards a More Evidence-Based Methodology*. New York: Routledge.

Tabery, J. (2004). 'Synthesizing Activities and Interactions in the Concept of a Mechanism'. *Philosophy of Science* 71.1: 1–15.

Wood, E. J. (2003) *Insurgent Collective Action and Civil War in El Salvador*. Cambridge: Cambridge University Press.

V

Research Narratives

When scientists write about their research, their narratives centre on their practices but reveal their beliefs about phenomena

12 Research Narratives and Narratives of Nature in Scientific Articles: How Scientists Familiarize Their Communities with New Approaches and Epistemic Objects

Robert Meunier

Abstract

The chapters in this volume show that narrative can be found on many levels and in many media in science. This contribution locates narratives in one of the most prominent forms of scientific literature in the twentieth century: the research article. It shows how in the experimental sciences accounts of natural processes and accounts of research activities both take the form of narratives, 'narratives of nature' and 'research narratives', respectively. For a hypothesis to enter the former or to be criticized, members of a scientific community need to grasp the research approach from which it emerges. The chapter argues that research narratives are designed to make readers familiar with an approach. Such narratives draw a path through epistemic scenes inhabited by a character representing the researchers. By stylistic means the researchers are construed as exemplars for members of the community, and their activities as exemplifying the approach to a shared problem.

12.1 Research Narratives and Narratives of Nature

In 1945, George Beadle, who was to receive the Nobel Prize in Physiology or Medicine in 1958, together with Edward Tatum, published a long review article on the state of biochemical genetics. In one section, entitled 'Eye pigments in insects', he summarized results stemming to a large extent from his own work, which he had initiated with Boris Ephrussi in 1935, preceding his collaboration with Tatum. Beadle and Ephrussi used the fruit fly *Drosophila melanogaster*, which at the time was already a well-established experimental organism. Their experiments, however, introduced a novel approach based on tissue transplants between flies carrying different combinations of mutations. The results of these and similar experiments, and further biochemical efforts to characterize the substances involved, led to the following account of the physiological process of the formation of brown eye pigment and the roles of genes therein:

Dietary tryptophan is the fly's initial precursor of the two postulated hormones. This is converted to alpha-oxytryptophan through a reaction controlled by the vermilion gene. A further oxidation to kynurenine occurs. [...] This is the so-called v^+ substance of Ephrussi and Beadle. This is still further oxidized to the cn^+ substance, which Kikkawa believes to be the chromogen of the brown pigment. The transformation of kynurenine to cn^+ substance is subject to the action of the normal allele of the cinnabar gene. (Beadle 1945: 34; references omitted)

This text constitutes a small narrative. It relates several events which occur in temporal order and are causally connected. The sequence has a beginning (the precursor is ingested), a middle (it is transformed in several reactions controlled by genes) and an end (the implied formation of brown pigment). Yet, this narrative does not recount particular events, but rather a type of event happening countless times in fruit flies (and similarly in many other insects); it is a generic narrative.[1]

In the natural sciences, such narratives are often found in review articles and textbooks, but also in summaries of the state of knowledge on a given subject in the introduction to research articles; they state what is taken as fact. Addressing scientific facts as narratives acknowledges that they are typically presented as complex and ordered accounts of a subject rather than single propositions. It is striking that no human agents, observers or cognizers are present in such narratives. They are accounts of events that are taken to happen 'in nature' when no researcher is intervening or even watching. Such narratives can thus be called 'narratives of nature'.[2]

Historians and philosophers of science no longer see the question of epistemology to be concerned with the truth of such knowledge claims alone, but also with the practices from which they emerge, and which enable, shape and delimit these claims. The references in Beadle's text make it clear that each proposition can be traced back to an episode of research. Narratives of nature emerge gradually from the research literature as facts accepted in a community. Accounts of the methods by which the knowledge was achieved are abandoned like ladders once the new state of knowledge is reached. The facts are turned into 'black boxes', which can, however, be reopened any time; methods are called into question when facts are challenged (Latour 1987).

To account for how a hypothesis derived from research eventually enters a 'narrative of nature', it is necessary to show how a hypothesis comes to be

---

[1] 'The world of the generic narrative [...] is not a unique world, but rather a class of worlds in which the activities and circumstances generally obtain. Any given event, agent, or object in a generic discourse actually stands for a class of such objects' (Polanyi 1982: 511).

[2] Myers (1990: 142) uses the expression in a related sense regarding popular science.

known and understood by members of a community in the first place. I will argue here that this requires peers to understand the research approach – which aligns a method and a problem, and in the context of which the hypothesis was formulated. Familiarization with the approach is achieved by using another type of narrative, realized primarily in research articles. The function of research articles is thus not only and primarily to convince readers that a hypothesis is supported by evidence, so that they will accept it.[3] Instead, by making readers familiar with the approach, the article enables them to understand how one gets in the position to formulate and support a hypothesis of this kind in the first place, its relevance regarding a problem recognized in the community, and the meaning of the terms used (i.e., to grasp the epistemic objects in question).[4]

An approach is a movement, it involves positioning oneself towards a phenomenon and accessing it from a particular direction and in a particular way.[5] The phenomenon, the experimental system employed for accessing it, the activities of intervention and observation afforded by the system, and the ways to make inferences from observations, including the recognition of invisible entities, make up what I call the 'practice-world' of researchers. The research article introduces the reader to this world and to the way that researchers position themselves by interpreting a problem pertaining to a phenomenon, to access the phenomenon, materially and cognitively, generate data and draw inference – in other words it makes the reader familiar with an approach. Only then can the hypothesis be understood; but it does not need to be accepted. Any criticism, refinement or amendment of the hypothesis is articulated in terms that are meaningful in the context of the approach and often involve the recreation of the approach by members of the community, introducing more or less substantial variation.[6]

In this chapter, I will show how research articles employ narrative to familiarize readers with an approach. Reporting the material (intervention and observation) and cognitive (inference) activities of researchers, research

---

[3] Crasnow (Chapter 11) argues for narrative processes as making evidence from data. Jajdelska (Chapter 18) explores an alternative means of familiarization in research articles, by means of narrative performativity.

[4] According to Rheinberger, experimental systems '"contain" the scientific objects in the double sense of this expression: they embed them, and through that very embracement, they restrict and constrain them', and thereby 'determine the realm of possible representations of an epistemic thing' (1997: 29). Approaches in the experimental sciences involve experimental systems, yet the notion is broader, referring to the ways an experimental system is used to address a problem and its output is interpreted.

[5] On a related notion of 'approach', see Waters (2004). On my account an approach is the equivalent for practice of what philosophers refer to as perspective regarding theoretical representation (e.g., Giere 2006).

[6] For experimental systems this has been referred to as 'differential reproduction' (Rheinberger 1997).

articles on the whole are narratives (even if they often contain non-narrative passages) and might be referred to as 'research narratives'. Like narratives of nature, research narratives are factual narratives, but in contrast to the former, they recount particular events, which happened at a specific site (e.g., a given laboratory) and a specific time; they are not generic. And yet, as will become clear, they do not present these events as unique either, but rather as exemplary.

In section 12.2, I will introduce several narratological concepts pertinent to the analysis. In 12.3, I will then trace back some of the elements of the narrative of nature above to an original research article by Beadle and Ephrussi, which I take to be representative of this genre in twentieth-century experimental life sciences. Section 12.4 will then return to the ways in which the particular implementation of an approach is rendered exemplary.[7]

12.2   Research Articles as Factual Narratives

Subsection 12.2.1 will argue that modern research articles are indeed narrative texts. As they are generally taken to be factual narratives, I will briefly address the question of how they relate to real-world events. Subsection 12.2.2 will clarify the relation of researchers in their double role as agents and authors, and as narrator and character. It will then relate these roles to the narratee and the implied and actual reader. I will also introduce two metaphors: narrative as path and narrating as guidance, to further characterize the relation of narrator and narratee.

12.2.1   *Research Articles are Factual Narratives*

When talking about narrative, one often thinks of fictional texts or accounts of personal experience.[8] Research articles might not meet common expectations about what a narrative is. Nonetheless, research articles should be seen as narratives. Before showing why, I will address some ways in which they depart from more typical narrative texts.

First, research articles have a unique structure in that they typically separate the accounts of various aspects of the same events. This partitioning of information is often realized in the canonical 'introduction, methods, results and discussion' (IMRaD) structure.[9] In the *Introduction*, researchers state where they see themselves standing in relation to various disciplines and theoretical

---

[7] Philosophers of science, like scientists, aim to make not only their case but also their approach (here a narratological approach) familiar and exemplary, such that the insights derived from it can be discussed and the categories employed be transferred to other cases (see, for example, Currie 2015).

[8] On the latter, see Hurwitz (Chapter 17).

[9] On the origin of the IMRaD structure, see Day (1989).

commitments, thereby positioning themselves towards a problem recognized in the community they address and motivating the activities to be narrated. The detailed description of the activities, including preparation, intervention and observation, is presented in the *Material and Methods* section collectively for all experimental events. The structured performance of these activities is reported in the *Results* section, albeit not necessarily according to their actual temporal order. Finally, the *Discussion* section recounts cognitive operations in which the material activities are revisited, often as involving entities which are inferred from patterns in the data.

Second, research articles tend to exhibit a characteristic style. As is often noted, they use impersonal language, i.e., various devices such as passive voice, adjectival participles, nominalization, abstract rhetors and impersonal pronouns to conceal the agent in an event (e.g., Harré 1990; Myers 1990). Furthermore, events are often reported in the present tense. These strategies give the impression of a generic narrative, even though (unlike a narrative of nature) the statements in fact refer to particular events. Such narratives are thus pseudo-generic, but in this way represent events as exemplary.

Taken together, these organizational and stylistic features result in the fact that research articles do not resemble other text types that are more often addressed in terms of narrative. And yet they should indeed be seen as narratives.

Most definitions of narrative or criteria for narrativity of a text include the notion that narratives relate connected events. The verb 'to relate' can be read in a double sense here: narratives recount the events and they also establish relationships between them. There is some dispute about the nature of the connections among events that lend themselves to being narrated – for example, whether connections need to be temporal or causal (Morgan and Wise 2017). It is, however, almost universally agreed that a mere assortment of event descriptions or a mere chronological list of events does not constitute a narrative. Another central criterion is the involvement of human-like or intelligent agents in the events. Again, further aspects of agency might be required, such as the representation of the mental life of the agents or the purposefulness of actions (Ryan 2007).

Depending on whether the latter condition is taken to be necessary, or how one interprets 'human-like', one might doubt the status of narratives of nature discussed above. Research articles, however, are clearly narratives in the light of these core criteria. They report connected events, and they report them as connected. Indeed, many of the events are temporally ordered, with previous determining subsequent events. Furthermore, the events involve the researchers as agents, and their actions are purposeful and accompanied by cognitive operations.

Although it has been observed that research articles often do not provide a faithful representation of the research process which they appear to report, research articles are not typically perceived as works of fiction either (Schickore 2008).[10] Instead, they are generally presented and perceived as factual narratives (Fludernik 2020). Accordingly, an account of factual narrative is required.

One of the most robust theoretical tenets of narratology is the distinction between story and discourse.[11] Seymour Chatman, for instance, states that

each narrative has two parts: a story (*histoire*), the content or chain of events (actions, happenings), plus what may be called the existents (characters, items of setting); and a discourse (*discours*), that is, the expression, the means by which the content is communicated. In simple terms, the story is the *what* in a narrative that is depicted, discourse the *how*. (Chatman 1978: 19)

It cannot be assumed, however, that in the case of factual narratives the real-world chain of events constitutes the story. If the observation of common mismatch between research process and report is accurate, then for research articles, at least, it is clear that the chain of events reconstructed from the discourse, the story, is not necessarily equivalent to the chain of events that make up the research process. The story as the sequence of events reconstructed from the discourse by the reader is a mental representation, as cognitive narratologists maintain (Ryan 2007). I will thus assume a semiotic model of factual narrative according to which the discourse invokes a story in the mind of the reader, and the narrative (discourse + story) represents real-world events, whether or not the events of the story fully match the represented events.[12] I will speak of the represented events as being part of a 'practice-world', however, to avoid false contrasts, as discourses and minds are, of course, part of reality, and to point out that these narratives represent only a fragment of the world which is inhabited by the actual researchers.

12.2.2   Communicating and Narrating

By putting their names in the title section, researchers as authors of scientific articles clearly assume responsibility for what they write, and they will be held accountable by others. Yet even if the narrator is identified with the author of these and other factual narratives, it cannot be equated with the author.[13] Authors will carefully craft the narrator and adorn it with properties which

---

[10] Such observations are based on lab ethnographies or the analysis of lab notebooks (e.g., Holmes, Renn and Rheinberger 2003; Knorr-Cetina 1981).
[11] See Hajek (Chapter 2).   [12] This model is thus Peircian, rather than Saussurean.
[13] Genette (1990) makes this equation.

they need not necessarily ascribe to themselves. In fact, as many articles – including Beadle and Ephrussi's – are co-authored, it would be challenging to construct a narrative voice that is faithful to the ways each of the authors perceives themselves or the group. In research articles, narrators are homodiegetic, i.e., they are also characters in the story (Genette 1980). Hence, by crafting the narrator, authors also craft the character of the researcher on the level of the story (for instance, as an able, attentive and accurate experimenter).

On the recipient side, the reader of a research article can be anyone, of course, even a philosopher of science looking at the text 80 years later to make it an example for narrative in science. There is also an implied reader, which can be inferred from paratextual as well as textual features (Iser 1978). Regarding the former, the journal in which an article is published is a key indicator. Textual features include the knowledge the authors take for granted – the kind of claims that do not need further justification or terminology, used without definition. The actual reader who matches the features of the implied reader is the addressee of the communicative act of the author.

Genette (1980) distinguishes the act of narrating from the discourse and the story. This act is performed by the narrator and is not part of the story; the addressee of this act can be called the 'narratee'. By creating the discourse, the author creates the voice of a narrator as if it (the narrator) had produced this discourse, and a narratee as the addressee implied in the discourse.[14] Thus the narratee cannot be equated with the reader addressed by the author. Furthermore, while the way the narratee is construed is informative of the way the implied reader is construed, these two categories need not necessarily overlap.

Based on the above model of factual narrative, I propose the following account of narrating. The narrator in the act of narrating represents the researchers in their role as authors in the precise sense that it is construed as having the same knowledge as the latter. The researcher-character, who is identified with the narrator, represents the researchers in their role as experimenters and reasoners in the practice-world. The narrator addresses the narratee to recount events in which it was involved as a character and which thus represent events in the practice-world of the researchers. A reader can cognitively and epistemically adopt the position of the narratee and thereby learn about these events. A reader who matches the implied reader will be more willing and able to do so. In this way, researchers as authors communicate information about the practice-world they inhabit as agents to a reader who might inhabit similar practice-worlds.

[14] For Genette (1990), narrating is prior to discourse and the author can perform the act of narrating directly in those cases where the author is equated with the narrator. See n. 22, below, for discussion of pronouns used in this chapter.

Narratologists routinely analyse differences regarding time (order, duration and frequency of events) between discourse and story.[15] Note, however, that if the story is distinguished from the practice-world events in factual narratives, the difference between these two regarding time is an entirely different issue. Take the order of events. The discourse might introduce events in the order B, A, C, while it can be inferred from the textual cues that the order in the story is A, B, C. The discourse then does not misrepresent the order – in fact, by means of the cues it does represent the events in the order A, B, C, and as the story is an effect of the discourse, the two levels cannot be compared independently. If the narrative (discourse + story) presents events in a given order A, B, C, while the practice-world order of events was in fact C, A, B, then this, instead, constitutes a mismatch (e.g., between research process and report). The above semiotic model maintains that the narrative still represents the practice-world events. By manipulating order, duration and frequency in the discourse, authors can create certain effects in the perception of the story. In the case of factual narrative, developing a story that misrepresents practice-world events in one aspect can help to highlight other important aspects of these events such that the overall representation might become even more adequate with respect to a given purpose.

The purpose of the research narrative, or so I argue, is to represent the practice-world events as an approach to a given problem. Seen from the perspective of the act of narrating, the discourse not only presents events which are reconstructed on the story level, but it consists of events of narrating. If the discourse introduces narrated events in the order B, A, C (including cues that indicate the order on the story level is A, B, C), then there will be three sequential events: narrating B, then A, and then C. The temporal order on the level of narrating might be employed to highlight an order of elements in the story world other than temporal (e.g., a conceptual order).[16]

On the level of narrating, the narrative might be described as a path through scenes in the story world which are considered in turn. By laying out a path, the narrator guides the narratee through the story-world. If this metaphor has a somewhat didactic ring, it is important to remember that it does not describe the relation of author and reader. The narratee in the research narrative is construed not so much as a learner who knows less about a subject but more as an apprentice who knows less about how to approach the subject. The narrator (who is also the researcher-character) will create a path connecting several diegetic scenes in which the character has certain beliefs, performs activities and observations, and reasons on their basis. The narratee qua guidee

---

[15] In a different way, Huss (Chapter 3) discusses the lining up of these different time patterns in terms of 'narrative closure'.

[16] This possibility is of particular importance for narratives in science (Morgan 2017).

is thus introduced to the epistemic possibilities of the approach. A reader willing and able to adopt the position of the narratee can thereby learn about the approach.

12.3   Familiarizing a Community with an Approach through Research Narratives

12.3.1   The Case: A Research Article on Physiological Genetics from the 1930s

I now turn to the work of George Beadle (1903–89) and Boris Ephrussi (1901–79) and in particular to one article, which can be analysed based on the considerations in 12.2. The article in question was published in the journal *Genetics* in 1936.[17] It was entitled 'The Differentiation of Eye Pigments in Drosophila as Studied by Transplantation' and reported research the authors had performed mainly in 1935, when Beadle, who was at Caltech at the time, visited Ephrussi in his lab at the Institut de Biologie Physico-Chimique, Paris.[18]

Leading up to Beadle's Nobel Prize-winning work with Tatum, which is usually associated with a the 'one gene – one enzyme hypothesis' and thus considered an important step in the history of genetics, the article is relatively well known, at least to historians of genetics, as well as philosophers of biology. While firmly embedded in the genetic discourse and practice of its time, it presents enough novelty to display clearly the work it takes to familiarize peers with a novel approach and the novel epistemic objects emerging from it. Finally, in employing the IMRaD structure and an impersonal style, it conforms to salient conventions of much scientific writing in twentieth-century life sciences. It is thus well suited for such an analysis.

Many geneticists at the time aimed to understand the physiological role of genes, an enterprise that was often referred to as 'physiological genetics'.[19] This was the kind of problem Beadle and Ephrussi set out to engage with. Their starting point was an observation made by Alfred Sturtevant. Sturtevant had studied genetic mosaics naturally occurring in *Drosophila* flies, that is, organisms which are composed of tissues with different genotypes.[20] In some flies it appeared that the eyes did not exhibit the eye colour that would be expected given their mutant genotypic constitution (indicated by other phenotypic

---

[17] The journal was founded in 1916. For the context of discipline formation, see Sapp (1987).

[18] On Beadle and Ephrussi's work, see Burian, Gayon and Zallen (1991); Harwood (1993); Kay (1993); Kohler (1994); Sapp (1987).

[19] Also 'developmental genetics'; see, for example, Harwood (1993). Beadle's (1945) 'biochemical genetics' came into use only in the 1940s and had a more limited meaning, referring to the study of genes in biochemical pathways.

[20] In this case, this was due to the loss of an X-chromosome in some cells early on in development.

markers), but rather the colour-phenotype associated with the normal (wild type) genotype present in other parts of the body. From this Sturtevant concluded that a substance might circulate in the body of the fly, affecting the development of the eye, and that the gene, which was mutated in the eye, but was functioning in other parts of the body, was involved in the production of this substance (Sturtevant 1932).

Beadle and Ephrussi developed an experimental system based on implanting larval structures that would give rise to the adult eye (imaginal eye discs) into host larvae. The procedure resulted in adult host flies which harboured an additional eye in their abdominal cavity. This allowed them to create mosaics artificially and thus in larger numbers, and to produce adequate experimental controls. They clearly began with Sturtevant's hypothesis regarding the existence of a circulating, gene-related substance. Furthermore, hypotheses about the nature of gene action, in particular, the idea that genes affected biochemical reactions (either because they were enzymes or because they played a role in their production) were common (Ravin 1977). Nonetheless, Beadle and Ephrussi's article did not frame the work as testing any specific hypothesis about the relation of these entities, but rather as exploratory. Their project aimed at producing evidence for the existence and interactions of further elements in the biochemical system.

The epistemic objects they dealt with were thus on the one hand a well-established one, the gene, of which, however, little was yet known regarding its physiological function in somatic contexts, and on the other hand the assumed circulating substances, which were presumably involved in physiological reactions and in some way connected to the action of genes. The article reported the approach through which they achieved material and cognitive access to these epistemic objects and thereby established novel concepts referring to them. The approach enabled the formulation of hypotheses pertaining to these objects.

12.3.2   The Analysis: The Research Narrative as Path through Epistemic Scenes

In the following, I will reconstruct the research article by Beadle and Ephrussi as a narrative. The narrative draws a path through several scenes in which the researcher-character performs material or cognitive activities in a story-world which in turn represents the practice-world of the researchers as experimenters. The researchers as authors construct the narrator to guide the narratee through these scenes in a way that enables an understanding of the epistemic possibilities of the approach they have developed and thus an understanding of the hypothesis put forward. I will identify four types of epistemic scenes (concerning what is known and what can be known through the approach), which roughly coincide with the canonical IMRaD sections.

*I   The Positioning Scene: Interpreting a Problem Shared
       by a Disciplinary Community*

Both the journal in which Beadle and Ephrussi published their article
(*Genetics*), as well as the things they take for granted, clearly indicate that
their text implies geneticists as readers, as opposed to, say, embryologists.

The article does not begin with a hypothesis to be tested, but with
a question or research problem to be explored, which pertains to the discipline of genetics, and more specifically to the subfield of physiological
genetics.[21]

Prominent among the problems confronting present day geneticists are those
concerning the nature of the action of specific genes – when, where and by what
mechanisms are they active in developmental processes? (Beadle and Ephrussi
1936: 225)

With respect to this question, an assessment is made of the state of research at
the time, which has a theoretical aspect (what is known or assumed about gene
action) and a methodological aspect (how the problem has been approached).
Regarding the former, it is asserted that 'relatively little has been done toward
answering [these questions]' (Beadle and Ephrussi 1936: 225). Regarding the
latter, advances that have been made are acknowledged:

Even so, promising beginnings are being made; from the gene end by the methods of
genetics, and from the character end by bio-chemical methods. (Beadle and Ephrussi
1936: 225)

However, a methodological obstacle to theoretical progress is identified in the
fact that those organisms, which are well-characterized genetically, are not
studied from a developmental perspective, and vice versa. It is suggested that
this impasse be confronted by studying developmental processes in
a genetically well-characterized organism (*Drosophila*), and in particular
regarding the formation of pigment in the eye, because many eye-colour
mutants were known in this species (and because of Sturtevant's previous
findings).

As these considerations are written in an impersonal style, one could see
them as considerations of the authors in the moment of writing. And yet they
are narrated as considerations of the researchers at the time of setting up the
project, as indicated by formulations such as this: 'Several facts have led us
to begin such a study' (Beadle and Ephrussi 1936: 225). As such, they are
events in the story-world (whatever was in fact considered in the practice-
world). They constitute the beginning of the story, the initial epistemic scene
in which the researcher-character ('we') – as a member of a discipline – finds

---

[21] On question-driven fields, see Love (2014); on exploratory research, see Burian (2007).

itself.[22] The narrator guides the narratee through the scene to let it understand how one can position oneself in the field characterized by certain problems and available methods, and to realize the advantages of the chosen approach.[23] This will resonate in particular with readers who are members of the community the authors belong to.

II   *The Methodology Scene: Having and Mastering an Approach*

The introduction of a new approach changes the situation in the field. It results in new possibilities for these researchers, and with them for everyone in their community. The new situation is characterized by the availability of the new experimental method, the new interpretation of the problem such that it can be addressed by the method, and the evidence and conclusions it affords. The narrator has already led the narratee to consider this new approach by setting it off against previous work in the *Introduction* section.

In the *Material and Methods* section, then, experimental events, consisting in applying a technique, are presented as generic, repeatable activities:

In brief, the desired organ or imaginal disc, removed from one larva, the donor, is drawn into a micro-pipette and injected into the body cavity of the host. As a rule, operations were made on larvae cultured at 25°C for three days after hatching from the eggs. (Beadle and Ephrussi 1936: 225–226)

In general, one function of this section can be to enable other researchers to reproduce the techniques in their own lab. In that sense, the text functions like a recipe (or 'protocol', in the language of experimental sciences). In this case, however, the detailed description of the technique has been relegated to an extra method article (Ephrussi and Beadle 1936). The information given in the *Material and Methods* section of the present article is possibly too sparse to allow for reproducing the experiments. This points to the fact that there must be another function: this section is similar to the exposition in a fictional text.[24] It introduces the reader to various elements ('existents') of the story, such as flies, fly larvae, donors, hosts, imaginal discs, various mutant lines and other things, and, furthermore, to the 'habitual' activities involving these elements performed by the researcher-character.

---

[22] The authors use an exclusive 'we' as narrative voice. The narrator/character acts as a single entity in the sense that the researchers are presented as interchangeable. To indicate this and to mark the narrator's status as a textual entity that is different from the actual persons, I will refer to the narrator/character with the third-person singular 'it'.

[23] On positioning, see Van Langenhove and Harré (1999). For another account that also puts ideas concerning positioning into relation with narrative see Berry (2021).

[24] 'It is the function of the exposition to introduce the reader into an unfamiliar world, the fictive world of the story, by providing him with the general and specific antecedents indispensable to the understanding of what happens in it' (Sternberg 1978: 1).

In the quotation, the first sentence uses the present tense. It is prescriptive in the sense of a protocol, but more importantly expresses the fact that the experiments can be performed by anyone who has the skills and access to the material. The second sentence is in the past tense, making it clear that the narrative nonetheless represents particular events when the researchers have performed these actions and indeed varied the conditions and found one that worked best. Following the contrastive presentation of the approach in the *Introduction*, the narrative in the *Material and Methods* section presents the character in a scene where it equips itself with a reliable method with which to approach the problem identified in the positioning scene.

III   *The Experimentation Scene: Addressing Questions Pertaining*
           to the Problem through the Approach

In the *Experimental Results* section, the narrative proceeds through questions which generate several epistemic scenes within the context of the broader disciplinary situation. These scenes are characterized by specific instances of ignorance (e.g., regarding the action of specific genes known through mutations) relative to the overarching research problem (gene action in general). These questions in turn have to be expressed in terms of the behaviours of the material in the context of the experimental interventions possible in the framework of the novel approach.

The path along which the narrator guides the narratee through these scenes is not fully determined by the temporal order in which the experiments were performed. Some questions can only be formulated if the data of previous experiments are obtained (or indeed only after conclusions are drawn from it, which are only presented in the *Discussion* section). But for many experiments the order in which they are performed is not relevant and hence also not represented in the text. The ordering created by the path is thus not always that of a sequence of events, but, instead, the intervention events are also ordered into series according to the logic of the experiments, in this case the combinatoric logic regarding donor and host genotype. The subsections have titles such as *Mutant eye discs in wild type hosts*, *Wild type discs in mutant hosts*, *Vermilion discs in mutant hosts*, etc.

For instance, the first subsection shows how the approach provides an assay to answer the question of which mutants are autonomous (i.e., when serving as donor, are not affected by the host tissue). The result that v and cn are the only exceptions, in that they are not autonomous, leads to a new epistemic scene. In the subsection *Vermilion discs in mutant hosts*, then, the narrative moves forward by means of a new question the researcher-character asks itself, and which can be addressed through the approach:

data should be considered which bear on the question of whether other eye color mutants have anything to do with this 'body-to-eye' phase of the v reaction [i.e., the influence of the host]. This question can be answered by implanting *v* eye discs into hosts which differ from wild type by various eye color mutants. Such data are given in table 3.

[*Table 3*]

These data show that, when implanted in certain mutant hosts ([*list of mutants*]), a *v* optic disc gives rise to a wild type eye; in others ([*list of mutants*]), it gives an eye with *v* pigmentation. (Beadle and Ephrussi 1936: 231–232)

Most of the researcher-character's activities of intervention (implanting) and observation (dissecting and comparing eye colours) are compressed into one sentence and relegated to the table. Again, the formulation in the present tense and the impersonal style suggest that for any researcher, at any time, these interventions would result in these observations. And yet these sentences clearly refer to particular events in the story. The table, for instance, lists the number of individuals tested. We learn, for instance, that a *v* disk has been implanted in a *bo* host only a single time, while it has been implanted in 18 *ca* hosts. By having the researcher-character note the regularities and notable exceptions in the data, the narrator enables the narratee to grasp what can be done with the experimental method within the approach.

IV   The Interpretation Scene: Formulating Hypotheses in the Context of the Problem and Approach

Already in the *Experimental Results* section, cognitive operations of the researcher-character are narrated:

From the data present above, it is seen that, in the cases of *cn* in wild type, *v* in wild type, and wild type in *ca*, the developing eye implant is influenced in its pigmentation by something that either comes or fails to come from some part or parts of the host. Just what this is, whether or not, for example, it is of the nature of a hormone, we cannot yet say. We shall therefore refer to it by the noncommittal term 'substance'. (Beadle and Ephrussi 1936: 232)

In this scene, the narratee is shown how the approach enables cognitive access to new epistemic objects through interpreting data resulting from past activities. It is in the context of the approach that the term 'substance' refers to new objects. It can then be used to formulate a new set of questions, which no longer concern the visible effects of the interventions in the materials, but the assumed entities which are not directly observable: '[I]s there only one substance? If not, are the different substances related and in what way? What is their relation to the genes concerned in their production?' (Beadle and Ephrussi 1936: 233).

These epistemic objects are thus introduced as objects of interaction, appearing when acting in the framework of the approach. For this purpose, in the

Discussion section, events reported in the *Experimental Results* section are revisited:

Since the pigmentation of a genetically v eye can be modified to v^+ by transplanting it to a host which supplies it with what may be called the v^+ substance, it follows that v differs from wild type by the absence of this substance. Evidently there is no change in the v eye itself which prevents its pigmentation from assuming wild type characteristics. It follows that the mutation $v^+ \rightarrow v$ has resulted in a change such that v^+ substance is no longer formed. (Beadle and Ephrussi 1936: 240)

The events of experimental intervention (implanting a v disk) are retold, but this time the unobservable events on the molecular level that are thought to link the intervention and observation made by the researcher are added. Yet the scene inhabited by the researcher-character is not one of experimentation but of reconsidering past experimental action. Together, the experimental scene, in which the narrator recounts what has been observed upon intervention, and the interpretation scene, which narrates the reconstruction by the character of what was actually happening on a hidden level, are akin to an 'epistemic plot'.[25] The narratee is led to understand the way activities in the context of the approach can be interpreted in terms of interactions with the epistemic objects.

12.4   Conclusion: Exemplification of an Approach, between the Particular and the Generic

If the hypothesized entities and relations in the research article are compared with the narrative of nature in the review article quoted above, then it is clear that some – for instance, regarding the roles of the v and cn genes – achieved the status of accepted facts. Other propositions never went beyond the status of 'preliminary hypothesis'. Regarding the relation of substances, Beadle and Ephrussi provide the following hypothesis:

Such an hypothesis assumes that the ca^+, v^+, and cn^+ substances are successive products in a chain reaction. The relations of these substances can be indicated in a simple diagrammatic way as follows:
   $\rightarrow ca^+$ substance $\rightarrow v^+$ substance $\rightarrow cn^+$ substance (Beadle and Ephrussi 1936: 243)

The entities and relations after the second arrow are conserved in the narrative of nature. For sure, Beadle and Ephrussi can claim to have discovered these substances and the relations holding among them and between the substances

---

[25] 'The trademark of the epistemic plot is the superposition of two stories: one constituted by the events that took place in the past, and the other by the investigation that leads to their discovery' (Ryan 2008: 7).

and genes.[26] But the details of the hypothesis do not matter much, nor which elements are conserved. When it turned out that the existence of an entity that would match their hypothesized ca^+ substance could be established, this by no means diminished the value of the work. To criticize the hypothesis on its own terms required understanding the approach from which it emerged. Further results of that sort would come from the application of a more or less substantially modified version of the approach. Indeed, the research (Clancy 1942) which led to the abandonment of the ca^+ substance, was 'undertaken in order to repeat and supplement the experiments of Beadle and Ephrussi' and '[t]ransplantation operations were performed by the method of Ephrussi and Beadle [1936]'. The author also added a novel technique for the 'extraction and measurement of the eye-color pigments' to the approach (Clancy 1942: 417, 419). Hence, amending Beadle and Ephrussi's hypothesis depended on understanding, applying and modifying their approach.

The approach to the problem faced by the discipline, rather than the hypothesis, was thus the main achievement of Beadle and Ephrussi's work. As stated right at the beginning of their article:

In this paper we shall present the detailed results of preliminary investigations [...] which we hope will serve to point out the lines along which further studies will be profitable. (Beadle and Ephrussi 1936: 225)

The actual process, the contingencies and detours are not the subject of the narrative. The activities are reported as they would have been performed if the researchers had known better from the beginning. This explains the common mismatch between research process and report. The result is an approach that works and that enables researchers to make certain kinds of claims. Understanding the approach is a condition for understanding the terms and the significance of the hypothesis, no matter how well supported it is by the evidence. Furthermore, it is this kind of knowledge researchers can employ to design new research projects (Meunier 2019). It is anticipated that further research 'along these lines' will lead to modifications of the theoretical claims. The purpose of the narrative is to make readers as members of the relevant community (geneticists) familiar with the approach, such that they understand 'some of the possibilities in the application of the method of transplantation' with regard to the shared problem of gene action (Beadle and Ephrussi 1936: 245). Accordingly, the hypotheses about these epistemic objects which might or might not enter the narrative of nature are not the only or even primary result.

---

[26] The actual distribution of credit is more complicated, because not only had Sturtevant anticipated the v^+ substance, but Alfred Kühn and collaborators had delivered similar results working with a different organism (Rheinberger 2000).

In order to present the approach as universally applicable to the problem faced by the community, the narrative takes on the character of a generic narrative, even though it is in fact about particular events. It is thus pseudo-generic. More positively, the particular events are presented as exemplary; the research article constitutes an exemplifying narrative.

A significant stylistic difference between research narratives and many other accounts of personal experience is the use of an impersonal style and the present tense. These literary devices remove 'indexicality' (Harré 1990). In sentences of the type 'when implanted into a x host, a y disk gives rise to a z eye', the researcher-character is hidden by omitting the pronoun, while the present-tense detaches the activities from time and site. On the level of narrating, this has the effect that the narratee, guided through the experimental scene as an apprentice, can occupy the vacant position of the agent and perceive the event from the character's point of view (or rather point of action). A reader can then adopt the narratee's and thereby the character's position.

Grammatically, the character is only referentially absent but performatively present as the agent of implantation. Hiding the character thus renders the narrated events universal experiences of an unspecified agent. However, the occasional use of 'we', reference to individual instances (flies), and the use of the past tense anchor the narrative in particular events experienced by the character. Semiotically, the character as a complex sign denotes Beadle and Ephrussi. In so far as their experience is represented by the narrative, they are construed as exemplars of researchers in their community, who could all have similar experiences when performing the approach exemplified in the activities in which Beadle and Ephrussi engaged.[27]

Members of the community can read the text as narrating what Beadle and Ephrussi did or as stating what can be done regarding the problem. This ambiguity is indeed necessary. An approach is seen as universally applicable to a type of problem, just like a hypothesis is seen as universally answering to a problem. But, an approach, unlike a hypothesis, is not justified; it is not shown to be true, but it is shown to work. This is achieved by guiding the narratee along a path through various epistemic scenes, to see that one can do these things because they have been done.

In conclusion, while understanding the terms and the significance of a hypothesis (and not least the degree to which it is supported by the evidence) through understanding the approach is the condition for members of the community to accept the hypothesis as fact and incorporate it into emerging narratives of nature, the primary result communicated through the research

---

[27] Kuhn's (1977) notion of 'exemplar', as one reading of his term 'paradigm', refers to theoretical solutions to a problem. Here, instead, the focus is on practices including both experimental techniques and reasoning strategies, which exemplify an approach.

narrative is the approach itself, as exemplified in the particular activities reported. Rendering the events generic, by stylistic means, helps members of the community to familiarize themselves with the approach as generally applicable to a shared problem.[28]

References

Beadle, G. W. (1945). 'Biochemical Genetics'. *Chemical Reviews* 37: 15–96.

Beadle, G. W., and B. Ephrussi (1936). 'The Differentiation of Eye Pigments in Drosophila as Studied by Transplantation'. *Genetics* 21.3: 225–247 www.ncbi.nlm.nih.gov/pmc/articles/PMC1208671/pdf/225.pdf.

Berry, Dominic (2021). 'Narrative and Epistemic Positioning: The Case of the Dandelion Pilot'. In Z. Pirtle, D. Tomblin and G. Madhaven, eds. *Engineering and Philosophy: Reimagining Technology and Social Progress*. Cham: Springer, 123–139.

Burian, R. M. (2007). 'On MicroRNA and the Need for Exploratory Experimentation in Post-Genomic Molecular Biology'. *History and Philosophy of the Life Sciences* 29.3: 285–311.

Burian, R. M., J. Gayon and D. T. Zallen (1991). 'Boris Ephrussi and the Synthesis of Genetics and Embryology'. In S. F. Gilbert, ed. *A Conceptual History of Modern Embryology*. New York: Plenum Press, 207–227.

Chatman, S. (1978). *Story and Discourse: Narrative Structure in Fiction and Film*. Ithaca, NY: Cornell University Press.

Clancy, C. W. (1942). 'The Development of Eye Colors in Drosophila Melanogaster: Further Studies on the Mutant Claret'. *Genetics* 27.4: 417–440.

Currie, A. (2015). 'Philosophy of Science and the Curse of the Case Study'. In C. Daly, ed. *The Palgrave Handbook of Philosophical Methods*. London: Palgrave Macmillan, 553–572.

Day, R. A. (1989). 'The Origins of the Scientific Paper: The IMRaD Format'. *American Medical Writers Association Journal* 4.2: 16–18.

Ephrussi, B., and G. W. Beadle (1936). 'A Technique of Transplantation for Drosophila'. *American Naturalist* 70.728: 218–225.

Fludernik, M. (2020). 'Factual Narration in Narratology'. In M. Fludernik and M. L. Ryan, eds. *Narrative Factuality: A Handbook*. Berlin: De Gruyter, 51–74.

Genette, G. (1980). *Narrative Discourse*. Trans. Jane E. Lewin. Ithaca, NY: Cornell University Press.

(1990). 'Fictional Narrative, Factual Narrative'. *Poetics Today* 11.4: 755–774.

Giere, R. N. (2006). *Scientific Perspectivism*. Chicago: University of Chicago Press.

Harré, R. (1990). 'Some Narrative Conventions of Scientific Discourse'. In C. Nash, ed. *Narrative in Culture: The Uses of Storytelling in the Sciences, Philosophy, and Literature*. London: Routledge, 99–118.

Harwood, J. (1993). *Styles of Scientific Thought: The German Genetics Community, 1900–1933*. Chicago: University of Chicago Press.

[28] The research was funded by the Deutsche Forschungsgemeinschaft (DFG, German Research Foundation) – 362545428. I thank the editors, reviewers, and participants of the authors' workshop for helpful feedback. *Narrative Science* book: This project has received funding from the European Research Council under the European Union's Horizon 2020 research and innovation programme (grant agreement No. 694732). www.narrative-science.org/.

Holmes, F. L., J. Renn and H.-J. Rheinberger (2003). *Reworking the Bench: Research Notebooks in the History of Science*. Dordrecht: Kluwer Academic Publishers.

Iser, W. (1978). *The Act of Reading: A Theory of Aesthetic Response*. Baltimore, MD: Johns Hopkins University Press.

Kay, L. E. (1993). *The Molecular Vision of Life: Caltech, the Rockefeller Foundation, and the Rise of the New Biology*. New York: Oxford University Press.

Knorr-Cetina, K. D. (1981). *The Manufacture of Knowledge: An Essay on the Constructivist and Contextual Nature of Science*. Oxford: Pergamon Press.

Kohler, R. E. (1994). *Lords of the Fly: Drosophila Genetics and the Experimental Life*. Chicago: University of Chicago Press.

Kuhn, T. S. (1977). *The Essential Tension: Selected Studies in Scientific Tradition and Change*. Chicago: University of Chicago Press.

Latour, B. (1987). *Science in Action: How to Follow Scientists and Engineers through Society*. Cambridge, MA: Harvard University Press.

Love, A. C. (2014). 'The Erotetic Organization of Developmental Biology'. In A. Minelli and T. Pradeu, eds. *Towards a Theory of Development*. Oxford: Oxford University Press, 33–55.

Meunier, R. (2019). 'Project Knowledge and Its Resituation in the Design of Research Projects: Seymour Benzer's Behavioral Genetics, 1965–1974'. *Studies in History and Philosophy of Science* 77: 39–53.

Morgan, Mary S. (2017). 'Narrative Ordering and Explanation'. *Studies in History and Philosophy of Science Part A* 62: 86–97.

Morgan, Mary S., and M. Norton Wise (2017). 'Narrative Science and Narrative Knowing: Introduction to Special Issue on Narrative Science'. *Studies in History and Philosophy of Science Part A* 62: 1–5.

Myers, G. (1990). *Writing Biology: Texts in the Social Construction of Scientific Knowledge*. Madison: University of Wisconsin Press.

Polanyi, L. (1982). 'Linguistic and Social Constraints on Storytelling'. *Journal of Pragmatics* 6.5: 509–524.

Ravin, A. W. (1977). 'The Gene as Catalyst; the Gene as Organism'. *Studies in the History of Biology* 1: 1–45.

Rheinberger, H.-J. (1997). *Toward a History of Epistemic Things: Synthesizing Proteins in the Test Tube*. Stanford, CA: Stanford University Press.

(2000). 'Ephestia: The Experimental Design of Alfred Kühn's Physiological Developmental Genetics'. *Journal of the History of Biology* 33.3: 535–576.

Ryan, M.-L. (2007). 'Toward a Definition of Narrative'. In D. Herman, ed. *The Cambridge Companion to Narrative*. Cambridge: Cambridge University Press, 22–35.

(2008). 'Interactive Narrative, Plot Types, and Interpersonal Relations'. In U. Spierling and N. Szilas, eds. *Interactive Storytelling*. Berlin: Springer, 6–13.

Sapp, J. (1987). *Beyond the Gene: Cytoplasmic Inheritance and the Struggle for Authority in Genetics*. New York: Oxford University Press.

Schickore, J. (2008). 'Doing Science, Writing Science'. *Philosophy of Science* 75.3: 323–343.

Sternberg, M. (1978). *Expositional Modes and Temporal Ordering in Fiction*. Baltimore, MD: Johns Hopkins University Press.

Sturtevant, A. H. (1932). 'The Use of Mosaics in the Study of the Developmental Effects of Genes'. In D. F. Jones, ed. *Proceedings of the Sixth International*

Congress of Genetics, Ithaca, New York, 1932. vol. 1. *Transactions and General Addresses*. Menasha, WI: Brooklyn Botanic Garden, 304–307.

Van Langenhove, L., and R. Harré (1999). 'Positioning and the Writing of Science'. In R. Harré and L. Van Langenhove, eds. *Positioning Theory: Moral Contexts of Intentional Action*. Oxford: Blackwell, 102–115.

Waters, C. K. (2004). 'What Was Classical Genetics?' *Studies in History and Philosophy of Science* 35.4: 783–809.

13 Thick and Thin Chemical Narratives

Mat Paskins

Abstract
This chapter introduces a distinction between two sorts of scientific narrative, modelled on Ted Porter's discussion of thick and thin description. In *thin* narratives, sequences of processes and experimental interventions are presented in a highly conventionalized form, their notation often assembled from a stock of familiar elements. *Thick* narratives, by contrast, offer a greater degree of context and contingency and may be attentive to social, environmental and other considerations. The distinction is discussed with examples from chemistry; I suggest that chemical reaction schemes, written to describe organic syntheses, are examples of thin narratives. But some chemists, as well as historians, geographers and sociologists who study chemistry, have expressed reservations about what such accounts leave out, and seek to develop modes for narrating chemical processes, experiments and impacts which can provide a thicker account.

13.1 Introduction

This chapter is about the role of narratives in chemistry. Recent studies by historians and philosophers of science have argued that narratives play an important part in shaping scientific explanations; narratives are not, according to this view, only concerned with rhetoric or communication, and not an added extra, but integral to the work of social and natural sciences. In Mary Morgan's concise definition, 'what narratives do above all else is create a productive order amongst materials with the purpose to answer why and how questions' (Morgan 2017: 86).

Notions of narrative are not alien to existing discussions of chemistry: most notably, the Nobel Prize-winning organic chemist Roald Hoffmann has argued that chemical findings should be given narrative form, and similar arguments are present (or at least implicit) in some chemical publications, process ontologies of chemistry and historians' and social scientists' critical accounts of chemistry. Despite their differences, these claims are based on a shared understanding of the purpose of narrative which goes beyond attention to productive order: they suggest that narratives should be used to challenge the conventional demarcations of chemical accounts and 'let the world back in' by incorporating contingencies, aspects of decision-making, social dynamics and the inter-actions between humans and chemical substances which are not usually included within the chemical literature. All continue to bring materials together, to answer questions – they are thus still narratives in Morgan's sense – but they also proceed contrastively, by trying to offer something beyond the conventions of writing in chemistry. These more capacious narratives contrast with the extremely terse form usually adopted by chemical publica-tions. I will call the conventional presentation of chemical findings, 'thin narratives', and the more capacious ones recommended by some chemists, philosophers and historians, 'thick narratives'.

My distinction between the thick and the thin is modelled on the anthropolo-gist Clifford Geertz's (1973) celebrated discussion of 'thick description'. Geertz gave the example of describing someone who was winking, first devel-oped by the philosopher Gilbert Ryle. We could describe a wink in physio-logical terms – through a very specific sequence of muscle contractions, or more simply in terms of what we observe directly. Or we could say something like, *the man winked conspiratorially, according to a cue we had agreed beforehand, and I was delighted.* The former confines its description to a single plane: that of observable physiological phenomena – Ryle (1947) called it a 'thin' description. The latter incorporates context and intentionality, which cannot just be read directly, but require additional elucidation and the incorporation of considerations behind the immediately observable. It is a 'thick' description. By extension, a thin narrative is a sequence or productive order, all of whose materials are presented as closely interrelated and condu-cing to the same purpose, and which can readily be transferred from one situation to another.[1] The thin narrative may also be presented in a formal language, which encodes relations and interactions between the entities involved in the narrative. A thick narrative, by contrast, is one which incorpor-ates more context and considerations which may not be directly related to the explanatory task at hand, and which may be more difficult to move around.

The distinction between thin and thick descriptions carries normative impli-cations. Geertz thought that anthropology needed thick descriptions; that its accounts would be incomplete and misleading without them. Similarly, the chemists and writers in chemistry who have called for the use of narrative form argue that understanding of chemical processes and chemists' decision-making will be impoverished without the incorporation of elements which are usually not found in works of chemistry. But the difference between the thick and the

---

[1] In remaining fixed when transferred between contexts, thin narratives contrast with the medical anecdotes studied by Hurwitz (Chapter 17).

thin has been understood in a much wider sense as well. The historian Ted Porter (2012) argues that the institutional and bureaucratic structures of modernity tend to privilege thin descriptions and to denigrate thick ones, and that natural sciences have been justified through an appeal to thinness, sometimes even changing their own thickets of practices and overlooking the persistence of skilled judgement in response to the pressure to offer thin descriptions.

I think that Porter is right to claim that thin descriptions (and thin narratives) are characteristic products of modernity, and that it has often been a chief aim of historical and sociological analysis to restore a measure of thickness. The views of chemistry discussed in this chapter are examples of arguments which have exactly this goal in mind. Nevertheless, Porter's view requires two qualifications. First, we should not give the impression that thin descriptions and narratives are impoverished, because this risks overlooking the functions which they serve, such as providing a condensed, unitary record of chemical reactions, or shared format for planning out new chemical syntheses. Those functions may come with considerable problems, but that does not imply they are unimportant, and indeed they are of considerable utility to working chemists.

Second, thickening can be seen as an end in its own right, an obvious good. But, as the examples discussed in this chapter indicate, different attempts to thicken a thin narrative can have rather divergent aspirations, incorporate details of different kinds, and also make significant omissions. As a result, even thick narratives can look somewhat thin if the goal is to provide a completely comprehensive account. This can be a strength, as long as thickening in itself is not seen as a way to escape the troubles of thinness, or a way to offer the 'whole story' which lurks behind the thin surface.

In this chapter, I describe and analyse thin and thick chemical narratives, using the example of synthetic reaction schemes linked to a 'classic' synthesis from the history of chemistry: Robert Robinson's 'one pot' production of tropinone, which was accomplished in 1917. In section 13.2, I present a twenty-first-century rendering of the tropinone reaction scheme, as well as its 1917 counterpart, and use work by the chemist-historian Pierre Laszlo to indicate some of the reasons that chemists may prefer to present their findings in such a thin form. Sections 13.3 and 13.4 contrast two kinds of arguments that conventional presentations of chemical results are deficient on the grounds of their thinness – those employed by chemists and those advocated by analysts of the science, respectively – and explore how such attempts played out in repeated retellings of Robinson's tropinone synthesis. This leads me, finally, to consider some implications of thinking in terms of thick and thin narratives for historical and philosophical writing about chemistry.

Before I do so, however, I want to introduce my historical case study of a thin chemical narrative which has repeatedly been thickened. The example is drawn from the career of celebrated organic chemist Robert Robinson. Born in Derbyshire in 1886, Robinson (d. 1975) would acquire a reputation as one of the foremost organic chemists of the first half of the twentieth century and become President of the Royal Society and an advisor to the British government on a range of chemical topics, including colonial development. In 1917, Robinson achieved a synthesis of the alkaloid tropinone that significantly simplified the previous multi-stage, and therefore highly inefficient, scheme. Robinson's scientific paper on the synthesis was published in the same year and detailed how he used counter-intuitive chemical starting products to produce tropinone at room temperature, and without any extremes of alkalinity or acidity. Furthermore, the process involved several reactions which led on from one another without requiring further intervention on the part of the chemist. These features of the synthesis led to its becoming one of the foundational works for Robinson's reputation as a significant synthetic chemist, and to its elevation to the status of a synthetic 'classic' – discussed in textbooks and cited as an inspiration by chemists even now (Medley and Movassaghi 2013). As we will see in section 13.3, Robinson's tropinone synthesis has been repeatedly retold by chemists, and was the subject of a sustained historical investigation by Robinson's one-time student, the Australian biochemist Arthur Birch.

13.2   Synthetic Reaction Schemes as Thin Narratives

Reaction schemes are one of the characteristic ways in which organic chemists plan and record their activities; it is therefore not surprising that Robinson's landmark publication on the one-pot synthesis of tropinone included such a scheme. Drawing on discussions by Robert Meunier (Chapter 12), Line Andersen (Chapter 19), Norton Wise (Chapter 22) and Andrew Hopkins (Chapter 4) from elsewhere in this volume, this section will discuss some of the features which make reaction schemes distinctive as thin narratives, as well as ways in which they are similar to scientific narratives found in other domains.

Figure 13.1 is taken from a 2013 reconsideration of Robinson's 'landmark' synthesis of tropinone and records a reaction scheme for the synthesis according to twenty-first-century conventions. Read in a clockwise direction, starting in the top left, the scheme shows the ways in which two starting products are subjected to various operations – diluted, reacted with other chemical substances, and so on – which change them into a series of intermediate forms, which gradually become more and more similar to the desired final product (tropinone – see molecule labelled 1 in Figure 13.1). The synthesis of complex natural products can involve many hundreds of separate stages, although this version of the tropinone synthesis only involves three intermediate stages. Indeed, from a chemist's point of view, what is striking about this reaction is

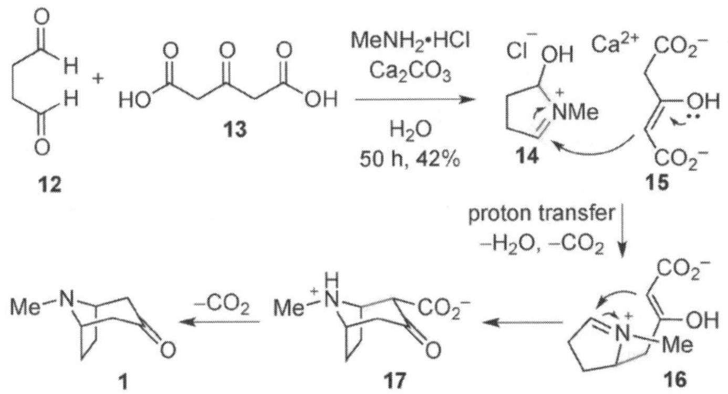

Figure 13.1 **Modern representation of Robinson's 'landmark' synthesis of tropinone**
Source: Medley and Movassaghi 2013: 10775–10777.

that a considerable amount of change happens in only a few stages. Each stage consists of one or several structural formulae: diagrams through which chemists represent both the composition of chemical substances and their spatial arrangement; knowledge of composition and structure helps chemists to construct explanations about how chemical substances will react with one another. Stages in the scheme occur in a particular sequence of reactions, where structural formulae indicate both the protagonists of the synthesis (the chemical substances which play a part in it) and the functions which these chemical substances can play. The transition between the different steps of the synthetic sequence is indicated by straight arrows, while the intermediary reactions are animated, so to speak, by the curved arrows that join together different chemical structures and show the movement of electrons. These curly arrows, which came into widespread use in the second and third decades of the twentieth century, allow the reaction sequence to offer an indication of what is happening at a molecular level to form the desired final chemical substance.

If the reaction scheme provides an ordered sequence of chemical events leading to a single goal (the end product), it is also important to note what the scheme does not show. It does not give an indication of what happens to any chemical substances which do not play a role in subsequent stages of the synthesis, and which are treated as waste products. Similarly, the scheme does not give any indication of the process by which the sequence was arrived at. It also presents a series of operations and reactions which may occur within an organism, or in a laboratory, as though they followed on naturally from each other – the role of the human chemist in performing the synthesis does not appear as distinct from the reactions of chemical substances.

Considered in this way, it makes sense to consider chemical reaction schemes as thin narratives: ordered sequences of chemical events conducing to a single, unified end, in which human intervention is flattened onto the same plane as chemical interactions. Moreover, the reaction scheme resembles a 'narrative of nature', in Robert Meunier's sense. As Meunier describes such a narrative (Chapter 12), it 'relates several events which occur in temporal order and are causally connected', and which is structured into a beginning, middle and end; like the narratives which Meunier discusses, the reaction scheme 'does not recount particular events, but rather a type of event happening countless times'. And the sequence appears to be self-evident: it does not foreground the role of a human experimenter or observer. In other ways, however, the sequence is rather unlike the examples which Meunier gives. It is told in a formal visual language (the structural formulae), which requires a chemical training to understand, rather than providing a neat compact set of events that are (potentially) intelligible to non-scientists. It is not that the reaction sequence cannot be paraphrased, or its events presented verbally; instead, a verbal paraphrase of the sequence of chemical events presented in the reaction scheme would be just as terse and technical as the reaction scheme, just as thin a narrative.

Here, for example, is one such verbal description (of a different synthetic reaction), presented by the chemist, historian and philosopher Pierre Laszlo:

L-Proline was esterified (12) by treating it with MeOH and thionyl chloride at 0°C, followed by Boc protection of secondary amine in dry tetrahydrofuran (THF) using triethyl amine as base at rt, furnishing (13), which on LAH reduction at 0°C in dry THF provided alcohol (14). (Singh et al. 2013; cited in Laszlo 2014: 101)[2]

Unpacking the meaning of this extremely terse sentence, Laszlo argues, relies on the implicit knowledge of the chemist. He attempts two glosses of this piece of 'chemese'. The first seeks to define the provenance of the chemical substances mentioned in the paper – indicating how they would be obtained – and a description of the verbs, suggesting what is turning into what.

The chemical recipient of this treatment is the amino acid proline, as the (natural) L-enantiomer. It can be bought from suppliers of laboratory chemicals. Its esterification means formation of an ester between its carboxylic COOH group and the simplest of alcohols, methanol (here written as MeOH), another commercial chemical, in the presence of thionyl chloride ($SOCl_2$), also commercial. The reaction scheme bears the instruction '0°C-rt, 4 h', in other words, 'dissolve proline and thionyl chloride in methanol, held in a cooling bath, made of water with floating ice cubes, at 0°C and let this mixture return to room temperature (rt) over four hours, before extracting the desired product'. (Laszlo 2014: 101)

---

[2] Note that each of the numbers indicates a structure in the reaction scheme.

Laszlo goes on to unpack the sentence's other implicit meanings, in a manner which draws them out towards the laboratory routines of the chemist:

[T]he stated 'room temperature' in fact has a meaning more elaborate than 'the temperature in the laboratory'. It means 'about 20°C', hence if the actual room temperature is markedly different, one ought to switch on either heating or air-conditioning. (Laszlo 2014: 102)

Laszlo's commentaries give one perspective from which to unpack the sentence, which works outward from the various materials employed in the experimental process to the routines of the laboratory and the chemist's view of her workflow and the conditions in which she is working. Different explications could be given. Laszlo's larger point is that the cognition of chemists involves associative processes, 'molecular polysemy', characterized by continually shifting horizons: new chemical discoveries add extra layers of association to the sentence's existing stock of substances by positing new relations between them. Sentences, such as Laszlo analyses, lack, even as an aspiration, an attempt to fix the meanings of their key terms.

The use of structural formulae and of the terse language of 'chemese' are the reasons that I think we should consider chemical syntheses, as typically presented, as thin narratives, even in their verbal form. The powerful and polysemous formal languages of organic chemistry provide a rich but also restrictive vocabulary for describing what has happened or can happen, in chemical terms – for keeping track of how chemical substances change and the reasons for thinking that they may be used to serve chemists' purposes. Chemists' use of diagrammatic sequences and of language bring accounts of chemical syntheses into a single plane, with all relevant chemical actions and events describable in the same terms. And structural formulae can be used not only to explain what has happened, to record synthetic achievements or to investigate synthetic pathways in living organisms; the formulae can also be used to plan novel syntheses, with the information encoded in the formulae giving a good idea of what approaches might or might not be workable within the laboratory.[3] On their own terms, such 'narratives of nature' are meant to be self-sufficient, a robust and portable sequence of events which can be unpacked by a skilled chemist.

My attempt to consider such terse and formalized sequences as narratives in their own right, however, also indicates their potential instability – reasons that others might call for them to be 'thickened'. Other chapters in this book have

---

[3] Structural formulae provide a unifying representation which in principle encompasses all synthetic possibilities and which thus allows chemists to generate myriad plans, each of which is an order of actions, a potential synthetic story, both human and chemical. In this way, they resemble the branching diagrams of narrativeworthiness discussed by Beatty (2017 and Chapter 20).

aligned narrative with the experiential dimension of interpreting a formalized sequence; this is the gist of Line Andersen's discussion of mathematical proofs (Chapter 19), which Norton Wise (Chapter 22) describes as follows: 'Reading a proof in experiential terms changes what looks to an outsider like a purely formal structure into a natural narrative for the reader; so too the experiential reading enriches the formal language of rigorous proof with the natural language of narrative, for it calls up meanings that the unaided formal language, lacking background and context, cannot convey'. If the opposition is drawn between formal language, on the one hand, and natural language, on the other, then thin narratives only become narrative when they are interpreted by a skilled reader, who is able to supply context and detail that may be absent from the plane of the formal representation itself. In the absence of a reader who possesses such 'scripts', reaction sequences cannot function as narratives. Even so – and I wish to insist on this – organic chemists do not simply animate the dry bones of their thin narratives with their competence, background knowledge and experience; chemists have also argued, explicitly, that the formal languages in which chemical research is presented provide an inadequate account of chemists' reasoning and the character of the interactions between chemical substances which they employ. I will discuss chemists' calls for thickening in the next section of this chapter.

For now, I wish to follow Meunier's lead and ask to what extent these thin chemical narratives might encode their origins in experimental research practices. Meunier (Chapter 12) emphasizes that each part of a narrative of nature 'can be traced back to an episode of research', with narratives of nature 'emerg[ing] gradually from the research literature as facts accepted in a community', with the experimental aspects of 'the methods by which the knowledge was achieved [...] abandoned like ladders once the new state of knowledge is reached'. Is something similar happening with the narrative of nature provided by the reaction scheme? The answer to this question is a qualified yes: indeed, the narrative of nature is related to past experimental work, but in organic chemical synthesis the experimental narrative retains a stronger presence in an organic chemical reaction scheme than would be the case for the biological narratives which Meunier examines.

We can see this with reference to Figure 13.2, which shows the tropinone reaction scheme as presented by Robinson in his 1917 publication. As before, the scheme bears features of a thin narrative: a sequence of chemical events leading to a single outcome (the tropinone molecule – see bottom right), presented in the formal language of structural formulae, with no explicit indication of the researcher's interventions or the laboratory context. But, if we compare Figure 13.2 with Figure 13.1, we note an important difference in the way the structural formulae are presented. As Laszlo (2001) remarks, to the eye of the present-day chemist, the structural formulae found in Figure 13.2 and

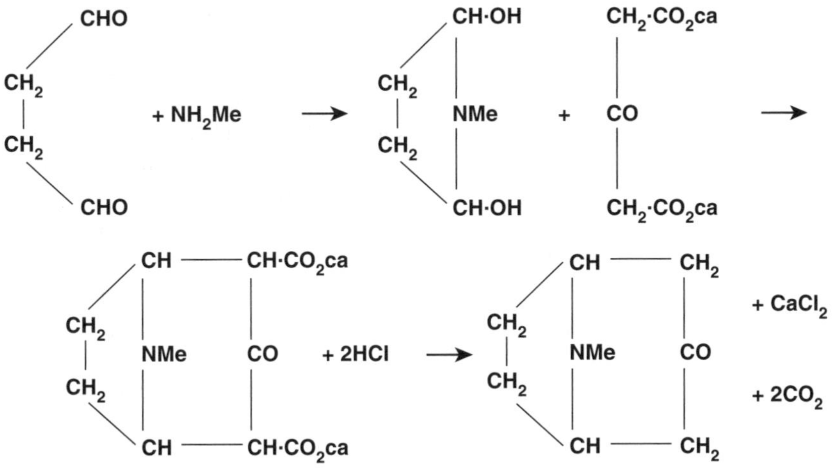

Figure 13.2 **Robinson's original representation of 'A Synthesis of Tropinone'**
Source: Robinson (1917: 762–768).

similar publications look like primitive attempts to capture the spatial arrangement of chemical substances. But this is a historical mirage. The way in which
late nineteenth- and early twentieth-century chemists used structural formulae,
Laszlo argues, was primarily to relate their experimental investigations to the
edifice of organic chemistry, to situate new findings in relation to existing work,
and to draw the map of relations between chemical substances. The formula
'spelled out to its proponent a historical account of how it came to be, of how it
had been slowly and carefully wrought. A formula was the sum total of the
work, of the practical operations, of the inter-relating to already known compounds, which had gone into its elucidation' (Laszlo 2001: 55). As such, the
formula amounted to a kind of 'condense[d] [. . .] narrative' (Laszlo 2001: xx),
whose history would need to be unpacked by a skilled chemist familiar with the
relevant literature.[4] In other words, the structural formulae of the late nineteenth and early twentieth centuries encoded what Meunier terms 'research
narratives' – although once again their narrative qualities were not obvious to
the non-specialist, and had to be unpacked. It is only on the basis of hindsight,
Laszlo says, that present-day chemists might see the structural formulae of the
late nineteenth and early twentieth centuries as continuous with those of
present-day chemistry. We might even say that these historical research narratives are so thin, bound so tightly into a single plane, that their practically and

---

[4] Similarly, geologists read seemingly descriptive statements as a temporal narrative of changes
   undergone by a particular feature (see Hopkins, Chapter 4).

epistemically significant details cannot easily be recovered by today's skilled practitioner in synthesis.

Before examining the different styles of thickening, I want to note some of the distinctive uses which a thin narrative could play in the hands of a chemist like Robinson. The same year that Robinson published his laboratory synthesis, he wrote and published a second paper proposing that he might have found a plausible pathway for the formation of alkaloids in living plants. This claim was absent from his first paper, which instead positioned tropinone as a precursor to a number of products of commercial and medical significance. Robinson's new claim relied on the reaction's status as a thin narrative. That is, it was a scheme that could be picked up from one context and inserted into another, without changing significantly. Contemporary textbooks show Robinson's speculations being reported respectfully, and alongside the proposals of other chemists; in the 1910s and 1920s, experimental methods were not available to trace the formation of chemical substances directly. This changed in the early 1930s, with the development of carbon tracing techniques; initially, Robinson's proposal appeared to have been borne out in practice, although subsequent experimental findings cast doubt on its correctness.

Robinson maintained his distance from experimental attempts to confirm his speculation and was even a little scornful of them. The Australian natural products chemist Arthur Birch, who was at one time Robinson's student, recalled that Robinson was reluctant to take 'pedestrian, even if obviously necessary steps beyond initial inspiration', and would even claim to be disappointed if his findings were confirmed. As a result, 'if Robinson correctly "conceived and envisaged" a reaction mechanism [...] he thought he had "proved" it' (Birch 1993: 282). For Robinson, the venturesome daring of the thin narratives of organic chemistry was all-important: a way to avoid becoming bogged down in the minutiae of subsequent development.

13.3   The Pot Thickens: Chemists' Claims

In this section, I will discuss some of the ways in which chemists have sought to thicken the thin narratives described in the previous section, beginning with arguments by the Nobel laureate, poet and playwright Roald Hoffmann. Then I will look at two other sorts of narrative thickening which chemists have employed, which proceed by emphasizing *contrastive* and *contingent* aspects of the chemical story.

Roald Hoffmann (2012: 88) argues that narrative gives a way to 'construct with ease an aesthetic of the complicated, by adumbrating reasons and causes [...] structuring a narrative to make up for the lack of simplicity'. In other words, the interactions between chemical substances which characterize chemical explanations and the decisions of human chemists which impact on chemical research programmes are highly particular, involving contingencies and speculations and evaluations in terms of human interest in order to make sense. Hoffmann aligns scientific narratives with literary ones on three grounds – a shared approach to temporality, causation and human interest – and he particularly emphasizes the greater narrative satisfactions which are often found in oral seminar presentations than in published scientific papers. In drawing his distinction between narrative and information, Hoffman quotes from the philosopher Walter Benjamin: information can only communicate novelty, whereas a story 'does not expend itself. It preserves and concentrates its strength and is capable of releasing it even after a long time' (Benjamin 1968: 81). In the terms which I am using in this chapter, Hoffmann offers a call for narrative thickening – for getting behind the surface of the conventional chemical article to explain the human dynamics and non-human particularities that have shaped chemical research. In Hoffmann's view, the role of narratives in chemistry should be taken seriously as a way for chemists to be clearer about how they actually think and work (as opposed to idealizations which would present chemistry as an affair of discovering universally applicable laws). Hoffmann's position has both descriptive and normative implications. He suggests that if we scratch the surface we will see that chemists *do* use narratives as a matter of course; but also that if chemists reflect on how they use narratives this will contribute to a better understanding of their work.

'Classic' syntheses, like Robinson's production of tropinone, come to take on the attributes of narratives in Hoffmann's sense. They are retold for their ingenuity and human interest, to motivate further inquiry, to suggest imitable problem-solving strategies and as part of chemists' professional memory; some chemists also argue that they are worth revisiting repeatedly to allow new lessons to be drawn. In this sense, they are more like stories than like information, in Hoffmann's terms. So, for example, the chemists Jonathan Medley and Mohammad Movassaghi (2013: 10775) wrote almost a hundred years after Robinson's initial synthesis that it had 'continue[d] to serve as an inspiration for the development of new and efficient strategies for complex molecule synthesis'. The tropinone synthesis has been retold by chemists on a number of different occasions over the past century, and these retellings have drawn out a variety of meanings from the synthesis and related it to subsequent chemical work in a number of different ways. In the process, chemists have used *contrasts* to emphasize different aspects of the synthesis, or tried to restore *contingent* historical details or aspects of context which would not be apparent from the elegant, but notably thin, reaction schemes discussed in the previous section.

A similar discontent to the one which Hoffmann expresses with the terseness of conventional chemical publications can be detected in some twentieth-century publications on organic chemical synthesis. Complex, multi-stage syntheses can take many researchers many years to achieve, but the final publication may ignore possible routes which were not taken, or which were successful but proved to be less efficient or in other ways less desirable than the final synthetic pathway. In an article from 1976, the chemist Ivan Ernest explicitly tries to challenge this tendency by reconstructing in some detail the plans which the research group made and the obstacles which led them to give up the approaches which had initially appeared promising. Rather than presenting the final synthesis as an edifice which could only adopt one form, this method of presentation emphasizes the chemists' decision-making, and the interaction between their plans and what they found in the laboratory. And rather than presenting the structural formulae of the reaction sequence simply as stepping stones towards a predetermined end, Ernest's article (1976) emphasizes that each stage of the synthesis should be considered as a node, a moment when several different decisions may be possible. Like Hoffmann's view of narrative in chemistry, Ernst's article emphasizes contingency and the human interest of chemical decision-making in the laboratory, giving a more complex and nuanced human story about what this kind of experimentation involves. In other respects, though, it does not diverge significantly from the conventional presentation of thin chemical narratives – it is still presented chiefly in the form of structural formulae, and its presentation is based chiefly on laying different routes alongside each other, giving additional clarity to the decisions made in the final synthesis by comparing it with paths not taken – what could have happened but did not. I call this *contrastive thickening* because it contributes to the scientific explanation by allowing for a contrast between the final decisions which were made and other paths which could have been taken. Every event in the narrative thus exists in the shadow of some other possibility; what did happen can be compared with what did not.

Beyond telling different ways in which things can happen, chemically, to allow the desired outcome to be reached, contrastive thickening also introduces a different way of thinking about the shape of the whole synthesis and what motivates the relations between its different stages. For example, when Robinson's reaction scheme for synthesizing tropinone is contrasted with that proposed by German organic chemist Richard Willstätter in 1887, contemporary chemists evoke notions of 'brute force' and an 'old style' of synthesis to describe Willstätter's approach. Robinson's scheme contrasts as a far more efficient experimental methodology, and the first glimpse of a more rational approach towards synthetic planning, which is based on starting with the final form of a molecule and then dividing it up.

Contrastive thickening tries to show that the final form of a chemical synthesis could have turned out differently, but does not make significant changes to the terse manner in which chemical syntheses are presented – Ernst's article is still narrated primarily in 'chemese'. *Contingent* thickening, in contrast, proceeds by fuller narration. For instance, Ernst's sense that conventional publications on synthesis failed to give the whole story was also cited as inspiration in the first volume of the book series *Strategies and Tactics in Organic Synthesis*, a collection of papers in which chemists were invited to reflect on the contingencies, human factors and tangled paths of their experimental work. The chapters adopt an avowedly narrative style, and emphasize the prolonged difficulty of synthetic work as well as its eventual achievements. Details include serendipitous discoveries in the chemical literature; and discussions of sequencing syntheses so that their more tricky or untested parts are not attempted at the end, putting previous work into jeopardy. These narrative approaches are intended to stir reflection on problem-solving, and how chemists do not rely on the formal language of structural formulae and planning primarily in their synthetic work. They also share with Hoffmann the goal of keeping chemists motivated and the less codifiable aspects of synthetic knowledge in clear view. The Harvard chemist E. J. Corey writes in his preface to the third volume of the series that

the book conveys much more of the history, trials, tribulations, surprise events (both negative and positive), and excitement of synthesis than can be found in the original publications of the chemical literature. One can even appreciate the personalities and the human elements that have shaped the realities of each story. But, above all, each of these chapters tells a tale of what is required for success when challenging problems are attacked at the frontiers of synthetic science. (Corey 1991: xv)

In Corey's view, it is easy to think of synthetic chemistry as 'mature' because it has grown more 'sophisticated and powerful' over the past two centuries. But the impression of maturity belies the fact 'that there is still much to be done' and that the 'chemistry of today will be viewed as primitive a century from now'. As such, it is important that 'accurate and clear accounts of the events and ideas of synthetic chemistry' should be available to the chemists of the future, lest they be misled into thinking that chemistry has become routine. Thickening, in this *contingent* form, reintroduces research narratives alongside the thin narratives of nature for the benefit of the discipline of chemistry: motivation and inculcation of junior researchers into the culture of synthetic research.

13.4   Analysts' Narratives: Processual and Contextual Thickening

I now want to discuss two other ways of thickening chemical narratives, which I will call *processual* and *contextual*. Whereas the thickenings discussed in section 13.3 have been developed by chemists themselves, accounts of processual and contextual thickening have been developed primarily by analysts of chemistry – philosophers, and historians and social scientists, respectively. The primary goal of these thicker accounts is not to offer a more complete record of laboratory activity in order to assist with chemists' own activities, but rather to move beyond the plane of the reaction in selecting what requires consideration in recounting chemical processes. Processual and contextual thickenings work to shift the focus of chemical narratives – calling into question the range of humans and non-humans who should be considered as the primary agents of chemistry, the actions and motivations which are held relevant and worthy of discussion, and the locations in which chemistry occurs. These ways of thickening, furthermore, open up the notion of chemical beginnings and endings by raising questions of how some chemical substances come to be available for chemists to study, and of what happens to chemical substances after chemists have finished using them.

To start with *processual* thickening, then. Some philosophers of chemistry have offered 'process ontologies', guided by the view that philosophy should give accounts of processes and the dynamic aspects of being. As the science of transformations of matter, chemistry can be treated in such dynamic terms, which also call into question the seeming fixity of the substances which chemists employ. These arguments proceed from two related claims: first, Gaston Bachelard's (1968: 45) view that the substances that chemistry studies require extensive purification, and hence 'true chemical substances are the products of technique rather than bodies found in reality'. In this view, the artificiality of chemical substances used in the laboratory circumscribes the types of stuff which are amenable to chemical analysis – samples taken from the messy world are therefore to be understood to the extent that they conform to what chemists can do with their artificial materials. The second claim is that, in the words of A. N. Whitehead (1978: 80), 'a molecule is a historic route of actual occasions; and such a route is an "event"'. What Whitehead meant was that the chemist's molecules arise from sequences of specific actions, whether constructed in the laboratory or found outside. So chemistry deals, above all, with processes – which may be occurring on different scales – rather than with fixed substances.

In his metaphysics of chemistry, the chemist Joseph Earley builds on these insights to claim that chemical substances are historically evolved, in the manner of other evolved systems, and have a vast array of potentials, but that in practice these are subject to considerable path dependencies, as '[e]very sample has a history (usually unknown and untold) that specifies its current context and limits the range of available futures' (Earley 2015: 226). In broad terms, Earley is observing that material history and institutional constraints matter for the definition of chemical substances – and this sounds like he is calling for thicker narratives of chemistry, which take these other factors into account. But, while his philosophical arguments can be read in this way, he also cautions that many of the relevant histories of chemical substances used in the laboratory are 'unknown and untold', and that to the extent their origins are unknowable it is not possible to construct narratives about them. This view suggests the need for a measure of caution concerning the extent to which narratives of chemistry can be thickened to incorporate all the relevant contributory historical factors. Although this chemical metaphysics might sound like an abstract warning, it touches on some of the factors which are described in the *Strategy and Tactics* research narratives – especially the impact on synthesis efforts of which materials happen to be locally obtainable.

I quoted from Birch's prolonged investigation into Robinson's synthesis earlier in this chapter; now I want to say a bit more about what he was trying to achieve and how attending to the contingent history of Robinson's materials helped him to do so. Birch had trained as an organic chemist but made his professional career as a biochemist. He was extremely sensitive to the differences in method and experimental technique between organic chemistry and biochemistry, and suspicious of attempts to claim that practitioners from the two fields could talk straightforwardly to each other, without taking such differences into account. As part of this wider argument, Birch condemned what he characterized as the mythology which had grown up around Robinson's synthesis – particularly the claim that Robinson had been inspired primarily by an attempt to imitate the natural process by which plants synthesize alkaloids. In an effort to challenge this narrative, Birch interrogated its chronology, drawing on both documentary and material evidence. He noted that Robinson had been interested in a somewhat similar synthesis some years earlier, as a result of a theoretical interest in the structure of alkaloid skeletons which he had developed in discussion with his colleague Arthur Lapworth. Robinson's initial experimental work for a one-pot type of synthesis, Birch showed, had taken place when he was based in Sydney. Birch even succeeded in tracking down the original bottle of one of the chemical reagents which Robinson had employed in his experiments. As a result, argued Birch, Robinson's motivations for attempting the tropinone synthesis could not be reduced straightforwardly to an attempt at bio-mimicry, and the synthesis should not be remembered as a precursor to a subsequent unification between organic chemistry and biochemistry. As Birch wrote, 'the chemist's natural products [. . .] tend to mark the diversity of organisms by their sporadic occurrences, whereas the biochemist's materials tend to represent the unity of living matter'; as a result 'the biochemists in a search for generalities have largely ignored the chemist's compounds' (Birch 1976: 224). Digging into Robinson's legacy, and locating it within the distinct material and processual culture of organic chemistry, gave Birch a way to demonstrate the tensions between different chemical subfields, their different ways of proceeding and the different entities which they considered. Birch also noted that Robinson's programme in Sydney may have been guided in part by the difficulty of obtaining chemical supplies in the early part of the twentieth century, and chemists' needs to improvise on the basis of the materials which were locally obtainable. As he noted, most chemical supplies 'came from abroad and normally might take up to six months to arrive for use', with the result that '[s]ynthetic programmes tended to be organized around what was already in the store,' and '[m]uch early Sydney work was on natural products which grow in the local Bush' (Birch 1993: 286).

The intention of Birch's historical narrative is to recover different conceptual and material sources for Robinson's synthetic work, and to caution against too close an equation between the practices of synthetic organic chemistry and those of biochemistry. He draws attention to the material constraints which bounded some of Robinson's synthetic decision-making, but which are absent from the published research narrative. In the process, he draws attention to the specificity of the molecular cast of characters involved in organic chemical synthesis. These moves all recall the aspirations of contingent thickening, described above; but they also suggest a wider set of material and conceptual constraints which might need to be incorporated into an account of how chemists make their decisions. These wider questions are consistent with the goals of processual thickening, even though their intent is not philosophical.

Like processual thickening, *contextual* thickening tries to give chemical narratives depth beyond the laboratory; but it goes beyond material and processual contingencies to explore how chemists' scientific activities might be informed by social, political, historical and environmental dynamics. This kind of thickening thus often shifts chemists away from the centre of accounts of chemistry in favour of other human users of chemical substances (the farmer who employs pesticides, the sunbather with her suntan lotion), the ways in which chemical substances interact with non-humans, and of the complex, ambivalent meanings associated with relations with chemical substances. In general, the goal of such studies is critical – to look beyond the way chemists think about their materials and the impacts of their activities, and to understand chemical substances not as 'isolated functional molecules', but rather in terms of 'extensive relations', as the historian Michelle Murphy puts it (2017: 496). What Murphy means is that chemists' own evaluations of the impacts of chemical substances are too limited and limiting, and are insufficiently attentive to the myriad roles which such substances play.

Again, some retellings of Robinson's tropinone synthesis enact a kind of contextual thickening, by showing that his work was not guided solely by scientific considerations, and nor by the material constraints identified by Birch. In Robinson's own memoir, written late in his life, he talked about what had been happening in his laboratory when he conducted experimental work for the tropinone synthesis at Liverpool University during the First World War. This was a time when the British government had taken a great interest in the utilization of chemical waste products, and the university's laboratories had been turned over to an effort to make pharmaceuticals from the chemical residues of manufacturing explosives. At Liverpool, they made large quantities of the painkillers beta-eucaine and novocaine by saturating acetone with ammonia; the process was improved by adding calcium chloride. Sludge produced by washing explosives with alcohol was brought from the TNT factory at Silvertown and kept in buckets underneath the laboratory benches. Robinson's colleague, the Reverend F. H. Gornall, derived useful intermediates from these wastes and analysed their chemical properties. According to Robinson (1976: 107), 'The improvisation of suitable apparatus required for this work, and the necessity for careful operation and control, was found to be a good substitute for the conventional courses'. Robinson was learning too, and by his own account returned to his own earlier experimental work from Sydney. Among the substances which the chemists sought to produce was atropine, an alkaloid which was closely related to tropinone and which was used in the treatment of people who had been exposed to poison gas. As Robinson recounted:

Atropine was in short supply during the First World War and the knowledge of this fact led me to recall that I had contemplated in Sydney a synthesis of psi-pelletierine from glutardialdehyde, methyl-amine and acetone. This idea was a possible extension of pseudo-base condensations and I realised, at Liverpool, that a synthesis of tropinone [...] might be effected in a similar manner, starting with succindialdehyde, and tropinone could probably be converted to atropine without difficulty. (Robinson 1976: 108)

There is no evidence that Robinson was able to produce significant quantities of atropine for the British war effort, and his synthetic technique would have been unable to produce large quantities of atropine in any case. But, in this telling, a part of his motivation for returning to this synthesis at this time was that the historical and institutional imperatives brought about by wartime restrictions made the pursuit of a highly efficient synthesis more desirable.

13.5   Conclusion: Unfinished Syntheses

This chapter has drawn a distinction between the use of thin and thick narratives in organic chemical synthesis. Thin narratives allow explanations to be given in a terse form which is portable and not dependent on a particular setting or set of historical circumstances; the four styles of thickening identified here all add depth to the apparent planar self-evidence of thin narratives by exploring the role of unsuccessful lines of research, contingencies, the processes by which substances become available for chemical inquiry, or the relations between chemical syntheses and wider historical, political, environmental and material contexts.

Read alongside the other chapters of this book, I hope that the discussion here clarifies some of the ways in which scientists use narratives. As with geological features, chemists sometimes revisit past synthetic achievements, open them up and unpack their implications. Some of these implications may not have been obvious when a synthesis was first conducted, and for this reason some classic syntheses have the inexhaustible, unfinished quality which Roald Hoffmann associates with stories, in contrast with information. Of course this attitude towards the potentials of past experimental work is not present among all chemists and is not applied to all syntheses. But when chemists do draw upon past experimental achievements or reflect on the ways in which the activities of chemistry ought to be documented, they talk quite often in terms of narrative, and with an explicit awareness of the shortcomings of conventional modes of chemical publication – with the sense that the terse formal languages of chemistry fall short in describing how chemists work and think. Much academic history and public discussion of chemical synthesis has focused on the ways in which synthetic decision-making can be made routine, guided by artificial intelligence and planned using the powerful 'paper tools' of chemical nomenclature and structural formulae. Although such an emphasis correctly identifies a major strand in the chemical synthesis of the last 60 years, it has also often been balanced (as in the writings of E. J. Corey, quoted above) with a sense of the abiding complexity and role of contingency which are involved in chemical syntheses. The suggestion here has been that thinking about the difference between thin and thick narratives is a way to preserve a sense of the significance of the two aspects of chemical synthesis.

In Ted Porter's contrast between thick and thin descriptions, which I quoted in the introduction to this chapter, thickness indicates the complex, contingent, often intractable world, whereas thinness stands for attempts to corral that world into predictable shape. As chemistry deals with processes which are often complex, contingent and intractable, it is perhaps unsurprising that alongside its very robust reaction schemes there should be repeated calls for thickening – ways to put the world back in. It is important to note, however, some of the differences, and possible overlaps, between the different styles of thickening which I have identified. Because processual and contextual thickening emerge chiefly from analysts' accounts and chemists' historical writings rather than chemical research publications, it is tempting to see them as offering different forms of narrative to those discussed in sections 13.2 or 13.3. I will first give some reasons that we might want to draw such a division, in terms of the familiar distinction between 'internalist' and 'externalist' accounts of science, and then indicate the reasons that even though this contrast is suggestive, it is also somewhat misleading. As a first approximation we might associate thin synthetic narratives, and chemists' own contrastive and contingent thickenings, with internalist accounts of chemistry, which seek to understand the development of the science in its own terms, according to its conceptual and experimental developments, but without reference to a larger historical context. Processual and contextual thickenings, on the other hand, might be aligned with an externalist view of the history of chemistry, which seeks to understand what chemists do and the significance of their narratives as emerging from and as feeding back into a wider array of political, social, material, philosophical and environmental considerations. We might therefore say that such externalist perspectives are proper to the narratives of philosophy, of social science, or of history – in other words, that the problems they seek to solve belong to these different disciplines, rather than to chemistry itself. So the information in Robinson's memoir might be useful to a historian constructing an account of the relations between wartime production restrictions and innovations in chemical research, but it would be much less likely to be of use to a chemist trying to develop a new synthesis.

Chemists' use of narratives also has implications for critical analyses of chemistry, especially the work associated with the recent 'chemical turn' by social scientists. The goal of these studies, exemplified earlier in this chapter by Michelle Murphy's work, has been to show the pervasive importance and ambivalent significance of chemical substances for both humans and non-humans, and chemicals' roles in sustaining ways of life as well as causing harms through pollution, poisoning and addiction. In the process, these studies have shifted focus away from laboratories and towards the myriad settings in which chemicals are found and play an active role. In taking the significance of chemistry away from the chemists, however, and displacing the locus of chemical study from laboratories, such studies may also fail to account for the distinctive terms in which chemists understand their science, and narrate their activities. Given the density and complexity of organic chemists' language, there are still few historical and social scientific studies which follow their distinctive practices of narrative ordering in significant detail, or which pursue the retellings of a single chemical synthesis, as I have tried to do here. There are not straightforward ways to incorporate the perspectives of the producers and users of thin chemical narratives into thicker accounts of chemistry without attempting to learn to speak 'chemese'; at the same time, chemists themselves sometimes argue for the need to give a fuller, thicker, account of their activity.[5]

[5] My most sincere thanks to Kim Hajek, without whose editorial acuity and moral support this chapter would not have been completed. *Narrative Science* book: This project has received funding from the European Research Council under the European Union's Horizon 2020 research and innovation programme (grant agreement No. 694732). www.narrative-science.org/.

References

Bachelard, G. (1984). *The Philosophy of No: A Philosophy of the New Scientific Mind.* New York: Orion Press.

Beatty, J. (2017). 'Narrative Possibility and Narrative Explanation'. *Studies in History and Philosophy of Science Part A* 62: 31–41.

Benjamin, W. (1968). 'The Storyteller: Reflections on the Works of Nikolai Leskov'. In *Illuminations.* Trans. Harry Zohn. New York: Schocken Books: 83–110.

Birch, A. J. (1976). 'Chance and Design in Biosynthesis'. *Interdisciplinary Science Reviews* 1.3: 215–239.

(1993). 'Investigating a Scientific Legend: The Tropinone Synthesis of Sir Robert Robinson, F.R.S.' *Notes and Records of the Royal Society of London* 47.2: 277–296.

Corey, E. J. (1991). 'Foreword'. In T. Lindberg, ed. *Strategies and Tactics in Organic Synthesis.* vol. 3. San Diego, CA: Academic Press, Inc.: xv–xvi.

Earley, J. (2015). 'How Properties Hold Together in Substances'. In G. Fisher and E. Scerri, eds. *Essays in the Philosophy of Chemistry.* Oxford: Oxford University Press: 199–234.

Ernest, I. (1976). 'Eine Prostaglandin-Synthese: Strategie und Wirklichkeit'. *Angewandte Chemie* 88.8: 244–252.

Geertz, C. (1973). *The Interpretation of Cultures.* New York: Basic Books.

Hoffmann, R. (2012). *Roald Hoffmann on the Philosophy, Art, and Science of Chemistry.* Oxford: Oxford University Press.

Laszlo, P. (2001). 'Conventionalities in Formula Writing'. In U. Klein, ed. *Tools and Modes of Representation in the Laboratory Sciences.* Dordrecht: Springer, 47–60.

(2014). 'Chemistry, Knowledge through Actions'. *HYLE: International Journal for Philosophy of Chemistry* 20.4: 93–119.

Medley, J. W., and M. Movassaghi (2013). 'Robinson's Landmark Synthesis of Tropinone'. *Chemical Communications* 49.92: 10775–10777.

Morgan, Mary S. (2017). 'Narrative Ordering and Explanation'. *Studies in History and Philosophy of Science Part A* 62: 86–97.

Murphy, M. (2017). 'Alterlife and Decolonial Chemical Relations'. *Cultural Anthropology* 32.4: 494–503.

Porter, T. M. (2012). 'Thin Description: Surface and Depth in Science and Science Studies'. *Osiris* 27: 209–226.

Robinson, R. (1917). 'LXIII. – A Synthesis of Tropinone'. *Journal of the Chemical Society, Transactions* 111: 762–768.

(1976). *Memoirs of a Minor Prophet: 70 Years of Organic Chemistry.* Oxford: Oxford University Press.

Ryle, G. (1947). *The Concept of Mind.* Chicago: University of Chicago Press.

Singh, R., M. K. Parai, S. Mondal and G. Panda (2013). 'Contiguous Generation of Quaternary and Tertiary Stereocenters: One-Pot Synthesis of Chroman-Fused S-Proline-Derived Chiral Oxazepinones'. *Synthetic Communications* 43.2: 253–259.

Whitehead, A. N. (1978). *Process and Reality.* New York: The Free Press.

14 Reporting on Plagues: Epidemiological Reasoning in the Early Twentieth Century

Lukas Engelmann

Abstract

The beginning of modern, twentieth-century epidemiology is usually associated with the introduction of mathematical approaches and formal methods to the field. However, since the late nineteenth century, the nascent field of epidemiology not only developed statistical instruments and stochastic models, but also relied on new forms of narrative to make its claims. This chapter will ask how chronologies of outbreaks, the increasing complexity of causal models and statistical and geographical representations were brought together in epidemiological reasoning. The chapter focuses on three outbreak reports from the third plague pandemic as critical examples. Reports grappled with the unexpected return of a devastating menace from the past, while inadvertently shaping the contours of a modern, scientific argument. Epidemiological reasoning emphasized historical dimensions and temporal structures of epidemics and integrated formalized approaches with empirical descriptions while contributing to the growing rejection of mono-causal explanations for epidemics.

14.1 Introduction

Epidemics make for powerful stories. Ever since Thucydides' account of the Plague of Athens, the epidemic story has joined the ranks of the grand tales of war, terror and devastation. Thucydides' account of the events of a plague in the Hellenic world also gave shape to a genre of writing that has since been copied, developed and expanded by countless witnesses to epidemic events in Western history. Since then, the epidemic narrative has contributed to the chronology through which an epidemic unfolds and has become the principal source to infer meaning and to make sense of epidemic crises. The same narrative has enabled authors to characterize the sweeping and limitless effects of epidemic events and to join aspects of natural phenomena, social conventions and cultural customs implicated in the distribution of plagues. The epidemic narrative, finally, has come to offer a generalized lesson, a common theme or an eternal truth, that the epidemic had laid bare (Page 1953; Wray 2004).

This chapter revisits the position of the epidemic narrative within a significant epistemological transformation. At the end of the nineteenth century, writing about epidemics shifted from an emphasis on storytelling to the production of methods and instruments to elevate an epidemic into the status of a scientific object. Written accounts of epidemic events were no longer judged upon their capacity to invoke lively pictures of terror or to excel in the inference of political lessons from tragic circumstances but were scrutinized within a field increasingly oriented towards a shared understanding of a scientific method.

As Olga Amsterdamska (2005; 2001) points out, epidemiology has a complicated history as a medical science. Without the tradition of the clinic and beyond the experimental and deductive methods of the laboratory, many of epidemiology's early protagonists turned to quantification to defend their work's status as a 'full-fledged science, no different in this respect from other scientific disciplines' (Amsterdamska 2005: 31). However, quantification and medical statistics were not the only resources required to establish the field's authority. Through boundary work, Amsterdamska shows how epidemiologists established epidemics as 'a collective phenomenon', as the field's 'special object' (Amsterdamska 2005: 42). In the quest to establish its unique scientific authority, the field came to rely 'on a wide range of widely used scientific methods' (Amsterdamska 2005: 42), which went far beyond statistics and quantification.

This chapter focuses on narrative reasoning as one of these methods deployed by epidemiologists to account for their special object at a time when the disciplinary boundaries of epidemiology were rather incongruent. The study of epidemics required a generalist dedication to historical accounts, a reliable understanding of medical classification, the capacity to account fluently for social dynamics, while maintaining expertise in the biological variables of contagion and infection. Epidemics were primarily medical events, as they constituted the multiplied occurrence of a specific disease, and most early epidemiologists approached the subject from the vantage point of their medical career. However, since Quetelet, even the medical profession had accepted that the aggregated occurrence of disease might not resemble the dynamics of the individual case.[1] The social body is, after all, not equivalent to the individual, and the spatial and temporal patterns of a series of cases in society followed discrete regularities (Armstrong 1983; Matthews 1995).

---

[1] In a series of publications in the 1830s and 1840s, Adolphe Quetelet aligned the theory of probability with statistical observation. His work on the 'average man' emphasized that any series of random individual acts and attributes will exhibit regularity and predictability, if their statistical summation is taken into account (Matthews 1995: 23).

Most of the historiography of epidemiology has looked to the quantification of epidemiological methods since the mid-nineteenth century to explain how this new object of concern took shape. Epidemics were represented in lists, tables, maps and diagrams to measure and calculate the dynamics of their waxing and waning, and medical statistics had become the dominant framework to envision the distribution of a disease within society (Magnello and Hardy 2002). Narrative reasoning, so the gospel of formal epidemiology goes, took on a secondary position, predominantly concerned with the interpretation and explanation of formalized expressions (Morabia 2013). With this chapter, I challenge the widely held assumption that narrative reasoning lost significance in the formation of a scientific method in epidemiology. Instead, I argue that narrative assumed a new epistemic function in the late nineteenth century, supporting the professional reorganization of the field and shaping what I call here 'epidemiological reasoning'.

This chapter will turn to outbreak reports of the third plague pandemic published between 1894 and 1904 to demonstrate how epidemiologists navigated the complexity of their 'special object'. To develop their account of epidemic events, the authors of the reports contributed to, engaged in and relied on epidemiological reasoning. The second section (14.2) will outline the nature of these reports and contextualize them within the field of medical and colonial reporting practices at the time. In the following sections, I will take in turn three aspects in which the reports' epidemiological reasoning advanced the constitution of epidemics as scientific objects. The third section (14.3) will return to the historical dimension of epidemics, asking how epidemiological reasoning has made epidemics 'known and understandable by revealing how, like a story, they "unfold" in time' (Morgan and Wise 2017: 2). In the fourth section (14.4), the focus will lie on the ordering capacity of epidemiological reasoning to produce epidemic configurations. Through narrative, the authors combine, or rather colligate (Morgan 2017), empirical descriptions, theoretical projections and a range of causal theories to capture the complex characteristics of the outbreak. In the fifth section (14.5) I will revisit the question of formalization in the evaluation of how lists, graphs and maps were positioned within an epidemiological reasoning dedicated to possibilities, conjecture, contradictions and contingency. Narrative is the technology which allows these reports to configure epidemics as more than just a multiplication of individual cases, more than just a result of social and environmental conditions and more than just the workings of a pathogen.

14.2   Outbreak Reports of the Third Plague Pandemic (1894–1952)

The production and circulation of outbreak reports was firmly grounded in a British administrative reporting practice: the Medical Officer of Health reports. As Anne Hardy has demonstrated in her extensive work on the 'epidemic streets', the reports of the medical officers of health were produced from the mid-1800s within a rationale of prevention, and established the provision of 'facts and faithful records about infectious disease' (Hardy 1993: 7). These reports were usually written with a focus on the range of diseases to be found in a specific district or city. Some diseases had also become the subject of dedicated reports during the nineteenth century, which compared and contrasted the occurrence of cholera or typhoid fever in different places (Whooley 2013; Steere-Williams 2020). However, only in the reporting on the third plague pandemic at the end of the nineteenth century do we see the emergence of a sizeable number of comparable reports.

Each of the over one hundred reports on plague was concerned with a city or a region, usually written after an outbreak had ended. A first look at these manuscripts shows them to be highly idiosyncratic pieces of writing, perhaps as much influenced by the authors' interests and professional expertise as by the specific local circumstances in which the outbreak occurred. However, comparing the range of reports published on outbreaks of bubonic plague between 1894 and 1904 allows for an appreciation of structural, stylistic and epistemological commonalities.[2] Over the course of the pandemic, reporting practices were neither discrete nor arbitrary; rather, authors tended to collect, copy, adapt and emulate their colleagues' work. The authors, who were local physicians, medical officers, public health officials or epidemiologists, would write their own account of a plague outbreak with a global audience of like-minded epidemiologists, medical officers and bacteriologists in mind. Archival provenance further suggests that these reports often circulated globally and followed the vectors of the epidemic. The occurrence of novel outbreaks in Buenos Aires or New South Wales appears to have prompted local health officers in these regions to collect outbreak reports from around the world to inform their actions and to adjust their narrative. A key function of the growing global collection of reports was to integrate each local outbreak into the expanding narrative of a global pandemic.

Comparable in form and style, the narrative genre of the epidemiological outbreak report resembles the medical genre of the clinical case report (Hess and Mendelsohn 2010).[3] Like clinicians, the authors of the reports practised epidemiology as an empirical art, dedicated to inductive reasoning and

---

[2] This work of comparison is ongoing in a collaborative project at the University of Edinburgh under the title Plague.TXT (www.ltg.ed.ac.uk/projects/plague-txt/). See Engelmann (2018); Casey et al. (2020).

[3] I have followed this comparison in another edited volume, in which I ask to what extent the outbreak report produces case-based knowledge and if epidemiological knowledge production should be seen as a form of casuistry. With an eye on both the comparability to the clinical case report as well as towards its function as an object of collection, comparison and generalization, I have asked if the genre could be understood as a paper technology (Engelmann 2021).

correlative modes of thinking. Unlike clinicians, the epidemiologist's scope was much less defined. Authors drew from history, clinical medicine, bacteriology, vital statistics, sanitary science, anthropology, sociology and demography. From Porto, San Francisco, Sydney to Hong Kong and Durban, the reports covered significant aspects of the location, ranging from climate trends, descriptions of the built environment, to the social and cultural analysis of populations in urban or rural communities.[4] These elements were bound together to constitute the epidemic narrative, tracing how the epidemic offered new ways of ordering what had appeared before as disparate sources and disconnected information.[5] With sections moving from questions of bacteriology to mortality rates, quarantine measures, outbreaks among rodents, to summaries of the longer history of plague, the narrative colligated disease, environment and population to let the epidemic emerge as a configuration of these coordinates. However, for this narrative to provide a formalized and ordered account of the epidemic – for it to become a scientific account – it was also punctuated with instruments of abstraction and formalization: tables, lists, graphs, and maps.

As such, reports are understood in this chapter as a peculiar global genre of epidemiological reasoning, which was ultimately concerned with producing a robust and global epidemiological definition of plague. This was not achieved just through cross-referencing and intertextual discussions of reports from different places. As a record of events, data and observations tied together by a single disease in a specific place, reports considered the local incident to shape the pandemic of plague as a global object of research.

The three reports discussed below, chosen from over one hundred written on the third plague pandemic, demonstrate three interlinked aspects of epidemiological reasoning: outbreak histories, epidemic configurations and visual formalizations. I have selected one of the first written reports on the emerging epidemic in Hong Kong in 1894, a second from the sprawling and fast-developing outbreak in Bombay, India, and a final one from a South African outbreak in Durban. All three epidemic events occurred within the confines of the British Empire at the time, and were subject to scrutiny, observation and reportage by officers under imperial British command. The reports on plague should therefore also be understood with regard to the long-standing forms of reporting carried out by British colonial officers. These forms included concerns of overseas administration; occasionally reports served as legal evidence for actions taken and they were instruments of stabilizing colonial hierarchies

---

[4] For a preliminary list of reports and locations, mostly limited to the English language, see 'Plague Dot Text': https://github.com/Edinburgh-LTG/PlagueDotTxt.

[5] This condensation of disparate and often polyphonic narratives resembles the way in which Bhattacharyya introduces the reconstruction of historical storms in the Bay of Bengal (Chapter 8).

of power and knowledge (Donato 2018). Reports and their destinations, the colonial archives, furnished the administration with knowledge to govern territory and populations, while establishing difference and hierarchy through 'epistemic violence' (Stoler 2010). Especially with regards to the governance of public health in British India, administrative reporting practices have been shown to have contributed substantially to the formation of common colonial tropes, such as the opacity of the colonial city as well as the pathogenicity of foreign territory. Reporting on outbreaks of plagues was therefore interlinked with evaluating plague's capacity to destabilize colonial rule and to provide evidence about how containment measures contributed to the reinstatement – but also the failure – of colonial power (Echenberg 2007). Ultimately, all reporting on outbreaks of plague in colonies and overseas territories was driven by utopian considerations of hygienic modernity (Rogaski 2004; Engelmann and Lynteris 2020), aiming to stabilize the increasingly fragile image of Europe as a place of immunity and security against epidemic risks.[6]

James Alfred Lowson, the author of the report on plague from Hong Kong in 1894, was a young Scottish doctor and acting superintendent of the civil hospital in Hong Kong by the age of 28 (Solomon 1997). He took on a key role in the outbreak, diagnosed some of the first cases and led early initiatives for rigorous measures to be put in place in the port and against the Chinese population. He remained on the sidelines of bacteriological fame, as the controversy between Kitasato and Yersin unfolded, both claiming to have first identified the bacterium responsible for the plague, later named yersinia pestis (Bibel and Chen 1976).

The second report is one of many written by the Bombay Plague Committee, which was at the time under the chairmanship of James McNabb Campbell (MacNabb Campbell and Mostyn 1898). The Scottish ethnologist had joined the Indian Civil Service in 1869 and served as collector, administrator and commissioner in the municipality of Bombay. In 1897, he succeeded Sir William Gatacre as the chairman of the Plague Committee to encourage cooperation, prevent further riots and contribute to the reinstatement of colonial rule (Evans 2018).

Ernest Hill, the author of the report on plague in Natal, was a member of both the Royal College of Physicians and the Royal College of Surgeons, and from 1897 was appointed health officer to the colony of Natal in South Africa. He authored a number of reports on the health challenges of the colony, notably on suicide as well as malaria outbreaks, and was reportedly involved in ambitious planning to introduce and establish vital statistics overseas (Wright 2006; Hill 1904).

---

[6] This prevailing sentiment is perhaps most clearly outlined by the French epidemiologists and the Pasteurian, Adrien Proust (1897).

14.3 Outbreak Histories

Until the early twentieth century, epidemiology had been a field intertwined with historical methods and narrative accounts. The historical geography of diseases, as exemplified by August Hirsch, had substantial influence on the development of formal accounts of epidemics (Hirsch 1883). Understanding the wider historical formation of a disease was, Hirsch and his contemporaries had argued, fundamental to anticipating which diseases were confined to certain geographies, which diseases occurred with seasonal regularity and how diseases corresponded to what Sydenham had called the epidemic constitution of societies (Susser and Stein 2009). History, in short, was what gave a form to epidemics, and it was historical narrative that enabled differentiation between smallpox, syphilis and phtysis (or tuberculosis) from plague and cholera (Mendelsohn 1998). Investigating the natural history of an epidemic disease was a powerful instrument of generalization and classification. Considering the origins, geographical distributions, stories of migration and relations to wars and famines offered a biographical form to diseases in the history of Western society (Rosenberg 1992b).[7]

It is therefore unsurprising to see most reports opening with some form of appreciation of the history of plague at large. Lowson, in his account of events in Hong Kong, even apologized for his limited access to relevant historical scholarship on the plague. However, revisiting what he had available in Hong Kong, he delved into a historiographical critique of the limited state of scholarship on Asiatic plague history. Knowledge of the historical geography of plague was for Lowson essential in considering how plague might have arrived in Hong Kong from Canton. The Cantonese outbreak had reportedly also begun in 1894, where plague might have been endemic for some time (Lowson 1895: 7).

Similarly, for the South African report, Ernest Hill dedicated his first chapter to the history of plague to emphasize three characteristics of the disease, known from extensive scholarship. He noted its 'indigenous' quality, as the epidemic appeared to persist historically in particular localities. Hill accounted for a predictable periodicity of outbreaks and included the fact that epidemics appear 'interchangeable' between men, rats and mice (Hill 1904: 5–6). From these generalized historical qualities, Hill inferred then a short history of the most recent outbreaks preceding the events in Natal, originating in Hong Kong and a series of outbreaks in India, Australia, and Africa to let the historical arc arrive finally in 1901 in the Cape Colony, from which the disease had most likely spread to Natal in 1903.

---

[7] Similar conclusions can be drawn when turning to the questions of origins, timing and dating in archaeology, as Teather argues in Chapter 6.

The report from India, however, does not refer to the recorded history of plague over previous centuries, but offers a different, perhaps more pertinent, account of outbreak history (MacNabb Campbell and Mostyn 1898). Where the large historical narratives of plague suggest generalization about plague, the repetitive chronologies, or what I call here outbreak history, emphasize a different register of historical reasoning. Without much preamble, the report gives a month-by-month overview of the development of the epidemic from July 1897 to April 1898. It continues on from previous reports that account for the development of the epidemic beginning in August 1896 in Bombay (Evans 2018; Gatacre 1897). Monthly summaries of the epidemic constitute by far the largest section of this report, and each monthly vignette cycles through aspects that the plague committee considered important to record over time in the epidemic diary. Rainfall, mortality and sickness, relief works, staffing, quarantines and migration of people in and out of Bombay were recorded monthly, each enclosed in a short narrative description. This entry for December 1897, for example, marks the beginning of the second outbreak and describes the reasoning for the 'segregation of contacts':

In early December the arrival of infected persons in Bombay, and in many attacks an increase of virulence and infectiousness, made it probable that at an early date the Plague would develop into an epidemic. To prepare for an increase in disease, two measures received the consideration of the Committee. These were the separation of Contacts, of the sick man's family, and the vacating of infected or un-wholesome houses, with the removal of the inmates to Health Camps. (MacNabb Campbell and Mostyn 1898: 12)

For some months, miscellaneous events such as riots or house inspections were added. But overall the author's choice of structure emphasized the temporal characteristics of the epidemic, which offers a sense of how circumstances, case numbers as well as reactive measures changed over time. August 1897 saw plenty of rainfall, with 15.59 inches, and a moderate number of 83 new plague cases. Relief works were required in August, as it was noted that 'the city was infested with numbers of starved idlers whose feeble condition, predisposing to plague, was a menace to public health' (MacNabb Campbell and Mostyn 1898: 4). Quarantine was established on sea routes to prevent importing plague and the total movement of people in and out of the city recorded an excess of 15,224 departures. In November of the same year, rainfall had been zero, while plague cases rose by a dramatic 661 cases. Relief works were in steep decline as movement of people into the city also continued to decrease.

The history of plague crafted in the Bombay report was structured to deliver a picture of the temporal dynamic of the epidemic within a complex configuration. Monthly summaries provide a granular view onto the variability of case numbers, of the changing climatic, social and political conditions in which plague emerged and thrived. This chronological reconstruction of the epidemic was entirely invested in the temporal dynamic of the outbreak.[8] Narrative gave a sense of the beginning, the waxing and the waning of the disease while integrating quantifiable indicators such as case numbers, rainfall and immigration as well as dense descriptions of what the committee perceived to be mitigating measures (poverty relief) and exacerbating circumstances (the immigration of homeless people).

Outbreak history, which considered the series of events that structured the local outbreak from its beginning to its end (if it had been reached), was a key component of reports of the third plague pandemic. Within epidemiological reasoning, this kind of temporal characterization was dislodged from the grand historical portraits of plague. While the latter were concerned with the settled story of what plague was, and how the contours of the epidemic's biography aligned and criss-crossed with sections of the history of the Western world, the former provided a lens for investigation and open-ended speculation. The historical arc of plague provided a hook, a larger, global narrative within which the report's account had to be situated, whereas the outbreak history offered the opportunity to bring the many facets that contributed to the local outbreak into a temporal order.

In contrast to that broad temporal arc, Lowson, in Hong Kong, dedicated only a small section explicitly to the 'time of the outbreak' (Lowson 1895: 30). A sense of the chronology of the Hong Kong outbreak can, however, be traced through all of Lowson's thematic sections. In his discussion of climatic influences, he reasoned on 'the increase of the disease after the rainy season' (Lowson 1895: 5), and in the 'Administrative' section, he provided day-by-day details on how staffing levels at the hospital were arranged and adapted to match the dynamic of the epidemic (Lowson 1895: 26). Lowson's section dedicated to statistics conveys a sense of the sudden growth and then quick slump in case numbers through June and July 1894. Overall, Lowson was eager to impart a picture of the Hong Kong plague as a sudden incident that emerged in April 1894 as 'people were reported fleeing from Canton on account of the plague' (Lowson 1895: 2) and which was expected to end with the strict observation of a list of recommendations provided by Lowson to improve the sanitary state of the city's worst dwellings (Lowson 1895: 26).

---

[8] To some extent, one might argue that these local micro-narratives of plague were aggregated across the outbreak reports to assemble and adjust the larger narratives of the temporal dynamic of plague. Through the reports, local narratives were entrusted with the epistemic capacity to shift the larger picture of how plague behaves, similar to how Darwin considered individual plant life cycles to inform broad arguments about plant movement (see Griffiths, Chapter 7), but not comparable to the seismologists who considered local incidents just to add facets to the larger narrative of the dynamics of earthquakes (see Miyake, Chapter 5).

Where Lowson let varied aspects of the outbreak chronology unfold in parallel, section by section, in the colony of Natal, Hill structured sections of his report very explicitly around the chronology of the outbreak, relating the 'origin', the 'course' as well as the 'spread' and the 'limitation' of the outbreak, each told in a dedicated section. The plague story of Natal began in the 'first weeks of December' 1902, when 'the disease was found to be causing a heavy mortality among rats over a roughly triangular area' at the Veterinary Compound (Hill 1904: 8). However, one month later, rats infected with plague were found in a produce store in the middle of Durban. Suspecting that the disease had been imported, Hill charted the epidemic distribution over time and space, as it spread to five or six further areas in Durban where it prevailed for some time. To characterize the temporal 'course of the epidemic' Hill utilized the metaphor of 'water spilt on a dry surface: a continuous forward progression with occasional branching off shoots, and now and again a return flow' (Hill 1904: 26). After detailed discussion of the relation between plague in rats and humans as well as white and (what Hill described as) native inhabitants, he closed his chronology with a detailed description of local measures put in place to control and end the outbreak.

14.4   Epidemic Configurations

The historian of medicine, Charles Rosenberg, identified two conceptual frameworks through which epidemics – his case was predominantly cholera – were explained until the late nineteenth century. The first, configuration, emphasized a systems view in which epidemics were explained as 'a unique configuration of circumstances' (Rosenberg 1992a: 295), each of which was given equal significance. Communal and social health was imagined as a balanced and integrated relationship between humankind and environmental constituents, in which epidemics appeared not only as the consequence, but also as the origin of disturbance, crisis and catastrophe. Rosenberg's second framework, contamination, prioritized particular identifiable causes for an epidemic event. Where configuration implies holistic concepts, the contamination perspective suggested a disordering element, a *causa vera*, suggestive of reductionist and mono-causal reasoning. As Rosenberg emphasizes, both of these themes have existed since antiquity in epidemiological reasoning, but it is particularly in the late nineteenth century, with the emergence of bacteriological science, that we can see a proliferation of these mutually resistant themes into polemical dichotomies.

Plague reports, however, did not neatly fit within this antagonism. Despite successful identification of the plague pathogen in 1894, and despite historiographical claims of a subsequent laboratory revolution (Cunningham 1992), the epidemic did not lend itself to reductionist attribution of cause and effect between seed and soil (Worboys 2000). As the previous section illustrates, understanding the puzzling configurations on the heel of the introduction of the contaminating pathogen was subject to much deliberation in plague reports. The texts' capacity to integrate questions of contamination and systems of configuration without adopting deterministic models is what I would propose here as a second advantage of epidemiological reasoning. Narrative was essential for a kind of reasoning that offered some breathing space around deterministic theories of cause and effect, while not resolving the question of cause altogether.[9]

This quality is perhaps best observed in the more speculative sections of the reports, where narrative reasoning enabled conjecture and allowed for contradiction. In stark contrast to the sober empirical tone of historical chronology, the reports engaged in intriguing ways with theories of distribution and transmission of plague. Many historians before me have shown that these factors were subject to heated global dispute (Echenberg 2007; Lynteris 2016). The return of plague as a global menace, no longer confined to historical periods as a 'medieval' disease, challenged as many convictions about 'hygienic modernity' (Rogaski 2004) as it supported spurious theories about racial superiority within colonial occupation and exploitation. Each of the authors of our three example reports offers a range of idiosyncratic theories attempting to arrange their observations within available causal concepts. To shed light on the circumstances under which plague moved through the communities of Hong Kong, Bombay and Natal, Lowson considered infection through the soil, the Bombay Plague Committee discussed the problem of infectious buildings, and Hill defended the rat as a probable vector of the disease.

The soil had been, as Christos Lynteris recently argued, a 'sanitary-bacteriological synthesis' (Lynteris 2017). Removed from traditional miasmatic understandings of contagion as an emanation from the ground, the soil became suspicious as a plausible source of infection as well as a reservoir for plague's pathogen. Lowson, who, like many of his contemporaries thought that implicating rats as the cause of plague was 'ridiculous' (Lowson 1895: 4), instead dedicated a full section to infection via the soil. The soil was a likely culprit, he argued, as it explained the geographically limited distribution of plague in the district of Taipingshan. With a vivid description of the living conditions of an area mostly occupied by impoverished Chinese labourers, Lowson drew attention to 'filth everywhere', 'overcrowding', the absence of 'light and ventilation' and basements with floors 'formed of filth-sodden soil' (Lowson 1895: 30).

---

[9] While not a taboo, the hesitancy and reluctance to engage in unambiguous causal theory in epidemiology resembles how Jajdelska (Chapter 18) outlines the resistance against determinant psychological inferences in narrative forms.

Lowson's account of the environmental configuration in Hong Kong was populated with vitriolic and racist descriptions of Chinese living conditions. Overcrowding, filth and the poor and damp state of houses, basements and stores were to him the driving factors of the plague, while latrines were particularly suspicious, as they were 'used by the bulk of the Chinese population'. The danger 'to every healthy person who went into the latrine' could be assessed through a quick 'glance' (Lowson 1895: 28). Remarkable here is not his relentless anti-Chinese sentiment, which was of course a common component of British rule in Hong Kong, but the seamless integration of bacteriological and sanitary perspectives into his reasoning.

'Predisposing causes', as Lowson qualified his perspective, assumed political urgency as authorities concerned themselves with the future of the district. As well as burning the district to the ground and destroying the squalid habitations, was it also necessary to remove an entire layer of soil? After consultations with bacteriologists and a series of experiments, Lowson came to the conclusion that the soil was innocent and that instead common sense should prevail. Resorting to his racist conviction, he concluded that the most 'potent factor in the spread of the epidemic' could be found in the 'filthy habits of the inhabitants' of Taipingshan (Lowson 1895: 32).

A similar reasoning about infective environments structured the writing of the plague committee in India. As a second line of defence, after patient cases had been relocated to hospitals and populations evacuated to quarantined camps, the remaining houses and buildings were perceived as a suspicious and potentially dangerous environment. This was reflected in extensive discussions about the need for thorough disinfection. Fire, as the report stated, was the 'only certain agent for the destruction of infective matter', but its application was considered too risky. Based on undisclosed experience from previous outbreaks, the local committee chose to use perchloride of mercury, followed by thorough lime-washing (MacNabb Campbell and Mostyn 1898: 65). Yet, evidence of the beneficial effect of such operations was difficult to obtain. Previously disinfected premises were, as the author states, not protected against the reintroduction of plague through vermin and people. Some disinfection officers had thrown into doubt the benefit of lime-washing, as bacteriologists had reportedly shown that bacteria thrive in alkaline environments, such as the one provided by hot slaked lime. Regardless, the committee held up against the contradictory perspective and insisted on continuing lime-washing operations, despite the death of 'two or three limewashing coolies', as its use following other means of disinfection was 'invaluable for sweetening and brightening up the rooms' (MacNabb Campbell and Mostyn 1898: 66).

Six years later, in Natal, Hill needed to consider a very different question when explaining the distribution of plague. The rat had by then become the most likely vector in the distribution of the disease, and observations of symptoms in rats were no longer the subject of myth but had moved to the centre of theories regarding the epidemic's aetiology. Accordingly, Hill discussed a series of cases, which seem to indicate clearly that plague in rats was a precursor to human cases. Examples of grain and produce stores and a railway locomotive shop and barracks were introduced as sceneries in which rat cadavers had been found, collected and tested positively for plague before cases in human occupants of the same structures were reported (Hill 1904: 77). However, eager to deliver a balanced view, Hill also offered cases 'of the opposite'. He reported on an employee at the same barracks, who, despite being contracted with the collection and destruction of rats in the premises, never once suffered from plague, and he included detailed description of rat-proofing constructions, encountering dozens of infected rats, which were 'carried out without any precaution, and yet for all that fortunately not one of the persons so engaged was attacked by the disease' (Hill 1904: 78).

14.5   Epidemiological Reasoning and Visual Formalization

Each of the reports contains formalized representations of epidemic outbreaks, such as tables, graphs and maps. With this third section, I return to the initial question of how we might position narrative reasoning within the more common perception of the field's trajectory towards quantification and mathematical formalization. In plague reports, medical statistics and maps take on a significant role to support, and at times to illustrate, narrative. Importantly, throughout the examples cited here, as well as within most of the remaining reports, little effort is given to the explanation and interpretation of statistical representations or spatial diagrams.

Lowson dedicated a short section to quantifiable data, which he entitled 'Statistical'. Rather than offering characterization and interpretation of the aggregated case numbers per hospital and along nationality and age groups, his writing was predominantly concerned with reasons that undermined the reliability of the listed numbers. In reference to a table of cases and mortality in different nationalities, he did not discuss or analyse the variable caseloads in the listed populations. Nor did he make any efforts to interpret the highly suggestive picture of mortality rates. Lowson did not use the visualized data to draw inferences, but the numbers appear to be listed to confirm the colonial framing of the outbreak as a Chinese issue, which had already been established through narrative. However, the table seems to have been still useful to Lowson as a rhetorical device to strengthen yet another colonial trope. The lack of reliable data, so he argued, was attributed to the invisible and unaccountable burial practices that emerged as a consequence of corpses left in the street.

The report from Bombay shared a similar agnosticism towards formal data in the characterization of the epidemic. While the repetitive structure of the chronological narrative, with its regular references to weather, migration and control measures, appears to adhere to a formal structure, there are no tables and lists within this lengthy description of a year of plague in the city. A substantial formalization of the epidemic's account, however, can be found in a separate set of documents that accompanied the report. The portfolio consists of a map of the island of Bombay, a second, similar map, now inscribed with detailed information on the epidemic, a complex chart of the epidemic's case rates as well as plans for a hospital and an ambulance. The first object of interest is the chart, in which daily plague mortality from June 1897 to April 1898 was plotted together with data on the usual mortality, temperature, population, humidity, velocity of wind, wind directions and clouds (see Figure 14.1). (For further details on data collection, see MacNabb Campbell and Mostyn 1898: 213–214.) According to the report authors,

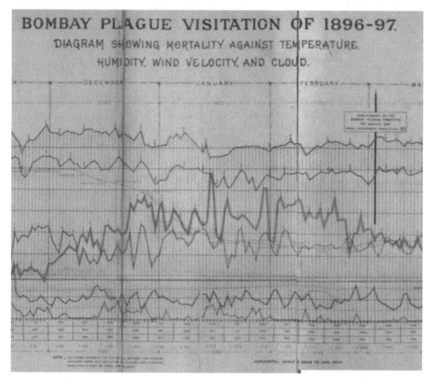

Figure 14.1 **Section of a chart provided by the Bombay Plague Committee for 1896–97**
The chart combines data on climatic factors and plague mortality rate to disprove spurious correlations.
Source: MacNabb Campbell and Mostyn (1898).

the chart was developed to mount further evidence against 'some of the theories freely advanced regarding the definite influence of temperature, humidity, wind and clouds on mortality' (MacNabb Campbell and Mostyn 1898: 214). Intriguingly, rather than an instrument of generalization, the chart takes on the opposite function, preventing misleading and simplistic causal theories about plague as a disease of climatic circumstances through the demonstration of the fallacy of correlations.

The second visual representation of interest attached to the report from Bombay is a 'progress map' of the epidemic (see Figure 14.2), plotting the course of the epidemic from September 1897 to the end of March in 1898. Each section of the city had been marked with a circle when it had become epidemic, and each of the circles was shaded to indicate in which months the outbreak occurred, based on granular data collection of 'actual cases from house to house'. While the report's authors saw the map as evidence for an improved overall picture of the epidemic compared with the previous year, its relation to the narrative account within the report requires a few further considerations.

First of all, the map was designed to reinstate the image of chronology previously developed in the narrative sections of the report. It illustrated inferences drawn in writing, rather than opening a new space of geographical exploration. Second, the map served to visualize the 'progress' of plague, invoking the image of sweeping coverage, in which the flow of contagion becomes as visible as the obstacles that were put in place to contain the epidemic.[10] Third, within the form of the administrative report, the map constitutes a remarkable picture of granular insight, which exposes the colonial urban space through the lens of its epidemic predisposition as a radical transparent, controlled and contained space (Shah 1995).

With maps like these, epidemiologists were able to deliver two-dimensional abstraction of the complex relations of a plague outbreak. As Tom Koch has written, such maps should not be read as representations of the outbreak, or as pictures of research results (Koch 2011). Rather, he emphasized their use to combine data and theories, to create a visual context in which theories could be tested. The 'progress map' enabled a theoretical exploration of the relationship between the temporal dynamic of the epidemic and its place, following the rationale outlined in the reports' chronology.

In South Africa, Hill used a quite similar map to combine temporal and spatial coordinates in his attempt to show the 'marked correspondence between rat plague and human plague' (Figure 14.3). His map demonstrated that in areas

[10] Despite the suggestion of progression through title and legend, the map does not offer a readable narrative in and of itself, if compared to the phylogenetic trees discussed by Kranke (Chapter 10). Rather, and in line with the function of aerial photography as introduced by Haines (Chapter 9), maps like these provide a space for the theoretical reconstruction of the temporal dynamic of the outbreak, while pointing at their formalization.

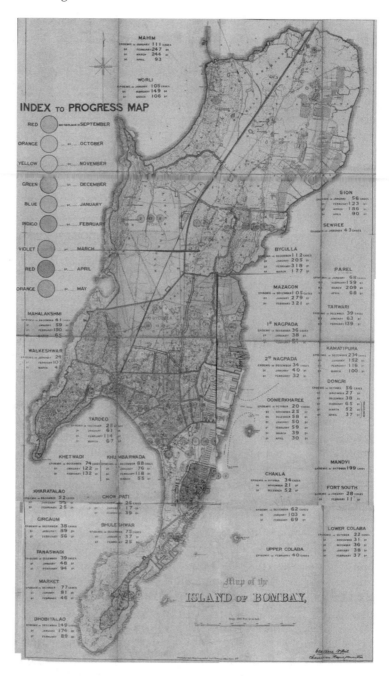

Figure 14.2 **A 'progress map' of the plague in Bombay in 1897 and 1898**
Circles indicate the temporal dynamic of the outbreak.
Source: MacNabb Campbell and Mostyn (1898).

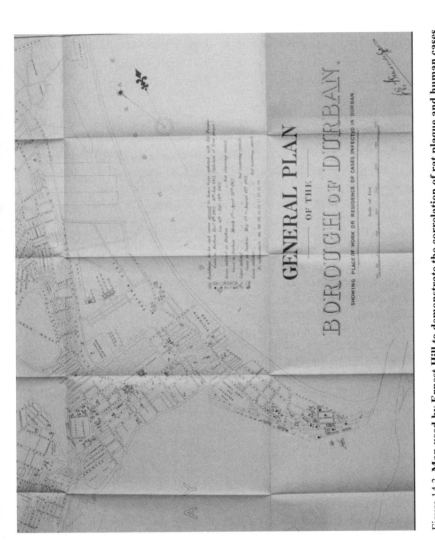

Figure 14.3 **Map used by Ernest Hill to demonstrate the correlation of rat plague and human cases, Durban 1903** Rat plague is indicated by the shaded buildings and human cases by dots. Source: Hill (1904).

where plague cases were rife, rats with plague had been found; while in areas without registered cases, rats were unaffected too (Hill 1904: 37). But this neatly mapped data could not ascribe a causal direction to the distribution of plague and infections between rats and humans, as many sceptics of the rodent-vector theory continued to argue. To support and indeed to strengthen the theory of the rat as a principal vector, Hill returned to narrative speculation about the professional occupation of human plague cases. Over 22 per cent of infected people were employed in grocery stores or stables where rat plague had been shown to reside. Here, in this focused line of argument, the map assumed a status of evidence to support his causal theory: as almost 50 per cent of cases stemmed from premises adjacent to or connected to such stores and stables, Hill concluded that 'the most important agency in the dissemination of plague was the rat' (Hill 1904: 39).

The tables used by Lowson in Hong Kong, as well as the charts and maps included in the reports from Bombay and Durban, have one aspect in common: they were used to illustrate, accompany and reinstate arguments and inferences already made in narrative form. The visualizations were not included to lift empirical observations up to a more generalizable state, nor were they used to replace the prevailing picture of uncertainty and conjecture with unambiguous representations of causal theories. All these authors raised doubts about the reliability of the data that went into the development of the tables, charts and maps and thus qualified the status of such visualizations as temporary, explora-tory and experimental rather than definitive.

Within epidemiological reasoning, this precarious status of formal represen-tation was neither derided nor seen as problematic. Particularly as these reports were concerned with the observation and explanation of an epidemic outbreak, their authors aimed to sustain the muddy ground between correlation and causation rather than to resolve the resulting account either into radical contin-gency or into simplistic mono-causality. The visualizations in this period maintained a dual position – as diagrams to formalize the temporal dynamic as well as street maps to visualize the epidemic on the ground (see Wise, Chapter 22). Narratives allowed the authors to convey a sense of correlation and causal implication, as they explained why and under which conditions a series of cases assume epidemic proportions. Narrative focused on the crucial questions, which at the same time were the most difficult to answer succinctly: how mortality rates were skewed by social behaviour, how the disease dynamic unfolded in relation to climatic or sanitary conditions, and if the parallel occurrence of diseases in rodents and humans emerges as causal theory if one considered professional occupation. The visual 'polemics' of graphs, maps and charts were not only mistrusted, but their misleading determinism required framing and containment within the possibilities that narrative conjecture raised.

14.6   Conclusion

In this chapter, I have revised perspectives on early twentieth-century epidemiology, which has been seen as a phase of quantification and medical statistics. Contrary to this historical account, I have introduced the outbreak report as a narrative genre and as a source to consider the emergence of epidemiological reasoning. In this narrative form, epidemics retain their character as complex phenomena, which could never fully be understood through the narrow lens of bacteriology, the limited perspective of the vital statistician or the diagnostic point of view of the clinician. Epidemiological reasoning set the groundwork for the development of epidemiology as a unique scientific practice, at a remove from the clinic and the laboratory, but dedicated to colligating an endless array of material, social and biological aspects.

Historical narration emphasizes the temporal nature of the epidemic as an object of research. The rhythm, patterns and dynamics of epidemics assume significance in the writing of the reporting authors, as they seek to account for the temporal shape of plague outbreaks. Crucially, epidemiological reasoning distinguishes between what would later be called the micro-histories of outbreaks and the macro-histories of disease biographies, to evaluate and to scrutinize their relations.

Beyond the sober empiricism of historiography and chronology, the reports also offer space for the negotiation of causal theory. Assumptions about contagious soil, spaces and rodents are often brought forward without robust justifications, strong experimental evidence or academic rigour. Rather, narrative epidemiological reasoning sustains the epidemic configuration as a series of disparate factors forced into relation by the epidemic event, allowing its authors to speculate about their correlation without losing sight of a probable causal inference.

The value of conjecture and the capacity to maintain uncertainty between correlation and causation assumes prominence when the narrative is contrasted with the blunt pictures of epidemics derived from tables, lists, graphs, charts and maps. Within the epidemiological reasoning of reports of the third plague pandemic, these representations of quantifiable aspects were framed in a rhetoric of unreliability and misleading mono-causality. Rather than instruments of standardization and generalization, visual formalizations took on a role of expressing theories, testing hypotheses and exploring spurious inferences.

As a practice of empirical observation, the reasoned argument about epidemics remains deeply indebted to the epidemic narrative as a form of storytelling. However, charged with the formalization of a scientific epidemiological discourse, the narrative in outbreak reports also begins to shape the epidemic as an object of knowledge structured by historical contingency, theoretical multiplicity and a rather hesitant formalization of causes and determinants. At the beginning of the twentieth century, it is in epidemiological reasoning, rather than in the formalization of medical statistics and mathematical formulae, where the epidemic emerges as a versatile point of reference to think through and beyond the boundaries of the clinic, the laboratory, the population, the city and an increasingly fragile colonial world order.[11]

References

Amsterdamska, O. (2001). 'Standardizing Epidemics: Infection, Inheritance, and Environment in Prewar Experimental Epidemiology'. In J.-P. Gaudillière and I. Löwy, eds. *Heredity and Infection: The History of Disease Transmission.* New York: Routledge, 135–180.

(2005). 'Demarcating Epidemiology'. *Science, Technology, and Human Values* 30.1: 17–51.

Armstrong, D. (1983). *Political Anatomy of the Body: Medical Knowledge in Britain in the Twentieth Century.* Cambridge: Cambridge University Press.

Bibel, D. J., and T. H. Chen (1976). 'Diagnosis of Plague: An Analysis of the Yersin–Kitasato Controversy'. *Bacteriological Reviews* 40.3: 633–651.

Casey, A., M. Bennett, R. Tobin, C. Grover et al. (2020). 'Plague Dot Text: Text Mining and Annotation of Outbreak Reports of the Third Plague Pandemic (1894–1952)'. *Journal of Data Mining & Digital Humanities* 7105. http://arxiv.org/abs/2002.01415.

Cunningham, A. (1992). 'Transforming Plague: The Laboratory and the Identity of Infectious Disease'. In A. Cunningham and P. Williams, eds. *The Laboratory Revolution in Medicine.* Cambridge: Cambridge University Press, 209–244.

Donato, M. P. (2018). 'Introduction: Archives, Record Keeping and Imperial Governance, 1500–1800'. *Journal of Early Modern History* 22.5: 311–326.

Echenberg, M. J. (2007). *Plague Ports: The Global Urban Impact of Bubonic Plague, 1894–1901.* New York: New York University Press.

Engelmann, L. (2018). 'Mapping Early Epidemiology: Concepts of Causality in Reports of the Third Plague Pandemic, 1894–1950'. In E. T. Ewing and K. Randall, eds. *Viral Networks: Connecting Digital Humanities and Medical History.* Blacksburg, VA: VT Publishing, 89–118. https://publishing.vt.edu/site/books/10.21061/viral-networks/.

(2021). 'Making a Model Plague: Paper Technologies and Epidemiological Casuistry in the Early Twentieth Century'. In C. Lynteris, ed. *Plague Image and Imagination*

[11] The thoughts explored in this chapter have benefited immensely from the encouraging interactions with the Narrative Science group. I am grateful in particular to the editors of this volume, Mary S. Morgan, Kim M. Hajek and Dominic J. Berry, for providing insightful and generous feedback on presentations, draft and edits. Many thanks also to the internal reviewers, Brian Hurwitz and M. Norton Wise as well as to all workshop participants for their very helpful criticism on an earlier version of this chapter. The author of this chapter acknowledges his own ERC Grant (grant agreement No. 947872). In addition, the *Narrative Science* book has received funding from the European Research Council under the European Union's Horizon 2020 research and innovation programme (grant agreement No. 694732). www.narrative-science.org/.

from Medieval to Modern Times. Medicine and Biomedical Sciences in Modern History. Cham: Springer, 235–266. https://link.springer.com/chapter/10.1007/978-3-030-72304-0_9.

Engelmann, L., and C. Lynteris (2020). *Sulphuric Utopias: A History of Maritime Fumigation*. Cambridge, MA: MIT Press.

Evans, N. H. A. (2018). 'The Disease Map and the City: Desire and Imitation in the Bombay Plague, 1896–1914'. In *Plague and the City*. London: Routledge, 116–138.

Gatacre, W. Forbes (1897). *Report on the Bubonic Plague in Bombay . . . 1896–97*. Bombay: Times of India Steam Press. http://archive.org/details/b24974523_0003.

Hardy, A. (1993). *The Epidemic Streets: Infectious Disease and the Rise of Preventive Medicine, 1856–1900*. Oxford: Clarendon Press.

Hess, V., and J. A. Mendelsohn (2010). 'Case and Series: Medical Knowledge and Paper Technology, 1600–1900'. *History of Science* 48.161: 287–314.

Hill, E. (1904). *Report on the Plague in Natal, 1902–3*. London: Cassell. http://archive.org/details/b2135392x.

Hirsch, A. (1883). *Handbook of Geographical and Historical Pathology*. London: New Sydenham Society.

Koch, T. (2011). *Disease Maps: Epidemics on the Ground*. Chicago: University of Chicago Press.

Lowson, J. A. (1895). *The Epidemic of Bubonic Plague in 1894: Medical Report*. Hong Kong: Noronha & Company. http://archive.org/details/b24398287.

Lynteris, C. (2016). *Ethnographic Plague: Configuring Disease on the Chinese–Russian Frontier*. Basingstoke: Palgrave Macmillan.

(2017). 'A "Suitable Soil": Plague's Urban Breeding Grounds at the Dawn of the Third Pandemic'. *Medical History* 61.3: 343–357.

MacNabb Campbell, J., and R. Mostyn (1898). *Report of the Bombay Plague Committee, Appointed by Government Resolution No. 1204/720P, on the Plague in Bombay, for the Period Extending from the 1st July 1897 to the 30th April 1898*. Bombay: Times of India Steam Press. http://archive.org/details/b24974535.

Magnello, E., and A. Hardy (2002). *The Road to Medical Statistics*. Amsterdam: Rodopi.

Matthews, J. Rosser (1995). *Quantification and the Quest for Medical Certainty*. Princeton, NJ: Princeton University Press.

Mendelsohn, J. A. (1998). 'From Eradication to Equilibrium: How Epidemics Became Complex after World War I'. In C. Lawrence and G. Weisz, eds. *Greater Than the Parts: Holism in Biomedicine, 1920–1950*. Oxford: Oxford University Press, 303–334.

Morabia, A. (2013). *A History of Epidemiologic Methods and Concepts*. Basel: Birkhäuser.

Morgan, Mary S. (2017). 'Narrative Ordering and Explanation'. *Studies in History and Philosophy of Science Part A* 62: 86–97.

Morgan, Mary S., and M. Norton Wise (2017). 'Narrative Science and Narrative Knowing: Introduction to Special Issue on Narrative Science'. *Studies in History and Philosophy of Science Part A* 62: 1–5.

Page, D. L. (1953). 'Thucydides' Description of the Great Plague at Athens'. *Classical Quarterly* 3.3–4: 97–119.

Proust, A. (1897). *La défense de l'Europe contre la peste et la conférence de Venise de 1897*. Paris: Masson. http://gallica.bnf.fr/ark:/12148/bpt6k4414581.

Rogaski, R. (2004). *Hygienic Modernity: Meanings of Health and Disease in Treaty-Port China*. Oakland: University of California Press.

Rosenberg, C. E. (1992a). *Explaining Epidemics and Other Studies in the History of Medicine*. Cambridge: Cambridge University Press.

　(1992b). 'Introduction: Framing Disease: Illness, Society, and History'. In C. E. Rosenberg and J. Golden, eds. *Framing Disease: Studies in Cultural History*. New Brunswick, NJ: Rutgers University Press, xiii–xxvi.

Shah, N. B. (1995). 'San Francisco's "Chinatown": Race and the Cultural Politics of Public Health, 1854–1952'. PhD thesis, University of Chicago.

Solomon, T. (1997). 'Hong Kong, 1894: The Role of James A. Lowson in the Controversial Discovery of the Plague Bacillus'. *The Lancet* 350.9070: 59–62.

Steere-Williams, J. (2020). *The Filth Disease: Typhoid Fever and the Practices of Epidemiology in Victorian England*. Woodbridge: Boydell & Brewer.

Stoler, A. L. (2010). *Along the Archival Grain: Epistemic Anxieties and Colonial Common Sense*. Princeton, NJ: Princeton University Press.

Susser, M., and Z. Stein (2009). *Eras in Epidemiology: The Evolution of Ideas*. New York: Oxford University Press.

Whooley, O. (2013). *Knowledge in the Time of Cholera: The Struggle over American Medicine in the Nineteenth Century*. Chicago: University of Chicago Press.

Worboys, M. (2000). *Spreading Germs: Disease Theories and Medical Practice in Britain, 1865–1900*. Cambridge: Cambridge University Press.

Wray, S. K. (2004). 'Boccaccio and the Doctors: Medicine and Compassion in the Face of Plague'. *Journal of Medieval History* 30.3: 301–322.

Wright, M. (2006). 'Public Health among the Lineaments of the Colonial State in Natal, 1901–1910'. *Journal of Natal and Zulu History* 24.1: 135–163.

15 The Politics of Representation: Narratives of Automation in Twentieth-Century American Mathematics

Stephanie Dick

Abstract

This chapter explores narratives that informed two influential attempts to automate and consolidate mathematics in large computing systems during the second half of the twentieth century – the QED system and the MACSYMA system. These narratives were both political (aligning the automation of mathematics with certain cultural values) and epistemic (each laid out a vision of what mathematics entailed such that it could and should be automated). These narratives united political and epistemic considerations especially with regards to *representation*: how will mathematical objects and procedures be translated into computer languages and operations and encoded in memory? How much freedom or conformity will be required of those who use and build these systems? MACSYMA and QED represented opposite approaches to these questions: preserving pluralism with a heterogeneous modular design vs requiring that all mathematics be translated into one shared root logic. The narratives explored here shaped, explained and justified the representational choices made in each system and aligned them with specific political and epistemic projects.

> If there is to be a bias, it is to be a bias towards universal agreement.
>
> *QED Manifesto*

15.1 Introduction

Automation is all about representation and representation is always a political project. In order to hand off a given task to a computer, that task must first be reconceived and reformalized as something that a computer can do, translated into its languages, its formalisms, its operations, encoded in its memory.[1] In service of those transformations, decisions have to be made about what is important, what will be lost in the translation, whose needs or goals will be prioritized. This chapter explores two influential attempts to automate and consolidate mathematics in the second half of the twentieth century – the QED system and the MACSYMA system – and the representational choices that constituted each: the languages of mathematics had to be translated into the languages and formalisms of computing; relatedly, mathematical procedures, like proof verification or algebraic simplification, had to be translated into computer-executable operations; and decisions had to be made about how best to formalize mathematics for automation, with what foundational logics, rules and premises.

MACSYMA and QED developers made very different representational choices and they used narratives to frame those choices. Marc Aidinoff has observed that historians often set out to unearth the 'hidden politics' of technological systems that are framed by their developers or users as value-neutral, objective, apolitical. He argues we should also 'listen to people when they tell us what, and who, they prioritized', we should attend to 'the political, as it lies on the surface of technology, as actors directly described it' (Aidinoff 2022). This chapter attempts to do just that by focusing on the narratives with which QED and MACSYMA were framed in order to make sense of the approaches to automation they represent, and the animating visions of mathematics and culture at work underneath.[2] These narratives were not just stories, extraneous and external to the systems. Nor were they post hoc, developed to explain choices that had already been made. They mapped directly onto and informed technical development and design decisions. They also mapped onto *practice* – the representational choices framed by these narratives corresponded with cognitive realities – how users would have to think about and do mathematics with these systems.[3]

---

[1] I use the term 'reformalism' to refer to the process of translating abstractions from one symbolic or material system to another – for example, the translation of logical relations from the symbol system in the pages of a logic text to encodings in computer memory (Dick 2014).

[2] Line Andersen (Chapter 19) explores the narrative qualities of formal demonstrations like mathematical proofs themselves. Here, I alternatively explore how seemingly 'external' narratives can *shape* and *direct* formalizing and reformalizing efforts within mathematics.

[3] In introducing this volume (Chapter 1), Mary Morgan proposes that we 'think of narrative as a "technology"', a 'general-purpose technology' at that, comparing it to steam power, electricity and computing. In fact, computers were not recognized as 'universal' or 'general purpose' in any obvious sense in their earliest decades of use. As we will see, narrative in fact played an outsized role in sounding out and establishing the limits and possibilities of computers, as well as explaining and justifying the decisions that researchers made while trying to make them useful in different domains.

As such, the narratives that framed each project were both *political* and *epistemic*.[4] They were foundational myths that advocated for the consolidation and automation of existing mathematical knowledge so that the computer could take over certain elements of mathematical labour – from algebraic simplification to proof checking – and in so doing open up new possibilities for knowledge-making. Mathematicians in the future, it was proposed, would be able to see new things, solve new problems and ask new questions with automated repositories of what was already known in hand.[5] Neither QED nor MACSYMA fulfilled their foundational myths, however. They were utopian narratives, at the intersection of political and epistemic imagination. Throughout the second half of the twentieth century, there was genuine uncertainty about what kind of tool the modern digital computer would turn out to be, what its epistemic and cultural limitations and possibilities were. The narratives explored here served to attribute meaning, possible futures and cultural values to mathematics as it would be made manifest in this new and undetermined technology.

15.2   Political Choices in Automation

The QED system, whose development began with an anonymously authored manifesto in 1994, was an attempt to combat the 'tower of Babel' its developers perceived in the automation of mathematics which had, throughout the 1970s and 1980s, involved a proliferation of 'incompatible reasoning systems and symbolic computation systems' that were inefficient, redundant, cacophonous, and that threatened mathematics' traditional claim to universal truth (*QED Manifesto* 1994: 242). The *QED Manifesto* accordingly called for the translation of mathematics into a single formal and computational system, 'that effectively represents all important mathematical knowledge and techniques' and that conforms 'to the highest standards of mathematical rigor, including the use of strict formality in the internal representation of knowledge and the use of mechanical methods to check proofs of the correctness of all entries in the

---

[4] Certain aspects of the reforming ambitions presented here, particularly the emphasis on communal standards, and the values which research aspires to, also resonate with the motivations and goals of synthetic biologists as discussed in Dominic Berry's chapter (Chapter 16).

[5] Developers of the MACSYMA system, for example, proposed that the system would serve as a *laboratory* within which mathematical scientists would experiment, even with procedures and operations in mathematics they would not know how to execute by hand. Mathematics, usually characterized as inhering in a logico-deductive 'style of reasoning', following Ian Hacking, would instead become increasingly empirical and experimental, and users would develop knowledge *of the system* and its capabilities and behaviours, rather than of the underlying mathematics (Hacking 1992). See also Huss (Chapter 3), for another case discussing the new possibilities afforded by computation. Automation and its mathematical discontents was a theme present in Lorraine Daston's narrative science public lecture 'Annihilating Time: The Coup D'Œil and the Limits of Narrative', given on 5 November 2019 at the London School of Economics: www.narrative-science.org/events-narrative-science-project-public-seminar-series.html.

system' (*QED Manifesto* 1994: 238). It was to be a 'monument', gathering together, verifying and unifying mathematics, the 'foremost creation of the human mind'. Writing in the wake of the Cold War, and amid the rise of American liberalism, the authors of the *Manifesto* proposed that the system would help 'overcome the degenerative effects of cultural relativism and nihilism' (*QED Manifesto* 1994: 239–240). They lamented the perceived loss of 'fundamental values' that the end of the Cold War and the rise of liberalism signalled and saw in mathematics a uniting and universalizing possibility.

QED would bring mathematics together by *making it all the same* – by formalizing it within one 'root logic', the same rules and foundations at work throughout. The *Manifesto* incorporated a narrative of 'Babel' and of the loss of shared cultural values in order to align their project with an ideological goal: they wanted to use the universality of mathematics in order to reinforce 'fundamental values', in the face of cultural difference. The home of the project was the Argonne National Laboratory (where some of the anonymous authors were based). This was an American government and military funded, Department of Energy hosted, effort to assert 'universal truth'. But their project highlights that 'the universality of mathematics' *is itself a construct*. QED would *make* mathematics universal, by demanding that different visions, approaches, logics and techniques be put into one formal and technological system. Anything that wasn't or couldn't be reformalized in this way would be 'outside of mathematics', excluded from the centralized system, from the monument to truth. The corresponding commitment to shared fundamental cultural values is similarly normative – values will only be universal and shared when everyone has been convinced (or forced) to adopt them.

The authors of the *Manifesto* were right about Babel in mathematics automation. Since the early 1960s, there had been a proliferation of attempts to automate different parts of mathematics, and the resulting systems did not conform to shared formal or computational specifications. Some of the 'cacophony' resulted from the fact that system developers were building from scratch and without collaboration or communication with other system developers. Some differences were the result of direct competition between them. But some of the formal and representational pluralism was done by design, including in the second case to be explored in this chapter.

The MACSYMA system, developed at Massachusetts Institute of Technology (MIT) between the mid-1960s and the early 1980s during the Cold War, was among the most influential early computer algebra systems. It was designed with multiple representational schemes, multiple logics, *on purpose*, because the developers believed this would make it more useful to practising mathematicians and mathematical scientists. MACSYMA, too, was meant to be a centralized, consolidated, automated repository of existing mathematical techniques – a toolkit mathematicians could use in order to spare themselves the time and effort of learning and executing those techniques for themselves. But MACSYMA developers believed that the best way to automate and consolidate mathematical knowledge was with as much heterogeneity and flexibility as possible. They wanted to bring mathematics together in pieces, stand-alone modules that each operated according to its own logic, its own internal design. This, they believed, would create a more accurate and more useful encoding of mathematical knowledge that would reflect and respect the pluralism of mathematical communities.

In an article explaining the representational choices one must make in the automation of mathematics, MACSYMA developers used political language. In a section called 'The Politics of Simplification', Joel Moses (a lead MACSYMA developer) described these choices in terms of how much freedom they afford the user, acknowledging that user freedom almost always adversely affects efficiency (Moses 1971). There are many different but equivalent ways that mathematical relations can be expressed, and mathematicians choose particular expressions because they are convenient to work with in a given context. But what is convenient for a mathematician on paper may not be efficient on the computer where very different constraints and economies, of memory and operations, are at stake.

For example, even simple addition can lead to trouble on the computer. Consider the sum of a series of numbers [1] $S = x_1 + \ldots + x_n$. In computers, numbers are typically stored in memory using a fixed number of bits, and for 'real numbers', a format called floating-point is used to represent them. However, floating-point schemes struggle to represent both very large and very small numbers. As such, for the purposes of automation when very large numbers may be involved, it might be simpler to work in 'log space' where the computer stores and operates on the logs of numbers rather than the numbers themselves, because they require less memory. **Incidentally, the capacity to simplify problems by calculating in 'log space' is what made tables of logarithms so valuable in the nineteenth century before automatic calculators.**

Expression [2] $\log\left(exp(x_1) + exp(x_2) + \ldots + exp(x_n)\right)$ calculates the same value as [1], but works in log space, and so is often more efficient for computation. If you want to compute the log-space representation of the sum of x_1 to x_n, you can convert out of log space (by exponentiating), compute the sum of the regular representation of the numbers, and then take the log again, as in [2]. But, on the computer, it can be even more efficient to represent this expression as

[3] $M + \log\left(exp(x_1 - M) + \ldots + exp(x_n - M)\right)$ where $M = \max(x_1, \ldots, x_n)$.

[2] and [3] are equivalent, but how could [3] possibly be more efficient than [2]? It has this extra term, M, added and subtracted throughout. [3] is called the 'log-sum-exp' trick and it is a way of computing the sum of a series of numbers in log space without having gigantic intermediate calculations that could

exhaust computer memory. While it complicates the *expression* by adding M, M simplifies the *computation* by ensuring that numbers are sufficiently small to be represented in available memory. But this way of looking at and working with sums may be counter-intuitive or difficult for a human user who may nonetheless be required to input expressions in this form or recognize and interpret them on the screen if sums have been implemented in this way in the system they are using. In this and so many other cases, what is easier and more efficient computationally may not be what is easiest for the mathematician.[6]

Typically, the more representational flexibility a user has, the more 'under the hood' processing needs to be implemented by developers to translate inputs into a form that the system was set up to manipulate. A 'user-friendly' system might allow a user to input simple expressions like [1] and, 'under the hood', the computer could convert them into the more computationally efficient forms in [2] or [3] before executing, and then convert back when displaying a result. But, these conversions also cost computing resources, so more rigid designs demand that the user become accustomed to working with, recognizing and generating computer-oriented representations themselves. This problem – how to implement and represent mathematical expressions and operations efficiently in memory, how users could input and work with mathematical expressions and operations, and how much work was needed to translate between the two – is a core problem for the automation of mathematics. These are the representational choices involved in any automation effort, and these are the choices MACSYMA developers framed through political narrative.

Moses surveyed the algebraic computing systems of the 1960s according to what he figured as the politics of their representational choices. There were the so-called 'radical systems' that could only 'handle a single, well-defined class of expressions. [. . .] This means that the system stands ready to make a major change in the representation of an expression written by a user in order to get that expression into the internal canonical form' (Moses 1971: 530). There was 'the new left', which 'arose in response to some of the difficulties experienced

---

[6] Throughout the history of mathematics, there is a related tension between aspirations fully to formalize mathematics, and a recognition of the convenience and productivity of working with informal, heuristic or more intuitive languages and representational systems. Formal systems can be cumbersome and tedious to work with, and actual mathematical practice tends not to adhere to strict formalization. Yet, without formalization, there is concern about the truth of mathematical conclusions and the foundations on which they rest (Livingston 1999). This tension was revisited in the context of computer automation (MacKenzie 2005) because computers require levels of formalization that may be unintuitive or difficult for human use. This tension also relates to Norton Wise's comments on the distinction between formal and natural language narratives in his afterword (finale) (Chapter 22). In this case, natural language narratives used to make sense of and ascribe cultural meaning to formalisms.

with radical systems' and which operated like a radical system but with some alternative algorithmic simplification mechanisms. There were 'the liberals', equipped with 'very general representations of expressions', the 'conservatives', who 'claim that one cannot design simplification rules which will be best for all occasions. Therefore, conservative systems provide little automatic simplification capabilities. Rather, they provide machinery whereby a user can build his own simplifier and change it when necessary' (Moses 1971: 532). There were also 'catholic' systems that used 'more than one representation for expressions and have more than one approach to simplification. The catholic approach is that if one technique does not work, another might, and the user should be able to switch from one representation and its related simplification facilities to another with ease' (Moses 1971: 532). MACSYMA was a catholic system, incorporating elements of liberal, radical and conservative representational choices – 'The designers of catholic systems emphasize the ability to solve a wide range of problems. They would like to give a user the ease of working with a liberal system, the efficiency and power of a radical system, and the attention to context of a conservative system. The problem with a catholic system is its size' (Moses 1971: 532). MACSYMA, with its catholic design, reflected a narrative that highlighted *horizontal management* – the system's modules operated independently of one another – and *pluralism* – each module operated according to its own representational schemes and internal logic (Martin and Fateman 1971).

Any attempt to encode and automate mathematics requires an answer to a host of representational questions – how should mathematical objects be stored in computer memory? What will be included and what will be excluded? How should human practice be translated into computer operations? Whose needs and perspectives will be prioritized – the user or the developer? How and how much should these processes and representations be made visible to the user on a screen or printout? How must users formulate their problems and objects of interest such that they can be input to the system? QED and MACSYMA were designed with different answers to this set of representational questions, both framed with politico-epistemic narratives. QED embodied a vision of mathematics as a source of universal, shared truth and 'fundamental values' in the face of scorned 'cultural relativism'. MACSYMA instead embodied a commitment to pluralism and flexibility in both mathematics and culture. These narratives flag the cognitive freedom or discipline that accompanies different approaches to automation – they describe how users must discipline their relationship to mathematics and mathematical representation in order to use a system effectively. They imagine a different role for computers in the production of mathematical knowledge, and different 'styles of reasoning' to accompany them (Hacking 1992).

15.3   From Political Choices to System Building

But how (and how well) do these narratives relate to on-the-ground realities of these projects? How free are the developers of technological systems to decide what their politics will be? What is highlighted and what is left out in these narratives? Jonnie Penn, a historian of artificial intelligence (AI) has demonstrated that, in spite of all of their self-proclaimed differences, early AI practitioners were in fact united by key underlying logics and values (Penn 2020). While they disagreed about how intelligence might be manifested in the machine, or what intelligence was, different approaches to AI were nonetheless united by many shared commitments – most notably, he identifies military and industrial logics and funding at work across them. For all their purported differences, they in fact agreed as much as they disagreed, especially about unspoken assumptions. Similarly, on the face of it, QED and MACSYMA embodied opposite approaches to the same problem – both projects aimed to centralize and automate mathematics, MACSYMA by preserving difference and adopting representational flexibility, QED by translating all of mathematics into one 'root logic' by unifying it. The narratives adopted by the developers of each system correspond to these opposing visions of automation. However, in spite of those differences, both systems shared a more fundamental belief that the consolidation and automation of mathematics was possible. They shared an underlying goal – to extract mathematical knowledge from people and communities and put it into the machine. To do so, both projects had to accommodate computers, whose limitations and possibilities constrained the epistemological and political values they could realize. The next sections offer a closer look at each automated system, the narratives that surrounded them and the practices that accompanied them.

15.3.1   MACSYMA

The MACSYMA system (for Project MAC Symbolic Manipulator) was developed under the auspices of Project MAC at MIT, beginning in the 1960s. The system was meant to offer automated versions of much of what mathematicians know and do: 'The system would know and be able to apply all of the straightforward techniques of mathematical analysis. In addition, it would be a storehouse of the knowledge accumulated about many specific problem areas' (Martin and Fateman 1971: 59). The system could multiply matrices, it could integrate, it could factor and simplify algebraic expressions, it could maximize and minimize functions and hundreds of other numeric and non-numeric operations. This automated repository of knowledge was meant to free mathematical scientists from 'routine mathematical chores', and free them even from the process of acquiring much mathematical knowledge for themselves (Engelman 1965: 413). With such a system at hand, one need only to know when different operations were useful in solving a particular problem, but not necessarily how to execute those operations by hand oneself. The system grew in popularity, especially among Defense Advanced Research Projects Agency (DARPA)-funded military, academic and industrial research centres throughout the 1960s and 1970s. The PDP-10 computer at MIT on which the system was housed could be accessed through the ARPANET and was, Moses recalled, one of the most popular nodes during the 1970s (Moses 2012: 4). MACSYMA grew popular enough, in fact, that by the mid-1970s, they shifted to a user consortium funding model rather than relying on DARPA funding alone. The initial consortium included the Department of Energy, NASA, the US Navy and Schlumberger, an oil and gas exploration company.[7] Universities and academic research labs continued to access the system freely until the early 1980s, when the system outgrew the development and maintenance capacities of the MIT team, and it was privatized (controversially) and licensed to Symbolics Inc.

MACSYMA was developed in explicit opposition to two other trends in artificial intelligence and automated mathematics research at the time, and these differences help to situate the developers' framing narratives. First, MACSYMA developers were critical of the 'symbolic' approach to AI which was largely characterized by an 'information processing' model of human intelligence in which minds took information as input and manipulated it according to a set of rules, and then output decisions, solutions, judgements, chess moves and other 'intelligent behavior' (Cordeschi 2002).

Following Allen Newell and Herbert Simon, AI researchers using this approach looked for the information-processing rules that governed different problem domains and set out to automate these. Newell and Simon's ultimate goal in this field was the development of a 'general problem solver' (GPS) – a computer program equipped with sufficiently general rules of reasoning that it could solve problems in any domain, by applying those rules in a top-down fashion to whatever symbolic input it was given (Newell, Shaw and Simon 1959). GPS was based on a 'theory of problem solving' that suggested 'very general systems of heuristics [...] that allows them to be applied to varying subject matters' (Newell, Shaw and Simon 1959: 2). The idea was that people do the same sorts of analysis and planning when they solve problems in chess, or in mathematics, or in governance alike, and that if you could identify and automate those 'heuristics', they could be successfully applied 'to deal with different subjects' (Newell, Shaw and Simon 1959: 6). Attempts to produce a general problem solver in this way, however, were fraught with failure and overpromise throughout the second half of the century.

---

[7] MIT Archives, Collection AC268, Boxes 22–24.

According to Moses, these failures were entirely unsurprising. He rejected both the belief that some one set of reasoning rules or heuristics was sufficient for problem-solving across domains, and the underlying vision of 'top down' control in automation. Reflecting in 2012, he wrote:

[...] I was increasingly concerned over the classic approach to AI in the 1950s, namely heuristic search, a top-down tree-structured approach to problem solving [...] There was Herb Simon [...] emphasizing a top-down hierarchical approach to organization. I could not understand why Americans were so enamored with what I considered an approach that would fail when systems became larger, more complex, and in need of greater flexibility. (Moses 2012 : 129)

Moses thought it was untenable to identify any set of top-down rules that would be effective in solving problems across domains in mathematics. He also believed that this was an inaccurate picture of how human minds work. He believed minds were modular as well, applying different tricks and methods here and there. He did not believe that there was a singular governing set of reasoning principles at work across all intelligent behaviour, not even in mathematics. The MACSYMA system was accordingly modular – one module to factor, another module to integrate, another module to find the Taylor expansion – and these modules did not operate according to a shared set of rules or a top-down governing principle. It fell to the user to chart a path through the available modules that would produce a solution to their problem, and this was based on experiment, intuition, trial and error.

Moses was born in Palestine in 1941 and found America to be more culturally homogeneous by comparison. He suggested that this cultural homogeneity explained the commitment to top-down hierarchical organizational structures, citing these as uniquely *American*. He believed that pluralist systems of organization had correlates both in other societies and in the branches of mathematics, and sought to reflect these in MACSYMA:

When I began reading the literature on Japanese management, I recognized ideas that I had used in [...] MACSYMA. There was an emphasis on abstraction and layered organizations as well as flexibility. These notions are present in abstract algebra. In particular, a hierarchy of field extension, called a tower in algebra, is a layered system. Such hierarchies are extremely flexible since one can have an infinite number of alternatives for the coefficients that arise in each lower layer. But why were such notions manifest in some societies and not so much in Anglo-Saxon countries? My answer is that these notions are closely related to the national culture, and countries where there are multiple dominant religions (e.g., China, Germany, India, and Japan) would tend to be more flexible than ones where there is one dominant religion. (Moses 2012)

Moses' interest in 'non-American' forms of organization informed his approach to automation and AI throughout his career. His critique of top-down control infrastructure was not just that, empirically, it was brittle and performed poorly, but also that it reproduced a commitment to homogeneity that he believed was characteristically American.

Moses recognized what historians of technology have long suggested – that culture and ideology can be reproduced in technical infrastructure – and the MACSYMA system was designed to reflect the political-technics of pluralistic places. MACSYMA's catholic modularity was intended to preserve pluralism, to allow for context, mixing radical, liberal and conservative elements. That modularity would, he believed, better meet the needs of mathematicians, avoid the brittleness and failings of top-down control hierarchies he perceived in other automation attempts and, he considered, in American culture overall.

15.3.2   QED

Where Moses sought to preserve pluralism in MACSYMA, the QED system, inaugurated in the 1990s, was meant to promote and even enshrine cultural homogeneity:

[P]erhaps the foremost motivation for the QED project is cultural. Mathematics is arguably the foremost creation of the human mind. The QED system will be an object of significant cultural character, demonstrably and physically expressing the staggering depth and power of mathematics. Like the great pyramids, the effort required may be great, but the rewards can be even more staggering than this effort. Mathematics is one of the most basic things that unites all people, and helps illuminate some of the most fundamental truths of nature, even of being itself. In the last one hundred years, many traditional cultural values of our civilization have taken a severe beating, and the advance of science has received no small blame for this beating. The QED system will provide a beautiful and compelling monument to the fundamental reality of truth. It will thus provide some antidote to the degenerative effects of cultural relativism and nihilism. (*QED Manifesto* 1994: 239–240)

The *QED Manifesto* was written by a collective of automated mathematics researchers, and anonymously published in the proceedings of the 1994 *Conference on Automated Deduction*, after the fashion of the mathematical collective called Nicholas Bourbaki.[8] Like Bourbaki, however, the *Manifesto* had a primary author – Robert Boyer, a professor of computer science, mathematics and philosophy at the University of Texas at Austin. Boyer had many collaborators Argonne, the institutional home of QED, which had also been an important site of automated mathematics research since the 1960s. Readers of the 1994 *Manifesto* were directed to email 'subscribe qed' to majordomo@msc.anl. gov in order to subscribe to the Argonne-supported qed@msc.anl.gov mailing

---

[8] 'Nicholas Bourbaki' was a pseudonym used by a group of primarily French mathematicians in the 1930s who collectively authored several texts aimed at modernizing mathematics through an emphasis on structure and abstraction (Corry 1998).

list. Argonne also hosted the first QED workshop, aimed at realizing the imagined project later in 1994.

Further reading of the *Manifesto* reveals which 'civilization' and whose values were perceived as under threat and in need of monumentalizing: they worked in the tradition of the European Enlightenment. The authors of the manifesto lamented the fact that 'the increase of mathematical knowledge during the last two hundred years has made the knowledge, let alone understanding of all, or even the most important, mathematical results something beyond the capacity of any human' (*QED Manifesto* 1994). In the late nineteenth century, during the so-called 'foundations crisis', similar concerns motivated efforts to consolidate and formalize mathematics, but in books and periodicals rather than computer systems (Corry 1998; Gray 2004). Logicians and philosophers like Giuseppe Peano, Gottlob Frege, Bertrand Russell and Alfred North Whitehead set out to develop logics whose premises and inference rules they hoped would be sufficient for the establishment of mathematical results from different fields, and they published lists of known theorems and proofs of foundational results within those systems. Their desire to consolidate emerged in part in response to concerns about the foundations of mathematics and the discovery of troubling paradoxes, but also in response to the professionalization and proliferation of mathematics, which developed distinct national cultures and schools during the nineteenth century.

If mathematics was to be the bedrock of 'universal truth', it wouldn't do for it to diversify, proliferate and divide in this way, threatening the Enlightenment narrative in which mathematics, and its nineteenth- and twentieth-century bedfellows *reason* and *rationality*, respectively, were the foundations for universal truth.[9] The *Manifesto* cites Aristotle on this point:

In the end, we take some things as inherently valuable in themselves. We believe that the construction, use, and even contemplation of the QED system will be one of these, over and against the practical values of such a system. In support of this line of thought, let us cite Aristotle, the Philosopher, the Father of Logic: That which is proper to each thing is by nature best and more pleasant for each thing; for man, therefore, the life according to reason is best and pleasantest, since reason more than anything is man. (*QED Manifesto* 1994: 240)

---

[9] The late nineteenth-century anxiety about mathematics has been called the 'Foundations Crisis' (Corry 1998; Gray 2004). In the European Enlightenment context, 'reason' was cast as a universal faculty equipped to produce objective knowledge and seek out truth, and yet simultaneously it was denied to colonized people, people of colour and women. Reason was deemed 'compromised' in Eastern thinkers – it was a 'universal' faculty that European men reserved only for themselves (Terrall 1999; Mazzotti 2012). Mathematics had a central role to play in this history, since it was associated with 'reasoning' itself, especially in the American Cold War context (Erickson et al. 2013; Phillips 2014).

The narrative that an antidote to cultural relativism was required, in the form of a monument to fundamental truth, participated in that century-old impulse to gather together and render immutable – by logic and consolidation – what is known in mathematics. The enlightenment commitment to 'reason' as the bedrock of truth, as an imagined 'universal' faculty, and of mathematics as its purest manifestation, were the values perceived as under threat by 'cultural relativism' and in need of reinforcement by QED. The commitment to reason, like the commitment to formalization, may seem in tension or at odds with the use of narrative tools, and yet, in the context of QED, they work in entangled ways. While acknowledging that there would be biases and disagreements in the implementation of the system, their belief in universalism was not swayed – 'If there is to be a bias, let it be a bias towards universal agreement' (*QED Manifesto* 1994: 241). This statement captures the tension and political fantasy that supported the project.

The late nineteenth- and early twentieth-century attempt to consolidate and fully formalize all of mathematics largely failed. While significant subsections of mathematics were subjected to successful axiomatization efforts, much of mathematics remained and remains unformalized. There were also the incompleteness and decision problem results of Kurt Gödel, Alonzo Church, and Alan Turing, which demonstrated that formalization has intrinsic limitations. There was similarly the fact that most formal systems were too obtuse for actual use in practice, and most research mathematicians did not work strictly within them.

Boyer and his co-authors on the *Manifesto* believed that the modern digital computer put the full formalization of mathematics back on the table. *Human* limitations had impeded earlier efforts, but these were limitations that the computer did not share – 'the advance of computing technology [has] provided the means for building a computing system that represents all important mathematical knowledge in an entirely rigorous and mechanically usable fashion' (*QED Manifesto* 1994). Where early twentieth-century efforts at consolidation and formalization had fallen short, computer automation, they believed, could succeed – 'The QED system we imagine will provide a means by which mathematicians and scientists can scan the entirety of mathematical knowledge for relevant results'. Mathematical knowledge would be redefined as that which was included in the system, and which adhered to its formal prescriptions, highlighting again that the field's 'universality' was *constructed* through inclusionary and exclusionary choices. Mathematicians would not need, they went on, 'minute comprehension of the details' of the knowledge they would find, use and build upon in the centralized database. In this way, human understanding of that knowledge was displaced in favour of machine-consolidation. Human understanding was further displaced by the

QED commitment to machine-verification. Results would be accepted, not if they were convincing to mathematicians, but if they were automatically verifiable by the system.

QED, as earlier projects projecting universalism in mathematics, largely failed to achieve its lofty goals. Although it led to the development of the Mizar library which currently holds the largest database of fully formalized and verified mathematical results, and projects are ongoing, no system has achieved the consolidation and automation they imagined.[10] The Manifesto itself pointed to numerous obstacles – 'social, psychological, political, and economic', not to mention technical and mathematical – that would need to be overcome (*QED Manifesto* 1994: 250). They imagined a vast number of people would be needed to achieve this project and suggested that credentialing systems and individualism in mathematics might also impede their vision (*QED Manifesto* 1994: 249). They noted even that QED should avoid 'any authorship or institutional affiliation' since these could undermine the universalism that QED sought to construct. Universalism would be the product of a particular social and labour organization, central planning, shifts in credentialling and motivations, as well as technical consolidation.

The *Manifesto* acknowledged that the establishment of leadership, and the cultivation of agreement about the priorities and plans that would guide the project, would be difficult. What they described, essentially, was a centrally planned economy – you need a central planner to make a centrally planned universal mathematics, to 'establish some "milestones" or some priority list of objectives', to 'outline which parts of mathematics should be added to the system and in what order. Simultaneously, an analysis of what sorts of cooperation and resources would be necessary to achieve the earlier goals should be performed' (*QED Manifesto* 1994: 249). The *Manifesto* proposed that, ideally, the 'root logic' with which mathematics would be represented in the system would be widely accepted: 'It is crucial that the "root logic" be a logic that is agreeable to all practicing mathematicians' (*QED Manifesto* 1994). However, they also acknowledged that no such 'root logic' was, as yet, universally accepted, and leadership and agreement would remain difficult. In practice, the QED project was guided by the perspectives of a small number of automated reasoning researchers and descendent efforts remain adjacent to both mainstream mathematics and computer science. In spite of continually running up against the realities of pluralism and individualism in mathematics, part of QED's foundational myth was that a 'root logic' could be established, that reasonable people would no doubt agree on it, and mathematical labour could be reorganized accordingly. The *Manifesto*'s acknowledgement of obstacles highlighted the fact that the unity and universalism of mathematics would have

---

[10] The Mizar Project: http://mizar.org/.

to be *constructed* – disagreements erased, a 'root logic' selected and then all of mathematics reformalized and implemented within it by labourers willing to eschew individual recognition for collaborative achievement. Although QED inspired significant efforts in this direction, no such fully formal, automatically verified, comprehensive consolidation of mathematics yet exists.

In spite of consistent failures, the *belief* that full formalization and consolidation of mathematics could be achieved, *just around the next corner*, with the next advancement, has been remarkably powerful and persistent in the history of mathematics. The authors of the *QED Manifesto* suggested that paper, pencil and human minds had simply been too limited for the task but the technological advances of computing had, by the mid-1990s, made it possible to achieve. Over the next several decades, mathematicians reflecting on the QED project proposed that it had failed because of limited interest and limited technical capacity but that *now* it might be possible. In 2007, Freek Wiedijk asked, 'Why the QED manifesto has not been a success (yet)', and concluded that 'I myself certainly believe that the QED system will come. If we do not blow up the world to a state that mathematics will not matter much anymore, then at some point in the future people will formalize most of their proofs routinely in the computer. And I expect that it will happen earlier than we now expect' (Wiedjik 2007: 132). In 2016, success still had not come, but Italian computer scientists Michael Kohlhase and Florian Rabe proposed that 'Even though [QED] never led to the concrete system, communal resource, or even joint research envisioned in the QED manifesto, the idea lives on and shapes the research agendas of a significant part of the community' (Kohlhase and Rabe 2016). Again, in 2014, Ittay Weiss proposed that 'two decades later it is safe to say the dream is not yet a reality'. But he, too, believed that success was just around the corner (Weiss 2014: 803). Weiss suggested a new approach to the complete automation of mathematics, which he named 'Mathropolis' – an imagined polity, *just over the next hill*, in which the monument to universal truth will be built, the pluralism of mathematics united in one formal system, the economy of mathematical labour centrally planned, the limited human mind and social vetting of truth replaced by the robust and reliable machine. His proposed system, named as a city, reflected the entanglement of politics, governance and epistemology at work within the QED project.

15.4   Conclusion

This vision – that mathematics will be fully consolidated, automated and formalized just around the next social or technical corner, that its universality will be made materially manifest – gained much traction in the late nineteenth and early twentieth centuries. Responding both to the discovery of several troubling paradoxes and to the proliferation of mathematical fields and centres of research, mathematicians around the turn of the twentieth century wanted to get all of mathematics into one place, they wanted to represent it all in the same formal system, the same symbolism and in the pages of one book. They were unable to do so, for formal, social and material reasons. With the perceived possibilities of modern digital computing in the 1960s and 1970s, many, including the developers of the MACSYMA system, believed that, *finally*, consolidation would be possible, especially through pluralism and horizontal management. It wasn't. Again, in the 1990s, the anonymous authors of the *QED Manifesto* proposed that *finally* the cost of computing and the intellectual will were such that it would be possible to gather up all of mathematics in one place, in one formal system. It wasn't. In revisiting the QED Manifesto two decades later, several mathematicians proposed that the time had *finally* come for the full and final consolidation of mathematics. It hadn't. This story – that mathematics will be fully unified, consolidated, formalized *just around the corner* – now that the conditions of past failures have been overcome – shapes whole research projects, and scaffolds belief in the universalism of mathematics.[11]

In spite of their different approaches to automation, and the different narratives that accompanied them, QED and MACYSMA both participated in that shared goal of consolidating mathematical knowledge and automating it, putting it in the machine. Moreover, both received initial funding from the same organizations – DARPA and the Office of Naval Research (ONR), especially. Both projects were undertaken at powerful hubs of military–industrial–academic research, MIT and Argonne National Laboratory, whose power grew out of the post-war American context. Both ascribed to ideologies of efficiency and logics of industrial planning in their imagining of automated mathematics, but to serve two different ideologies. Both projects rested on the belief that, whether pluralistically or not, knowledge could be extracted from human knowers, that it could and should be 'put into the machine'. And both set out to redefine, transform and encode mathematical knowledge with computer-oriented representations and processes. QED and MACSYMA have more in common than their framing narratives may suggest.

MACSYMA was meant to preserve pluralism and empower mathematicians for new programs of problem-solving. It was meant to free time and energy for new questions and explorations by handing over much mathematical labour to the machine. However, the freedom afforded by MACSYMA required users to work with and within highly disciplined and often counter-intuitive computer-oriented representational schemes, and that freedom cultivated dependency on the system, once a user came to rely on the system for the execution of techniques they did not themselves understand (Dick 2020). The developers conceded the

---

[11] Further chapters in this volume making use of the scaffold notion include Kranke (Chapter 10), Teather (Chapter 6) and Miyake (Chapter 5). On examples of background knowledge active in cases, see Crasnow (Chapter 11).

point that MACSYMA required mathematicians to reconceive what they know for the purpose of automation, and even encouraged users to transform their own knowledge into automated modules for inclusion in the system. The modularity that was meant to serve a pluralistic and modular vision of mathematical practice also made it easier for mathematicians to take what they knew and 'put it in the machine'. Users could contribute to a SHARE Directory – an ever-growing repository of new modules, user-generated, that expanded the system's capabilities and made more 'knowledge' available to more people. The claim that MACSYMA freed mathematicians and that it preserved pluralism of practice betrayed the fact that incredible accommodation to the machine was first required and that the system was primarily useful and usable to elite and defence-funded institutions. When MACSYMA was privatized in 1981, and licensed to Symbolics Inc., the users who had worked so hard to learn and accommodate and even contribute to the system were then transformed into a set of buyers in a market who had to now pay for the privilege of consuming the goods they had in part made themselves. MACSYMA wasn't the materialization of freedom and pluralism that its narrative suggests.

Lewis Mumford cautioned, in opposition to strong theories of social construction, that there are technological systems that cannot be aligned with any politics whatever, but rather operate according to fundamental logics that cannot be overcome through creative use, alternative intention or new narrative. Mumford suggested that computers are essentially authoritarian technics, centralized command and control technologies, no matter how often people have tried to align them with democracy, freedom, counter-culture and pluralism (Mumford 1964; Turner 2008). Even if one doesn't accept Mumford's analysis in its entirety, it would still be safe to suggest that no American militarily funded effort to extract knowledge from knowers and communities and make it efficiently and automatically available to defence-funded research institutions, can be aligned with the politics of pluralism.

Both QED and MACSYMA were supposed to serve a dual purpose. First, both were meant to automate mathematics, and in this they differed – the former meant to automate by representing all of mathematics in a shared 'root logic', the latter, automating mathematics modularly, attempting to preserve logical and methodological pluralism, as well as offer users flexibility. In this difference, the narratives the developers attached to the projects suit. But both projects were also meant to consolidate all of mathematical knowledge, efficiently and automatedly. Both entailed and in fact *celebrated* the displacement of human understanding – users need not understand that which the system can do. For MACYSMA, users would be spared the need to learn mathematical techniques for themselves because of having an automated system available to execute them instead. In QED, the fundamentally social project of establishing mathematical truth was displaced in favour of automatically verified results. Both entailed theories of knowledge that did not require a subjective knower, only a machine encoding. And in this regard both displace human understanding, social processes and the pluralism these entail. And both projects consolidated resources and decision-making power, as well as the automated mathematical knowledge itself, in the hands of a small number of institutions, also limiting pluralism. Both projects also minimize the productive capacity of friction, miscommunication, disagreement, misunderstanding and difference. While MACSYMA preserved *logical* pluralism in its modularity, all modules still had to accommodate the constraints of a single arbiter: the PDP-10 computer on which they ran. We might call this *computational-pluralism*, and it was only as plural as those constraints permit. The politics of technology go beyond the technical design choices made within them to include the context in which they are developed, who pays for them, profits from them, and how much freedom or discipline users and contributors have in their engagement with technical systems.

In these histories of mathematics automation, narratives map onto design and implementation decisions, they acknowledge the representational choices involved in accommodating the machine and the user and they reflect beliefs about mathematics' relationship to culture. But the narratives that developers use to frame their technological systems may also serve to direct our gaze away from certain institutional realities and unspoken assumptions. These epistemic–political narratives highlight entanglements between mathematics and culture, and conformity and freedom, in the representational choices that automation always involves.[12]

References

Aidinoff, M. (2022). 'Centrists against the Center: The "Jeffersonian" Internet as Policy and Politics'. In J. Abbate and S. Dick, eds. *Abstractions and Embodiments: New Histories of Computing and Society.* Baltimore, MD: Johns Hopkins University Press.

Cordeschi, R. (2002). *The Discovery of the Artificial: Behavior, Mind, and Machines Before and Beyond Cybernetics.* New York: Springer.

Corry, L. (1998). 'The Origins of Eternal Truth in Modern Mathematics: Hilbert to Bourbaki and Beyond'. *Science in Context* 12: 137–193.

Dick, S. A. (2014). 'After Math: (Re)configuring Minds, Computers, and Proof in the Postwar United States'. PhD thesis, Harvard University.

(2015). 'Of Models and Machines: Implementing Bounded Rationality'. *Isis* 106.3: 623–634.

(2020). 'Coded Conduct: Making MACSYMA Users and the Automation of Mathematics'. *BJHS Themes* 5: 205–224.

[12]   *Narrative Science* book: This project has received funding from the European Research Council under the European Union's Horizon 2020 research and innovation programme (grant agreement No. 694732). www.narrative-science.org/.

Engelman, C. (1965). 'MATHLAB: A Program for On-Line Machine Assistance in Symbolic Computations'. *AFIPS (American Federation of Information Processing Societies) Conference Proceedings*. vol. 27. *Fall Joint Computer Conference*. London: Macmillan: 413–423.

Erickson, P., J. L. Klein, L. Daston, R. Lemov et al. (2013). *How Reason Almost Lost Its Mind: The Strange Career of Cold War Rationality*. Chicago: University of Chicago Press.

Gray, J. (2004). 'Anxiety and Abstraction in Nineteenth-Century Mathematics'. *Science in Context* 17.2: 23–47.

Hacking, I. (1992). '"Style" for Historians and Philosophers'. *Studies in the History and Philosophy of Science A* 23.1: 1–20.

Kohlhase, M., and F. Rabe (2016). 'QED Reloaded: Towards a Pluralistic Formal Library of Mathematical Knowledge'. *Journal of Formalized Reasoning* 9.1: 201–234.

Livingston, E. (1999). 'Cultures of Proving'. *Social Studies of Science* 26.6: 867–888.

MacKenzie, D. 2005. 'Computing and the Cultures of Proving'. *Philosophical Transactions of the Royal Society A* 363: 2335–2350.

Martin, W. A., and R. J. Fateman (1971). 'The MACSYMA System'. In *Proceedings of the Second ACM Symposium on Symbolic and Algebraic Manipulation*. New York: Association for Computing Machinery: 59–75.

Mazzotti, M. (2012). *The World of Maria Agnesi, Mathematician of God*. Baltimore, MD: Johns Hopkins University Press.

Moses, J. (1971). 'Algebraic Simplification: A Guide for the Perplexed'. *Communications of the ACM* 14.8: 527–537.

(2012). 'MACSYMA: A Personal History'. *Journal of Symbolic Computation* 47.2: 123–130.

Mumford, L. (1964). 'Authoritarian and Democratic Technics'. *Technology and Culture* 5.1: 1–8.

Newell, A., J. C. Shaw and H. Simon (1959). 'Report on a General Problem-Solving Program'. Revised version. *RAND Technical Report P-1584*. typescript. http://bit savers.informatik.uni-stuttgart.de/pdf/rand/ipl/P-1584_Report_On_A_General_P roblem-Solving_Program_Feb59.pdf.

Penn, J. (2020). 'Inventing Intelligence: On the History of Complex Information Processing and Artificial Intelligence in the United States in the Mid-Twentieth Century', PhD thesis, University of Cambridge. www.repository.cam.ac.uk/handle/1810/315976.

Phillips, C. (2014). *The New Math: A Political History*. Chicago: University of Chicago Press.

QED Manifesto (1994). In 'Automated Deduction: CADE 12'. *Lecture Notes in Artificial Intelligence* 814: 238–251.

Terrall, M. (1999). 'Metaphysics, Mathematics, and the Gendering of Science in Eighteenth Century France'. In W. Clark, J. Golinski and S. Schaffer, eds. *The Sciences in Enlightened Europe*. Chicago: University of Chicago Press, 246–271.

Turner, F. (2008). *From Counterculture to Cyberculture : Steward Brand, the Whole Earth Network, and the Rise of Digital Utopianism*. Chicago: University of Chicago Press.

Weiss, I. (2014). 'The QED Manifesto after Two Decades: Version 2.0.' *Journal of Software* 11.8: 803–815.

Wiedijk, F. (2007). 'The QED Manifesto Revisited'. *Studies in Logic, Grammar, and Rhetoric* 10.23: 121–133.

16 Chronicle, Genealogy and Narrative: Understanding Synthetic Biology in the Image of Historiography

Dominic J. Berry

Abstract

Where some chapters in this volume find narrative in the phenomena addressed by scientists, or in their reporting and representational practices, or in their argumentation and reasoning, this chapter finds narrative at the level of field and subfield formation. It does so through the history of historiography and philosophy of history, particularly the work of scholars who have differentiated the many forms of historical knowledge. Focusing on just three – the chronicle, the genealogy and the narrative – the chapter explains how these means for making historical knowledge might be made to cover knowledge-making in the sciences. The first half of the chapter develops this analytical approach, while the second applies it to the case of synthetic biology. By taking narrative's epistemic significances more seriously we arrive at a new way to explain scientific change over time.

16.1   Introduction

In this chapter, the history of historiography and the philosophy of history is brought to the aid of the history and philosophy of science (Uebel 2017; Roth 2020; Virmajoki 2020; Kuukkanen 2012). Narrative has sometimes been taken to define historical knowledge, and to define it in contrast with scientific knowledge. The Narrative Science Project undermines this contrast (Morgan and Wise 2017; Morgan 2017; Wise 2017; Cristalli 2019; Griesemer 1996). If narrative is a constitutive feature of scientific knowledge then perhaps the making of historical and scientific knowledge is more similar than has otherwise been assumed or allowed. For historians and philosophers who have investigated the so-called historical sciences, most prominently geology, palaeontology, evolutionary biology and natural history (Currie 2018; Cleland 2011; Rudwick 1985; Gallie 1955; Richards 1992; Hubálek 2021), or who have attended to science's archival practices (Daston 2017; Strasser 2019; Leonelli 2016), such similarities might already seem obvious.[1] But the approach taken here extends beyond these bounds.

Not only is it the case that scientific knowledge contains more narrative than has been appreciated, but historical knowledge contains less. Historical knowledge has existed in many forms, not all of which are indebted to narrative. Chronicles and genealogies are among the most well-known alternatives which do not assimilate to narratives, although they may possess narrativity. If it is the case that within historiography we recognize narrative as only one part of our epistemic apparatus (working with chronicles and genealogies), and we also find narrative at work in science, then perhaps there is something about the relations between chronicle, genealogy and narrative within historiography that might be illuminating within the sciences. This chapter argues that this is indeed the case. It does so through an analogy between, on the one hand, the making of new historiographical fields and subfields, and, on the other, the making of new scientific fields and subfields. It argues that the process is relational, with field-forming choices taken by individual historians and scientists being made to a considerable extent through reflection on the apparent field-forming choices made by others. The content of these choices tracks the terrain of chronicle, genealogy and narrative. Sections 16.2 and 16.3 acclimatize the reader to thinking with these three forms of knowledge within historiography. Section 16.4 applies them to a case study in the sciences.[2]

16.2   Three Forms of Historical Knowledge

Chronicles are some of the earliest known examples of historical writing and thought (Breisach 2007; Aurell 2004). While their variety of contents and styles is considerable, they can be grouped together thanks to their sharing some key exaggerated features. A convenient example is included in the University of Leeds digital collection of Medieval Manuscripts, which are freely available online.[3] The 'Anonimalle Chronicle', which can be found there, is a fourteenth-century manuscript which exemplifies key features of a chronicle (Childs and Taylor 1991). A chronicle can be eclectic, but establishes rough terms for what it will include, bounded by some geographical or temporal limit. It records people and events deemed important. For example, in the case of the Anonimalle, this includes the Peasants' Revolt. There is little

---

[1] A number of the chapters in this volume pursue relevant examples, including Hopkins (Chapter 4), Miyake (Chapter 5), Teather (Chapter 6) and Huss (Chapter 3).

[2] The political and epistemic tensions present in the automation of mathematics (see Dick, Chapter 15) exhibit a number of striking parallels with the case of synthetic biology presented here.

[3] Medieval Manuscripts Guide: https://library.leeds.ac.uk/special-collections/collection/707%3c/int_u.

thematic or argumentative ordering, as chronicles are mainly organized according to sequences and chronology (Spiegel 2016). While there may be some evidence of forward referencing from past events to future ones, so that there is room for some overarching narrativity, these features are muted (Pollard 1938). At different times, and in different cultures, what it has meant to produce a factual account, and the means by which a chronicle's evidences and descriptions have been assessed as reliable, have varied considerably. Today, key distinctions between historiographical approaches very often hinge on changes in the chronicle. For instance, while feminist historiography has inspired many significant and ongoing changes, the most fundamental has been recognition that the chronicles of history have been drawn ridiculously narrowly. The same can be said for those urging for global history-making, or environmental history-making or animal history. To boil things down, we can say that making a chronicle concerns choices of relevance and irrelevance, facing epistemic constraints of the present and the absent.

When it comes to genealogy, the most complete digitized work in the Leeds online collection is the 'Biblical and genealogical chronicle from Adam and Eve to Louis XI of France'. This fifteenth-century manuscript includes a genealogical tree of the pedigrees of French kings and their descendants. It achieves this record both in tables and through tree diagrams showing these relations.[4] While there are many earlier examples of genealogical working and thinking, in Europe it was not until the twelfth and thirteenth centuries that this form came to be developed into a prose genre in its own right, its primary function being to define and legitimize particular lines of descent and their authority (Spiegel 1983; 1993). A historical genealogy finds ways to pick out certain objects that it can follow over time, objects bounded by some privileging rationale. The choice of ending point will have a direct and immediate effect on the overall message or moral, a choice which the genealogical author is considered responsible for. Genealogy was given a new lease of life in the second half of the twentieth century, adopted by a large and diverse set of historians and sociologists who found it could be fruitfully applied to histories of concepts and ideas. This mode is most commonly associated with Foucault, although his own broader debts in arriving at genealogy are worth remembering, as are alternative approaches to genealogical history (Roth 1981; Bevir 2008; Prescott-Couch 2015). In addition, many publics commit themselves to making genealogies, be it of their DNA, or of their own family history, all of which has become big business (Nelson 2008; Tutton and Prainsack 2011). Often in genealogical research it is the finding of connections that matters over and above any explaining which those connections might achieve. In

---

[4] On thinking through such visualizations and their genealogical significance, see Kranke (Chapter 10).

contemporary historiography, genealogies often distinguish themselves from narrative histories (discussed shortly), by explicitly resisting the latter's epistemic expectations and genre conventions, particularly by denying closure. In its defining features, genealogy concerns choices of following and unfollowing, facing epistemic constraints of what it is possible or impossible to follow.

As with chronicles and genealogies, there is a wide range of different examples of narrative histories (Momigliano 1990; Levine 1987; Bentley 1999). Nevertheless, while some form of *narrativity* is present in chronicles and genealogies, that *narrative itself* can be recognized as requiring its own care and attention within historical epistemology allows us to mark out a third distinct form of historical knowledge (White 1987). I cannot account for the multiple potential origins of this form's emergence, although it presumably occurred in piecemeal fashion somewhere between *The Iliad* and Braudel's *The Mediterranean*. This form brings together a range of evidence to serve an argument or set of arguments, organized in the form of a narrative or set of narratives. A narrative has to know its end before it begins, and its terms are bounded by the questions it pursues. The motivations and justifications which take it from beginning to end (or from the end back to the beginning) are drawn from some present-centred interests which help to determine its informational order (we need to know X before we get to Y if we are to truly appreciate or agree to Z). Sometimes the written account will be narrated much like an unfolding novel, other times it is intended for a narrative to be read into it. Even when leaning into the grandiose and the rhetorical, their ambitions remain factual. At times the presence of narrative in historical knowledge, as offering something too much like fantasy or storytelling, has been contentious (White 1984; Spiegel 2007), but today it remains the dominant and preferred form of historical knowledge, facing little meaningful scepticism. Those who recognize narrative as providing a means of explanation in its own right can hold any rhetorical or storytelling features at arm's length.[5] To boil its key features down: narrative concerns choices of beginnings and endings, and makes connections – explicit or implicit – between elements of evidence which constitute an argument about, or give an explanation of, their subject.

16.3   The Analytical Apparatus: Six Elements of Historical Knowledge

The three forms are not incompatible with one another, and most examples of historical knowing and understanding will contain aspects of all three. A narrative history necessarily adopts some chronicles and not others, while treating some

---

[5] For a chapter explicitly discussing narrative explanation, see Beatty (Chapter 20). For a defence of their explanatory power against the 'Just-so story' charge, see Olmos (Chapter 21). On the importance of storytelling, see Jadjelska (Chapter 18).

objects genealogically and not others. A chronicle will necessarily serve some genealogical interests better than others and be more amenable to some narrative syntheses and not others. A genealogy necessarily includes some chronicles and not others and occupies some narrative worlds more than others. Having described them, I argue that the making of fields and subfields of history, and indeed the making of any given historian's identity as a historian, is achieved by the combining of different choices concerning the chronicles, genealogies and narratives that one adopts or rejects. This understanding is partially inspired by Gabrielle Spiegel, particularly her work on the 'social logic of the text' (1990), and earlier work that I completed with Paolo Palladino on biological time (Berry and Palladino 2019). Some of these choices are aesthetic, others political, others correspond to competing epistemic goals and values. Other aspects of these choices concern the kind of time in which one wishes to situate one's research objects and audiences. Different fields, subfields and historians assess the value of historical knowledge encoded in these three forms according to their own criteria. This process is relational: seeing in what one person or group is doing an excellence or an excess, and in some other person or group something improper or deficient. It is relational because these assessments help to motivate and justify change (or stasis) in oneself. When two or more historians arrive at the same or similar evaluative criteria, or when they share an emphasis on the importance of one or the other forms for a particular topic, we may discern the beginnings of a subfield.

It is this state of affairs within historiography, which, so this chapter argues, is paralleled in the sciences.[6] However, the descriptions of the three forms provided thus far has been too general for the purposes of making analogies between history and science. We need smaller focal points.

The six elements running down the rows of Table 16.1 have been intuited from reading history of historiography, narrative theory and philosophy of history. They concern ways in which the chronicle, the genealogy and the narrative are distinguishable. Some of these six elements are taken quite directly from the existing work of other scholars, and these debts will be clear in citations. However, the gloss which each is given serves the unique aims of this chapter. Section 16.4 applies these elements analogically to a case in the sciences.

1. *Means of construction.* When a chronicle is being compiled, one is primarily faced with choices of inclusion or exclusion, determined by whatever criterion one has adopted. When a genealogy is being composed, even if the key figures or subjects are picked out, the primary choices one faces concern which to follow when, and which to cease following when. As for

---

[6] Chris Mellingwood's doctoral thesis (2018), which concerned (among other things) the making of new 'amphibious' selves in synthetic biology through contrasts with other perceived selves elsewhere in biology and engineering, helped inspire this analysis.

Table 16.1 *Reading history of historiography, narrative theory and philosophy of history*

	Chronicle	Genealogy	Narrative
1. Means of construction	include/ exclude	follow/unfollow	begin/end
2. Means of ordering	chronology	material overlap	presentism
3. Likely (or available) modes of narrativity	embryonic/ deferred	multiple/braided	underlying/figural
4. Reflexivity	low	medium	high
5. Ending	arbitrary	dependent on pursued material/s	dependent on pursued question/s
6. Orientation to the world	by universal/ god's time	by persistence and loss	by argument

the construction of a narrative, the most essential manoeuvre for the building of a narrative world concerns when to start (and why), and when to stop (and why), two decisions which are really one.

2. *Means of ordering.* The question of ordering has helped motivate the Narrative Science Project from the outset (Morgan 2017). The organizer of a chronicle works under the expectation that each entry will be ordered chronologically. Chronology may also matter for the organizer of a genealogy, but this will be mixed with a selectivity towards a particular object, the phenomenon of interest, that which is being traced over time. The term 'material overlap' – which I use to describe this means of ordering – I take from Griesemer (2000), which he introduced to help characterize what is interesting about evolutionary dynamics in particular, but which I think is extendable to anything lineal. When it comes to narrative, despite very clearly important concerns which often guard historians against presentism, the producer of a narrative will be inescapably presentist, and indeed that presentism contributes to a narrative's value. Their materials will therefore be ordered according to their argumentative ambitions in the present.

3. *Likely modes of narrativity.* The six modes of narrativity that run across this row are all directly taken from Ryan 1992, who also lists many more.[7] I have chosen the six which best help differentiate the three forms. To explain them very briefly, an *embryonic narrative* has some of the most important features for narrative-making without any identifiable plot (the historical chronicle is one of Ryan's illustrative examples). A *deferred narrative* provides no narrative of its own but intervenes into something which might eventually

---

[7] I thank my colleague Kim Hajek for suggesting I consider Marie-Laure Ryan's modes of narrativity in the course of this research.

become one (the example given is that of a newspaper report). *Multiple narrativity* keeps many plots in play at once but without requiring that they be related or interact (the suggested examples are *The Decameron* and *The Arabian Nights*). *Braided narrativity* might have plots which interact, join, depart from one another, etc. (Ryan's preferred examples are family sagas and soap operas). *Underlying narrativity* is read into some source material without being stated in explicit narrative form (examples are offered from everyday life, such as witnessing a fight and interpreting it as the outcome of some longer set of events).[8] Historians commonly use the latter mode of narrativity in an effort to create distance between the materials presented and their own preferred narrative, or when they are attempting to delay the selection of one narrative over alternatives. Last, *figural narrativity*, which again arises outside of any explicitly stated narrative, and occurs when some source or other conjures up in our minds a stand-in, a figure, of one kind or another. One of Ryan's preferred examples is the making of nation states into characters on a global historical stage.

4. *Reflexivity*. This row places the three on a scale, from low reflexivity to high. A chronicle requires little reflexivity on the part of the chronicler once the criteria for selection are established. As such, it might be better to say that the reflexivity of the chronicler is required prior to the making of the chronicle, rather than it being an explicit feature of the account. A genealogy requires a little more reflexivity because the choices concerning what to follow or unfollow, and when, cannot be specified prior to the composition, but will be more dependent on author choices. Last, a narrative history expects a very reflexive and self-conscious author, even when an externalizing 'all-seeing' voice is adopted.

5. *Ending*. A chronicle is not building to some ending or other but ends at some arbitrary point. It could easily continue, without any effect on its overall structure or significance, but it simply does not. A genealogy can only last as long as do the materials it is addressing. It could continue, provided the object continues, and different ending points can produce different lessons. The ending of a narrative history, meanwhile, will have been baked into it from the beginning, because it needs to be coherent with the questions it pursued (Roth 2017; Morgan 2017).

6. *Orientation to the world*. This last distinguishing element is the most difficult to explain concisely. Very briefly, when we pick up either a chronicle, a genealogy, or a narrative history, we are also picking up the relationship between reader and world which each generates or assumes. This element is similar to 'rhetorical structure' as conceived in genre theory

---

[8] On the widespread presence and usefulness of these world readings in science, which can be interpreted as 'scripts', see Hopkins (Chapter 4), Andersen (Chapter 19) and Hajek (Chapter 2).

(Frow 2015), establishing the posture of the audience.[9] Chronicles, by and large, are written such that they are set against the backdrop of 'all time', or 'God's time'. 'These are the chronicles of X', some disembodied voice impresses upon us, 'and they were recorded because they are important'. Genealogies instead orientate audiences by facts of existence, i.e., that some things which once were are no more, other things which have been are still, and still others that have not yet been, might. Finally, narrative histories orientate a reader between three points: (A) the world of the narrative; (B) the world as it has been known to the reader, and (C) an argument which is taken to describe and explain B through A.[10] Narrative history is a large and complex modelling exercise.

Having explained these elements, Table 16.1 can now be used both as a diagnostic for detecting the presence of the three forms of historical knowledge and as a means by which to more clearly distinguish between them. These elements will now be applied analogically to a case in the sciences – more specifically, the development of the field of synthetic biology.

16.4   Narrative in the Sciences: The Case of Synthetic Biology

Synthetic biology is, among other things, an epistemic programme of reform. This programme is sometimes 'imposed' on the biological sciences by engineering-trained outsiders, but just as often is pursued by biologists and biochemists from within (O'Malley 2009). The creation of this field, or subfield, has been relational in the sense that synthetic biologists explain their own aims and ambitions largely through comparison and contrast with alternative existing fields (the primary alternatives being molecular biology and engineering). On my terms, this is the normal state of affairs for all areas of science and historiography, which make and remake themselves by these relational claims and choices on a daily basis. But in the case of synthetic biology the process has been particularly pronounced and instructively explicit. Indeed, there are a vast array of things which synthetic biology seeks to, and often *has* to, distinguish itself from, in order to marshal any autonomy. The list of competitors ready to swallow it up include biochemistry, systems biology, genetics, microbiology, data-centric biology, molecular biology, developmental biology, biochemical engineering and biotechnology. Indeed, it might not make sense to conceive of synthetic biology as existing outside of the parallel development of some of these alternatives, particularly systems biology (which it has grown up alongside),

---

[9] For a case building more thoroughly on genre theory, see Griffiths (Chapter 7).
[10] For an account demonstrating the importance of distinguishing narratives of research and narratives of nature, see Meunier (Chapter 12).

each dealing with 'epistemic competition' in similar ways, although emphasizing different aspects of biological knowledge (Gross, Kranke and
Meunier 2019). Nor has synthetic biology entirely settled in one identity
or another, with both biological engineering and design biology being
alternatives sometimes adopted by practitioners and institutions.

This chapter explains the emergence of synthetic biology not only as a direct
result of the influence of key charismatic individuals (Campos 2013), nor only
thanks to the availability of novel experimental commodities (Berry 2019), nor
only by attachment to the aspirations of national and international patrons
(Schyfter and Calvert 2015), nor only as a product of techno-futurist venture
capital (Raimbault, Cointet and Joly 2016), alongside all the other candidate
features of importance which scholars have already addressed, but also as the
bringing together of a set of epistemic choices made relative to other subfields.
The epistemic choices in question track the terrain of chronicle, genealogy and
narrative.

Figure 16.1 contains most of the essential features one might need in order to
illustrate the epistemology of synthetic biology. From the point of view of this
chapter, the knowledge production and interpretation practices found in this
image exemplify all six of the scientific analogues for the elements of historical
knowledge explained above. It is an exemplary image of synthetic biology, and
was intended to be, published as it was in a PhD thesis completed in
a laboratory dedicated to bringing plants to synthetic biology and synthetic
biology to plants (Pollak Williamson 2017).[11] It was produced to test predictions concerning the relative strengths of different promoters (lengths of DNA
that raise the rate at which some other DNA in the cell gets transcribed, to
ensure it gets expressed) in different regions of plant tissue. Such promoters are
prototypical 'parts' for synthetic biologists. What parts are is sometimes
a fraught question, but are in general lengths of DNA with some characterized
and specified function. Here 'characterized' means that data has been generated
describing their behaviour in one or several biological, chemical or biochemical contexts. The promoter parts in question were being tested for inclusion in
a new registry of standardized parts, which would enable more plant scientists
to work with the organism in question, *Marchantia polymorpha*, as a model
organism (Delmans, Pollak Williamson and Haseloff 2017). The combining of
different radiating or fluorescing reporters with different microscope technologies provides some of the most important historical background to this

---

[11] I became familiar with this case during short periods of laboratory observation conducted in
the Haseloff Lab at the University of Cambridge while I was employed on the Engineering
Life project at the University of Edinburgh. I am very grateful to Bernardo Pollak
Williamson for discussing his research project with me at length and remain grateful to
Professor Haseloff for permitting me access to his facility. For more on the Engineering Life
project, see www.stis.ed.ac.uk/engineeringlife.

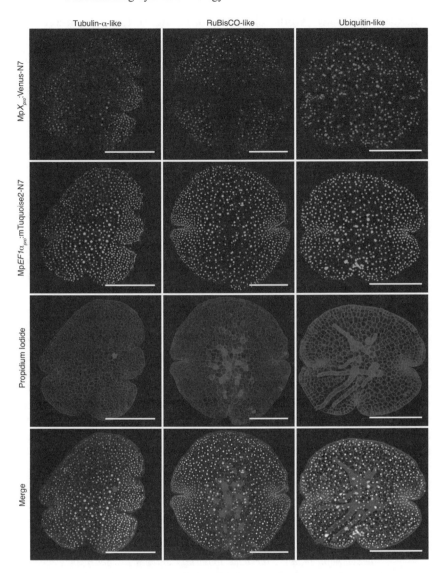

Figure 16.1 **Patterns of expression of different promoters transferred to three plants**
Columns 1–3 show that their expression is localized to different areas of plant tissue, visualized by monitoring two distinct fluorescent proteins (rows 1–2) attached to those promoters. Images taken through confocal microscopy, including one taken against a dye-stained background, which illuminates distinct plant cells (row 3), all overlaid in a composite image (row 4).
Source: Pollak Williamson (2017: 84).

particular case study (Worliczek 2020). In Figure 16.1, we see fluorescent proteins (FPs), which are the source of the dots that you can see in rows 1, 2 and 4, combined with confocal microscopy (the particular microscope technology which took these images).[12] Fluorescence in row 3 is produced by a dye. Figure 16.1 is also demonstrating a particular method, known as 'ratiometry'. All of these features (standardized parts, registries, fluorescent proteins, confocal microscopy and ratiometry) will be explained in more detail below.

For some scholars, my decision to treat synthetic biology as a relatively well-defined field or subfield with a distinctive collective epistemic culture would be problematic, on the grounds that it is not quite so distinctive as it thinks itself to be and is substantially coextensive with existing fields of biological research. In this respect, my attention to confocal microscopy in combination with FPs will not be an effective way to explain or argue for the particular emergence of synthetic biology, because the tools and methods that I am focusing on matter far more widely than in synthetic biology alone, right through molecular and developmental biology (Baxter 2019). For instance, Hannah Landecker has already recognized the importance of live cell imaging techniques (as seen in Figure 16.1) throughout the biological sciences, and attended to the novel epistemic perspective which they enable (Landecker 2012). Likewise, regarding the emphasis I will place on improving data quality and management: these characteristics have been studied extensively by Sabina Leonelli (2016) in the wider phenomena of 'data-centric' biology. But, for myself, the point is that whatever significance these methods and representations might have throughout the biological sciences they are nevertheless regularly *claimed on behalf of* synthetic biology – and not entirely illegitimately, thanks to reasons that correspond to their epistemic choices and preferences. Of course, the extent to which actors' epistemic choices and preferences are practised coherently, and the extent to which actors live up to their own self-image, are always important questions. But they simply fall outside the analytical bounds of this particular chapter.

16.4.1   Means of Construction

For synthetic biology, the most important choices in this respect concern inclusion and exclusion. Synthetic biology pushes for the rigour and standardization of engineering, rejecting the artisanal choices of molecular biology. As such, it emphasizes the <u>chronicle</u> form of knowledge.

---

[12] For those reading the printed version, the dots in Row 1 are red, and they are green in Row 2. In Row 3 the plant tissue has been dyed blue. All three layers are merged by being overlaid in Row 4.

Synthetic biologists often start with worries and complaints that too many molecular biologists have been recording too many insignificant details, with insufficient rigour, too idiosyncratically, too selfishly, for too long. On their estimation, the terms for deciding what should be *included and excluded* in shared records have not been attended to with sufficient care and scrutiny, and so they want to increase the relevance and value of what is recorded. These claims and goals are analogous to choices in historiography regarding what to incorporate into the chronicle. The promoter parts used in Figure 16.1, and the parts registry to which they were submitted, are a useful icon to think with in this context, particularly as discussion surrounding them has developed considerably since the early 2000s when they were first posited as necessary (Frow and Calvert 2013; Stavrianakis and Bennet 2014). The idea of a 'parts registry' (with an attendant physical repository for samples) is partially built on the back of earlier national and international infrastructures for the sharing of materials and services in biology, such as GenBank. But parts registries are also novel to synthetic biology in ways which are directly related to its dissatisfaction with molecular biology. The quality of a part's characterization data is intended to be higher, and collected more rigorously and in more standardized ways than has been common in molecular biology, excluding anything esoteric or artisanal. On my terms, the application of the skills, technologies and methods which constitute parts, is intended to expand, and systematically to improve on, biology's chronicle, in contradistinction with the repository and data collection practices performed by others. One should imagine historians choosing to replace or alter existing chronicles on the grounds that they were made inexpertly, or under the guidance of misleading prejudices.

16.4.2   Means of Ordering

Synthetic biology uses new techniques and technologies to see through multiple scales of life during processes of material overlap, avoiding the retrospective stitching together of dead bits and pieces which it considers characteristic of molecular biology. As such, it emphasizes the genealogical form of knowledge.

Images like those in Figure 16.1 are taken to evidence the synthetic biologist's *particular* powers of observation, precision, and control over their materials as those organisms and materials undergo processes of material overlap. This is all the more so when used to demonstrate the desired activity of a molecular part, which in turn evidences their competence in designing and making. These technologies for visualization provide biological scientists finer-grained detail in the investigation of phenomena as they happen *over time*, at both molecular and phenotypic levels simultaneously, all while the cell, tissue or organism in question is still alive. On my terms, this concerns synthetic biology's and molecular biology's means of ordering. Both groups prize observing objects undergoing processes of material overlap at molecular and phenotypic scales simultaneously, but (so it is argued) only synthetic biologists have prioritized developing truly reliable means for doing so. By contrast, so it is said, molecular biologists addressing questions of biological development in relation to molecular and phenotypic scales have worked more indirectly, often retrospectively piecing together developmental sequences from dead matter, or addressing molecular and phenotypic levels separately rather than simultaneously (Ankeny 2001; Schürch 2017). On my terms, synthetic biology claims to arrive at the same kind of genealogical knowledge sought by molecular biology, but better and more reliably. One should imagine historians finding apparent evidentiary smoking guns at the centre of developing historical phenomena, or finding new potential paths of connection.

16.4.3   Likely (or Available) Modes of Narrativity

Where molecular biology measures its success by greater or lesser incorporation into the narrative of evolution, synthetic biology measures success in making simpler and more finely tuned systems, more like engineering than those found in nature. As such, it once again emphasizes the <u>chronicle</u> form of knowledge.

The next feature of Figure 16.1 that we will discuss, its use of ratiometry, exemplifies the mode of narrativity which synthetic biology prefers. Massimiliano Simons has identified the emergence of 'postcomplex' life sciences in the twenty first century. This refers to 'sciences [that] do not imply a denial of the complexity of nature at the experimental level, but rather [...] desire to transcend it' (Simons 2019: 151). This is a very helpful way to understand what synthetic biologists are getting up to in general, and with ratiometry in particular.

In biology, ratiometry is a practice that was first developed in the biomedical sciences as an improvement on earlier fluorescence-based diagnostic and observational techniques (Haidekker and Theodorakis 2016). Because living cells and tissues are so context sensitive, and subject to multiple complex influences, actors in biomedicine began to develop dyes that fluoresce at two different wavelengths of light. Measuring both frequencies provided a check on the overall biochemical context, while also gathering the actual reaction data which one is interested in. This is because the second fluorescence measurement can be used as a constant reference point. In Figure 16.1, the reference signal is found in Row 2. If the reference point behaves bizarrely, one has a reason to question the validity of the experiment and the data it yields. If, however, the reference point behaves well, one can gather even more precise data concerning the reaction of interest by monitoring the ratio between the two outputs, hence 'ratiometrics'. While not all synthetic biologists use ratiometry, it is nevertheless precisely the kind of effort which the field celebrates and can be appreciated as forming part of their productive response to the 'problem' of biological 'noise' (Knuuttila and Loettgers 2014). Interestingly, the practice and the term 'ratiometric measurement' originated outside of biology and biomedicine, within electrical engineering, where electrical rather than optical signals were used (Holloway and Nwaoha 2013). The precision of the ratiometric results arrived at increases the chance that whatever design constraints are impinging on this biological context will become easier to spot, if not now then at some point in the future. This is an *embryonic or deferred narrativity*, more akin to what one finds in a *chronicle*, which synthetic biology prefers over and above the more figural or underlying evolutionary and biochemical narratives, which drive, and are prized in, molecular biology. In making this kind of choice, existing outside of evolutionary narrative-making, synthetic biology is by no means alone (Love 2018). One should imagine historians increasing sceptical pressure on certain evidences and standards of assessment, which often requires or affords a postponement on overarching conclusions.

16.4.4   Reflexivity

Synthetic biology interprets molecular biology as possessing low reflexivity because the latter has rarely considered the authors of protocols, metadata and other foundational sources as worthy of recognition. By contrast, synthetic biology increases authorial pride over protocols, metadata and so on. As such, it emphasizes the <u>narrative</u> form of knowledge.

At the same time as improving the materials used, synthetic biology also promises to improve the ways they are used, by prioritizing thoughtful, planned and well-managed sharing. The ambition is not only to increase the number of parts available for the synthetic biologist to work with, but to improve their capacity to work with them by also collecting as much useful data and metadata concerning their use as is feasible (McLaughlin et al. 2018). Fluorescent proteins, and the parts they are used to characterize, are themselves expected to be created more reflexively in synthetic biology than they have been in molecular biology through the adoption of some explicit standard, which, it is hoped, will ensure their compatibility with other parts that have been made according to the same standard (Peccoud et al. 2008). Sabina Leonelli's research on the broader phenomena of data-centric biology is important here, and data 'curators', the class of experts permeating the biological sciences on whom Leonelli focuses, are particularly pronounced in synthetic biology (Leonelli 2016). These claims and goals emphasize authorial reflexivity. On my terms, by emphasizing greater reflexivity concerning authorship, synthetic biology is emphasizing the value of *narrative knowledge* for the field, in contrast with those fields outside of it. One should imagine curators and archivists developing international standards for description, the making of definitive translations of historical texts or the making of more widely accessible historical archives.

16.4.5   Ending

> *Where molecular biology seeks to find what has been and why, synthetic biology arrives at what can be, and (hopefully) an eventual understanding as to why, ending as engineering often does with something that works. As such, it emphasizes the <u>genealogical</u> form of knowledge.*

Synthetic biology's preferences for narrativity (section 16.4.3) are coextensive with its preferred ending points, which are often the demonstrations of what they can now make with their materials rather than necessarily a reflection on what questions they might answer (Schyfter 2013). Whatever complexity is present in the cells of Figure 16.1, and whatever overall biochemical or evolutionary narrative which these findings might contribute to, this research programme cuts through all of that, in order to produce a simpler, immediate and fine-grained picture of a system of protein expression, one which might be further refined or made tuneable, as in engineering. Of course, this does not stop researchers also considering evolutionary significances, but these are simply not a requirement for the field. On my terms, such endings most closely resemble genealogical knowledge, because parts and constructs *could* always be developed further, characterized further, put to work in more places, etc. But after they have been shown effective in at least one or two places the synthetic biologist can choose to stop. Nor are they required to place them in evolutionary context. These then are their preferred kinds of ending, which are demonstrative ones. One should imagine historians engaging in re-enactments, or developing new uses for old sources, or new methods by which to study sources, or simply finding new sources. These are all sufficient, as useful and valuable endings, in their own right.

16.4.6   Orientation to the World

> *Where molecular biology illuminates evolutionary lineages, and the persistence or loss of forms over time, synthetic biology renders organisms in the image of its argument for engineering. As such, it once again emphasizes the <u>narrative</u> form of knowledge.*

When it comes to how synthetic biology orientates an audience to the world, design principles are key. This is a topic which Sara Green has addressed through the concept of 'constraint-based generality' in the closely aligned field of systems biology. 'Constraint-based generality makes possible the identification of general principles underpinning a class of systems exhibiting similar structural or dynamic patterns' (Green 2015: 635). Where systems biologists attempt to find and understand these patterns in extant biological organisms and systems, synthetic biologists look for structural patterns not only in existing systems but also in the parts which they make (Koskinen 2017). When submitting parts to the registry and attempting to standardize the ways in which these parts are characterized, synthetic biologists are building up biology's recognized design space, one part at a time. This is the world which synthetic biology orientates a reader to – some possible designed one rather than a merely evolutionary and natural historical one (Keller 2009; Knuutilla and Koskinen 2021). Molecular biology therefore asserts its authority through contrast with an 'actual' genealogy of evolution and natural history. Interestingly, one of the ways in which Michel Morange allows that synthetic biology might prove liberating for evolutionary biology is by undermining its otherwise uniformitarian tendencies when it comes to the actual narratives of evolution (Morange 2009: 374).[13] As synthetic biology's constructs serve as justification for the approach and embody their argument, I interpret synthetic biology as producing an orientation to the world which is much more like *narrative history* rather than chronology or genealogy. Just as historians orientate the audience towards the world through their historical model, so do synthetic biologists orientate the audience to the world through their designed synthetic one.

16.5   Conclusion

In this chapter, I have offered six elements distinguishing three forms of knowledge, arguing that choices concerning these six elements are means by which fields and subfields define and differentiate themselves in relation to other fields and subfields, regarding their understanding of the world and their practices. While the three forms in question were derived from analyses of historiography, the chapter argues that they also undergird knowledge-making in the sciences. Table 16.1 lists these elements and the forms of knowledge they most typically align with, turning the table into a diagnostic tool. I have used the table to analyse features of synthetic biology present within Figure 16.1, illustrating how these six elements are located in practice in a scientific case study of a field's formation and self-understanding of its knowledge-making. We come to appreciate that synthetic biology differentiates itself by sometimes emphasizing 'chronicular'

---

[13] On the prevalence of uniformitarian thinking in the case of geological narrative-making, see Hopkins (Chapter 4).

knowledge, other times genealogical knowledge, other times narrative knowledge. When it comes to its means of construction and narrativity, it emphasizes the chronicle. When it comes to its research endings and preferred ordering of phenomena, it emphasizes genealogy. When it comes to its level of reflexivity and orientation to the world, it emphasizes the narrative form. If the overall picture has been effectively communicated, the reader will be able to recognize ways in which the process of making new scientific fields and their knowledge claims turns on similar considerations to those which arise in the process of making new historical fields and their knowledge claims. The aim has been to advance understanding in at least two directions.

First, the historians and scientists who populate any given period are rarely considered together and are typically treated as requiring different and distinct analytical apparatus. But this need not be the case. The kinds of analysis which some historians of science already make concerning the narrative, genre and literary conventions which have mattered for science throughout time (Pomata 2018; Buckland 2013; Beer 1983; Dear 1991) could be good starting points for such an approach. Aspects of these accounts will be illuminating for the history of science and historiography alike. As such, a future direction which historians and philosophers could take would involve looking across the waxing and waning of modes of narrativity, or preferred endings, or means of construction, etc., of a given period in science in tandem with historiographical fashions concerning the same, in the search for shared patterns of epistemic change. Such shared patterns might direct us towards underlying cultural shifts.

Second, this chapter has established a new framing for examining the formation of fields and subfields in both history and science. This approach prioritizes actors' categories without remaining beholden to them. It is dynamic, as the six elements of knowledge formation described here can be applied imaginatively to a wide variety of aspects of scientific and historical life – be they publishing norms, experimental practices, representational preferences, intellectual property norms or what they study and how they study it. The differences between chronicles, genealogies and narratives, and the different times when we wish to emphasize the significance of the one or the other, are worth bringing to bear on more cases, particularly where clashing 'narratives' are believed to be in play, as between synthetic biology and its immediately adjacent subfields.[14]

---

[14] I am deeply grateful to Mary S. Morgan, who pushed this chapter through many rounds of revision, and to Jane Calvert, Sara Green, Paolo Palladino, Massimiliano Simons and my internal reviewers, Lukas Engelmann and Norton Wise, all of whom gave me extensive, insightful and critical feedback, which greatly improved every section of the chapter. I am also very grateful to my colleagues Kim Hajek, Andrew Hopkins and Robert Meunier, in all our collaborative efforts over the past few years. My thanks also to the participants at the *Narrative Science* book workshop, particularly Sabina Leonelli, Teru Miyake, Sharon Crasnow, Devin Griffiths and Elizabeth Haines, who all engaged with these ideas generously and critically.

References

Ankeny, R. A. (2001). 'The Natural History of Caenorhabditis Elegans Research'. *Nature Reviews Genetics* 2.6: 474–479.

Aurell, J. (2004). 'From Genealogies to Chronicles: The Power of the Form in Medieval Catalan Historiography'. *Viator* 36: 235–264.

Baxter, J. (2019). 'How Biological Technology Should Inform the Causal Selection Debate'. *Philosophy, Theory, and Practice in Biology* 11.2: 1–17.

Beer, G. (1983). *Darwin's Plots: Evolutionary Narrative in Darwin, George Eliot and Nineteenth-Century Fiction*. Cambridge: Cambridge University Press.

Bentley, M. (1999). *Modern Historiography: An Introduction*. London: Routledge.

Berry, Dominic J. (2019). 'Making DNA and Its Becoming an Experimental Commodity'. *History and Technology* 35.4: 374–404.

Berry, Dominic J., and P. Palladino (2019). 'Life, Time, and the Organism: Temporal Registers in the Construction of Life Forms'. *Journal of the History of Biology* 52: 223–243.

Bevir, M. (2008). 'What Is Genealogy?' *Journal of the Philosophy of History* 2.3: 263–275.

Breisach, E. (2007). *Historiography: Ancient, Medieval, and Modern*. 3rd edn. Chicago: University of Chicago Press.

Buckland, A. (2013). *Novel Science: Fiction and the Invention of Nineteenth-Century Geology*. Chicago: University of Chicago Press.

Campos, L. (2013). 'Outsiders and In-Laws: Drew Endy and the Case of Synthetic Biology'. In O. Harman and M. Dietrich, eds. *Outsider Scientists: Routes to Innovation in Biology*. Chicago: University of Chicago Press, 331–348.

Childs, W. R., and J. Taylor, eds. (1991). *The Anonimalle Chronicle 1307 to 1334: From Brotherton Collection MS 29*. Cambridge: Cambridge University Press.

Cleland, C. E. (2011). 'Prediction and Explanation in Historical Natural Science'. *British Journal for the Philosophy of Science* 62.3: 551–582.

Cristalli, C. (2019). 'Narrative Explanations in Integrated History and Philosophy of Science'. In E. Herring, K. M. Jones, K. S. Kiprijanov and L. M. Sellars, eds. *The Past, Present, and Future of Integrated History and Philosophy of Science*. Abingdon: Routledge, 72–87.

Currie, A. (2018). *Rock, Bone, and Ruin: An Optimist's Guide to the Historical Sciences*. Cambridge, MA: MIT Press.

Daston, L. (2017). *Science in the Archives: Pasts, Presents, Futures*, Chicago: University of Chicago Press.

Dear, P., ed. (1991). *The Literary Structure of Scientific Argument*. Philadelphia: University of Pennsylvania Press.

Delmans, M., B. Pollak Williamson and J. Haseloff (2016). 'MarpoDB: An Open Registry for Marchanthia Polymorpha Genetic Parts'. *Plant and Cell Physiology*, 58.1: e51–9. https://academic.oup.com/pcp/article/58/1/e5/2919392.

Frow, E., and J. Calvert (2013). '"Can Simple Biological Systems Be Built from Standardized Interchangeable Parts?" Negotiating Biology and Engineering in a Synthetic Biology Competition'. *Engineering Studies* 5.1: 42–58.

Narrative Science book: This project has received funding from the European Research Council under the European Union's Horizon 2020 research and innovation programme (grant agreement No. 694732). www.narrative-science.org/.

Frow, J. (2015). *Genre*. London: Routledge.

Gallie, W. B. (1955). 'Explanations in History and the Genetic Sciences'. *Mind* 64.254: 160–180.

Green, S. (2015). 'Revisiting Generality in Biology: Systems Biology and the Quest for Design Principles'. *Biology and Philosophy* 30: 629–652.

Griesemer, J. R. (1996). 'Some Concepts of Historical Science'. *Memorie della Societàitaliana di scienze naturali e del Museo civico di storia naturale di Milano* 27: 60–69.

(2000). 'The Units of Evolutionary Transition'. *Selection* 1.1: 67–80.

Gross, F., N. Kranke and R. Meunier (2019). 'Pluralization through Epistemic Competition: Scientific Change in Times of Data-Intensive Biology'. *History and Philosophy of the Life Sciences* 41.1. 1–29.

Haidekker, M. A., and E. A. Theodorakis (2016). 'Ratiometric Mechanosensitive Fluorescent Dyes: Design and Applications'. *Journal of Materials Chemistry C* 14: 2707–2718.

Holloway, M., and C. Nwaoha (2013). *Dictionary of Industrial Terms*. Beverly, MA: Scrivener Publishing.

Hubálek, M. (2021). 'A Brief (Hi)Story of Just-So Stories in Evolutionary Science'. *Philosophy of the Social Sciences* 51.5: 447–468.

Keller, E. F. (2009). 'What Does Synthetic Biology Have to Do with Biology?' *BioSocieties* 4.2–3: 291–302.

Knuuttila, T., and R. Koskinen (2021). 'Synthetic Fictions: Turning Imagined Biological Systems into Concrete Ones'. *Synthese* 198: 8233–8250.

Knuuttila, T., and A. Loettgers (2014). 'Varieties of Noise: Analogical Reasoning in Synthetic Biology'. *Studies in History and Philosophy of Science A* 48: 76–88.

Koskinen, R. (2017). 'Synthetic Biology and the Search for Alternative Genetic Systems: Taking How-Possibly Models Seriously'. *European Journal for Philosophy of Science* 7.3: 493–506.

Kuukkanen, J.-M. (2012). 'The Missing Narrativist Turn in the Historiography of Science'. *History and Theory* 51.3: 340–363.

Landecker, H. (2012). 'The Life of Movement: From Microcinematography to Live-Cell Imaging'. *Journal of Visual Culture* 11.3: 378–399.

Love, A. C. (2018). 'Individuation, Individuality, and Experimental Practice in Developmental Biology'. In O. Bueno, R.-L. Chen and M. B. Fagan, eds. *Individuation, Process, and Scientific Practices*. Oxford: Oxford University Press.

Leonelli, S. (2016). *Data-Centric Biology: A Philosophical Study*. Chicago: University of Chicago Press.

Levine, J. M. (1987). *Humanism and History: Origins of Modern English Historiography*. Ithaca, NY: Cornell University Press.

McLaughlin, J. A., C. J. Myers, Z. Zundel, G. Mısırlı et al. (2018). 'SynBioHub: A Standards-Enabled Design Repository for Synthetic Biology'. *ACS Synthetic Biology* 7.2: 682–688.

Mellingwood, C. (2018). 'Amphibious Researchers: Working with Laboratory Automation in Synthetic Biology'. PhD thesis, University of Edinburgh.

Momigliano, A. (1990). *The Classical Foundations of Modern Historiography*. Berkeley: University of California Press.

Morange, M. (2009). 'Synthetic Biology: A Bridge between Functional and Evolutionary Biology'. *Biological Theory* 4: 368–377.

Morgan, Mary S. (2017). 'Narrative Ordering and Explanation'. *Studies in History and Philosophy of Science Part A* 62: 86–97.

Morgan, Mary S., and M. Norton Wise (2017). 'Narrative Science and Narrative Knowing: Introduction to Special Issue on Narrative Science'. *Studies in History and Philosophy of Science Part A* 62: 1–5.

Nelson, A. (2008). 'Bio Science: Genetic Genealogy Testing and the Pursuit of African Ancestry'. *Social Studies of Science* 38.5: 759–783.

O'Malley, M. A. (2009). 'Making Knowledge in Synthetic Biology: Design Meets Kludge'. *Biological Theory* 4: 378–389.

Peccoud, J., M. F. Blauvelt, Y. Cai, K. L. Cooper et al. (2008). 'Targeted Development of Registries of Biological Parts'. *PLoS One* 3.7: e2671.

Pollak Williamson, Bernardo (2017). 'Frameworks for Reprogramming Early Diverging Land Plants'. PhD thesis, University of Cambridge.

Pollard, A. F. (1938). 'The Authorship and Value of the "Anonimalle Chronicle"'. *English Historical Review* 53: 577–605.

Pomata, G. (2018). 'The Medical Case Narrative in Pre-Modern Europe and China: Comparative History of an Epistemic Genre'. In C. Ginzburg with L. Biasiori, eds. *A Historical Approach to Casuistry: Norms and Exceptions in a Comparative Perspective*. London: Bloomsbury, 15–46.

Prescott-Couch, A. (2015). 'Genealogy and the Structure of Interpretation'. *Journal of Nietzsche Studies* 46.2: 239–247.

Raimbault, B., J.-P. Cointet and P.-B. Joly (2016). 'Mapping the Emergence of Synthetic Biology'. *PloS ONE* 11.9: e0161522.

Richards, R. J. (1992). 'The Structure of Narrative Explanation in History and Science'. In M. Nitecki and D. Nitecki, eds. *History and Evolution*. New York: State University of New York Press, 19–54.

Roth, M. S. (1981). 'Foucault's "History of the Present"'. *History and Theory* 20.1: 32–46.

Roth, P. A. (2017). 'Essentially Narrative Explanations'. *Studies in History and Philosophy of Science Part A* 62: 42–50.

(2020). *The Philosophical Structure of Historical Explanation*. Evanston, IL: Northwestern University Press.

Rudwick, M. J. S. (1985). *The Great Devonian Controversy: the Shaping of Scientific Knowledge among Gentlemanly Specialists*. Chicago: University of Chicago Press.

Ryan, M.-L. (1992). 'The Modes of Narrativity and Their Visual Metaphors'. *Style* 26.3: 368–387.

Schürch, C. (2017). 'How Mechanisms Explain Interfiled Cooperation: Biological–Chemical Study of Plant Growth Hormones in Utrecht and Pasadena, 1930–1938'. *History and Philosophy of the Life Sciences* 39.16: 1–26.

Schyfter, P. (2013). 'How a "Drive to Make" Shapes Synthetic Biology'. *Studies in History and Philosophy of Biological and Biomedical Sciences* 44.4: 632–640.

Schyfter, P., and J. Calvert (2015). 'Intentions, Expectations and Institutions: Engineering the Future of Synthetic Biology in the USA and the UK'. *Science as Culture* 24: 359–383.

Simons, M. (2019). 'The Raven and the Trojan Horse: Constructing Nature in Synthetic Biology'. PhD thesis, KU Leuven.

Spiegel, G. M. (1983). 'Genealogy: Form and Function in Medieval Historical Narrative'. *History and Theory* 22.1: 43–53.

(1990). 'History, Historicism, and the Social Logic of the Text in the Middle Ages'. *Speculum* 65.1: 59–86.

(1993). *Romancing the Past: The Rise of Vernacular Prose Historiography in Thirteenth-Century France*. Berkeley: University of California Press.

(2007). 'Revising the Past/Revisiting the Present: How Change Happens In Historiography'. *History and Theory* 46.4: 1–19.

(2016). 'Structures of Time in Medieval Historiography'. *Medieval History Journal* 19.1: 21–33.

Stavrianakis, A., and G. Bennett (2014). 'Science and Fabrications: On Synthetic Biology'. *BioSocieties* 9: 219–223.

Strasser, B. J. (2019). *Collecting Experiments: Making Big Data Biology*. Chicago: University of Chicago Press.

Tutton, R., and B. Prainsack (2011). 'Enterprising or Altruistic Selves? Making Up Research Subjects in Genetics Research'. *Sociology of Health and Illness* 33.7: 1081–1095.

Uebel, T. (2017). 'Philosophy of History and History of Philosophy of Science'. *HOPOS: Journal for the History of Philosophy of Science* 7.1: 1–30.

Virmajoki, V. (2020). 'What Should We Require from an Account of Explanation in Historiography?' *Journal of the Philosophy of History*: 1–32 (published online ahead of print). https://doi.org/10.1163/18722636-12341446doi.org/10.1163/18722 636–12341446.

White, H. (1984). 'The Question of Narrative in Contemporary Historical Theory'. *History and Theory* 23.1: 1–33.

(1987). *The Content of the Form: Narrative Discourse and Historical Representation*. Baltimore, MD: Johns Hopkins University Press.

Wise, M. Norton (2017). 'On the Narrative Form of Simulations'. *Studies in History and Philosophy of Science Part A* 62: 74–85.

Worliczek, H. L. (2020). 'Wege zu einer Molekularisierten Bildgebung: Eine Geschichte der Immunfluoreszenzmikroskopie als visuelles Erkenntnisinstrument der modernen Zellbiologie (1959–1980)'. PhD thesis, University of Vienna.

Zimmer, M. (2015). *Illuminating Disease: An Introduction to Green Fluorescent Proteins*. Oxford: Oxford University Press.

VI

Narrative Sensibility and Argument

When narrative acts as a site for reasoning

17   Anecdotes: Epistemic Switching in Medical Narratives

Brian Hurwitz

Abstract

This chapter examines the narrative and epistemic coordinates that underpin the way anecdotes notice and reason about the situations they recount. A selection of anecdotes from medical publications over the last two-and-a-half centuries is examined for the way they marshal their materials, present and reframe information, entertain explanations and bring situations and conceptual frameworks for their understanding into close relationship with each other. Through swapping perspectives on their objects of contemplation in heuristics centring on observational and conceptual vantage points, anecdotes bring before their audience's eyes scenes viewed in a new light. By changing the domains in which phenomena become explicable, anecdotes effect 'epistemic switches' – for example, from a biomedical to a sociopsychological domain or from a non-medical to a pathological one. In highly abbreviated tales of the unexpected, their 'nutshell narratives' persuade prime audiences that prior assumptions and understandings require to be adjusted in the light of experience.

> [F]rom Procopius's *Anekdota* onward [. . .] anecdotes have run counter to the order of imperial authorization.   Peter Fenves, *Arresting Language*, 153.

> The anecdote is a narration that claims to present (whether true or not, verifiable or not) a historical event, usually a single event detached from other events.
> Paul Fleming, 'The Perfect Story: Anecdote and Exemplarity in Linnaeus and Blumenberg', 74.

17.1   Introduction

This chapter examines the narrative means anecdotes employ to marshal observations and entertain interpretations of them. It also examines the way anecdotes point to explanations of the situations they recount. A term of long lineage that originally invoked 'a revelation of events previously undivulged' (Burke 2012: 60), anecdote in the eighteenth century gained wider currency as a striking account of an incident in a person's life (Johnson 1773); informative, entertaining and often memorable, it took the form of a bon mot on notable happenings or repartee that praised and mocked human accomplishments (Adams 1789).[1]

The anecdotes considered here partake in some of the features of such closely worked accounts; however, the main focus of attention will be on their modes of noticing and narrative reasoning and how they bring situations and conceptual frameworks for understanding them into close relationship with each other. In *An Essay Concerning Human Understanding* (1690), John Locke recounted an anecdote concerning the behaviour of water, which contained a striking twist:

> As it happened to a Dutch ambassador, who entertaining the King of Siam with the particularities of Holland, which he was inquisitive after, amongst other things told him that the water in his country would sometimes, in cold weather, be so hard that men walked upon it, and that it would bear an elephant, if he were there. To which the king replied, 'HITHERTO *I* HAVE BELIEVED THE STRANGE THINGS YOU HAVE TOLD ME, BECAUSE *I* LOOK UPON YOU AS A SOBER FAIR MAN, BUT NOW *I* AM SURE YOU LIE'. (Locke 1690: vol. 2, book IV, chap. 15, para. 5)

That the surface of water could support the weight of a human being proved too fantastical a particularity for the king to accept. Although a 'fair man', the Ambassador failed to overcome what Bernard Williams called the 'positional disadvantage' (Williams 2004: 55) of his royal interlocutor, a disadvantage that was geographical and above all perspectival.[2] To persuade the king of its truth – never having ventured into cold climates – the Ambassador would have to have reframed his notion of the fluid. His failure to do so stemmed from the overly narrow way he approached the task. Whether it was water's capacity to solidify or the nature of the solid once formed that was inconceivable, the unfamiliar object to the king was ice (not water). Had the Ambassador sought to expand the range of the king's conception of its physical states, he might have had more success.[3] In the event, reliance solely on his own testimony did nothing to transform the intellectual frame the king brought to the fluid. The anecdote

---

[1] John McCumber explicates the etymology of the term as something '*not* given out [...]' originally a young woman not yet married' (McCumber 2009: 58). Procopius's exposé of Justinian and Theodora's sixth-century CE rule is the earliest extant text bearing the title *Anekdota*. Its divulgences were taken to grant anecdotes a role in writing history 'from the inside' and 'from below' (Burke 2012: 61). The historian Antoine Varillas applauded 'Anecdoto-graphers [...] who draw a Picture through Conversation and Witness', but argued for their selective deployment only if they illuminated 'peculiar Connexion[s]' and 'notable Events' (Varillas, 1686: unpaginated).

[2] Where the balance of competing testimonies favoured an informant's claims, Locke counted witnesses with integrity and skill as credible sources of knowledge. In the face of such testimony, this anecdote can be read as a caution against 'extreme incredulity' (Daston 1988: 307).

[3] The Ambassador might have approached the task by pointing to changes with temperature in the state of substances known in Siam, such as water to steam and solidification of hot cheese and molten metal; and by advising the king of activities conducted on ice in cold climates, such as

offered a pointed example of how the credibility of a witness's claims turns on their conformity with the experiential and conceptual frames of an audience (Shapin 1995: 244).

In one of the earliest treatments of anecdotes in English, Isaac D'Israeli found this 'species of composition' (Anon. 1799: 181) to comprise 'minute notices of human nature, and of human learning' (D'Israeli 1793: 80–81). Impressed by its special kind of realism, the twentieth-century literary scholar Jürgen Hein noted the propensity of anecdotes to present 'a differentiated incident' and a 'dramatically shaped action or saying [...] through a social story-telling situation' (Hein 1981: 15–18).[4] As a stripped-down account claiming import and significance beyond the particulars recounted, the literary scholar Joel Fineman found anecdotes provide 'pointed, referential access to the real' in structured formats that carry 'peculiar and eventful narrative force' (Fineman 1989: 67), a forcefulness, I argue, that derives from epistemic vantage points, narrative branchings, reversals and punchlines (Beatty 2016; 2017).

The anecdotes to be discussed appeared in medical publications over the past two-and-a-half centuries. Selected for their narrative and epistemic articulations and the way they connect healthcare scenarios to general propositions and frame-works, these accounts effect 'epistemic switches', a term Richard Wollheim coined in relation to a thought experiment of A. J. Ayer's, concerning a man whose knowledge of something was initially based on the testimony of others, until he found the source of it – unchanged in content – in his own memory. For Wollheim (1979: 199), 'epistemic switch' referred to a change in the reasons for a true belief. I adapt it here to include a sudden change of evidential base, which alters the explanatory level of a phenomenon, by changing the domain in which it is explained from, for example, a biomedical to a socio-psychological domain or from a normal to a pathological one. By swapping perspectives on their objects of contemplation, anecdotes bring before the mind's eye scenes viewed in a new light.

17.2   Epistemic Effects of Medical Anecdotes

A dialogue published in *The Lancet* in 1824 headed 'Anecdote', involved Michel-Philippe Bouvart (1717–87), a French physician to Louis XV's court, and his patient, a Marquis recovering from a severe illness:

'Good day to you, Mr. Bouvart, I feel quite in spirits, and think my fever has left me.' – 'I am sure of it,' replied the doctor, 'the very first expression you used convinced me of it.' – 'Pray explain yourself.' – 'Nothing more easy. In the first days of your illness,

---

ice-skating and frost fairs. Had he indicated that ice was not a bizarre particularity but a temperature-dependent property of water he could have helped the king escape his experiential and conceptual landscape.

[4] Translated and cited in Marion C. Moeser, *The Anecdote in Mark, the Classical World and the Rabbis* (London: Sheffield Academic Press, 2002), 32–33.

when your life was in danger, I was your *dearest friend*; as you began to get better, I was your *good Bouvart*; and now I am Mr. Bouvart: depend upon it you are quite recovered'. *Ward's* [*sic*] *Nugæ Chirurgicæ* (Anon. 1824: 256)

The proposition advanced is that bedside salutations can be medically note-worthy. In response to the Marquis's declaration of buoyant spirits, the doctor positioned his greetings – hitherto peripheral to the Marquis – as central to his clinical assessment of him and claimed that his expressions of affiliation were a function of the disruptive effects of illness. The contention could have been based on an antecedent generalization: that familiarity of greeting varies inversely with recuperation, recovery restoring social and inter-personal distance. But how well established was such a generalization? Was it grounded in Bouvart's personal observations – in contemporary terms, a small, biased sample likely to engender overgeneralization – or in an accepted maxim or hunch, entertained for suppositional purposes? Whatever its source and standing, the claim put a spotlight on social exchanges in clinical practice, elevating their notice to a level of prognostic significance. In proposing that greeting style carried meaning beyond simple etiquette, the dialogue showed how clinical attention could move from biomedical considerations to a domain of socio-psychological observation that encompassed interpersonal aspects of doctoring.

James Wood finds anecdotes in the Enlightenment played an important role in advocating the naturalistic study of customs (Wood 2019). From this per-spective, what the Marquis relayed to Bouvart – regarding his spirits and lack of fever – was less important to the medical assessment of him than the Marquis's recourse to plain 'Mr.', which Bouvart located within a series 'dearest', 'good' and 'Mr.', a pattern that pointed to recovery. Even before the Marquis disclosed how he was feeling, the pattern supplied Bouvart with foreknowledge of his well-being.

Within Bouvart's diagnostic process, an 'epistemic switch' takes place, which grants greeting style an evidential value. The scenario shows how medical consideration moves away from reported feelings and symptoms to focus on inadvertently expressed greetings which match a pattern of signs or clues; cognitively, the switch persuades readers that social and interpersonal relations could be a source of valuable information.

In creating a 'picture through conversation and witness' (Varillas 1686: unpaginated), the anecdote revealed a doctor drawing out meaning from social interactions which otherwise appeared incidental and mundane. Whatever doubts we may harbour today about its embedded generalization, the acuity of Bouvart's discernment lends the account plausibility. If we remain sceptical about its central claim, the dialogue prompts less specific considerations, such as that other facets of healthcare exchange could vary with illness; concerning what sort of courtesy it is that intensifies at moments of extreme dependency; and about the levelling of birth and social rank by serious illness. Read in these ways, the anecdote brings a plurality of possibilities to the fore (Paskins and Morgan 2019), including radical and ironical meanings.

Thomas Wakley, founder editor of *The Lancet*, took the dialogue not from its earlier source in *L'Esprit des journaux*, but from William Wadd's *Nugæ Chirurgicæ*, published in London in 1824 (Wadd 1824: 199). The volume offered a miscellany of medical portraits in which Wadd announced his intention of 'blend[ing] the "utile" with the "dulce", the learned and the ignorant, the regulars and the irregulars [. . .] with the Republic of Medicine' (Wadd 1824: i), and of reforming the hierarchies and outdated medical practices of the *Ancien Régime*.[5] By interlinking the social world and health – even in the case of a Marquis – the anecdote captured something of the rationalist universalism of post-Revolutionary medicine, and the cultural authority of medical men. That this socio-political context was occluded in *The Lancet*'s retelling of the dialogue is characteristic of the anecdotal form, which the literary scholar Paul Fleming takes to be 'a discrete isolated narrative' that lacks 'chronological connection to any surrounding narration of events' (Fleming 2011: 74). Although anecdotes may cite news reports, witness accounts and informants' memories as sources for events recounted, the literary scholar André Jolles found their defining characteristic to be a pronounced capacity to condense the 'flow of events', enabling diverse elements of a situation to be grasped in a 'bound factuality' (Jolles 2017: 161–174). As well as its sources, the anecdote's credibility for Jolles also derived from its capacity to realize events and people 'concretely', qualities it shares with the clinical case and its capacity to delineate particularities and reason about them (Hurwitz 2017).

In 1985, *The Lancet*'s 'In England Now' column recounted an anecdote about a man who upturned the conceptual framework of an X-ray report:

How often have we stumbled on a diagnosis by accident and then taken credit for a good examination? Like the man whose chest we X-rayed routinely to find a peculiar thin line stretching from one lung to the other through the mediastinum. Technical fault, said the report, but lungs normal. Only when the patient commented on the number of technical errors his X-rays caused, did the penny drop. A thoracotomy revealed a bicycle spoke lying within the chest cavity, a relic of a road accident some thirty years previously. It produced some pleural pain and an interesting publication. (Anon. 1985: 381)

There is something glib and 'off pat' about this account of how a patient was briefly recognized as a situated knower after he remarked how commonly

---

[5] Although Wadd saw in Bouvart a doctor able to impress upon an aristocrat how dependent he'd been on his physician, Bouvart in fact seems always to have been identified with the *Ancien Régime* (Brockliss and Jones 1997: 475, 637).

a 'technical fault' appeared in his chest X-ray reports. The information made it a priori unlikely that yet another technical fault accounted for the X-ray appearance. His disclosure on the contrary suggested that the 'peculiar thin line' denoted a persistent lung abnormality. The patient's mention of 'pleural pain' and a history of road accident constituted a decisive first-person vantage point from which the diagnosis inverted: no longer could it invoke 'technical fault though lungs normal' but instead 'adequate image and lungs *abnormal*', an epistemic switch that led to the removal of a spoke from his lungs!

The change in what the image signified was mediated through the patient's attentiveness, memory and agency. His 'context-dependent knowledge' (Flyvbjerg 2006: 221) switched the interpretation of the X-ray from that of an artefact to that of pathology and led to publication of the image. Even so, the anecdote's punchline implied minimal acknowledgement of the patient's pivotal role in the diagnosis: although his vantage point changed the explanatory part of the explanation – the explanans – that applied to what required explanation – the explanandum – his role in the change appears not to have featured in the report.

An epistemic switch that removed the explanans of a phenomenon altogether occurred in an anecdote recounted by the general practitioner, writer and broadcaster Michael O'Donnell, in 2016:

A paediatrician travelling on a train grew intrigued by the state of an infant girl cradled in the arms of a woman who sat opposite him. He could see there was something abnormal about the child but couldn't diagnose what it was. He carefully watched the respiratory movements of her chest, listened to the timbre of an occasional cry and whimper, observed the way she moved her limbs. But he remained baffled.

He described in *The Lancet* how 20 minutes passed before the penny dropped. The child had a condition that rarely came his way. She was perfectly healthy. (O'Donnell 2016: 10)

Once the object of observational interest – 'the state of an infant girl' – had fallen outside the doctor's frame of reference, the classificatory gaze wielded at the start of the encounter proved nugatory by the end of it. The story is about 'uncontrolled observation' and the medicalization of everyday life, through imposition of a narrow conceptual frame on a non-medical context. The change of object, from a sick infant requiring pathological explanation, to one needing no explanation, amounted to a radical reframing that effected an epistemic switch. But, as the punchline makes clear, the apparent rarity of such an encounter for the doctor – even on a train – carried its own compensation: a report in an international journal. In responding to what was unfamiliar by attempting to diagnose it – and only later appreciating that what he thought was 'something abnormal' was in fact a healthy child – the doctor fell victim to the insidious power of medicalization. Through publication of a case in which the medical framework had no grip – a case of nothing – the doctor's mindset was rewarded and seemingly left intact his drive to deploy it again on the next train journey.

Despite its brevity, O'Donnell's anecdote is quite a layered story whose meanings emerge as the framing shifts.[6] A prototype for this effect is suggested by Bernard Williams's discussion of how differences in spatiotemporal observational position justify tellings, on the grounds that tellers were there when the audience was not (Williams 2004: 55). The situated and conversational aspects of such grounds were noted by ethnographer and sociologist Paul Atkinson, who observed anecdotes recounted by UK and US haematologists during clinical ward rounds in the 1970s:

> One of the most striking characteristics of speech exchange in daily rounds is the deployment of personal narratives and reminiscences on the part of senior physicians [. . . who] claim, tacitly, the right to tell stories and to relate medical knowledge back to their biographical experiences. The justification for the story as evidence does not derive from the warrant of textbooks, journals or other sources of biomedical science. (Atkinson 1995: 137–139)

Caught between personal, experiential accounts of medical phenomena and formal, impersonal accounts of medical knowledge, anecdotes occupy a contested space. On the one hand, they offer experience-based insights into healthcare phenomena, on the other, insights based on haphazardly collected observations. Within this space, anecdotes are viewed as anachronistic reports, 'the enemy of objective, dispassionate observation [. . .] riddled with bias, faulty memory and "foolish optimism"' (Campo 2006).

Despite such sampling problems and the lowly position anecdotes occupy in hierarchies of evidential credibility (Murad et al. 2016: 125–126), they continue to be published in medical journals. Since the *British Medical Journal* (*BMJ*) initiated its Minerva column in the 1880s and its subsequent 'Literary Notes' section a few decades later, it has carried snippets of anecdotal information, anonymous experienced-based testimonies, mirrored by *The Lancet*'s 'In England Now' column, and by its more recent 'Uses of Error' section, which featured personal accounts of medical mistakes.

Anecdotes remain prominent features of contemporary medicine's written and oral culture, a circumstance that led the literary scholar Kathryn Montgomery Hunter to insist that '[s]omething so pervasive and so contrary to medicine's scientific ideal [. . .] must have a function in the everyday business of medicine' (Hunter 1991: 70). Within the schema of a 'nutshell narrative' (Morgan, Chapter 1), these functions include delineating modes of

---

[6] Despite their brevity and portability, anecdotes do not always align with the 'thin descriptions' and narratives of the chemical reactions Paskins discusses in Chapter 13; some constitute thicker narratives.

noticing that effect epistemic switches that prime audiences to adjust their prior understandings of situations in the light of experience.

17.3   Colligating Medical Anecdotes

Anecdotes played a distinctive role in William Withering's discovery of the therapeutic value of the foxglove in the treatment of dropsy, a condition that caused bodily swelling. At the head of his *Account of the Foxglove* (1785), Withering featured several anecdotes, foremost of which was the following:

> In the year 1775, my opinion was asked concerning a family receipt for the cure of the dropsy. I was told that it had long been kept a secret by an old woman in Shropshire, who had sometimes made cures after the more regular practitioners had failed. I was informed also, that the effects produced were violent vomiting and purging; for the diuretic effects seemed to have been overlooked. This medicine was composed of twenty or more different herbs; but it was not very difficult for one conversant in these subjects, to perceive, that the active herb could be no other than the Foxglove. (Withering 1785: 2)

With this anecdote Withering simultaneously proclaimed and reframed the therapeutic value of the remedy. The account depicted more than a sudden apprehension of the active ingredient of 'a family receipt', a recipe handed down from one generation to another; it also relayed his realization that diuresis accounted for its efficacy, not the 'violent vomiting and purging' the remedy also provoked, which had long been accepted as beneficial for dropsy. Withering bolstered this aperçu with further anecdotes: he 'knew of a woman in the neighbourhood of Warwick' and another 'in Yorkshire' who possessed a similar 'receipt', and relayed a 'circumstance' told to him by his 'truly valuable and respectable friend, Dr. Ash' that the Principal of Brasenose College, Oxford, had been cured of dropsy 'by an empirical exhibition of the root of the Foxglove, after some of the first physicians of the age had declared they could do no more for him' (Withering 1785: 3).

The source of these causal claims was lay experience,[7] which proved rich in anecdotal remedies for dropsy in accounts that were not themselves case reports but testimonies which emerged 'from the empirical usages and experience of the populace' (Withering 1785: 1). At a time when such anecdotes featured in advertisements for quack remedies and false nostrums, Withering was well aware of the precarity of their credibility: 'There are men who will hardly admit of any thing which an author advances in support of a favorite medicine, and I allow they may have some cause for their hesitation' (Withering 1785: viii). Contrary to advertising testimonies – which were countered as 'cases of cures never performed, and copies of affidavits never sworn to'

---

[7] Withering may not have been the first to identify the foxglove as the remedy's active ingredient; he records its use in Edinburgh in 1777 (Withering 1785: xx).

(Adair 1790: 256)[8] – Withering drew attention to people harmed by the remedy in an anecdote that could not be so easily questioned or dismissed:

I recollect about two years ago being desired to visit a travelling Yorkshire tradesman. I found him incessantly vomiting, his vision indistinct, his pulse forty in a minute. Upon enquiry it came out, that his wife had stewed a large handful of green Foxglove leaves in half a pint of water, and given him the liquor, which he drank at one draught, in order to cure him of an asthmatic affection. This good woman knew the medicine of her country, but not the dose of it, for her husband narrowly escaped with his life. (Withering 1785: 9–10)

Withering's first anecdote contained more than a simple flash of inspiration: in positing the remedy's active ingredient and mechanism of action, he narrowed the explanandum and refined the type of explanans that accounted for its benefits, which no longer centred on purgings. The change fuelled the need to develop a different treatment regimen, based on careful dose adjustment, which Withering set out in a hierarchy of some 150 case histories that precisely reflected his capacity to vouch for their details.[9] Whereas the anecdote that announced the discovery invoked observations by others, Withering's cases focused on his own observations, and how he maximized the remedy's benefits and minimized its harms. By standardizing the use of extracts and adjusting doses, he set out the candidate 'causal relations' and 'manipulable facts' that Rachel Ankeny identifies as some of the key contributions cases make to medical knowledge (Ankeny 2014: 1009).

Eighteenth-century case reports invoked intricate descriptions taken to reflect their authors' powers of perception, which helped underpin the verisimilitude claimed for their accounts (Da Costa 2002). Hess and Mendelsohn note that cases increasingly were treated as collections of observable data linked to specific individuals, based on information excerpted from their histories (Hess and Mendelsohn 2010; 2014). However, unlike cases, anecdotes remained as solitary micro-narratives which arose from fleeting, unformalized aspects of medical practice, demonstrated by the anecdote published in the BMJ in 1881, which turned on the perils of making inferences across cases:

The following tale has lately been reported. [...] An epidemic of typhoid fever broke out in a small village in the South of France. A locksmith fell ill, and called in the local medical man, who came, prescribed, and went away. The next day, during his

---

[8] Cited in R. Porter, *Health for Sale: Quackery in England, 1660–1850* (Manchester: Manchester University Press, 1989).

[9] In his *Account*, people assessed and treated at home by Withering were given pride of place, followed by those in hospital supervised by others for some of their stay; next came patients referred to him by other doctors whose complete course of treatment he did not observe; and lastly, those not seen by him but reported by others.

usual rounds, he called at the locksmith's, and asked his wife after the health of the interesting patient. She replied: 'Ah, sir, only imagine, whilst I went to fetch the medicine, my husband ate two pickled herrings and a dish of bean salad.' 'Good heavens! Then he is . . . .' 'Quite well, doctor. He went to work this morning as usual, and is as well as possible.' 'That is extraordinary', exclaimed the doctor; 'what a wonderful remedy for typhoid; I must make a note of it.' And he accordingly entered in his note-book: 'Typhoid fever: tried remedy, two pickled herrings and bean salad.' Two days afterwards, a bricklayer was attacked by the same disorder. 'Take', said the same doctor who was consulted, 'two pickled herrings and a dish of bean salad. I will come again to-morrow.' To-morrow-alas! the bricklayer was dead. The doctor, taking a logical view of his experimental method, again entered in the famous note-book. 'Typhoid fever. Remedy. pickled herrings and bean salad. Good for locksmiths, bad for bricklayers.' (Anon. 1881: 248)

To attribute a difference in outcome to a difference in occupation, the same treatment given to the locksmith would have to have been taken by the bricklayer. But whatever the medical man prescribed for the locksmith, assuming he took it, which is not certain, the doctor assumed it played no role in his unexpected recovery. In his surprise at the locksmith's recovery, the doctor purloined his contingent dietary likes as treatment for the next case of typhoid. As a burlesque on clinical reasoning and the doctor's 'experimental method', the account relayed a scenario built on the perils of inferring causation within and across individual cases. In striving for, and falling comically short of, a generalizable possibility (Hurwitz 2017), the anecdote parodies how easily certainty and uncertainty can be elided concerning the cause of a clinical outcome. It also prefigures how unreliable patient and practitioner testimonies would come to be considered in the following century.

17.4   Quirky, Anecdotal Testimony

In the second half of the twentieth century, testimony rooted in patient and practitioner experiences came to be viewed as profoundly unreliable. The contemporary clinical researchers Murray Enkin and Alejandro Jadad defined the anecdotal as 'any type of information informally gained, either from personal or clinical experience, one's own or that of others, in contradistinction to evidence generated by formal research studies' (Enkin and Jadad 1998: 963). The physician Mark Crislip put the position succinctly:

Anecdotes are how patients transmit the particulars of their disease to their health care providers. The medical history, as taken from the patient, is an extended anecdote, from which the particulars of the disease have to be extracted. (Crislip 2008)

On this account, patients' concerns and fears – the whole symptomatic realm – are viewed as anecdotes and placed at the very centre of practice. Steven Novella, an evidence-based neurologist, classed all 'uncontrolled subjective observations' as anecdotal, although in the context of discovery he argued that they may still be useful:

Many medical discoveries started as anecdotal observations. But then those observations have to be tested with controlled observations or experiments – and most anecdotal observations will turn out to be wrong or misleading, because they are quirky and uncontrolled. (Novella 2010)

Yet the editors of *The Lancet* singled out anecdotes as the lingua franca of clinical exchange, information and learning, claiming that:

Clinicians learn from anecdotes – stories they heard at medical school, stories they tell each other, and stories their patients tell them. This is an efficient way to grasp new knowledge – even the most obscure hints and warnings can be made memorable if tagged to real people and actual events. (Bignall and Horton 1995)

Less formalized than case reports, although consonant with Withering's stance, the editors treat anecdotes as potential sources of knowledge 'if tagged to real people and actual events'. Without some reliance on anecdotal testimony it is difficult to see how therapeutic substances could ever have been developed. Jeff Aronson, a clinical pharmacologist, observes how commonly contemporary research and development of new drugs 'start with an anecdotal report of some sort' (Aronson 2008: xxxi), which provokes further tests of efficacy and toxicity in animal and human tissues, Phase 1 trials, then trials in larger, more representative human populations.

In 1992, fragmentary, unshaped anecdotal accounts concerning a quirky effect of a drug prompted the pharmaceutical company Pfizer to upturn the rationale of its testing programme on sildenafil citrate, a chemical it had synthesized in the 1980s as a possible treatment for high blood pressure and angina. Six years later, the company began marketing sildenafil under the brand name Viagra (Tozzi and Hopkins 2017; Dunzendorfer 2004).

Sildenafil was the product of a 'rational drug design program' (Terrett et al 1996: 1819), targeted at inhibiting a family of enzymes believed to play a part in increasing thrombosis and vascular resistance. Synthesized in Pfizer's UK Discovery Chemistry and Biology Laboratories, the drug reached Phase 1 trials in 1992 in eight healthy South Wales miners. Twenty-five years later, David Brown, a Pfizer chemist, recalled the response to a routine question at the end of the study to the whole group of volunteers: 'Is there anything else you noticed you want to report?'

One of the men put up his hand and said, 'Well, I seemed to have more erections during the night than normal,' and all the others kind of smiled and said, 'So did we.' (Tozzi and Hopkins 2017)

Although Brown's recollection reconstructed the occasion as a moment of revelation, Ian Osterloh, who was in charge of research in Pfizer's laboratories, reported that: 'At the time, no one really thought, "This is fantastic, this is great news, we're really onto something here. We must switch the direction of this program"'. Nick Terrett, another Pfizer researcher, recalled that when asked about side effects of the drug, volunteers said: '"Yeah, I've got a headache. I feel a bit dizzy" and some added "I've got erections"', but these did not register as more significant than other side effects, such as backache, throbbing and stomach upsets (Friend 2017: 480–481). Osterloh recounted Pfizer's initial response:

None of us at Pfizer thought much of this side effect at the time. I remember thinking that even if it did work, who would want to take a drug on a Wednesday to get an erection on a Saturday? So we pushed on with the angina studies. (Osterloh 2015)

But when intravenous sildenafil showed little cardiovascular effect in volunteers, hopes of a tangible anti-anginal benefit faded and the three-year development programme faced closure (Jackson et al. 1999). Notwithstanding the costs already incurred, reports voiced as 'I *seemed* to have more erections', which were not 'thought much of' by Pfizer, and likely biases from a combination of embarrassment, self-censorship and embellishment in testimonies that could have become 'improved in the telling' (Gross 2006: xii) – 'all the others kind of smiled' (Tozzi and Hopkins 2017) – the company credited the miners' reports with sufficient warrant to commit additional funding to further investigate sildenafil.

Testimony takes place within a context framed by a variety of practices and institutions that affect both its content and the level of warrant it purports to deliver. To interpret and assess it requires sensitivity to such contextual factors. Rather than concluding that testimony is in general warranted or that only the testimony of informants who are known to be reliable is warranted, we assess testimony in light of [...] not just who is talking and what she is saying, but also what is at issue, what is being assumed about the facts, the circumstances, the testifier, the audience, and the cognitive context. [...] A testifier can transmit no more warrant than she has. But her audience may have epistemic resources that she lacks [...] [which] is why testimony can be a vehicle for the advancement, not just the dissemination, of understanding. (Elgin 2002: 307)

To consider the anecdotal as solely comprised of claims is to miss the cognitive and other contexts in which anecdotes arise, are received, and made sense of. To the anecdotes he heard in the 1770s, Withering deployed highly developed epistemic resources not only for the historical and biographical purposes of announcing how he happened upon the discovery, but in granting anecdotes warrant for the remedy's therapeutic and harmful effects. In respect of the bicycle spoke in the lung anecdote, the information the patient provided proved decisive evidence against the likelihood of another faulty X-ray, which threw doubt on the reliability of the diagnosis. In regard to the sildenafil testimonies, the epistemic resources of Pfizer's researchers appeared initially under-developed;[10] it was not that they doubted the truth of the miners' reports, but the erectile effect was unexpected and their testimonies stood outside the anti-anginal framework in which the drug had been developed. At a time when there were no oral therapies for impotence and the biochemistry of tumescence was only partially understood, the testimonies lacked a framework in which to articulate the effect, which was medically and culturally unsituated (Giere 2006; Massimi 2018).

Consistent with Fleming's view that anecdotes can only be collected and 'not sewn together into a single story' (Fleming 2011: 74), Withering attempted no summation of his anecdotes. Although they carried evidential warrant, each stood alone in his *Account*. Pfizer attempted no summation of the anecdotal testimonies of the Welsh miners as their reports at best could only have been counted as a 'pre-cursor to [scientific] evidence' (DeWald 2013). Nevertheless, they proved sufficient to persuade Pfizer to turn around its research programme, a switch that would garner the company $30 billion of revenue from sildenafil sales in the first two decades of the twenty-first century (Statista 2020).

Pfizer attempted no summation of the testimony of the volunteers who subsequently joined its sildenafil trials: instead, a different type of evidence was developed that reflected the experiences of men on and off the drug. This evidence pertained to several components of the erectile process – speed of onset, hardness and duration, capacity to penetrate – gauged from answers to questionnaires and records of measuring devices in studies that recruited over 4,000 men worldwide (Eardley et al. 2002). Key to these studies was the development of an International Index of Erectile Function, a self-completable measure that took little account of emotional, inter-relational, non-erectile aspects of sex (Burnett 2020), but proved sensitive and specific enough to detect treatment- and dose-related erectile effects within and between drug and placebo (Rosen et al. 1997; Goldstein et al. 1998).

Once the biochemical and physiological mechanisms underlying tumescence had been fully elucidated, the effects of sildenafil could be fitted into a causal account of human erection (Baier 2019) and the voices of the miners could be overlaid by a set of multidimensional scores of thousands of male heterosexual performances, on and off the drug. In place of their anecdotal testimonies stood a matrix of standardized, combinable, self-reported scores, averaged for separable components of the erectile process, partitioned by age, dosage and timing, severity and cause of erectile dysfunction.

---

[10] The epistemic resources called on by anecdotes might be compared to the creative cognitive and affective processes that enable reconceptualization of mental models of the world (see Jajdelska, Chapter 18).

According to the philosopher Elizabeth Fricker, the paradigm of testimony is a face-to-face encounter in which expression of a knowledgeable belief is vouched for by a combination of the speaker's trustworthiness and 'choice of words' (Fricker 2006: 594), a model that fits the testimony of Withering's informants, the bicycle patient and Welsh miners. However, as we have seen, the miners' natural language testimonies came to be revoiced in the 'de-anecdotalized', formal language of Pfizer's subsequent trial participants. In place of their 'choice of words' – which was never elicited – were scored responses to the closed questions of impotence researchers, expressed in data points derived from Lickert scales and channelled into a quantitative construct of erectile effectiveness.[11] These standardized and manipulable scores spoke the quantized language of mean differences in effect size between drug and placebo, and took the place of the anecdotal accounts of the miners.

17.5   Conclusions

In examining how anecdotes handle medical materials, size up situations and gain attention, the aim has not been to advocate a presumptive right for their claims to be believed. It has been to delineate the epistemic shifts interwoven in their narratives, which confer an immediacy of engagement with what is recounted.[12] In the Greco-Roman period, anecdotes provided a view of political power profoundly at odds with authorized accounts (Fenves 2001: 153); in the Renaissance, they became 'the principal register of the unexpected and hence of the encounter with difference' (Greenblatt 1991: 2–3), and in questioning the social order they recruited 'the unreliable, eccentric, and the improper' (Patterson 1997: 160). In the eighteenth century, their 'fragmentary, eye-witness character' fitted them to a 'variety of verbal practices, both oral and written, both popular and cultivated: the joke or the tall story; the jewel-like short narrative, with its witty punchline [. . .] usually containing a moral lesson' (Gossman 2003: 149–150).

Such tropes are evident in the dramatic and humorous nature of anecdotes in the modern era, which may overshadow their epistemic coordinates. Michael O'Donnell valued anecdotes for their quick-witted formulation of 'uncertainties, paradoxes, life-affirming surprises and black comedy' (O'Donnell 2013: i). Within this heterogeneity, a variety of anecdote can be delineated, which draws together observations and dialogues from scattered zones of healthcare experiences, in a cognitively and affectively rousing narrative. It opens on a situation that quickly faces challenge from a new vantage point and

---

[11] For discussion of the differing natures of natural and formal languages, see Wise (Chapter 22).

[12] The sense of immediacy derives in part from epistemic switching that invokes not only a change in episteme, but also a switch of speaker and the voices heard in healthcare situations. See Hajek (Chapter 2) and Wise (Chapter 22).

perspective, which introduce other views that make different sense of the situation, and lead to a degree of resolution.

Within this minimalist format, altered appearances and understandings based on first-person perspectives immediately create unanticipated effects that can severely test epistemic resources. To make sense of the mental processes believed to govern the ability to place the parts of a familiar situation in context, the psychologist Francesca Happé argued that 'the best way to convey what is meant by the tentative and exploratory notion of central coherence is to give an anecdote':

> A clinician testing a bright autistic boy presented him with a toy bed, and asked the child to name the parts. The child correctly labelled the bed, mattress, and quilt. The clinician then pointed to the pillow and asked, 'And what is this?'. The boy replied, 'It's a piece of ravioli'. (Happé 1995: 173–174)

Happé emphasized that the boy was neither visually impaired nor joking and that his 'clinician commented that the pillow did indeed look like a piece of ravioli' (Happé 1991: 174). His uncanny response challenged researchers to explain how a cognitive process that correctly identified some of the accoutrements of a toy bed discounted dissimilarities in texture, function and socio-cultural context of a key component of the ensemble, allowing its place to be supplanted by a piece of pasta. The anecdote made the explanatory potential of weak central coherence the vantage point from which the boy's response became explicable.

Walter Benjamin argued that anecdotes develop a 'pathos of nearness', which brings distant happenings spatiotemporally closer to us (Benjamin 1999: S1a, 3).[13] It manifests by making apparent and proximate contours, patterns, frameworks and views of situations hitherto outside of the field of vision. Bernard Williams argued that an 'epistemic division of labour' (Williams 2004: 43) underpins viewpoint and perspective in the role observational position plays in the look of things. Although most of the anecdotes we have examined are not first-order accounts of sensory-based observations, but retellings of the noticings of others, Williams's topological argument illuminates how observations made from different angles, conceptual, figurative, first-person vantage points and hypothetical positions, create different views of their objects of interest.[14]

In the context of a broadly observational and inferential practice such as medicine, anecdotes are the culturally mediated schemata that alert

---

[13] Cited in R. S. Lehman, *Impossible Modernism: T. S. Eliot, Walter Benjamin, and the Critique of Historical Reason* (Stanford, CA: Stanford University Press, 2016), 145–146.

[14] By extension, Williams's point encompasses Maxwell's advocacy of different ways of looking in elucidating the nature of electro-magnetic force lines; see Wise (Chapter 22). For how different views and perspectives on the same geological formation have given rise to different explanatory narratives, see Hopkins (Chapter 4).

audiences to the importance of epistemic vigilance, by demonstrating adjustment of prior understanding of a situation in the light of vantage points and experiences emergent from them (Herman 1999: 25).[15] By bringing observational and conceptual frameworks into close relation with each other, anecdotes bring to the fore scenarios viewed in a new light before their audiences. Those we have encountered include a pattern in bedside greetings; a radical change in diagnosis and treatment; a switch of diagnostic mindset; a change in the mode of action of a remedy; and the intimation that a tentatively voiced side effect pointed to an unrecognized therapeutic effect.

Kathryn Hunter argues that 'medicine both scorns the anecdotal and provides for the careful reporting of single instances' (Hunter 1991: 118). Unlike case reports, which have become highly regulated medical accounts, anecdotes remain informally patrolled schema, cast in a vernacular language that has less recourse to technical and formal terminology than cases. Where anecdotes set out their own observations and descriptions, they do so in a register more humanly voiced in the idioms of conversation and hearsay than that of cases. Epistemic switches, comprehensibility and human-centred focus are interwoven features of anecdotal narratives and contribute to their pungency. Despite the increasingly formalized nature of bioscientific discourse, anecdotes retain a continuing presence in medical culture. They demonstrate how suffused with vantage point and perspective medical understanding is and how dependent medical knowledge remains on the speakers and voices heard in healthcare.[16]

Bibliography

Adair, J. M. (1790). *Essays on Fashionable Diseases*. London: T. P. Bateman, 256.

Adams, J. (1789). *Anecdotes, Bons-mots, and Characteristic Traits of the Greatest Princes, Politicians, Philosophers, Orators, and Wits of Modern Times*. Dublin: Printed for Messrs. H. Chamberlaine, P. Byrne, etc.

Ankeny, R. A. (2014). 'The Overlooked Role of Cases in Causal Attribution in Medicine'. *Philosophy of Science* 81.5: 999–1011.

---

[15] To adopt Mary Morgan's terms, these schemata arise from a narrative-making practice that works and reworks healthcare materials to make new sense of them, in ways that can have ontological implications (Morgan, Chapter 1).

[16] Many thanks to Mary Morgan, Kim Hajek, Mat Paskins, John Heywood, Jeremy Howick, Jens Brockmeier, Anna Elsner, Iain Bamforth, Graham Matthews, David Griffith, James Le Fanu, John Launer, Jacek Mostwin, Max Saunders, Neil Vickers, Mathias Wirth, Martina Zimmermann and Ruth Richardson, and to participants of NS workshops for their critical comments at various stages of the preparation of this chapter. *Narrative Science* book: This project has received funding from the European Research Council under the European Union's Horizon 2020 research and innovation programme (grant agreement No. 694732). www.narrative-science.org/.

Anon. (1799). '[Review of] *Biographiana. By the Compiler of "Anecdotes of Distinguished Persons"'. The London Review, and Biographia Literaria*. vol. 1. London: Vernor & Hood.

Anon. (1824). 'Anecdote'. *The Lancet* 34.2: 256.

Anon. (1881). 'New Remedies'. *British Medical Journal* (6 August): 248.

Anon. (1985). 'In England Now'. *The Lancet* (17 August): 381.

Aronson, J. K. (2008). *Side Effects of Drugs Annual: A Worldwide Yearly Survey of New Data and Trends in Adverse Drug Reactions*. Amsterdam: Elsevier.

Atkinson, P. (1995). *Medical Talk and Medical Work: The Liturgy of the Clinic*. London: Sage.

Baier, S. (2019). 'Drug Discovery Graph, Hoffmann-La Roche AG'. In M. Paskins and Mary S. Morgan, eds. *Anthology I: Case Book of Narrative Science*. London: London School of Economics, 48–49. www.narrative-science.org/uploads/3/1/7/6/31762379/narrative_science_anthology_complete_final_draft_9_11_2019.pdf.

Beatty, J. (2016). 'What Are Narratives Good For?' *Studies in History and Philosophy of Science Part C* 58: 33–40.

   (2017). 'Narrative Possibility and Narrative Explanation'. *Studies in History and Philosophy of Science Part A* 62: 31–41.

Benjamin, W. (1999). *The Arcades Project*. Trans. Howard Eiland and Kevin McLaughlin. Cambridge, MA: Harvard University Press, SA,1a,3. https://monoskop.org/images/e/e4/Benjamin_Walter_The_Arcades_Project.pdf.

Bignall, J., and R. Horton (1995). 'Learning from Stories: *The Lancet*'s Case Reports'. *The Lancet* 346: 1246.

Brockliss, L., and C. Jones (1997). *The Medical World of Early Modern France*. Oxford: Oxford University Press.

Burke, P. (2012). 'The Rise of "Secret History"'. In C. Emden and D. Midgley, eds. *Changing Perceptions of the Public Sphere*. New York: Berghahn Books, 57–72.

Burnett, A. L. (2020) 'Commentary RE: The International Index of Erectile Function (IIEF): A Multidimensional Scale for Assessment of Erectile Dysfunction'. *Urology* 145: 308–309.

Campo, R. (2006). '"Anecdotal Evidence": Why Narratives Matter to Medical Practice'. *PLoS Med* 3.10: e 423. https://journals.plos.org/plosmedicine/article?id=10.1371/journal.pmed.0030423.

Crislip, M. (2008). 'A Budget of Anecdotes'. *Science-Based Medicine* (25 September). https://sciencebasedmedicine.org/a-budget-of-anecdotes/.

D'Israeli, I. (1793). *A Dissertation on Anecdotes; by the Author of Curiosities of Literature*. London: Kearsley & Murray.

Da Costa, P. F. (2002). 'The Making of Extraordinary Facts: Authentication of Singularities of Nature at the Royal Society of London in the First Half of the Eighteenth Century'. *Studies in History and Philosophy of Science Part A* 33.2: 265–288.

Daston, L. (1988). *Classical Probability in the Enlightenment*. Princeton, NJ: Princeton University Press.

DeWald, J. (2013). 'Anecdotes and Science': Part 2. *Skeptoid Blog*. https://skeptoid.com /blog/2013/02/15/anecdotes-and-science-part-2/.

Dunzendorfer, U., ed. (2004). *Sildenafil*. Frankfurt am Main: Springer.

Eardley, I., P. Ellis, M. Boolell and M. Wulff (2002). 'Onset and Duration of Action of Sildenafil for the Treatment of Erectile Dysfunction'. *British Journal of Clinical Pharmacology* 53.1: 61S–65S.

Elgin, C. (2002). 'Take It from Me: The Epistemological Status of Testimony'. *Philosophy and Phenomenological Research* 65.2: 291–308.

Enkin, M., and A. Jadad (1998). 'Using Anecdotal Information in Evidence-Based Health Care: Heresy or Necessity?' *Annals of Oncology* 9.9: 963–966.

Fenves, P. (2001). *Arresting Language*. Stanford, CA: Stanford University Press.

Fineman, J. (1989). 'The History of the Anecdote: Fact and Fiction'. In H. A. Veeser, ed. *The New Historicism*. New York: Routledge, 49–87.

Fleming, P. (2011). 'The Perfect Story: Anecdote and Exemplarity in Linnaeus and Blumenberg'. *Thesis Eleven* 104.1: 72–86.

Flyvbjerg, B. (2006). 'Five Misunderstandings about Case-Study Research'. *Qualitative Inquiry* 12.2: 219–245.

Fricker, E. (2006). 'Second-Hand Knowledge'. *Philosophy and Phenomenological Research* 73.3: 592–618.

Friend, D. (2017). *The Naughty Nineties*. New York: Hachette Book Group.

Giere, R. N. (2006). *Scientific Perspectivism*. Chicago: University of Chicago Press.

Goldstein, I. , T. F. Lue, H. Padma-Nathan, R. C. Rosen et al. (1998). 'Oral Sildenafil in the Treatment of Erectile Dysfunction'. *New England Journal of Medicine* 338: 1397–1404.

Gossman, L. (2003). 'Anecdote and History'. *History and Theory* 42: 143–168.

Greenblatt, S. (1991) *Marvelous Possessions: The Wonder of the New World*. Chicago: University of Chicago Press.

Gross, J. ed. (2006). *The New Oxford Book of Literary Anecdotes*. Oxford: Oxford University Press.

Happé, F. (1995). 'Theory of Mind and Communication in Autism'. PhD thesis, University College London.

Hein, J. (1981). 'Die Anekdote'. In O. Knörrich, ed. *Formen der Literatur in Einzeldarstellungen*. Stuttgart: Alfred Kröner, 15–18.

Herman. D. (1999). 'Parables of Narrative Imagining'. *Diacritics* 29.1: 20–36.

Hess, V. J., and A. Mendelsohn (2010). 'Case and Series: Medical Knowledge and Paper Technologies, 1600–1900'. *History of Science* 48: 287–314.

(2014). 'Sauvages' Paperwork: How Disease Classification Arose from Scholarly Note-Taking'. *Early Science and Medicine* 19.5: 471–503.

Hunter, K. M. (1991). *Doctors' Stories: The Narrative Structure of Medical Knowledge*. Princeton, NJ: Princeton University Press, 1991.

Hurwitz, B. (2017). 'Narrative Constructs in Modern Clinical Case Reporting'. *Studies in History and Philosophy of Science Part A* 62: 65–73.

Jackson G., N. Benjamin, N. Jackson and M. J. Allen (1999). 'Effects of Sildenafil Citrate on Human Hemodynamics'. *American Journal of Cardiology* 83: 13C–20C.

Johnson, Samuel (1773). *A Dictionary of the English Language*. 4th folio edn. London: Printed for W. Strahan; J. and F. Rivington; J. Hinton; etc.

Jolles, A. (2017). *Simple Forms*. Trans. Peter Schwartz. London: Verso.

Katzenstein, L. (2001). *Viagra*. New York: Medical Information Press.

Locke, J. (1690). *An Essay Concerning Human Understanding, in Four Books*. 2nd edn. London. www.gutenberg.org/cache/epub/10616/pg10616-images.html.

McCumber, J. (2009). 'To Be Is to Be an Anecdote: Hegel and the Therapeutic Absolute'. *SubStance* 38.1: 56–65.

Massimi, M. (2018). 'Perspectivism'. In J. Saatsi, ed. *The Routledge Handbook of Scientific Realism*. London: Routledge, 164–175.

Murad, M. H., N. Asi, M. Alsawa and F. Alahdab (2016). 'New Evidence Pyramid'. *Evidence Based Medicine* 21.4: 125–127.

National Institute for Health Care Excellence (2018). 'Chronic Heart Failure in Adults: Diagnosis and Management' [NICE guideline 106]. www.nice.org.uk/guidance/ng106.

Novella, S. (2010). 'The Context of Anecdotes and Anomalies', *Neurologica Blog, History of Science/Medicine, Skepticism*. https://theness.com/neurologicablog/index.php/the-context-of-anecdotes-and-anomalies/.

O'Donnell, M. (2013). *The Barefaced Doctor*. Leicester: Matador.

(2016). *Medicine's Strangest Cases*. London: Portico.

Osterloh, I. (2015). 'How I Discovered Viagra'. *Cosmos Newsletter* (27 April). https://cosmosmagazine.com/biology/how-i-discovered-viagra.

Paskins, M., and Morgan, Mary S. (2019). 'Preface'. In M. Paskins and Mary S. Morgan, eds. *Anthology I: Case Book of Narrative Science*. London: London School of Economics, 7–9. www.narrative-science.org/uploads/3/1/7/6/31762379/narrative_science_anthology_complete_final_draft_9_11_2019.pdf.

Patterson, A. (1997). 'Foul, His Wife, the Mayor, and Foul's Mare: The Power of Anecdote in Tudor Historiography'. In Donald R. Kelley and David Harris Sacks, eds. *The Historical Imagination in Early Modern Britain: History, Rhetoric, and Fiction, 1500–1800*. Cambridge: Cambridge University Press, 159–178.

Rando, D. (2011). *Modernist Fiction and News: Representing Experience in the Early Twentieth Century*. New York: Palgrave Macmillan.

Riley, D. S, M. S. Barber, G. S. Kienle, J. K. Aronson et al. (2017). 'CARE Guidelines for Case Reports: Explanation and Elaboration Document'. *Journal of Clinical Epidemiology* 89: 218–235.

Rosen, R. C., A. Riley, G. Wagner, I. H. Osterloh et al. (1997). 'The International Index of Erectile Dysfunction (IIEF): A Multidimensional Scale for Assessment of Erectile Dysfunction'. *Urology* 49.6: 822–830.

Shapin, S. (1995). *A Social History of Truth: Civility and Science in Seventeenth-Century England*. Chicago: Chicago University Press.

Statista (2020). 'Worldwide Revenue of Pfizer's Viagra from 2003 to 2019'. www.statista.com/statistics/264827/pfizers-worldwide-viagra-revenue-since-2003/.

Terrett, N. K., A. S. Bell, D. Brown and P. Ellis (1996). 'Sildenafil (Viagra^{TM}), a Potent and Selective Inhibitor of Type 5 cGMP Phosphodiesterase with Utility for the

Treatment of Male Erectile Dysfunction'. *Bioorganic and Medicinal Chemistry Letters* 6.15: 1819–1824.

Tozzi, J., and J. S. Hopkins. (2017). 'The Little Blue Pill: An Oral History of Viagra'. *Bloomberg*. www.bloomberg.com/news/features/2017-12-11/the-little-blue-pill-an-oral-history-of-viagra.

Varillas, A. (1686). 'The Author's Preface'. In *'Ανεκδοτα Έτερουριακα. Or, the Secret History of the House of Medicis*. Trans. Ferrand Spence. London: R. Bentley and S. Magnes.

Wadd, W. (1824). *Nugæ Chirurgicæ*. London: John Nichols & Son.

Williams, B. (2004). *Truth and Truthfulness: An Essay in Genealogy*. Princeton, NJ: Princeton University Press.

Withering, W. (1785). *An Account of the Foxglove and Some of Its Medical Uses: With Practical Remarks on Dropsy, and Other Diseases*. Birmingham: M. Swinney.

Wollheim, R. (1979). 'Memory, Experiential Memory and Personal Identity'. In G. F. Macdonald, ed. *Perception and Identity: Essays Presented to A. J. Ayer*. London: Macmillan, 186–233.

Wood, J. (2019). *Anecdotes of Enlightenment*. Charlottesville: University of Virginia Press.

18   Narrative Performance and the 'Taboo on Causal Inference': A Case Study of Conceptual Remodelling and Implicit Causation

Elspeth Jajdelska

Abstract

Storytelling can be understood as a performative social event that instantiates a specific relationship between storyteller and audience. This relationship supports inferences of narrative causation in hearers, both locally (episode x caused episode y) and globally (repeated patterns of causation at a more abstract level). This applies to passages of performative speech in a narrative event that are non-narrative, such as description or digression. Scientific writing is often conceived as non-performative and impersonal, with causation expressed explicitly. However, I suggest in this chapter that discourse of this kind can make the task of configuring global patterns of causation more difficult. Performative narrative discourse, on the other hand, offers support for readers in the task of remodelling existing theoretical causal structures through reconceptualization. I illustrate this argument through an analysis of narrative and non-narrative performative discourse in the field of cognitive psychology.

18.1   Introduction

Research on live storytelling illustrates how a performative storyteller holds the floor, as the audience lends them the authority to control who speaks, on which topics, when and for how long. Storytellers are also granted some authority over the organization of space, movement within it and acceptable behaviours. It is this wholeness of the social event that supports a shared attribution of underlying causes linking one event in the narrative sequence to the next (Lwin 2010; Goffman 1974). This willingness to cede, temporarily, authority to a storyteller leaves traces in written narrative, for example in readers' willingness to trust narrative voices (on general 'truth bias', see Gilbert, Tafarodi and Maone 1993; in narrative in particular, see Yanal 1999). In what follows,

I suggest that experience of narrative performativity in a science paper can influence readers' trust not just in the authors' factual authenticity (e.g., is the data accurate?) but in the more contingent *interpretation* of facts. I view narration as entwined with performance, such that even non-narrative episodes of performative speech in an event that is narrative overall contribute to the inference of causes. In this light, I show how causal modelling is affected by local instances in science writing of both narrative and non-narrative performativity, arguing that together they incite the reader to construct an implicit and new narrative model relating a set of cognitive concepts.[1]

I use 'model' here in the sense developed by philosophers, psychologists and computer scientists in reference to predictive theories of mind organized around Bayesian probability. Here, perception is a continual revision of predictions through error correction, a dynamic network of shifting probabilities giving rise to implicit, and continually changing, causal relationships through a process of reconceptualization (Clark 2013).

'Reconceptualization' is used in the sense employed by Churchland and Boden in their accounts of creativity in the context of connectionist, and more recently neural network, approaches. In this light, concepts are not the brittle units of modular approaches to mind (Fodor), but 'flexible, distributed representations' (Kiefer and Pulvermüller 2012: 805). As a result they can be reconfigured by experience, or by repeated iterations of thought, as in Churchland's example of Newton reconceptualizing the moon, from something like a ball bearing in a track to something like a projectile that has been hurled. Reconceptualization in this sense involves reconfiguring a concept's causal profile and its relationship to the rest of the mind's world model.

Explicit and implicit mental causal modelling of this kind is found across scientific disciplines, and the process of reconceptualization which can be involved in this, and particularly in moments of scientific 'breakthrough', has been identified as a key creative component of scientific thinking. Examples include the identification of the Benzene ring, or the theory of gravity, mentioned above (Boden 2004; Churchland 2012). I have argued elsewhere that narrative fiction engages readers in a comparable process of creative reconceptualization (Jajdelska 2019). At the same time, causal modelling is foundational to the comprehension and experience of narrative performance, whether live or mediated by silent reading of narrative texts. Anthropologists, for example, have shown how causal links between story episodes are largely a matter of audience/reader inference, in response to the experience of the text as a whole. As a result, a single event sequence

---

[1] See Andersen (Chapter 19), who also investigates the work of readers in comprehension (of mathematical proofs rather than cognitive theory), in terms of contextual support from scripts rather than performativity.

can be associated in different performances with radically different under-lying causal models (Bauman 1986; Trompf, Gough and Eckhart 1988). Narrative performance, then, has the potential to generate creative causal inference in support of scientific remodelling.

The analysis below illustrates this idea, using a theory paper in scientific psychology by two cognitive neuroscientists. Scientific psychology is a good place to start this investigation, as it has a particularly anxious relationship with causal attribution. In part, this relates to its origins, which cohered in the twentieth century around research methods 'institutionalised across the varied communities of experimental, animal, educational, social, clinical, and applied psychologists' (Flis 2018). But it also relates to the dual status of its practi-tioners as simultaneously agents (as researchers) and objects (as minds) of study (Smith 2007). Smyth, for example, has shown that, compared to other scientific disciplines, scientific psychology 'behaves in textbooks as if there are grounds for not trusting its statements' (Smyth 2001: 392). The current repli-cation crisis in the discipline (Wingen, Berkessel and Englich 2020) is arguably symptomatic of these problems, as is an intense focus on methods rather than 'ways for initial selection and identification of relevant phenomena', which are 'in comparison [to methods] underdeveloped' (Flis 2019: 167) and a 'seriously limited' 'conceptual analysis of psychological phenomena' (Flis 2018: 160).

As a result, the discipline suffers from what Grosz and colleagues identify as 'a taboo' against what might be considered a 'central goal of research', that is aiming for 'explicit causal inference' to describe the world (Grosz, Rohrer and Thoemmes 2020). This fear of explicit causal inference can explain the 'encyclopedic incrementalism' of writing in the discipline, adding more and more discrete pieces of knowledge to the field. The promise that this could 'finish' psychology – produce comprehensive knowledge or at least 'organized theory' is one it 'just could not deliver' (Flis 2018: 31; Bazerman 1987; 1988). If the genre of the 'APA [American Psychological Association] article' has ended up in a cul-de-sac, the Barrett and Bar article I consider here is an interesting example of experimentation with a different genre, one that alter-nates with the APA conventions described by Flis, drawing on Bazerman. I argue that this paper's elements of narrative performativity, and of performa-tive language more generally in support of narrative inference, enable compre-hension of the underlying causal claim of the paper: that emotion is a cause of perception rather than just an effect.

My analysis is informed by the work of pioneering folklorist and anthro-pologist of verbal art, Richard Bauman (1975; 1986; Bauman and Briggs 1990). For Bauman, performativity is a measure of how far a speaker and audience understand their relationship to be one of evaluation of the speaker by the audience, not by reference to content, accuracy or informativeness, for example, but in relation to the speaker's 'communicative competence'

(Bauman and Briggs 1990: 66). In this respect, performativity is embedded in the social event, and is also scalar: speech can be more or less performative, not just either/or. Texts emerging through performative speech are likely to be characterized by greater or lesser degrees of poetic patterning and meta-textual features (features that draw attention to the artful status of the text itself). In the article analysed here, I suggest that both narrative and non-narrative performative sections are quite starkly distinguished from those that are non-performative, but that the former nonetheless play a vital role in explicating the non-performative sections by supporting an implicit narrative involving causal relations.

I chose the paper by Barrett and Bar because I have learned from it, and cited it in my own work. If the analysis appears critical at any point this is in the context of a critique of a scholarly practice rather than a critique of the authors. It is also a comparatively rare, although far from unique, example of this approach of blending the performative with the APA genre.[2] This technique also suggests unexpected overlaps between scientific psychology and those social scientists who successfully put 'elements into relation to each other when they appear in opposition' in publications, such that social scientists have 'good reason to think that they should fit together – in some way or other – rendering disparate and even oppositional matters into a narrative explanation' (Morgan 2017: 90). In this case it is not just content that is juxtaposed in productively puzzling ways, but modes of performance, or genre.

18.2　Analysis

'See It with Feeling: Affective Predictions during Object Perception', by psychologist Lisa Feldman Barrett and neuroscientist Moshe Bar (Barrett and Bar 2009), reviews existing findings in the fields of neuroscience and cognitive psychology to 'develop the hypothesis that the brain's ability to see in the present incorporates a representation of the affective impact of those visual sensations in the past'. In some ways, the combination of title and subtitle here captures precisely the sense of a 'taboo' on causal inference identified above by Grosz et al. The title conveys a relationship between emotion and perception of an environmental stimulus, in which each mutually influences the other; if we 'see with feeling' then it is not obvious that we could see without feeling. The causal relationship between the two in this scenario therefore departs from earlier theories in both behaviourist and later cognitive traditions, which 'assumed that affect occurred after object perception and in reaction to it' (Barrett and Bar 2009: 1328, citing Arnold 1960). One way to figure this is

---

[2] The paper by Barrett and Bar I discuss here was not in fact published in an APA journal, which may be why it is unusually performative for an article on scientific psychology.

as two titles for two distinctive narrations: the first more performative, the second less so. This division between two genres potentially allows for one to do the work of the reconceptualization, and the other to supply the evidence.

The authors' moves into and out of the stance of a narrative performer throughout the paper enable a relationship with the reader which can support creative reconceptualization of emotion and perception, and thereby allow the emergence of the more global theories which Flis identifies as missing from the project of psychology writing in the APA model. It should be noted that 'narration' here refers to the stance of *performing* a narrative, although the content of the paper itself does not provide the event sequence that can give the article as a whole the status of narrative. However, this performative stance of narration, supplemented by non-narrative moments of verbal performance, is, I suggest, crucial to the implicit narrative of temporally organized causal relations between emotion and perception which the article points towards.[3]

The authors' abstract in full will help with my further analysis below:

See it with feeling: affective predictions during object perception
People see with feeling. We 'gaze', 'behold', stare', 'gape' and 'glare'. In this paper, we develop the hypothesis that the brain's ability to see in the present incorporates a representation of the affective impact of those visual sensations in the past. This representation makes up part of the brain's prediction of what the visual sensations stand for in the present, including how to act on them in the near future. The affective prediction hypothesis implies that responses signalling an object's salience, relevance or value do not occur as a separate step after the object is identified. Instead, affective responses support vision from the very moment that visual stimulation begins. (Barrett and Bar 2009: 1325)

18.2.1   One Title, Two Modes

Returning to the title, it captures the contribution of local performative elements that are not explicitly narrative to the implicit narrative model developed by readers throughout the article: 'See it with feeling: affective predictions during object perception' (Barrett and Bar 2009: 1325). The first part of the sentence uses wordplay ('say/see' recalling 'say it with feeling', a stereotypical instruction to, for example, drama students), recalling Bauman's characterization of performance as drawing 'special attention to and heightened awareness of the act of expression' (Bauman 1975: 293). It also introduces the counterintuitive idea that there is no seeing, in the non-metaphorical sense of visual perception, without feelings. We may believe that sometimes we simply see

---

[3] See Meunier (Chapter 12), who also considers narrative as a means to familiarize new concepts, in his case through the narrator's relation to the narratee.

a scene or object in a disengaged or neutral way, but that, it is suggested, is not in fact the case. The performative section of the title, therefore, is also introducing the article's most counter-intuitive idea, one, moreover, which commits the authors to a broader theory of emotion and cognition. Performative text – that is text subject to evaluation for competence in delivery, rather than content or information – here enables a bolder claim than might otherwise seem proper to the APA article genre. Performativity may act here, then, as a creative training ground for what may be a new idea: that emotion contributes to perception rather than following it. In this respect, while the performative passages in the paper, as we will see, move in and out of narration, they cumulatively support a cohesive implicit narrative process in the reader of causal attributions in a model of mind and perception.

A different, less performative, relationship with the reader is constituted immediately after the colon: 'affective predictions during object perception'. 'Feeling' and 'affect' both have specialized meanings in cognitive science. 'Affect' generally refers to 'a state characterized by emotional feeling rather than rational thinking', a state which generally involves 'arousal', or 'corresponding bodily reaction (but not necessarily action)'. 'Feeling' can be used in neuroscience as a synonym for 'emotion', but it is often treated as 'one component' of emotion, 'the proprioceptive representation' of emotional 'bodily changes' (Sander and Scherer 2009). 'Feeling' however, unlike 'affect', also has a role in non-expert discourse, as a synonym for 'emotion', but also as a conceptual metaphor. Conceptual, or basic, metaphors are metaphors whose status we barely notice, as they are essential to everyday discourse. They take a range of verbal forms, cross languages and cultures, and emerge from fundamental aspects of embodied experience. HAPPINESS IS UP, for example, can be seen in expressions like 'cheer up', and relates, it is argued, to the upright, bipedal status of the human body. In the case of feeling, the everyday, non-specialist equation of feeling and emotion maps the experience of physical pain or pleasure onto mental or emotional pain or pleasure, through the basic metaphor EMOTIONAL EFFECT IS PHYSICAL CONTACT (Lakoff and Johnson 1980: 18; 50).

The first, more performative, part of the title, then, introduces readers to a new and surprising idea, but via appeal to a fundamental and familiar one. In using a term like 'feeling', moreover, with a place (although not quite the same place) in both expert and non-expert discourse, the abrupt disjunction between the more performative first half and the less performative second half is ameliorated. The second half is marked by a switch from the Germanic vocabulary associated in English with the everyday to the Latinate vocabulary associated with formal and abstract speech (Bar-Ilan and Berman 2007). This formal/informal contrast does not tell us how performative a stretch of discourse might be; both can be used for specific effects in both more and less performative contexts. However, formal speech does imply greater social distance between speaker and hearer (Brown and Levinson 1987). The second part of the sentence is thereby abruptly distinguished again from the social event of narrative performance implied by the first part.

This is not just a case of restating the case in discourse more suited to the evaluation of accuracy rather than to the performative criterion of competence in delivery. The case itself is somewhat altered.[4] As discussed above, 'affect', unlike 'feeling', is specialized to psychology and neuroscience, distinct from the wider range of concepts connected by everyday uses of 'feeling'. 'Perception', on the other hand, is *broader* in some ways than 'seeing', covering all modalities, not just vision, as well as processes to which we do not have conscious access. But in other ways 'perception' is *narrower*, generally excluding the easy slippage in non-expert discourse between literal and metaphorical senses of 'see' (such as 'I see what you mean'). It is explicitly applied here only to 'objects', not, for example, to scenes. As an appendix to the imperative main clause before the colon, no verb is required, which allows this noun phrase to evade any time-locked claims. These would specify a starting point for the prediction and a causal relationship between prediction and perception, such as 'affective predictions *occur* during object perception' (my italics).

The second part of the title, then, both makes a different claim and establishes a different relationship with readers. The claim is hedged in the way Smyth characterizes as typical of scientific psychology papers, whereas the claim in the performative first half is not. Because it is less performative, it is also more accountable to its audience for accuracy. It is more formal (which is often, although not always, associated with an expectation of objectively described, accurate information). It claims that one aspect of 'feeling' as understood in non-expert discourse, that is 'affect', contributes to predictions. These predictions are described as happening during object perception; it is implicit that the process of prediction is in fact equivalent to the process of perception, with prediction/perception ending at the point where prediction error from the environment is minimal enough to be ignored (Clark 2013).

This pattern of an easy yet abrupt shift between performative (either narrative or with narrative implications) and informative relationships with the reader persists throughout the article. This might be seen as a symptom of the dual nature of science as a social practice and science as a methodology, which seeks to overcome the distortions of personal subjectivity by establishing objective facts. The oscillation between kinds of discourse is a way of keeping

---

[4] In the afterword (finale) to this volume (Chapter 22), Wise discusses the division of labour between different kinds of language, such as formal and natural.

both aspects in play.[5] The readers of a scientific article are both part of a social community of practice, which can bond through verbal performance, including jokes (Reimegård 2014), and committed to looking at the world, as best they can, as though from an extra-human perspective (Reiss and Sprenger 2017).

On this understanding, the reader's relationships between more and less performative authorial voices capture an abrupt disjunction between their functions. The more performative voice instantiates science as a social practice, while the less performative one enacts science as a method. While the two can sit side by side, it might be assumed they need not, indeed should not, interact. This interpretation relies on a strict separation between the social and methodological aspects of scientific practice, an interpretation with long-standing challenges from history, philosophy and social science (e.g., see Barnes and Bloor 1982; Shapin 1994; and Latour 1987). On these views, scientific sociability and scientific methods are intimately entangled. More recently, some psychologists, confronted with the replication crisis discussed in the introduction, have turned to the social aspect of science to explain some failures in methodology (Ritchie 2020). From this perspective, the coexistence of more and less performative voices might be symptomatic of a problem. In this view, the performative voice instantiates readers as a social group sharing a broad paradigm of cognition. The less performative voice is then interpreted within that broad paradigm, leaving readers less critical of both paradigm and evidence as each is interpreted as potential confirmation for the other.

I suggest an alternative explanation for the function of the performative mode. The performative mode, I argue, does indeed engage readers in building a cognitive theory. But the theory is itself an *implicit narrative* embedded in the performative text, and emerges, I will suggest, from a process of reconceptualization and remodelling implicit causation in the world. It is this narrative remodelling and reconceptualization which lets them identify a causal mechanism in the data despite the 'near taboo on causal inference' identified by Grosz et al., above (Jajdelska 2019). Causal inference in an unfamiliar scientific paradigm, then, can be a creative act. And while performative narration in the article may be dispersed in fragments rather than performed in a sustained social event by, for example, a Homeric rhapsode or a West African griot, it can serve some of the same functions: reshaping our world models in ways that let us see otherwise opaque causal structures. Put differently, performative narration allows for the scrutiny of concepts, causal inference and broader theory building which the genre of scientific psychology writing, as discussed by Bazerman and Flis earlier, inhibits.

---

[5] See Engelmann (Chapter 14), who also makes a case for narrative as a way to keep alternative causal explanations for a phenomenon in play.

18.2.2   Narrative Performance and Reader Trust

The opening paragraph of the 'Introduction' (following the abstract) immediately instantiates a voice with a narrator's authority to hold the floor:

Michael May lost the ability to see when he was 3 years old, after an accident destroyed his left eye and damaged his right cornea. Some 40 years later, Mr May received a corneal transplant that restored his brain's ability to absorb normal visual input from the world (Fine et al. 2003). With the hardware finally working, Mr May saw only simple movements, colours and shapes rather than, as most people do, a world of faces and objects and scenes. [. . .] As time passed, and Mr May gained experience with the visual world in context, he slowly became fluent in vision. (Barrett and Bar 2009: 1325)

This is a narrative of the following event sequence in Bauman's terms: accident; destruction or damage to eyes; transplant; simplified vision; full vision. Like the verbal performers in Bauman's studies, in performance, the authors, and in turn hearers, attribute causal links to this sequence: the accident caused the destruction and damage; the transplant 'restored his brain's ability to absorb normal visual input from the world'; 'experience with the visual world in context' caused his later fluency 'in vision'. These causal attributions may seem uncontroversial but they are not inevitable. The transplant could have been described as a restoration of 'the ability of the cornea to react to rays of light', for example, making the story one of simpler physical components of low-level cognition. 'He slowly became fluent in vision', could have been, for example, 'He learned to match visual data to the data higher up in processing streams of other modalities', again modelling vision as a mechanical process, rather than a complex skill drawing on multiple domains of knowledge and processing, such as 'fluency' in a language.

The authors' narrative phrasing of causal links, then, models a world in which agents interact with their environment by acting on it, and adjusting their expectations through the feedback from their actions. In other words, narrative performance here generates a model which will support their hypothesis: 'When the brain receives new sensory input from the world in the present, it generates a hypothesis based on what it knows from the past to guide recognition and action in the immediate future' (Barrett and Bar 2009: 1325). The narrative performance of the story of Mr. May, then, enables readers to model the article's hypothesis not just by providing a case study, but by modelling a specific causal structure for long enough to establish a hypothesis prior to assessing it.

This short verbal narrative gives way to material which cannot, without effort at least, be characterized as narrative. However, the *authority* of a narrator may persist through the use of textual features found in verbal art and performance, and it is this performative-authoritative role which supports broader narrative remodelling in these passages throughout the article. For example, the narrative to non-narrative transition sees a continued use of declarative sentences and clauses, a continuity which carries authority in the narrative domain over to the domain of neuroscience:

> As time passed, and Mr May gained experience with the visual world in context, he slowly became fluent in vision. [. . .]
>   What Mr May did not know is that sighted people automatically make the guesses he was forced to make with effort. (Barrett and Bar 2009: 1325; my emphasis)

It seems unlikely here that the second sentence, which opens the paragraph after the narration ends and which does not form an event in a sequence, leads readers to believe that the authors have interviewed Mr May.[6] More likely, they accept the claim as a continuation of the authority granted to verbal performers. In both sentences, for example, there are verbal cues to this performative status. Both begin with information available to narrators but not, typically, to readers: that time passed before Mr May experienced change, and what he did or did not know at particular points in the narrative. Compare, for example, these words from novelist Ali Smith's third-person narrator of *Autumn*: 'It is still July'. Readers have neither independent means to establish that it is July, nor any reason to question the claim. Similarly, continuing the performative narrative voice into a non-narrative paragraph might make readers less vigilant about broad, and not always consensus, positions on the nature of vision, and narrower claims about what a specific individual knew about vision, than they would be otherwise.

In the next paragraph, the speaking voice takes a step away from the narration of Michael May's experience, but maintains the authority associated with it through declarative statements: 'External sensations do not occur in a vacuum, but in a context of internal sensations from the body that holds the brain'. The claim here is no longer a simple one premised on a narrator's privileged access to information ('What did Mr May know about cognition, and when?'). Here, there is an opposition between a model of external sensations 'in a vacuum' compared to 'a context of internal sensations from the body that holds the brain'. Rhetorically, the force of this is to create a binary choice between a model hard even to conceive of (external sensations in a vacuum) and a holistic model intertwining the world ('external'), the body ('internal'), and cognition ('the brain'). This sense of a non-modular, dynamic system is reinforced by the selective use in this paragraph of verbs in the present continuous (-ING):

> The sensory–motor patterns being stored for future use include. . .
> In addition to directly experiencing changes in their breathing. . .
> In addition to learning that the sounds of a voice come from moving lips . . .
>   (Barrett and Bar 2009: 1325; my emphasis)

---

[6] Meunier (Chapter 12) looks at the distinction between a sequence of actions performed in a lab, and how that sequence is represented in the corresponding research article's narrative; despite the mismatch between the two, the implication is that all events recounted have in fact been performed.

The present continuous is associated with a range of rhetorical effects, including the narrative voice of the 'historic present' (Wolfson 1982). Here it marks a move away from the narrative of Mr May, but a move which preserves the authority of the performative, narrative voice, just as it preserves and develops the implicit world model established in the narrative section. Stylistic choices between declarative or present continuous sentences, or the contrast between a more and less persuasive model of cognition, can be seen as rhetorical choices. All writers must make choices of this kind, and all of those choices will have some kind of effect on the reader. But, independently of rhetorical persuasion, artful expression and narration create a performative discourse, helping the reader to conceive of the world in a way that lets them understand the causal structure underlying the hypothesis, separately from accepting the hypothesis itself. The effect of performative language, I suggest, can be to support, rather than manipulate, the reader in a creative process of reconceptualization in order to assess the hypothesis by first understanding it. In this respect, then, the elements of narrative performance in the article provide an effective way around the 'taboo on causal inference' problem in scientific psychological writing, but one which does not confront that problem directly and therefore might not be sustainable in the field's discourse as a whole.

18.3   Creativity as Reconceptualization

At the heart of the article is the case for emotion and perception as having an entwined rather than a causally sequential relationship, at least at sub-personal speeds and levels of consciousness. As with modelling the relationship of mind and world as dynamic and holistic rather than linear and modular, this case requires some reconceptualization of emotion. For readers who trained in psychology under the influence of Jerry Fodor's work, emotion might be not just *habitually* distinguished from perception, but *by definition* distinguished from perception. Fodor's modular account of cognition drew inspiration from Turing's identification of a minimal mechanism that can do cognitive work automatically (Turing 1936). In a modular system, processing streams for different modalities, and for language, remain distinct until processing is advanced, at which points the information from each module is blended in 'central processes'. Modularity does not allow the kind of dynamic, holistic interaction of emotion, action and perception that Barrett and Bar outline (Fodor 1983). As Barrett and Bar explain, 'Affective responses were ignored in cognitive science altogether' (Barrett and Bar 2009: 1328). To even assess the coherence of their hypothesis, therefore, some readers may need first to reconceptualize (in Boden's and Churchland's sense described above) perception, affect and emotion. The performative, and either implicit or explicitly narrative, voice established at key points of the article, I argue, supports this reconceptualizing process in readers, in part through engaging aesthetic emotions.

Take the following extracts from the second and third paragraphs of the article:

This is how people learn that the sounds of a voice come from moving lips on a face, that the red and green blotches on a round Macintosh apple are associated with a certain tartness of taste, and that the snowy white petals of an English rose have a velvety texture and hold a certain fragrance. [. . .] [T]hey learn that they enjoy tartly flavoured red and green (Macintosh) apples or the milder tasting yellow apples (Golden Delicious); and they learn whether or not they prefer the strong fragrance of white English roses over the milder smell of a deep red American Beauty, or even the smell of roses over lilies. (Barrett and Bar 2009: 1325)

The authors explain here that perception is contextual, and that the context includes the perceiver's early affective responses to the target stimulus. To make this point, they draw on readers' prior experience not only of apples, roses and lilies, but also on their capacity for aesthetic emotion, a form of emotion likely to be accessible to consciousness, not least because the stimulus itself, that is the artefact (in this case a passage of text), draws attention to itself as an object of conscious attention (Miall and Kuiken 1994; 1999). Aesthetic effects are also associated with performativity (Bauman and Briggs 1990).

The authors make a number of aesthetically directed choices here. First of all, the objects they choose (apples, roses, lilies and their colours, tastes and smells) are all 'motifs' which have been identified by folklorists in folk and fairy tales. Motifs migrate between different tale types with a power that is independent of narrative context (Aarne, Thompson and Uther 2004; Thompson 1955–58). This status as standalone objects, divorced in mental imagery from scenes, plays a role in the vividness with which they can be imagined, creating an undiluted focus on aesthetic affordances in relation to taste, sight and smell. The focus on sight in relation to taste and smell also has the potential to heighten vividness through synaesthetic effects (Jajdelska et al. 2010; Scarry 2001; Jajdelska 2019: 570–572).

Aesthetic emotions have been variously categorized (see, for example, Fingerhut and Prinz 2020; Brown and Dissanayake 2009; Miall and Kuiken 1994). A common thread, however, is their association with higher-order processing, that is processing which is more accessible to conscious report (Brown and Dissanayake 2009: 51–52; Fingerhut and Prinz 2020). Aestheticized verbal descriptions may both evoke emotions and make readers aware of the emotions' source. These descriptions of apples and roses potentially *enact* in readers the concept of emotion as intrinsic to cognition, and thereby help them to understand the article's otherwise counter-intuitive hypothesis.

The enactment, insofar as it occurs, would not be evidence in support of the hypothesis, which concerns not 'emotion' in general, but affect in particular:

We have proposed that the medial [orbitofrontal cortex] (OFC) participates in an initial phase of affective prediction ('what is the relevance of this class of objects for me'), whereas the lateral OFC provides more subordinate-level and contextually relevant affective prediction ('what is the relevance of this particular object in this particular context for me at this particular moment in time'). (Barrett and Bar 2009: 1331)

Instead, the experience of mental imagery intertwining sensual experience with aesthetic emotion allows those readers for whom emotion and perception are distinct by definition to reconceptualize both and then potentially to evaluate this causal relationship.

18.4   From Reconceptualization to Causation

Reconceptualizing emotion is a first step to successfully identifying and understanding the dynamic network of causal interactions implicit both in the paper's hypothesis and in the implicit narrative identified by performative discourse throughout. A second step is to develop an understanding of the causal links with greater precision. Here too the move between more and less performative voices may help. The more performative material metaphorically introduces agency into sub-personal processes. The most prominent example of this is the status of the brain, and/or the body as a whole, which appears in the paper's varied discourses as at some points an objectively viewed system, lacking free will, and at others as an agent pursuing an identifiable set of goals whose parameters are established by evolution. The degree to which people can be understood as free agents is a question involving considerable philosophical effort, informed in recent years by findings in cognitive science (Vierkant 2017). The authors of the paper are not, either implicitly or explicitly, expressing a position in this debate. Instead, I suggest that they are moving between more and less performative discourses in ways that allow readers to imagine and reimagine their hypothesis in different lights, one narrative and performative and one non-narrative. In doing so, readers are supported in modelling causation through narrative discourse and then importing this into non-narrative discourse.

From the perspective of folklore, anthropology and to some extent the psychology of memory, any given narration's underlying structure is defined not by its causal links but by its sequence of events (Aarne, Thompson and Uther 2004; Bauman 1986; Bartlett 1995). In this view, for any given event sequence underlying a narrative performance, causation is built in only at the point of narration; causation is not an intrinsic quality of the sequence itself, but emerges in the audience and only through performative narration, and a single event sequence can be narrated on different occasions with different causal links. A critical element of this process, building causality into an event sequence through narration and performance, is attributing agency.[7] In the most influential psychological account of narrative production and comprehension of recent decades, for example, 'change in character's goals', a signature aspect of agency, is one of the small number of key dimensions that hearers/readers monitor consistently in order to comprehend and recall a story (Zwaan and Radvansky 1998; Zacks, Speer and Reynolds 2009). Mythical event sequences, which explain or make sacred features of the world that arose independently of humans, attribute agency either to the rocks, rivers and mountains themselves, or to divine beings who can control them (Doty 2000: 74–76).

Where an event sequence does not feature an easily recognizable agent, then, the narration attributes agency to non-agents, as in this example, continuing from the previous section on roses and apples:

They learn whether or not they prefer the strong fragrance of white English roses over the milder smell of a deep red American Beauty, or even the smell of roses over lilies. When the brain detects visual sensations from the eye in the present moment, and tries to interpret them by generating a prediction about what those visual sensations refer to or stand for in the world, it uses not only previously encountered patterns of sound, touch, smell and tastes, as well as semantic knowledge. It also uses affective representations – prior experiences of how those external sensations have influenced internal sensations from the body. Often these representations reach subjective awareness and are experienced as affective feelings, but they need not. (Barrett and Bar 2009: 1325)

This passage opens with the synaesthetic, and aesthetic, evocation of experiencing roses and lilies at the level of the person as agent, with the sensual processing of the moment linked to that of the past, and to high-level concepts which can potentially be expressed in language to others ('I prefer English roses'). Having indicated for the reader the relevant event structure by evoking aesthetic emotions in the reader, the passage then *re*narrates the story (or story component of the wider, implicit narrative generated by performative discourse across the paper), attributing agency this time to 'the brain'. The brain as agent has goals and priorities which are distinct from those of the person who weighs English roses against American ones. The goals of the brain as agent include 'interpreting visual sensations' and the method includes 'generating a prediction' of the world drawing on previous experiences, semantic

---

[7] Engelmann (Chapter 14) provides a persuasive account of the need to attribute agency, as well as the difficulties in doing so, in plague narratives.

knowledge and, critically for the paper's hypothesis, 'affective representations' drawn from past experience of the relationship between external and internal sensations, all of which can operate below the level of 'subjective awareness'.

While the person as agent has a (momentary) goal of 'establishing preferences', the brain as agent has the (longer-term) goal of 'interpreting sensations'. In the second case, as the paper's author and reader are certainly aware, 'the brain's goal' is a metaphor, mapping something like, 'has evolved in such a way that responses to past stimuli statistically shape present responses to stimuli' onto 'interprets current stimuli in the light of past ones'. One effect of this extended metaphor is, as with the aesthetic effects of apple and rose verbal imagery, to support readers in developing a new model not just of the causal relationships between affect and relation perception, but of the broader network of causal relationships among body, brain and world, a network in which all three are continuously and dynamically interacting in ways that cannot be captured by a unidirectional flow from stimulus to brain to action. The goal of identifying our preferred fruits and flowers, with which we are already familiar from our own conscious experience, can then be mapped onto a less familiar, sub-personal goal of making predictions to support optimal actions, and from there to a modified account of cognition more broadly. Performative discourse, then, even when not explicitly narrative, contributes to global narration across the paper through reconceptualization and its associated causal attributions.

The same movement – between recognizable personal experience and sub-personal cognition – can be seen in reverse in this passage:

With back and forth between visual cortex and the areas of the prefrontal cortex (via the direct projections that connect them), a finer level of categorization is achieved until a precise representation of the object is finally constructed. Like an artist who successively creates a representation of objects by applying smaller and smaller pieces of paint to represent the light of different colours and intensities, the brain gradually adds high spatial frequency information until a specific object is consciously seen. (Barrett and Bar 2009: 1328)

Here, the authors open without attributing agency to the brain, using the passive voice ('categorization is achieved', 'the object is finally constructed') to avoid attributions of agency altogether. The comparison to a supposed artist creating an ever more refined representation reframes the process *with* an agent, and then finally the brain returns, this time *as* agent, one whose goal is to refine a representation, using different kinds of inputs, including affect, to the point where it enters consciousness. We start with observable data about activity in specific brain regions under specific conditions (provided through citation), and the passive voice mutes any causal claims about the causal relationship between that activity and the subsequent representation. The simile of the artist then creates explicitly a causal relationship, which is parallel, although not identical, to the implicit causal relationship in the observable data: 'the artist creates a representation' with the goal of making an object, where, implicitly, the activity of relevant brain regions makes a representation with the goal of supporting appropriate action. Finally the causally neutral data sequence and the causally specific simile are brought together in the brain as agent, whose goal is to make a representation available to consciousness. The shifts between different levels of more and less performative and narrative discourse allow readers not just to identify the causal relationship that the authors embed in their hypothesis, but to model processes of cognition in a way that lets them assess the plausibility of that hypothesis.

18.5   Conclusion

The article discussed in this chapter was not written in response to the current replication crisis in scientific psychology, but it does implicitly address the absence of theory and conceptual work identified by Flis as a problem in the discipline (Flis 2018: 163, and elsewhere). It does so by alternating between the genres of narrative performance and of APA approved explication. The first enables the generation of theory in the reader, and the second articulates this theory in non-performative and non-narrative ways. In this way the paper maintains the division between trustworthy 'scientific psychology writing' and potentially less trustworthy engagement of creative imagination in readers, but still manages to look at the key topics in theorized ways. This points to a useful role for mixed discourses in scientific writing more generally. However, whether in psychology or less troubled areas of science, an acknowledgement of the role that performative narrative discourse plays in the work of theory, challenging the sense that this kind of narrative is merely an entertaining experiment with lower status forms of discourse, a holiday from rigour, can contribute to our broader theory-building, reconceptualization and remodelling.[8]

Bibliography

Aarne, A., S. Thompson and H.-G. Uther (2004). *The Types of International Folktales: A Classification and Bibliography, Based on the System of Antti Aarne and Stith Thompson*. Helsinki: Academia Scientiarum Fennica.
Arnold, M. B. (1960). *Emotion and Personality*. New York: Columbia University Press.
Austin, J. L. (1962). *How to Do Things with Words*. Oxford: Oxford University Press.

---

[8] *Narrative Science* book: This project has received funding from the European Research Council under the European Union's Horizon 2020 research and innovation programme (grant agreement No. 694732). www.narrative-science.org/.

Bahls, P., A. Mecklenburg-Faenger, M. Scott-Copses and C. Warnick (2011). 'Proofs and Persuasion: A Cross-Disciplinary Analysis of Math Students' Writing'. *Across the Disciplines* 8.1: 1–20.

Bar-Ilan, L., and T. Berman (2007). 'Developing Register Differentiation: The Latinate–Germanic Divide in English'. *Linguistics* 45.1: 1–35.

Barnes, B., and D. Bloor (1982). 'Relativism, Rationalism, and the Sociology of Knowledge'. In M. Hollis and S. Lukes, eds. *Rationality and Relativism*. Oxford: Blackwell: 21–47.

Barrett, L. F., and M. Bar (2009). 'See It with Feeling: Affective Predictions during Object Perception'. *Philosophical Transactions of the Royal Society. Series B* 364.1521: 1325–1334.

Bartlett, F. (1995 [1932]). *Remembering: A Study in Experimental and Social Psychology*. Cambridge: Cambridge University Press.

Bauman, R. (1975). 'Verbal Art as Performance'. *American Anthropologist* 77.2: 290–311.

(1986). *Story, Performance and Event: Contextual Studies of Oral Narrative*. Cambridge: Cambridge University Press.

Bauman, R., and C. L. Briggs (1990). 'Poetics and Performance as Critical Perspectives on Language and Social Life'. *Annual Review of Anthropology* 19: 59–88.

Bazerman, C. (1987). 'Codifying the Social Scientific Style: The APA Publication Manual as Behaviorist Rhetoric'. In J. S. Nelson, A. Megill and D. N. McCloskey, eds. *The Rhetoric of the Human Sciences: Language and Argument in Scholarship and Public Affairs*. Madison: University of Wisconsin Press: 257–277.

(1988). *Shaping Written Knowledge: The Genre and Activity of the Experimental Article in Science*. Madison: University of Wisconsin Press.

Boden, M. (2004 [1990]). *The Creative Mind: Myths and Mechanisms*. London: Routledge.

Brown, P., and S. C. Levinson (1987). *Politeness: Some Universals in Language Usage*. Cambridge: Cambridge University Press.

Brown, S., and E. Dissanayake (2009). 'The Arts Are More than Aesthetics: Neuroaesthetics as Narrow Aesthetics'. In M. Skov, O. Vartanian, C. Martindale and A. Berleant, eds. *Neuroaesthetics*. London: Routledge: 43–57.

Butler, J. (1990). *Gender Trouble: Feminism and the Subversion of Identity*. New York: Routledge.

Churchland, P. M. (2012). *Plato's Camera: How the Physical Brain Captures a Landscape of Abstract Universals*. Cambridge, MA: MIT Press.

Clark, A. (2013). 'Whatever Next? Predictive Brains, Situated Agents and the Future of Cognitive Science'. *Behavior and Brain Science* 36: 191–253.

De Pierris, G., and M. Friedman (2018). 'Kant and Hume on Causality'. In Edward N. Zalta, ed. *The Stanford Encyclopedia of Philosophy*. https://plato.stanford.edu/archives/win2018/entries/kant-hume-causality/.

Doty, W. (2000 [1986]). *Mythography: The Study of Myths and Rituals*. Tuscaloosa: University of Alabama Press.

Fine, I., A. R. Wade, A. A. Brewer, M. G. May et al. (2003). 'Long-Term Deprivation Affects Visual Perception and Cortex'. *Nature Neuroscience* 6.9: 909–910.

Fingerhut, J., and J. Prinz (2020). 'Aesthetic Emotions Reconsidered'. *Monist* 103: 223–239.

Flis, Ivan. (2018). 'Discipline through Method: Recent History and Philosophy of Scientific Psychology'. PhD thesis, Utrecht University.

(2019). 'Psychologists Psychologizing Scientific Psychology: An Epistemological Reading of the Replication Crisis'. *Theory and Psychology* 29.2: 158–181.

Fodor, J. (1983). *The Modularity of Mind: An Essay on Faculty Psychology.* Cambridge, MA: MIT Press.

Fugelsang, J., M. Roser, P. Corballis, M. Gazzaniga and K. Dunbar (2005). 'Brain Mechanisms Underlying Perceptual Causality'. *Cognitive Brain Research* 24: 41–47.

Gigerenzer, G. (1991), 'From Tools to Theories: A Heuristic of Discovery in Cognitive Psychology'. *Psychological Review* 98.2: 254–267.

Gilbert, D., R. Tafarodi and P. Maone (1993). 'You Can't Not Believe Everything You Read'. *Journal of Personality and Social Psychology* 65: 221–233.

Glymour, C. (2003). 'Learning, Prediction and Causal Bayes Nets'. *Trends in Cognitive Sciences* 7.1: 43–48.

Goffman, E. (1974). *Frame Analysis: An Essay on the Organisation of Experience.* Boston: North Eastern University Press.

Gopnik, A. (2009). *The Philosophical Baby: What Children's Minds Teach Us about Truth, Love and the Meaning of Life.* London: The Bodley Head.

Grosz, M., J. M. Rohrer and F. Thoemmes (2020). 'The Taboo against Explicit Causal Inference in Nonexperimental Psychology'. *Perspectives on Psychological Science* 15.5: 1243–1255.

Jajdelska, E. (2019). 'The Flow of Narrative in the Mind Unmoored: An Account of Narrative Processing'. *Philosophical Psychology* 32.4: 560–583.

Jajdelska, E., C. Butler, S. Kelly, A. McNeill et al. (2010). 'Crying, Moving, and Keeping It Whole: What Makes Literary Description Vivid?' *Poetics Today: International Journal for Theory and Analysis of Literature and Communication* 31.3: 433–463.

Jajdelska, E., M. Anderson, C. Butler, N. Fabb et al. (2019). 'Picture This: A Review of Research Relating to Narrative Processing by Moving Image Versus Language'. *Frontiers in Psychology* 10: 1161.

Kiefer, M., and F. Pulvermüller (2012). 'Conceptual Representations in Mind and Brain: Theoretical Developments, Current Evidence and Future Directions'. *Cortex* 48.7: 805–825.

Kilner, J., K. Friston and C. Frith (2007). 'Predictive Coding: An Account of the Mirror Neuron System'. *Cognitive Processing* 8.3: 159–166.

Lakoff, G., and M. Johnson (1980). *Metaphors We Live By.* Chicago: University of Chicago Press.

Latour, B. (1987). *Science in Action.* Cambridge, MA: Harvard University Press.

Lwin, S. (2010). 'Capturing the Dynamics of Narrative Development in an Oral Storytelling Performance: A Multimodal Perspective'. *Language and Literature* 19.4: 357–377.

Miall, D., and D. Kuiken (1994). 'Foregrounding, Defamiliarization and Affect: Response to Literary Stories'. *Poetics* 22.5: 389–407.

(1999). 'What Is Literariness? Three Components of Literary Reading'. *Discourse Processes* 28.2: 121–138.

Morgan, Mary S. (2017). 'Narrative Ordering and Explanation'. *Studies in History and Philosophy of Science Part A* 62: 86–97.

Pearl, J. (1995). 'Causal Diagrams for Empirical Research'. *Biometrika* 82.4: 669–710.

Pearl, J., and D. Mackenzie (2018). *The Book of Why: The New Science of Cause and Effect*. London: Penguin.

Reimegård, L. (2014). 'Here Comes the Story of the Dylan Fans'. *Karolinska Institutet News*. https://news.ki.se/here-comes-the-story-of-the-dylan-fans.

Reiss, J., and J. Sprenger (2017). 'Scientific Objectivity'. In Edward N. Zalta, ed. *The Stanford Encyclopedia of Philosophy*. https://plato.stanford.edu/archives/win2017/entries/scientific-objectivity/.

Ritchie, S. (2020). *Science Fictions: Exposing Fraud, Bias, Negligence and Hype in Science*. London: Random House.

Sander, D., and K. Scherer (2009). *Oxford Companion to Emotion and the Affective Sciences*. Oxford: Oxford University Press.

Scarry, E. (2001 [1999]). *Dreaming by the Book*. Princeton, NJ: Princeton University Press.

Shapin, S. (1994). *A Social History of Truth: Civility and Science in Seventeenth-Century England*. Chicago: University of Chicago Press.

Smith, L. (1986). *Behaviorism and Logical Positivism: A Reassessment of the Alliance*. Stanford, CA: Stanford University Press.

Smith, R. (2007). *Being Human: Historical Knowledge and the Creation of Human Nature*. New York: Columbia University Press.

Smyth, M. (2001). 'Certainty and Uncertainty Sciences: Marking the Boundaries of Psychology in Introductory Textbooks'. *Social Studies of Science* 31.3: 389–416.

Thompson, S. (1955–58). *Motif-Index of Folk-Literature: a Classification of Narrative Elements in Folktales, Ballads, Myths, Fables, Mediaeval Romances, Exempla, Fabliaux, Jest-Books, and Local Legends*. Revised edn. Bloomington: Indiana University Press.

Trompf, G., J. Gough and O. Eckhart (1988). 'Western Folktales in Changing Melanesia'. *Folklore* 99.2: 204–220.

Turing, A. (1936). 'On Computable Numbers, with an Application to the Entscheidungsproblem'. *Proceedings of the London Mathematical Society* 42.1: 230–265.

Vierkant, T. (2017). 'Does Science Show That We Lack Free Will?' In D. Prichard and M. Harris, eds. *Philosophy, Science and Religion for Everyone*. Abingdon: Routledge: 140–148.

Vigen, T. (2015). *Spurious Correlations*. New York: Hachette Books.

Weisberg, D., and A. Gopnik (2013). 'Pretense, Counterfactuals, and Bayesian Causal Models: Why What Is Not Real Really Matters'. *Cognitive Science* 37.7: 1368–1381.

Wetzel, Linda. (2018). 'Types and Tokens'. In Edward N. Zalta, ed. *The Stanford Encyclopedia of Philosophy*. https://plato.stanford.edu/archives/fall2018/entries/types-tokens/.

Wingen, T., J. Berkessel and B. Englich (2020). 'No Replication, No Trust? How Low Replicability Influences Trust in Psychology'. *Social Psychological and Personality Science* 11.4: 454–463.

Wolfson, N. (1982). *CHP: The Conversational Historic Present in American English Narrative*. Dordrecht: Foris Publications.

Yanal, R. (1999). *Paradoxes of Emotion and Fiction*. University Park: Pennsylvania State University Press.

Yuill, N., and J. Oakhill (1991). *Children's Problems in Text Comprehension: An Experimental Investigation*. Cambridge: Cambridge University Press.

Zacks, J. M., N. K. Speer and J. R. Reynolds (2009). 'Segmentation in Reading and Film Comprehension'. *Journal of Experimental Psychology* 138.2: 307–327.

Zwaan, R. A., and G. A. Radvansky (1998). 'Situation Models in Language Comprehension and Memory'. *Psychological Bulletin* 123.2: 162–185.

19   Reading Mathematical Proofs as Narratives

Line Edslev Andersen

Abstract

Mathematical proofs and narratives may seem to be opposites. Indeed, deductive arguments have been highlighted as clear examples of non-narrative sequences by narrative theorists. I claim that there are important similarities between mathematical proofs and narrative texts. Narrative texts are *read* in a quite distinct way, and I argue that mathematical proofs are often read like narrative texts by research mathematicians. In this way, narratives play an important role in mathematical knowledge-making. My argument draws on recent empirical data on how mathematicians read proofs. Furthermore, my examination of mathematical proofs and narratives provides an account of what it means for research mathematicians to *understand* mathematical proofs.

19.1   Introduction

Mathematicians sometimes emphasize the major role of *inductive* reasoning in mathematics (see, for example, Borwein and Bailey 2003). Results in mathematics are usually tested in reliable ways before a *deductive* mathematical proof of the results is produced. For example, the famous Riemann hypothesis has been verified in billions of instances. But results in mathematics are established by deductive mathematical proofs, and the Riemann hypothesis will remain just that, a hypothesis, until a deductive mathematical proof has been produced. Historically, deductive mathematical proof has become the ultimate method of justification in mathematics. For this reason, proofs play a very central role in mathematical research practice.

Mathematical proofs and narratives may seem to be opposites. Indeed, deductive arguments have been highlighted as clear examples of non-narrative sequences by narrative theorists (see, for example, Bruner 1987: 11–14; Herman 2009: 157). By contrast, scholars interested in the nature of mathematical proof have occasionally conceived of mathematical proofs as narratives (Doxiadis 2012; Robinson 1991: 269; Thomas 2007). Their accounts mainly focus on the similarities between mathematical proofs *as written* and narrative texts.

I similarly compare mathematical proofs with narratives but take a different approach.[1] Narrative texts are read in a quite distinct way, and I argue that mathematical proofs are often *read* like narrative texts by research mathematicians.[2] Hence, my account of mathematical proofs and narratives mainly focuses on the relationship between mathematical proofs and their readers rather than their writers.[3] The argument constitutes an independent argument for conceiving of mathematical proofs as comparable with narratives, and sheds new light on the reading of mathematical proofs. My argument draws on recent empirical data on how mathematicians read proofs.

Furthermore, my account of mathematical proofs and narratives provides an account of what it means for research mathematicians to *understand* mathematical proofs. This account of proof understanding appears to have implications for how we should conceive of proof validation and for how proofs should be taught.

19.2   Reading Proof as Narrative

Before I say more about how proofs are read, I should say a few words about how proofs are *made* and *presented*. In recent years, philosophers have developed accounts of how we should conceive of a mathematical proof as a sequence of inferential *actions* performed by an agent on various objects, such as propositions, diagrams and mental images (De Toffoli and Giardino 2015; Larvor 2012; Netz 1999: 51–56; Tanswell 2017a; 2017b: 144–153). For a vivid image of a proof involving inferential actions performed on diagrams, one may watch one of the many videos on YouTube of different diagrammatic proofs of the Pythagorean Theorem (or see Tanswell 2017a: 223–225).

If a proof is a sequence of actions, we may conceive of a proof *as written*, as found, for example, in a research article or a textbook, as a telling of a sequence of actions. Hence, a written proof is a telling of how something happened. A written proof is a telling of a sequence of actions performed on mathematical objects by an agent with the aim of proving a given proposition (in line with Hamami and Morris 2020). This implies that a proof as written is a *narrative* in at least a minimal sense. A narrative is often minimally characterized as a telling of a sequence of particular events or actions involving humans or humanlike characters with particular goals (see, for example, Sanford and Emmott 2012: 1–5; Toolan 2001: 4–8). To be more precise, a written proof is

---

[1] On the role of narrative in scientific reasoning more broadly, see Morgan (Chapter 1).

[2] In the introductory chapters in this volume, Morgan (Chapter 1) and Hajek (Chapter 2) distinguish between two broad senses of narrative in relation to science: narrative representation and narrative reasoning. My argument is about narrative reasoning.

[3] For a broader account of the relationship between narration in science and the reader, see Hajek (Chapter 2).

a telling by the author of a sequence of actions on mathematical objects performed by an agent, usually by the author of the proof herself, with the aim of establishing a particular theorem (this is in line with Doxiadis 2012: 330–331, on Euclidean proofs).

I ask readers to have something like this picture of proofs in mind as they read on. But rather than focusing on the narrative features of written proofs, the aim of this chapter is to argue that written proofs can be and often are *read* like narratives by research mathematicians. A narrative is not only characterized by features of presentation. Narratives are also characterized by how they are read or heard. In fact, narratives have a quite distinct relationship with their readers or auditors. This relationship is sometimes described by narrative theorists as having two key features (see, for example, Bruner 1991: 11–13; Herman 1997).

The first key feature of the relationship between narratives and their readers or auditors is that narratives are interpreted by their readers or auditors against the background of patterns of belief and expectation with respect to how events or actions of the kinds represented in the narrative usually take place. The background patterns of belief and expectation are called *scripts* by theorists like David Herman and they stem from the prior experience of the readers.[4] Hence, scripts describe *standard* sequences of events or actions against which a *particular* narrative is read. The narrative cues readers to activate the scripts.[5]

Consider, for example, the following narrative or part of a narrative: 'John went to Bill's birthday party. He watched Bill open his presents. John ate the cake and left' (adapted from Schank and Abelson 1977: 39). We read this particular narrative against our birthday party script which describes standard sequences of events and actions that usually take place at birthday parties in our experience. About similar examples of narrative sequences, Herman (1997: 1051) writes: 'I can make an astonishing number of inferences about the situations and participants – fill in the blanks of the stories, so to speak – because the sequences unfold against the backdrop of the familiar birthday-party script'. For example, when we read the narrative about John against our birthday party script, we can fill in sequences of actions such as John congratulating Bill upon arriving at his house, John giving Bill a present, and Bill tearing the wrapping paper before opening the box.

This conception of how narratives are read or heard implies that readers *reconstruct* a detailed story from a narrative text with less detail. It is important to note that readers reconstruct the actions behind the narrative text with the way events or actions usually occur based on the reader's prior experience. This is to say that scripts vary across readers whose prior experience is substantially

---

[4] I am grateful to Kim M. Hajek for alerting me to the research on scripts and its potential for work on mathematical proofs.
[5] On the value of the notion of script to the Narrative Science Project, see Hajek (Chapter 2).

different. Different readers may also make different reconstructions when they have similar scripts available if they make different assumptions about the context of a particular narrative. Thus a reader who assumes John and Bill to be children would fill in a sequence where Bill's mother lit candles on the cake, Bill blew out the candles, then his mother cut the cake and handed a slice to John. Another reader, assuming Bill to be an older man, might imagine there to be no candles, and Bill cutting the cake himself. An author who wanted to specify who cut the cake would need to write it out explicitly as part of the narrative sequence, not just rely on the birthday party script to do that job.

The process by which readers or auditors activate their prior experience captured in scripts when reading or hearing a narrative is sometimes described as the process by which they come to *understand* the narrative. I will return to the topic of understanding towards the end of this chapter.

The second key feature of the relationship between narratives and their readers or auditors is that readers or auditors are usually *surprised* by parts of the narratives.[6] The surprises are surprises against the backdrop of scripts. They are breaches of the scripts. In other words, the unusual or surprising aspects of narratives are unusual and surprising against the backdrop of the scripts. When the narratives convey something unexpected relative to existing scripts, the narratives can feed into new scripts. Hence, a narrative unfolds against the backdrop of scripts but also contributes to the creation of new scripts. Existing scripts are exploited to generate new scripts.

For example, we may add to the birthday party narrative featuring John and Bill that Bill's cat tried to lick John's hand, attracted to the cheese that had oozed out as John bit into the cake. There is a breach of the birthday party script here, as we expect a birthday cake to be made of flour, sugar, eggs and so on, not cheese. The breach of the birthday party script may lead us to reconsider the new trend of using cheese rounds as cakes, and when we on several occasions have read or heard narratives where scripts are breached by involving cheese rounds as cakes, we may be led to new scripts about usual sequences of actions in which a cake made of cheese rounds is served.

The two key features of the relationship between narratives and their readers or auditors are closely related. Herman (1997) describes the relationship thus: 'Stories stand in a certain relation to what their readers and auditors know, focussing attention on the unusual and remarkable against a backdrop made up of patterns of belief and expectation. Telling narratives is a certain way of reconciling emergent with prior knowledge' (Herman 1997: 1048). Focusing

---

[6] Netz has written on narrative and narrative surprises in mathematics. He writes about the narrative structure of Greek mathematical treatises with a particular focus on narrative surprises (Netz 2009: 80–91). See also Hurwitz (Chapter 17) on narrative surprises in medical anecdotes.

on social science research, Morgan (2017) similarly emphasizes the importance of narrative in framing and resolving puzzles.

I claim that research mathematicians often read proofs as narratives. Hence, I claim that proofs are often read as narratives in addition to having the presentational features of narratives by being a telling of a sequence of actions. More specifically, proofs are often interpreted by their readers against the background of scripts. The scripts are about how different kinds of results or sub-results are usually proved, about standard sequences of proving actions; and the scripts are based on the experience with proofs of the research mathematicians.[7] Furthermore, the readers focus attention on the breaches of the scripts, on that which appears unusual or surprising against the background of the scripts.

For example, consider the following telling of a sequence of actions aimed at proving the formula $1 + 3 + 5 + \ldots + (2n - 1) = n^2$, which holds for any natural number n. I begin by showing that the formula holds for n = 1. $((2 \times 1) - 1)$ equals 1, which, in turn, equals 1^2 and thus the formula holds for n = 1. I now show that *if* the formula holds for some value $n = n_0$, *then* it also holds for $n_0 + 1$. Hence, I make the assumption that the formula holds for $n = n_0$, that is, I make the assumption that $1 + 3 + 5 + \ldots + (2n_0 - 1) = n_0^2$. Given this assumption, I have to show that the formula holds for $n = n_0 + 1$, that is, I have to show that $1 + 3 + 5 + \ldots + (2n_0 - 1) + (2(n_0 + 1) - 1) = (n_0 + 1)^2$. Using the assumption, I get that $1 + 3 + 5 + \ldots + (2n_0 - 1) + (2(n_0 + 1) - 1) = n_0^2 + (2(n_0 + 1) - 1)$. And, simplifying, I get $n_0^2 + (2(n_0 + 1) - 1) = n_0^2 + 2n_0 + 2 - 1 = n_0^2 + 2n_0 + 1 = (n_0 + 1)^2$. In short, $1 + 3 + 5 + \ldots + (2n_0 - 1) + (2(n_0 + 1) - 1) = (n_0 + 1)^2$, which is what I wanted.

If we have some experience reading proofs by mathematical induction we may interpret the telling of the sequence of actions against the background of what we may call the proof by mathematical induction script, that is, against the standard sequence of actions performed in proofs by mathematical induction. We will then see that the particular sequence of actions taken in the example follows the script. In fact, in this example, we need to read the telling of the sequence of actions against the background of this script or pattern of actions in order to see that the formula has been proved. An analogous point can be made if we consider the following narrative: 'John went over to Bill's house. He watched Bill open his presents. John ate the cake and left' (adapted from Schank and Abelson 1977: 39). Readers will see that the description fits the birthday party script and can fill in the important 'detail' that Bill is having a birthday party. In short, relying on scripts, readers recognize that a standard proof by mathematical induction is taking place without being told so explicitly

---

[7] Hopkins (Chapter 4) similarly emphasizes the role of the training and experience of geologists in how they read narrative texts in geology.

and, similarly, readers recognize that a birthday party is taking place without being told so explicitly.

In the particular case of the example of the proof by mathematical induction, there are no surprises when we read the telling of the sequence of actions against the background of the proof by mathematical induction script. This is to say that the sequence of actions in the example proceeds entirely as we expect given our experience with how proofs by mathematical induction are usually carried out. But we would probably have been surprised or paid special attention if an unusual idea were used to show that the formula holds for $n = n_0 + 1$ or if the formula were of a kind where we would not expect the formula to be provable by mathematical induction.

It is worth emphasizing that my claim that proofs are often read as narratives is a claim about how research mathematicians read proofs. My discussion of the proof by mathematical induction thus provides a simplified illustration of how research mathematicians read proofs, since I speak about how we, who are not research mathematicians, would read the proof by mathematical induction. Most of us possess only few and simple scripts about standard sequences of actions in mathematical proofs.[8]

19.3   Evidence

In this section I draw on recent interview studies with research mathematicians about how they read proofs. The interview data is consistent with the claim that mathematicians can and often do rely on scripts when they read proofs, and that they focus attention on the breaches of the scripts, on that which appears unusual or surprising in the proofs against the background of scripts.

It is important to note that how mathematicians read proofs may well have changed substantially over time. And even if we assume that mathematicians now tend to read proofs as narratives and always have tended to read proofs as narratives, the scripts they have used will have changed over time. For

---

[8] My account of how proofs are read focuses on *actions* on the part of the authors and the readers of proofs. Previous accounts of mathematical proofs that focus on the actions on the part of the readers of proofs have conceived of a proof as a *recipe* of sorts for how to prove a proposition (Larvor 2012: 725–726; Sundholm 2012; and Tanswell 2017b). They claim that a proof as written is a recipe for how to execute an actual proof. Reading a proof is like reading a recipe and the readers are supposed to follow the recipe and perform steps of the actual proof as they read the proof recipe. My account of how proofs are read as narratives and the account of proofs as recipes are not necessarily inconsistent. It is possible that the two accounts capture different aspects of proof reading. In any case, the two accounts emphasize different kinds of actions on the part of the readers. When we conceive of a proof as a recipe, the action on the part of the readers of performing steps in the proof is emphasized. By contrast, when we conceive of reading a proof as reading a narrative, then the action on the part of the readers of connecting steps in the proof to scripts, of recognizing the steps performed by the author of the proof as instances of scripts, is emphasized.

example, the fact that mathematicians have used different criteria for mathematical proof across cultural and historical contexts means that they will have used different scripts. I here rely on research in the history and philosophy of mathematics which has uncovered how mathematical proofs as they occur in mathematical practice are context-sensitive. For example, the level of rigour required for mathematical proof has varied across time and discipline.

19.3.1 Reliance on Scripts

Various recent interview studies with research mathematicians suggest that mathematicians read proofs against the background of what they know from their experience with proofs. For example, these studies suggest that mathematicians have beliefs and expectations about which methods and techniques work in which situations and on which sorts of mathematical objects, and that mathematicians rely on these beliefs and expectations as they read proofs. The mathematicians seem to see recognizable patterns of action in the sense of standard sequences of proving actions in the proofs. In this sense, proofs seem to unfold against the backdrop of scripts about standard sequences of proving actions. In sum, mathematicians, as they read proofs, seem to rely on scripts about which methods and techniques work in which situations and on which sorts of mathematical objects.

For example, based on their interviews with mathematicians, Weber and Mejía-Ramos (2011: 340) suggest that mathematicians, when they read proofs, 'might encapsulate strings of derivations into a short collection of methods and determine whether these methods would allow one to deduce the claim that was proven'. Whether the methods will work to prove a result is something they judge based on their experience. Weber and Mejía-Ramos note (2011: 340) that Konior (1993) provides further data to support their claim. Konior reports on the analyses of several hundred proofs. Konior found that a written proof often contains cues that indicate to the reader how to separate the proof into parts and what methods were being used in each part. For example, a part of a proof may begin with: 'We have to define a one-to-one mapping g of X onto Y' (Konior 1993: 255). In this way, a proof seems to cue readers to activate their scripts about which methodological moves work when.

Andersen (2020) has interviewed mathematicians about their proof reading practices when they act as referees for mathematics journals. Based on the interviews, she similarly suggests that mathematicians read proofs against their experience concerning which approaches work to prove different kinds of results. Mathematicians appear to have reliable intuitions based on experience about which type of approach can typically be used to prove a sub-result of this or that type. The beliefs and expectations the interviewees have about proving actions seem to correspond to what we here call scripts.

A study by Andersen, Johansen and Sørensen (2019) indicates that the scripts may in part be provided by the main text preceding a proof in an article. The study reports on interviews with a supervisor and his PhD student. In line with the interviews referenced above, the supervisor described how he studies the 'general pattern' or flow of a proof and sees if it is recognizable or instead raises 'red flags', which can be interpreted to mean that he studies whether the proof follows the scripts or there are points at which a script is breached. He pays special attention 'if some of this pattern recognition raises a red flag or indeed gives any hint of unease or alienation' (Andersen, Johansen and Sørensen 2019: 11). His expectations with respect to the flow of the proof are sometimes informed by the main text of the article presenting the proof. He emphasized how an article may provide examples of how a mathematical object behaves in different situations before presenting proofs establishing results about the object in question. The examples may then influence the expectations of the supervisor with respect to how the results can be proved.

The interview data is consistent with my claim that mathematicians can and often do rely on scripts based on their experience as mathematicians when they read proofs. The interview data does not shed light on the question of what concrete scripts that play a role in mathematical practice look like exactly. This is a question for future research. Note that we would expect that the parts of proofs that follow the scripts are *not* the parts readers focus attention on, exactly because there are no breaches of the scripts. This is supported by Andersen, Johansen and Sørensen (2019) and Andersen (2020), whose interviews suggest that mathematicians do not thoroughly read the parts of a proof that unfold the way they would expect.

19.3.2   Breaches of Scripts

Sometimes something unusual happens in a proof. In a number of interview studies, mathematicians describe how they pause and pay close attention when they read a proof and encounter something 'surprising' (Andersen 2020: 238) or something 'strange' or 'odd' (Weber 2008: 448). Or how they pay close attention when a part of a proof is 'suspicious' (Müller-Hill 2011: 307–308, 327–328) or raises a 'red flag' (Andersen, Johansen and Sørensen 2019: 11). My argument above offers an interpretation of the parts of proofs the mathematicians describe here. I argued that the *usual* moves in proofs can be interpreted as the moves that follow the scripts about standard sequences of proving actions. The *unusual* moves that mathematicians describe that make them pause and pay close attention when they are reading proofs should then be interpreted as the moves that do not follow the scripts about standard sequences of proving actions. The unusual moves are breaches of the scripts.

As described in section 19.2, 'Reading Proof as Narrative', above, when narratives convey something unexpected relative to existing scripts, the narratives feed into new scripts. In the case of mathematical proofs, we may ask how unusual moves contribute to the creation of new scripts. Moves that may be unusual at one point in time may become standard moves at a later point in time because they have been shown to work in various proofs.[9] Before new moves can turn into standard moves, the new moves must be carefully checked in the proofs that use them. This probably involves careful attention to detail and filling intentional gaps with extra details.[10] It has previously been claimed that readers of mathematical proofs commonly fill intentional gaps in the proof and thus engage in a kind of *reconstruction* of the proof (see, for example, Fallis 2003; Netz 2009: 71–80; Rav 1999).[11] It is worth adding that how the readers perceive the *narrator* or the author of the proof may affect how thoroughly they check unusual moves. Readers may be more thorough if the author is a PhD student than if the author is an experienced mathematician (Andersen 2017: 184–187; Inglis and Mejía-Ramos 2009; Mejía-Ramos and Weber 2014: 165–168).

When proofs are read as narratives, reading proofs really involves two kinds of *reconstruction* of the proofs on the part of the readers, one of which is the kind of reconstruction of proofs that has previously attracted attention from philosophers. Readers of proofs engage in the kind of reconstruction that has been discussed previously when they *fail to see* what is going on, when they cannot follow a step in a proof, that is, when they cannot see why B follows from A as the author claims. The readers will then insert extra steps between A and B. As just mentioned, this kind of reconstruction probably plays a role where breaches of scripts occur and in establishing *new* scripts. But readers also engage in a kind of reconstruction when they *recognize* what is going on, when they recognize a move in a proof as an instance of a script for a standard way of proving this sort of result. The details they insert in the proof are provided by

[9] This process is similar to the process described in Morgan (2005: 324) of how a 'surprising behaviour pattern' observed in an experiment in economics may turn into a 'genuine behaviour pattern' over time, 'after many experimental replications with many subjects and with slight variations in the experimental design'.

[10] Jajdelska (Chapter 18) demonstrates a different way in which readers of research articles are led to accept unusual ideas presented in the articles: through narrative performativity. Meunier (Chapter 12) demonstrates how readers of research articles can be made familiar with new methods and epistemic objects by being guided through a narrative sequence by the authors.

[11] For example, Rav (1999) describes his experience with reading proofs. He writes that, when one reads a proof, it 'often happens – as everyone knows too well – that one arrives at an impasse, not seeing why a certain claim q is to follow from claim p, as its author affirms'. Thus, 'one picks up paper and pencil and tries to fill in the gaps', both by reflecting 'on the background theory [and] the meaning of the terms' and by 'using one's general knowledge of the topic' (Rav 1999: 14). Sørensen, Danielsen and Andersen (2019) provide an account of how this kind of reader engagement can be taught to students as an aspect of proof.

the *existing* script. Hence, in both kinds of reconstruction details are filled in by the readers but the details are different in kind and are inserted in different parts of the proof.

19.4   Understanding Proofs

The present account of how proofs are read provides a way of thinking about mathematical understanding, more specifically the understanding of proofs.[12] Among cognitive scientists, understanding is commonly characterized as 'a process by which people match what they see and hear to pre-stored groupings of actions that they have already experienced' (Schank and Abelson 1977: 67; quoted in Herman 1997: 1048). In particular, coming to understand a narrative is the process by which narratives are interpreted by their readers or auditors against the background of scripts about how events or actions of the kinds represented in the narrative usually take place. Hence, the process by which readers or auditors come to understand a narrative is the process by which they use scripts to reconstruct the narrative. Consider again the narrative: 'John went over to Bill's house. He watched Bill open his presents. John ate the cake and left' (adapted from Schank and Abelson 1977: 39). This narrative does not make much sense if we do not interpret the narrative against our knowledge of how birthday parties usually take place. We come to understand the narrative by reading the narrative against our birthday party script. Thus, we envision the guests arriving at Bill's house, the guests each giving Bill a present, and Bill tearing the wrapping paper.

When the narratives convey something unexpected relative to existing scripts, the readers can fail to understand. For example, in the case of the narrative where Bill's cat tried to lick John's hand, the readers may fail to understand how John ended up with cheese on his hand when he ate cake, since, according to their birthday party script, a birthday cake is made of flour, sugar and eggs, not cheese. But, when the narratives convey something unexpected relative to existing scripts, the narratives can also contribute to the creation of new scripts and thus new 'models for understanding' (Herman 1997: 1056). Hence, when the readers see that Bill's birthday cake is made of cheese rounds, and on a number of other occasions have read narratives where scripts are breached by involving cheese rounds as cakes, they may be led to new scripts about usual sequences of actions in which a cake made of cheese rounds is served.

---

[12] Avigad (2008) argues that we must consider mathematical understanding of different things, such as theories, theorems and proofs, separately. Sandborg (1997: 140–141) discusses the difference between mathematical understanding of theorems and proofs.

If proofs are read as narratives, as I suggest, the process by which mathematicians come to understand proofs is the process by which they use scripts to reconstruct the proofs. Thus, the process by which mathematicians come to understand a proof involves recognizing the moves in the proof as instances of scripts about standard sequences of proving actions.[13] Hence, the process by which mathematicians come to understand the proof of the formula $1 + 3 + 5 + \ldots + (2n - 1) = n^2$ I gave earlier involves recognizing my moves in the proof as instances of the proof by mathematical induction script.

Consider the following useful analogy suggested by Norton Wise between coming to understand a proof and coming to understand how to frame a new roof. Coming to understand how to frame a new roof requires experience with patterns of roof framing in many particular instances. Considering only how each part of the new roof framing is placed is not enough for understanding how the stability of the whole emerges. Similarly, going through the steps of a given proof is not enough for understanding the proof but requires experience with patterns of proving in many particular instances. In other words, understanding the proof requires scripts about standard sequences of proving actions.

The present account of proof understanding can explain why mathematicians emphasize that one may have verified every logical step of a proof and still not have understood the proof. Poincaré makes this point. A mathematician may have 'examined [the elementary] operations one after the other and ascertained that each is correct' and still not have 'grasped the real meaning' of the proof (Poincaré 1958: 217–218). Feferman similarly notes that, 'It is possible to go through the steps of a given proof and not understand the proof itself', and adds that understanding the proof is 'a special kind of insight into how and why the proof works' (Feferman 2012: 372; quoted in Folina 2018: 136).

While verification is not a form of understanding, the opposite may be true. Understanding may be a form of verification. Research on proof reading tends to focus on the validation of proofs rather than the understanding of proofs. But the present account of mathematical understanding seems to suggest that there is a strong connection between the understanding and validation of proofs (in line with Dutilh Novaes 2018; and Mejía-Ramos and Weber 2014). The present account of understanding indicates that mathematicians *understand* proofs through action pattern recognition, which is the same kind of action pattern recognition that previous studies, based on interviews and a survey with mathematicians, suggest that mathematicians use to *validate* proofs (Andersen 2020; Andersen, Johansen and Sørensen 2019; Mejía-Ramos and Weber 2014; Weber and Mejía-Ramos 2011). Hence, proof understanding may be a form of proof validation.

---

[13] By contrast, Cellucci (2015) argues that understanding a proof consists in seeing how the different parts of the proof fit together.

I end this section by briefly considering how the present account of mathematical understanding is relevant to mathematics education, to the teaching of proofs. As noted by Weber and Mejía-Ramos, 'a goal of many research programs is to lead students to think and behave more like mathematicians with respect to proof' (2011: 330). This raises the question of how students may read and come to understand proofs in a way that is similar to how mathematicians read and come to understand proofs. Needless to say, students will always have different scripts available to them than research mathematicians do. And presumably it is tempting for students to read proofs word by word without at all engaging in the kind of reconstruction of the proof scripts that mathematicians engage in. But the picture of proof reading presented in this chapter suggests that students cannot come to understand proofs this way. Not even mathematicians come to understand proofs by reading them only word by word rather than against the backdrop of scripts. Hence, if we want to teach students to read and come to understand proofs in a way that is similar to how mathematicians read and come to understand proofs, the teaching of scripts about standard sequences of proving actions is important. For example, it is valuable to teach students about the sorts of results that can be proved by mathematical induction and the commonalities between different proofs by mathematical induction.

19.5   Conclusion

Focusing on how mathematical proofs are *read*, I have argued that mathematical proofs can be and often are read like narratives by research mathematicians. Mathematicians read proofs as narratives when they read proofs against the backdrop of experience-based scripts about standard sequences of proving actions. They focus attention on the breaches of the scripts, on that which appears unusual or surprising against the backdrop of the scripts. The account I have defended of how proofs are read as narratives also provides an account of how to conceive of proof *understanding*, which is a topic that has received very little attention in the literature. On this account of proof understanding, a process by which mathematicians come to understand proofs is the process by which they relate proofs to scripts about standard sequences of proving actions.[14]

[14] I would like to thank editors Kim Hajek and Mary Morgan, referees Norton Wise and an anonymous referee, as well as Yacin Hamami and K. Brad Wray for highly valuable comments on earlier versions of this chapter. I would also like to thank the editors, Mary Morgan, Dominic Berry and Kim Hajek, for their dedication and excellent work in preparing this volume. The research for this paper was funded by K. Brad Wray's grant from the Aarhus University Research Foundation, AUFF-E-2017-FLS-7–3. *Narrative Science* book: This project has received funding from the European Research Council under the European Union's Horizon

References

Andersen, L. E. (2017). 'On the Nature and Role of Peer Review in Mathematics'. *Accountability in Research* 24.3: 177–192.

(2020). 'Acceptable Gaps in Mathematical Proofs'. *Synthese* 197: 233–247.

Andersen, L. E., M. W. Johansen and H. K. Sørensen (2019). 'Mathematicians Writing for Mathematicians'. *Synthese* 198: 6233–6250.

Avigad, J. (2008). 'Understanding Proofs'. In P. Mancosu, ed. *The Philosophy of Mathematical Practice*. New York: Oxford University Press, 317–353.

Borwein, J., and D. Bailey (2003). *Mathematics by Experiment: Plausible Reasoning in the 21st Century*. Wellesley, MA: A. K. Peters.

Bruner, J. (1987). *Actual Minds, Possible Worlds*. Cambridge, MA: Harvard University Press.

(1991). 'The Narrative Construction of Reality'. *Critical Inquiry* 18.1: 1–21.

Cellucci, C. (2015). 'Mathematical Beauty, Understanding, and Discovery'. *Foundations of Science* 20: 339–355.

De Toffoli, S., and V. Giardino (2015). 'An Inquiry into the Practice of Proving in Low-Dimensional Topology'. In G. Lolli, M. Panza and G. Venturi, eds. *From Logic to Practice: Italian Studies in the Philosophy of Mathematics*. Cham: Springer, 315–336.

Doxiadis, A. (2012). 'A Streetcar Named (Among Other Things) Proof'. In A. Doxiadis and B. Mazur, eds. *Circles Disturbed: The Interplay of Mathematics and Narrative*. Princeton, NJ: Princeton University Press, 281–288.

Dutilh Novaes, C. (2018). 'A Dialogical Conception of Explanation in Mathematical Proofs'. In P. Ernest, ed. *The Philosophy of Mathematics Education Today*. Cham: Springer, 81–98.

Fallis, D. (2003). 'Intentional Gaps in Mathematical Proofs'. *Synthese* 134.1–2: 45–69.

Feferman, S. (2012). 'And so on . . . : Reasoning with Infinite Diagrams'. *Synthese* 186: 371–386.

Folina, J. (2018). 'Towards a Better Understanding of Mathematical Understanding'. In M. Piazza and G. Pulcini, eds. *Truth, Existence and Explanation*. Cham: Springer, 121–146.

Hamami, Y., and Rebecca L. Morris (2020). 'Plans and Planning in Mathematical Proofs'. *Review of Symbolic Logic*: 1–36.

Herman, D. (1997). 'Scripts, Sequences, and Stories: Elements of a Postclassical Narratology'. *PMLA* 112.5: 1046–1059.

(2009). *Basic Elements of Narrative*. Malden, MA: Wiley-Blackwell.

Inglis, M., and Mejía-Ramos, Juan P. (2009). 'The Effect of Authority on the Persuasiveness of Mathematical Arguments'. *Cognition and Instruction* 27.1: 25–50.

Konior, J. (1993). 'Research into the Construction of Mathematical Texts'. *Educational Studies in Mathematics* 24: 251–256.

Larvor, B. (2012). 'How to Think about Informal Proofs'. *Synthese* 187.2: 715–730.

Mejía-Ramos, J. P., and K. Weber (2014). 'Why and How Mathematicians Read Proofs: Further Evidence from a Survey Study'. *Educational Studies in Mathematics* 85.2: 161–173.

2020 research and innovation programme (grant agreement No. 694732). www.narrative-science.org/.

Morgan, Mary S. (2005). 'Experiments Versus Models: New Phenomena, Inference and Surprise'. *Journal of Economic Methodology* 12.2: 317–329.

(2017). 'Narrative Ordering and Explanation'. *Studies in History and Philosophy of Science Part A* 62: 86–97.

Müller-Hill, E. (2011). 'Die epistemische Rolle formalisierbarer mathematischer Beweise'. PhD thesis, Rheinische Friedrich-Wilhelms-Universität. https://hdl.handle.net/20.500.11811/4850.

Netz, R. (1999). *The Shaping of Deduction in Greek Mathematics: A Study in Cognitive History*. Cambridge: Cambridge University Press.

(2009). *Ludic Proof: Greek Mathematics and the Alexandrian Aesthetic*. New York: Cambridge University Press.

Poincaré, H. (1958). *The Value of Science*. Cambridge: Cambridge University Press.

Rav, Y. (1999). 'Why Do We Prove Theorems?' *Philosophia Mathematica* 7.1: 5–41.

Robinson, J. A. (1991). 'Formal and Informal Proofs'. In R. S. Boyer, ed. *Automated Reasoning*. Dordrecht: Kluwer Academic Publishers, 267–282.

Sandborg, D. A. (1997). 'Explanation in Mathematical Practice'. Dissertation, University of Pittsburgh.

Sanford, A. J., and C. Emmott (2012). *Mind, Brain and Narrative*. Cambridge: Cambridge University Press.

Schank, R. C., and R. P. Abelson (1977). *Scripts, Plans, Goals, and Understanding: An Inquiry into Human Knowledge Structures*. Hillsdale, NJ: Erlbaum.

Sørensen, H. K., K. Danielsen and L. E. Andersen (2019). 'Teaching Reader Engagement as an Aspect of Proof'. *ZDM: The International Journal on Mathematics Education* 51.5: 835–844.

Sundholm, G. (2012). '"Inference Versus Consequence" Revisited: Inference, Consequence, Conditional, Implication'. *Synthese* 187.3: 943–956.

Tanswell, F. (2017a). 'Playing with LEGO® and Proving Theorems'. In W. Irwin and R. T. Cook, eds. *LEGO® and Philosophy: Constructing Reality Brick by Brick*. Hoboken, NJ: John Wiley and Sons, 217–226.

(2017b). 'Proof, Rigour and Informality: A Virtue Account of Mathematical Knowledge'. PhD thesis, University of St Andrews. https://research-repository.st-andrews.ac.uk/handle/10023/10249.

Thomas, R. S. D. (2007). 'The Comparison of Mathematics with Narrative'. In B. Van Kerkhove and J. P. Van Bendegem, eds. *Perspectives on Mathematical Practices*. Dordrecht: Springer, 43–59.

Toolan, M. (2001). *Narrative: A Critical Linguistic Introduction*. 2nd edn. New York: Routledge.

Weber, K. (2008). 'How Mathematicians Determine If an Argument Is a Valid Proof'. *Journal for Research in Mathematics Education* 39.4: 431–459.

Weber, K., and J. P. Mejía-Ramos (2011). 'Why and How Mathematicians Read Proofs: An Exploratory Study'. *Educational Studies in Mathematics* 76.3: 329–344.

20   Narrative Solutions to a Common Evolutionary Problem

John Beatty

Abstract

To give a Darwinian explanation of the traits of a species, it is not enough to show that the traits are appropriate for the environments inhabited. One must also show that the traits in question are more appropriate than the (presumed) ancestral traits from which they are derived. But one must go further still. Even if there is no question that the derived traits are more appropriate, one must still specify the sequence of modifications leading from the ancestral to the derived traits, each step of which is fitness-enhancing. How better – indeed, how else – than by a narrative? I illustrate these points through the evolution of flatfish eyes. This is part of an ongoing project concerning what narratives are good for, what narratives do better than non-narrative arguments: in short, why we need narratives.

20.1   Introduction

Sometimes, in order to understand an occurrence, we need to know what happened prior to that. (Yes, that sounds obvious, and yet . . . .)

And sometimes it's not enough to know what happened immediately prior to that. We need a *backstory* that rewinds time to some event in the more distant past, and then takes us forward through events that (1) were not foreseeable from the starting point, and (2) were consequential for the outcome of interest, (3) in the order in which they occurred and not just any order. Such a backstory is *narrative-worthy* – a narrative is just right for the occasion – as I will explain later.

I'll illustrate these points with a common problem from evolutionary biology. Or rather, I'll rely on the common problem in order to introduce/motivate the need for a narrative solution. Evolutionary explanations sometimes (often?) invoke only circumstances contemporaneous with the phenomena to be explained – no backstory, indeed, *atemporal* evolutionary reasoning, as odd as that may sound. This might be satisfactory in some contexts. But it is not satisfactory in other contexts, like those I'll discuss, where narrative-worthy backstories are called for.

20.2   Darwinian Assumptions: Successive, Slight Modifications

Flatfishes – halibut, turbot, sole, others – live horizontally/flat on the sea floor. They differ from vertical fishes in appropriate ways. Most notably, instead of having one eye on each side of their head, they have both eyes on one side, the topside, making it easier to see their prey and watch out for predators. Flatfishes are commonly coloured in a way that camouflages them against their background. Usually, just the topside is camouflaged; the unseen bottom is pigmentless (Figures 20.1 and 20.2).

It is common to account for the traits of a species – like the lifestyle, anatomy and coloration of flatfishes – in terms of their appropriateness, the ways in which they individually, and in combination, enhance the fitness of their possessors. The possession of fitness-enhancing traits is what we expect from evolution by natural selection.

This kind of reasoning is odd, though, when you think about it: it amounts to an evolutionary account of the present that does not invoke the past, just prevailing circumstances. It's an atemporal evolutionary account.

It's a common enough manner of reasoning to have been called out for criticism by Stephen Gould and Richard Lewontin (1979). They dubbed it the 'Panglossian paradigm', after Voltaire's Dr. Pangloss, for whom everything in

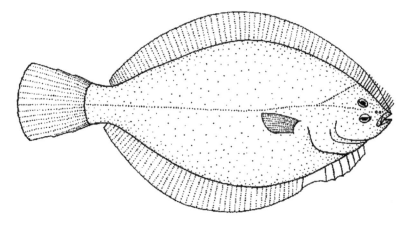

Figure 20.1 **Flatfish (flounder) topside**
Rhombosolea leporina (Yellowbelly flounder)
Source: This illustration is licensed under the Creative Commons Attribution
ShareAlike 1.0 license. The author is Dr Tony Ayling. The illustration was originally
published in Tony Ayling and Geoffrey Cox, *Guide to the Sea Fishes of New Zealand*
(Auckland: William Collins Publishers, 1982) https://commons.wikimedia.org/wiki/
File:Rhombosolea_leporina_(Yellowbelly_flounder).gif.

Figure 20.2 **Still life by Jan van Kessel the Elder**
Jan van Kessel the Elder, 1626–79.
Source: Wikimedia Commons.

the world was maximally appropriate. Thus, species are maximally adapted to their environments. But, to account for the traits of species entirely in terms of their '*current* utility', Gould and Lewontin objected, is to pretend that adaptation does not take time. Such accounts reflect the untenable assumption of '*immediate* adaptation' to whatever environment a species inhabits, the '*immediate* work of natural selection'.[1]

Surely the targets of Gould and Lewontin's critique did not really believe that species instantaneously adapt to their environments. But perhaps they assumed, along with John Maynard Smith, that 'most populations have had time to come close to the *optimum* for the environment in which they live' (Maynard Smith 1993: 11–12; my emphasis; but see also Maynard Smith et al. 1985). Had there been *in*sufficient time for populations to 'come close to the optimum for the environment in which they live', then one could not make sense of their traits without taking into consideration the ancestral starting points from which they had not completely departed. But, as it (supposedly) happens, there's no need to bother with the past; the present is enough.

---

[1] Olmos (Chapter 21) unpacks the logic of Gould and Lewontin's argument in detail, finding that their criticisms do not bear only on the narrative nature of adaptationist accounts, but on other aspects of them as well.

On the Panglossian paradigm, there is another, related respect in which history is irrelevant. It has to do with the equilibrating character of an optimizing process like evolution by natural selection – or, rather, as evolution by natural selection is commonly conceived. Consider an analogy. We find a marble lying in the bottom of a bowl. How did this come about? We need only take into account the prevailing properties of the marble and the bowl, and the principles that govern this mini-universe. The past is largely irrelevant, since the marble would have rolled around and eventually come to rest there, no matter where it started. Similarly, one might think that a species (marble) evolves by natural selection in its environment (bowl) until it attains *the* optimum combination of traits, the equilibrium point where it then rests, no matter its starting point. One need only take into account the prevailing circumstances.

Darwin himself could not have reasoned persuasively in this manner. His case for evolution by natural selection vs special creation depended on linking the present to the past. For instance, it makes better sense to attribute imperfect but satisfactory traits – like the wonky but workable placement of flatfish eyes – to the trial and error modification of an imagined ancestor, in this case with an eye on each side, than to an all-knowing and benevolent creator who engineers each species from scratch.

Evolutionary biologists today are rarely concerned to dispatch special creation. But, insofar as they are *Darwinians*, they have other sceptics to contend with, and in doing so they have other reasons for looking beyond the present into the past. And here's (at least one reason) why.

To make sense of the traits of a species, a Darwinian should be able to go back in time to an ancestor of that species, and then forward to the species in question. But, in going forward from the ancestor, the good Darwinian cannot rely on an all-at-once modification. It should be possible to specify a *sequence of slight modifications* that would lead to the descendant. As Darwin acknowledged:

> If it could be demonstrated that any complex organ existed which could not possibly have been formed by numerous, successive, slight modifications, my theory would absolutely break down. (Darwin 1859: 189)

I would just add the following friendly amendment (this is after all what he meant):

> If it could be demonstrated that any complex organ existed which could not possibly have been formed by numerous, successive, slight modifications, *each of which increases fitness*, my theory would absolutely break down.

There are times when Darwinians do not hold themselves – and are not held – to these standards, presumably on the grounds that the ancestral and gradual intermediate stages of evolution are not difficult to fathom and are perhaps not worth the worry. But there are also cases where it is not at all clear that there is a backstory that meets these standards, and the challenge is to provide one.

20.3   Plausible Orderings of Modifications

St. George Jackson Mivart, for one, challenged Darwin on his own terms (Mivart 1871).[2] And flatfish eyes (among other examples, see further) served him well in this regard. To give a Darwinian account of their eyes, by Darwin's own criteria, it is not sufficient to demonstrate the usefulness of that arrangement at present. One must also propose a sequence of slight modifications leading from an ancestor with one eye on each side to descendants with both eyes on one side, each step of which increases fitness.

What's a plausible sequence? Surely not by slight displacements of one eye *through* the skull to the other side! Surely it would involve slight displacements of one eye *over the top* of the skull to the other side. But that leaves unanswered how the initial and early displacements could have been fitness enhancing. He imagines a fish lying flat on its side with one eye in the sand. What's the advantage of having the lower eye only slightly closer to the top of the skull? It's still in the sand. How can the initial migrations of the eye have been anything but injurious, given the skull/eye-socket reconstructions involved?

Another instance which may be cited is the asymmetrical condition of the heads of the flat-fishes (Pleuronectidæ), such as the sole, the flounder, the brill, the turbot, &c. In all these fishes the two eyes, which in the young are situated as usual one on each side, come to be placed, in the adult, both on the same side of the head. If this condition had appeared at once, if in the hypothetically fortunate common ancestor of these fishes an eye had suddenly become thus transferred, then the perpetuation of such a transformation by the action of 'Natural Selection' is conceivable enough. Such sudden changes, however, are not those favoured by the Darwinian theory [. . .] But if this is not so, if the transit was gradual, then how such transit of one eye a minute fraction of the journey towards the other side of the head could benefit the individual is indeed far from clear. It seems, even, that such an incipient transformation must rather have been injurious. (Mivart 1871: 37–38)

Mivart generalized the problem and gave it a name: 'the incompetency of "natural selection" to account for the incipient stages of [ultimately] useful structures' (Mivart 1871: 23). It has since been shortened to 'the problem of incipient stages'.[3] As he put the point:

'Natural Selection,' simply and by itself, is potent to explain the maintenance or the further extension and development of favourable variations, which are at once

---

[2] A nice introduction to Mivart's challenge and the evolution of flatfish eye placement is Zimmer (2008).

[3] In his work on the evolution of leaf mimicry in butterflies, Suzuki (2017) includes updates on most of the problematic cases of incipient stages that Mivart raised, including flatfishes.

sufficiently considerable to be useful from the first to the individual possessing them. But Natural Selection utterly fails to account for the conservation and development of the minute and rudimentary beginnings, the slight and infinitesimal commencements of structures, however useful those structures may afterward become. (Mivart 1871: 23)

In addition to flatfish eyes, he illustrated the problem with other traits like the giraffe's neck, vertebrate limbs and mimicry. And mammary glands:

Is it conceivable that the young of any animal was ever saved from destruction by accidentally sucking a drop of scarcely nutritious fluid from an accidentally hypertrophied cutaneous gland of its mother? (Mivart 1871: 47)

For these and other reasons, Mivart inferred that new species arise not gradually, but with 'suddenness'.

Not only are there good reasons against the acceptance of the exclusive operation of 'Natural Selection' as the one means of specific origination, but there are difficulties in the way of accounting for such origination by the sole action of modifications which are infinitesimal and minute, whether fortuitous or not.

Arguments may yet be advanced in favour of the view that new species have from time to time manifested themselves with suddenness, and by modifications appearing at once [...] the species remaining stable in the intervals of such modifications. (Mivart 1871: 97)

Darwin well understood and appreciated the difficulty that Mivart had raised, and in the 6th and final edition of *On the Origin of Species* he devoted considerable space to the problem. Referring to this and other criticisms, Darwin wrote:

A distinguished zoologist, Mr. St. George Mivart, has recently collected all the objections which have ever been advanced by myself and others against the theory of natural selection, as propounded by Mr. Wallace and myself, and has illustrated them with admirable art and force. When thus marshalled, they make a formidable array. (Darwin 1872: 176)

Darwin responded to a variety of Mivart's objections. But he took most seriously, and spent the most time responding to, the problem of incipient stages (1872: 177–190), including a solution to the problem of flatfish eyes. He affirmed that the trajectory of evolution involved the eye moving over the top of the skull, onto the other side. The maturation of flatfishes provided the evidence. According to August Malm (1867), flatfish larvae swim vertically and – appropriately under the circumstances – have one eye on each side of their head. But, as they develop, one eye begins to migrate towards and then over the top of the skull, to the side that becomes the topside of the horizontal, bottom-dwelling adult, as shown in the sequence in Figure 20.3. The migration of the eye during the development of individual flatfishes, together with their

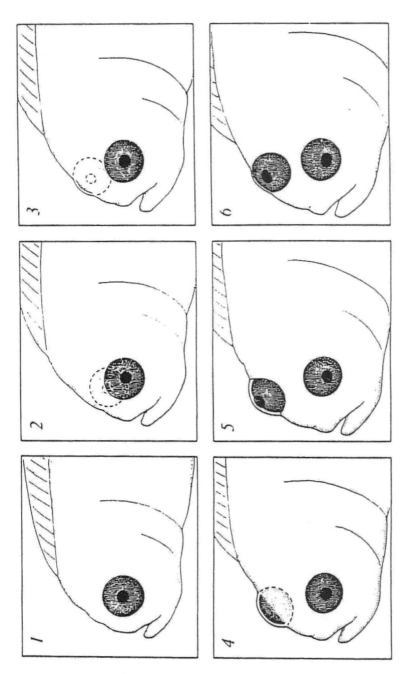

Figure 20.3 A depiction of eye migration in starry flounder larvae, that also illustrates Darwin's suggested evolutionary account of the flatfish eye.
Source: Policansky (1982).

change in orientation from vertical to horizontal, reflects the trajectory of their evolution. Why else would flatfishes undergo that course of development, other than because they were descendants of vertical, symmetrically eyed fishes?

But that leaves unanswered how the initial migration of the one eye, and early extensions of that migration, could have been fitness-enhancing, which had been Mivart's main puzzle. Darwin didn't entirely capitulate on this, but partly/largely. He attributed the initial, slight migration of the eye, and then early extensions of that migration, not to evolution by natural selection, but rather to Lamarckian inheritance of acquired characters. He took it on authority from Malm that very young flatfishes lying on their sides on the sea floor, before eye migration is complete, strain to see with their bottom eye

[. . .] and they do this so vigorously that they eye is pressed hard against the upper part of the orbit [socket]. The forehead between the eyes consequently becomes, as could be plainly seen, temporarily contracted in breadth. (Darwin 1872: 187)

Suppose this forced displacement of the eye, so as to see better, was inherited by the next generation, who also strained to see. Resulting in still further displacement of their eyes. The further, forced displacement was also inherited. This took place generation after generation until the eye made its way far enough around the skull that it was sufficiently out of the sand, at which point its migration to the other side was maintained and extended to the present state by natural selection. It was a largely Lamarckian, only partly Darwinian solution.

We thus see that the first stages of the transit of the eye from one side of the head to the other, which Mr. Mivart considers would be injurious, may be attributed to the habit, no doubt beneficial to the individual and to the species, of endeavouring to look upwards with both eyes, whilst resting on one side at the bottom. We may also attribute to the inherited effects of use the fact of the mouth in several kinds of flat-fish being bent towards the lower surface. (Darwin 1872: 187–188)

Darwin further explained the lack of pigment on the bottom of flatfishes in terms of the Lamarckian notion that disuse of a trait, over many generations, leads to its loss (Darwin 1872: 188).

Somewhat tangentially, Lamarck had offered his own explanation of flatfish eye placement. The ancestors of flatfishes fed in very shallow waters along shorelines, he supposed, waters so shallow that they had to lie flat on their sides. '[T]his requirement has forced one of their eyes to undergo a sort of displacement, and to assume the very remarkable position found in the soles, turbots, dabs, etc.' (Lamarck 1914: 120).

The all-at-once modification, as applied to flatfish evolution, and many other problematic cases, was developed in great detail by Richard Goldschmidt (1940) and had considerable influence well into the twentieth century and, in one form or another, to this day. Here is a thin version of his thinking, in connection with flatfish eyes:

In a former paper (Goldschmidt 1933) I used the term 'hopeful monster' to express the idea that mutants producing monstrosities may have played a considerable role in macroevolution. A monstrosity appearing in a single genetic step might permit the occupation of a new environmental niche and thus produce a new type in one step [...]. A fish undergoing a mutation which made for a distortion of the skull carrying both eyes to one side of the body is a monster. The same mutant in a much compressed form of fish living near the bottom of the sea produced a hopeful monster, as it enabled the species to take to the life upon the sandy bottom of the ocean, as exemplified by the flounders. (Goldschmidt 1940: 390–391; and see 1933: 545)

The question of flatfish eyes is often posed as one that pits a Goldschmidtian (and, to be fair, Mivart-inspired) solution against a Darwinian approach – for example, in Thomas Frazzetta's (2012) review: 'Flatfishes, Turtles, and Bolyerine Snakes: Evolution by Small Steps or Large, or Both?'.

The problem of flatfish eyes, from a Darwinian point of view, seems to have eluded even the master Darwinian communicator, Richard Dawkins. No, Dawkins doesn't go Lamarckian, nor Goldschmidtian. But he pulls up short of going fully Darwinian. He makes the Darwinian point that it is more reasonable to attribute asymmetric flatfish eyes to the modification of a symmetrically eyed ancestor than to special creation. Surely an intelligent designer would have created flatfishes more in the manner of skates and rays, flattened from top to bottom, with both eyes on top, rather than flattened from side to side and requiring the migration of one eye to the other side.

Even though its [the flatfish's] evolutionary course was eventually destined to lead it into the complicated and probably costly distortions involved in having two eyes on one side, even though the skate way of being a flat fish might *ultimately* have been the best design for bony fish too, the would-be intermediates that set out along this evolutionary pathway apparently did less well in the short term than their rivals lying on their side. (Dawkins 1986: 92–93)

Yes, 'apparently' in the lineages that beget flatfishes, lying flat on one side with a migrating eye prevailed over flattening from top to bottom. But how could the admittedly 'complicated and probably costly distortions' of the intermediate stages have been sufficiently advantageous to be selected for?

Interestingly, there have been recent – for the first time – fossil findings of flatfishes with intermediate stages of eye migration (Friedman 2008). But there is still no generally accepted, functional account of the fitness contributions of the early steps. The most promising clue is a fact about some flatfishes, maybe many or all, that has been known for quite a long time although not considered until recently in this connection (Olla, Wicklund and Wilk 1969; Stickney, White and Miller 1973; Friedman 2008: 211; Frazzetta 2012: 33). That is, adult flatfish (flounders in this case) sometimes use their dorsal and anal fins (what would have been the top and bottom fins in their ancestors but are right- and left-side fins in flounders) to prop themselves up, raising their heads to better see above them and lunge at their prey.

Now if early flatfishes could raise their heads in this way, and in the process raise their lower eyes out of the sand, then a slight migration of the lower eye could have allowed slightly better vision than having it face straight down and may have been selected for. And a further extension of the migration would be advantageous and selected for. And so on and so on, the entire eye migration thus being due to evolution by natural selection.

The sequence here is crucial. Head elevation for lunging is not only adaptive in combination with other flatfish traits. Its position in the sequence of evolutionary events, prior to eye migration, is what makes eye migration adaptive and hence evolutionarily possible.

An account of flatfish eye placement in terms of its current usefulness is not wrong, but it is possibly misleading, and in any case sorely incomplete, prompting the sort of objection raised by Mivart. A Darwinian explanation requires a backstory – back to an ancestor that had symmetrically placed eyes; and then forward from there, through a careful – just the right, so to speak – sequence of stages.

The problem of incipient stages – requiring a backstory – arises for various reasons. In the case of flatfish eye migration, it has to do with so-called 'epistatic' interactions, where the fitness contribution of a trait depends on the presence or absence of other traits. In the case of 'sign epistasis', the fitness contribution of a trait is positive or negative (plus 'sign' or minus 'sign') or zero depending on the presence or absence of another trait (Weinrich, Watson and Chao 2005; Weinrich et al. 2006; Poelwijk et al. 2007). For example, eye migration is fitness-enhancing in combination with head elevation, but fitness-neutral, or more likely fitness-diminishing alone. This particular kind of epistatic interaction results in there being multiple, sequential pathways to an optimal outcome, some of which are traversable by natural selection and some not. Which is to say, again, that it is not enough to attribute even highly adaptive traits to natural selection without also positing an ancestor and a carefully sequenced route from the ancestor to the descendant in question.

Consider another example of the problem of incipient stages that also points to the importance of backstory, but that arises and is resolved in a different way – different from the epistasis case. It concerns the evolution of wings. Mivart is often said to have asked 'What use is half a wing?' Mivart did not say that (at least not in the text regularly cited), and that does not sound like his manner of posing the more general problem. Stephen Gould (1991) had a better way of putting Mivart's point: the

'5 percent of a wing' problem. How could evolution by natural selection of slight modifications lead from wingless, flightless ancestors to winged, flying descendants? How could the miniscule, incipient wing-lets have been sufficiently useful for flying in order to be favoured by natural selection? Mivart concluded, 'It is difficult [...] to believe that the Avian limb was developed in any other way than by a comparatively sudden modification of a marked and important kind' (1871: 107). Which, again, violates Darwin's 'successive, slight modifications' constraint.

The most promising solution in this case is one that Darwin himself proposed, and illustrated with the case of wings.[4] The basic idea is that the incipient stages of the trait in question were useful in a (perhaps very different) way than the later stages. 'In considering transitions of organs, it is so important to bear in mind the probability of conversion from one function to another' (Darwin 1872: 183).

The structures that were eventually modified for flight might have served a variety of other functions, depending on the animal in question (e.g., insects vs birds) and depending on issues of scale. Darwin himself suggested that the thoracic wings of some insects might be modifications of parts originally related to respiration. Wings of insects and birds might, in their incipient stages, have served a variety of aerodynamic uses other than flight, like gliding and altitude control during descent. Narratives of the evolution of bird flight generally begin with the modification of four-legged ancestors into bipedal descendants, followed by modification of the forelimbs into wings. On some accounts, the initial modifications improved running and jumping (e.g., by improving balance). On other accounts, the modification followed tree climbing, and served aerodynamic uses related to descent mentioned above.

20.4   When Narratives Are Worthwhile

The evolution of flatfish eyes, wings and many other traits are narrative-worthy, in a sense I'll now explain. But first: I'm going to rely on a fairly minimal view of narratives, namely that they relate what happened, one event at a time. In this regard I'm following the lead of narrative theorists who adopt similarly minimal views of what counts as a narrative and who

---

[4] Gould 1991 and Brandon 1990 are nice analyses of the issues involved in the stepwise evolution of wings. Both focus on the now-classic work of Kingsolver and Koehl (1985; see also 1994) on the evolution of insect wings. Gould puts it in the context of Mivart's problem of incipient stages, and Darwin's solution of functional shift, while Brandon uses it to illustrate the character of explanations. Garner, Taylor and Thomas (1999) includes a useful presentation of the main theories of the evolution of avian flight and the sequences of trait acquisition that the alternative theories require.

concede that almost anything is narratable, but who deny that everything narratable is worth narrating. Some narratives are pointless. As William Labov famously commented:

Pointless stories are met (in English) with the withering rejoinder, 'So what?' Every good narrator is continually warding off this question; when his narrative is over, it should be unthinkable for a bystander to say, 'So what?'

There are a great many ways in which the point of a narrative can be conveyed – in which the speaker signals to the listener why he is telling it. To identify the evaluative portion of a narrative, it is necessary to know why this narrative – or any narrative – is felt to be tellable. (Labov 1972: 366, 370)

Labov's term 'tellable' has become the state of the art (narrative theory) term for what I prefer to call narrative-worthy. But I like his term 'pointless stories'. What is a narrative-worthy as opposed to a pointless story? The criteria that I offer may just be a few of many criteria for narrative-worthiness. Perhaps there are stories worth narrating that do not meet the following criteria but are worth telling for other reasons.

For starters, I'd say narratives are good for situations where we don't know – on the basis of what has already happened, and general principles – what will happen subsequently, and we need to be told.[5]

To clarify, this does not render pointless all of those stories where the narrator begins with the ending. Most historical narratives, in both civil history and natural history, begin with the outcome, and the narrator then proceeds to tell how it came about. Rather, the criterion calls into question the need for narrating how an outcome came about, when the outcome was already foreseeable from the initial events.[6]

An example of a situation where narratives are not particularly useful – where they do not serve this basic function – involves equilibrating/optimizing processes like the one discussed earlier of a marble coming to rest in the bottom of a bowl. Why narrate its trajectory – 'it was there, then it was there, then there' – if we can derive from the start where the marble will end up? And regardless of where it started from.

Similarly, why narrate the evolution of a species in an environment if we 'know'/suppose that it will eventually reach its predictable optimal state given that environment. And regardless of where it started from?

Whereas to make sense of flatfish eyes, wings, etc. in terms of evolution by natural selection, one must provide a backstory – back to a presumed ancestor, and then the sequence of stages moving forward. And these stages were hardly guaranteed by what preceded them. They could hardly be derived from past

---

[5] Crasnow (Chapter 11), links such narrative-worthiness to the work of tracing and casing.

[6] Andersen (Chapter 19), uses the notion of 'scripts' to argue that mathematicians skip precisely such foreseeable sequences when reading mathematical proofs.

circumstances. Flatfish head elevation was hardly foreseeable from the point at which their ancestors first lay flat on their sides on the sea floor. It was certainly not foreseeable by generations of naturalists who contemplated the evolution of flatfish eye placement. Nor, if one were to start the backstory earlier (as I'll discuss shortly) would it be predictable that predatory fishes inhabiting the sea floor would adopt the behaviour of lying flat on their sides, given that many bottom-dwelling predators (e.g., groupers) never have.

Consider the prominence of narratives in Darwin's work, and what makes them so worthwhile. Their employment, and their value reflect in part Darwin's view that, outside of gradual adaptation to environmental circumstances, there is nothing inevitable in the history of life.

I believe in no fixed law of development, causing all the inhabitants of a country to change abruptly, or simultaneously, or to an equal degree. The process of modification must be extremely slow. The variability of each species is quite independent of that of all others. Whether such variability be taken advantage of by natural selection, and whether the variations be accumulated to a greater or lesser amount, thus causing a greater or lesser amount of modification in the varying species, depends on many complex contingencies, – on the variability being of a beneficial nature, on the power of intercrossing, on the rate of breeding, on the slowly changing physical conditions of the country, and more especially on the nature of the other inhabitants with which the varying species comes into competition. (Darwin 1859: 314)

[I]f we must marvel, let it be at our presumption in imagining for a moment that we understand the many complex contingencies, on which the existence of each species depends. (Darwin 1859: 322)

These 'many', unforeseeable 'complex contingencies' need to be added to each evolutionary narrative, in the order they arise.

So, again, a good occasion for a narrative is when we don't know what will happen next and need to be told. But that still leaves room for a lot of narratives not worth telling, pointless. The events worth including are not just those that we would not have foreseen otherwise. They should also be consequential. Otherwise what is the point of including them in the narrative?

Note that, in the last of the Darwin quotes above, he refers not only to the 'many complex contingencies' that arise in the course of the evolution of each species, but those contingencies 'on which the existence of each species depends', i.e., which are consequential for the evolution of each species. These two facts about evolutionary history correspond to two reasons why narratives are so appropriate for making sense of evolutionary outcomes: the unpredictability of the events narrated, and their consequential character.[7]

---

[7] Elsewhere (Beatty 2006; 2016; 2017), I have discussed these two criteria of narrative-worthiness in terms of the events narrated being *contingent* (or *contingent* per se) – unpredictable, matters of chance – and in terms of the narrative outcome being *contingent upon* – dependent upon – those

Philosopher of history William Gallie stressed the importance of the two criteria in his reader-centric view of what makes a narrative 'worth following'. In the first paragraph below he stresses the otherwise unpredictable elements that we rely on narratives to supply. In the second paragraph, he stresses that the events in a worthwhile narrative are consequential for the outcome.[8] Generally, in narratives,

[...] there is a dominant sense of alternative possibilities: events in train are felt to admit of different possible outcomes – particularly those events that count [...] that deserve to be recorded, that could be the pivot of a good story. [...] [S]ide by side with this there is the recognition that many events, or aspects of events, are predictable either exactly or approximately. But, although recognised, this predictable aspect of life is, so to speak, recessive or in shadow. It is in contrast to the generally recognised realm of predictable uniformities that the unpredictable developments of a story stand out, as worth making a story of, and as worth following. [Or, in other words, using the terminology of footnote 7 (which I said I was trying to avoid, but I just can't help myself), the *contingent* (or *contingent* per se) developments are 'worth making a story of'.]

[O]f [still] greater importance for stories than the predictability relation between events is the converse relation which enables us to see, not indeed that some earlier event necessitated a later one, but that a later event required, as its necessary condition [i.e., that it was *contingent upon*], some earlier one. (Gallie 1964: 26; my emphasis)

I like to represent these two features of narrative-worthy stories with a branching tree of possibilities (Figure 20.4). In this world, the occurrence of event A leaves open the possibility of either B1 or B2. The occurrence of B1 leaves open the possibility of either O1 or O2 and forecloses the possibility of B2 and along with it O3 and O4. A–B1–O2 is one possible history in this world, A–B2–O4 another. There are multiple possible histories in this world; only one can come to pass.

Let's say it was A–B1–O2. B1 was not derivable from A; B2 might have occurred instead. Moreover, B1 was consequential – it made a difference; had B2 occurred instead, O2 would not have occurred. In the literature on narrative theory, events like B1 are often referred to as turning points or branch points (Beatty 2016: 36–37 and references therein). As far as evolutionary narratives are concerned, 'A' stands for the ancestral state with which the backstory begins, and the 'O's' stand for alternative evolutionary outcomes.

---

events. I have focused on those two uses of the term 'contingent', and the significant differences between them, because the terminology is ubiquitous in the biological literature, and because the two uses can be and are conflated. This point has been pretty well received; Griffiths (Chapter 7) uses it to articulate the differences between the Darwins' plant research and that of Julius Sachs. Nonetheless, here I am trying to see what good or ill comes from dropping that terminology in favour of the language I have substituted in the text above.

[8] Hajek (Chapter 2) proposes, by contrast, that the consequence of events in scientific narratives can also derive from meta-diegetic considerations.

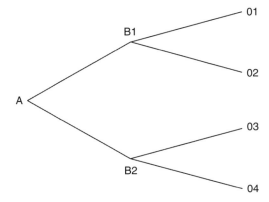

Figure 20.4 **Branching-tree representation of narrative-worthy stories**

The diagram helps to show how the sequence A–B1–O2 counts as an explanation of O2 on the prominent 'counterfactual difference-making' conception of explanation. As James Woodward expresses the basic idea:

An explanation ought to be such that it enables us to see what sort of difference it would have made for the explanandum if the factors cited in the explanans had been different in various possible ways (Woodward 2003: 11)

[A] common element in many forms of explanation, both causal and non-causal, is that they must answer what-if-things-had-been-different questions. (Woodward 2003: 221)

The occurrence of B1 helps to explain O2, in the sense that, had B1 not occurred (had B2 occurred instead), then O2 would not have resulted. Whether B1 or B2 occurs makes a difference.

There is a case to be made that worthwhile narratives include, at least implicitly, what did not occur as well as what did, at least some of the counterfactual as well as the factual sequences of events. But I won't press that case here (see Beatty 2016; 2017). At the very least, to see the worth of a narrative is to consider what did not happen and thereby see that there were consequential turning points, which, again, contributes to the explanatory character of the narrative.

Figure 20.5 shows a branching time representation of flatfish evolution. The acquisition of head elevation is a turning point that was not inevitable given past events, and that was consequential for the outcome. The order of events here is crucial. It is not enough to consider the three traits in question purely contemporaneously. Yes, they work well together, but that does not explain their presence. The acquisition of the trait, lying flat, made possible the evolution of head elevation for lunging, which in turn made possible the evolution of eye migration.

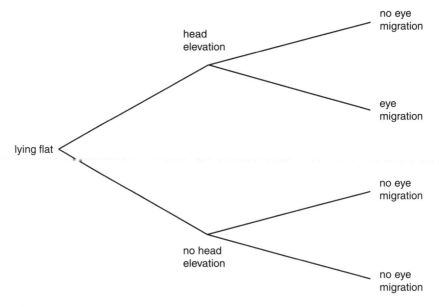

Figure 20.5 **Branching time representation of flatfish evolution**

20.5   Conclusion

In order to understand an occurrence, we sometimes need a backstory that rewinds time to some event in the more distant past, and then takes us forward through events that (1) were not foreseeable from the starting point, and (2) were consequential for the outcome of interest, (3) in the order in which they occurred and not just any order. We need a backstory that is narrative-worthy in these respects. Such a backstory is explanatory.

I'll end with a question that may have occurred to you already, namely how far back should the backstory go? I'm not sure there is a definitive answer to this. But surely it depends in part on the question being asked, or what counts as puzzling.

Gerd von Wahlert rewound the flatfish evolution clock back beyond their ancestors' horizontal lifestyle – the point at which the narratives above begin – to their more distant ancestors' vertical lifestyle. According to his narrative, the ancestors then evolved to rest/'sleep' lying flat. (Yes, some fishes rest/sleep on their sides.)[9] And subsequently evolved a horizontal lifestyle. He proposed this

---

[9] Aquarium owners may be familiar with the 'beds' or 'hammocks' or 'pads' that can be attached to the glass so that they can see their pet 'betta' fish napping, often lying on their sides.

as a solution to the puzzle of flatfish eye placement, and a way to avoid Goldschmidt's hopeful monster scenario.

The flatfish are usually cited not only as a paradigm of adaptation to benthonic life but frequently as a case of an unexplainable major evolutionary step; they are referred to by Goldschmidt as owing their origin to a 'hopeful monster'. Analysis of their structure and their habits has, however, revealed a simpler story (Wahlert 1961). A shift from an upright to a horizontal sleeping position occurred in the symmetrical ancestors of the flatfish; sleeping on either side is done in some of the present-day symmetrical acanthopterygians, such as triggerfish and wrasses. If this sleeping position were maintained as a resting or hiding position after the animal awoke, a shift of the eye from the blind towards the upper-most side would be an advantageous modification. The shift of the eye on the blind side to the margin of the head would enable the fish to scan the waters above it with binocular vision (Wahlert 1965: 290).

But von Wahlert's suggestion hardly solves – hardly addresses – the questionable adaptive value of the initial stages of eye migration, and that Goldschmidt (and Mivart) tried to circumvent by invoking an all-at-once transformation. On the other hand, starting with the deeper ancestral state of a vertical lifestyle, as von Wahlert does, followed by the evolution of horizontal *resting*, does seem a promising solution to a different puzzle, namely how flatfishes acquired a horizontal *lifestyle* in small steps each of which was selectively favoured. That is, they spent more and more waking time in what was previously just a resting posture, taking more and more advantage of that less conspicuous and motionless position to avoid predators and surprise prey.

And this has the elements of a worthwhile narrative. The acquisition of horizontal resting was hardly guaranteed. Indeed, von Wahlert offers no suggestion as to how it came about. His narrative discloses what was not foreseeable, a basic function of a worthwhile narrative. Moreover, once disclosed, the acquisition of horizontal resting serves as a counterfactual difference-maker in his narrative; it is consequential. Consider the counterfactual alternative: that horizontal resting was not acquired prior to acquisition of a horizontal lifestyle. It is not at all clear how the small steps from a vertical lifestyle directly to a horizontal lifestyle could be advantageous. What could be the advantage of tilting just slightly from vertical to horizontal?

As for the gradual transformation from a vertical lifestyle to horizontal resting, well, any suggestions?[10]

[10] I am very grateful for the opportunity to participate in the Narrative Science Project. Thank you Mary, Kim and Dominic, and thanks to fellow participants for their feedback and wealth of perspectives. I'm very lucky to join you in this volume. I was also lucky to contribute to the 2017 special issue of *Studies in History and Philosophy of Science, Part A*, devoted to scientific narratives, and edited by Mary and Norton Wise. *Narrative Science* book: This project has received funding from the European Research Council under the European Union's Horizon

References

Beatty, J. (2006). 'Replaying Life's Tape'. *Journal of Philosophy* 103.7: 336–362.

(2016). 'What Are Narratives Good For?' *Studies in History and Philosophy of Science Part C* 58: 33–40.

(2017). 'Narrative Possibility and Narrative Explanation'. *Studies in History and Philosophy of Science Part A* 62: 31–41.

Brandon, R. (1990). *Adaptation and Environment*. Princeton, NJ: Princeton University Press.

Darwin, C. (1859). *On the Origin of Species by Means of Natural Selection, or the Preservation of Favoured Races in the Struggle for Life*. 1st edn. London: John Murray.

(1872). *On the Origin of Species by Means of of Natural Selection, or the Preservation of Favoured Races in the Struggle for Life*. 6th edn. London: John Murray.

Dawkins, R. (1986). *The Blind Watchmaker: Why the Evidence of Evolution Reveals a Universe without Design*. New York: W.W. Norton.

Frazzetta, T. H. (2012). 'Flatfishes, Turtles, and Bolyerine Snakes: Evolution by Small Steps or Large, or Both?' *Evolutionary Biology* 39.1: 30–60.

Friedman, M. (2008). 'The Evolutionary Origin of Flatfish Asymmetry'. *Nature* 454: 209–212.

Gallie, W. B. (1964). *Philosophy and the Historical Understanding*. New York: Schocken.

Garner, J. P., G. K. Taylor and A. L. R. Thomas (1999). 'On the Origins of Birds: The Sequence of Character Acquisition in the Evolution of Avian Flight'. *Philosophical Transactions of the Royal Society of London. Series B* 266.1425: 1259–1266.

Goldschmidt, R. (1933). 'Some Aspects of Evolution'. *Science* 78.2033: 539–547.

(1940). *The Material Basis of Evolution*. New Haven, CT: Yale University Press.

Gould, S. J. (1991). 'Not Necessarily a Wing'. In *Bully for Brontosaurus: Reflections in Natural History*. New York: W.W. Norton, 139–153.

Gould, S. J., and R. C. Lewontin. (1979). 'The Spandrels of San Marco and the Panglossian Paradigm: A Critique of the Adaptationist Programme'. *Proceedings of the Royal Society of London. Series B* 205.1161: 581–598.

Kingsolver, J., and M. A. R. Koehl. (1985). 'Aerodynamics, Thermoregulation, and the Evolution of Insect Wings: Differential Scaling and Evolutionary Change'. *Evolution* 39.3: 488–504.

(1994). 'Selective Factors in the Evolution of Insect Wings'. *Annual Review of Entomology* 39: 425–451.

Labov, W. (1972). *Language in the Inner City*. Philadelphia: University of Pennsylvania Press.

Lamarck, J.-B. (1914 [1809]). *Zoological Philosophy*. Trans. Hugh Samuel Roger Elliott. London: Macmillan.

Malm, A. W. (1867). 'Bidrag til kannedorn af Pleuronektoidernas utveckling och byggnad'. *Kongliga Svenska Vetenskaps-Akademiens Handlingar* 7: 128.

2020 research and innovation programme (grant agreement No. 694732). www.narrative-science.org/.

Maynard Smith, J. (1993). *The Theory of Evolution*. 3rd edn. Cambridge: Cambridge University Press.

Maynard Smith, J., R. Burian, S. Kauffman, P. Alberch et al. (1985). 'Developmental Constraints and Evolution: A Perspective from the Mountain Lake Conference on Development and Evolution'. *Quarterly Review of Biology* 60.3: 265–287.

Mivart, St. George Jackson (1871). *On the Genesis of Species*. New York: Appleton & Co.

Olla, B. L., R. Wicklund and S. Wilk (1969). 'Behavior of Winter Flounder in a Natural Habitat'. *Transactions of the American Fisheries Society* 98.4: 717–720.

Poelwijk, F. J., D. J. Kiviet, D. M. Weinreich and S. J. Tans (2007). 'Empirical Fitness Landscapes Reveal Accessible Evolutionary Paths'. *Nature* 445: 383–386.

Policansky, D. (1982). 'The Asymmetry of Flounders'. *Scientific American* 246.5: 116–123.

Stickney, R. R., D. B. White and D. Miller (1973). 'Observations of Fin Use in Relation to Feeding and Resting Behavior in Flatfishes (Pleuronectiformes)'. *Copeia* 1: 154–156.

Suzuki, T. K. (2017). 'On the Origin of Complex Adaptive Traits: Progress since the Darwin vs. Mivart Debate'. *Journal of Experimental Zoology B* 328: 304–320.

Wahlert, G. von (1961). 'Die Entstehung der Plattfische durch ökologischen Funktionswechsel'. *Zoologische Jahrbücher. Abteilung für Systematik, Geographie und Biologie der Tiere* 89: 1–42.

(1965). 'The Role of Ecological Factors in the Origin of Higher Levels of Organization'. *Systematic Biology* 14.3: 288–300.

Weinrich, D. M., N. F. Delaney, M. A. DePristo and D. L. Hartl (2006). 'Darwinian Evolution Can Follow Only Very Few Mutational Paths to Fitter Proteins'. *Science* 312.5770: 111–114.

Weinreich, D. M., R. A. Watson and L. Chao (2005). 'Sign Epistasis and Genetic Constraint on Evolutionary Trajectories'. *Evolution* 59.6: 1165–1174.

Woodward, J. (2003). *Making Things Happen: A Theory of Causal Explanation*. Oxford: Oxford University Press.

Zimmer, C. (2008). 'The Evolution of Extraordinary Eyes: The Cases of Flatfishes and Stalk-Eyed Flies'. *Evolution: Education and Outreach* 1.4: 487–492.

21 Just-so What?

Paula Olmos

Abstract

This chapter examines the criteria exposed by Stephen Jay Gould's original paper on *just-so stories* to sustain such a charge. I show that Gould's concerns were neither directed to narrative explanations nor were they ineluctably linked to their narrative quality. Then I analyse how advocates of narrative science have met the challenge. I identify two basic defensive approaches: the vindication of explanatory narratives in cases where the historical, contingent and causally complex nature of the phenomena demand a narrative approach and an *unveiling* strategy showing how there's a narrative *behind* each law-like generalization or nomological explanatory formula. The chapter's concentration on the argumentative moves of the discussants helps clarify their positions. Moreover, the argumentative quality of their object of study (scientific reason-giving practices) is also emphasized. I claim that the *dialectical requirement* of openness to collective survey and discussion is what may prevent *just-so charges* for any kind of explanatory model.

21.1   Introduction

The recent interest shown by philosophers of science and scholars in related fields concerning the narrative qualities of our scientific explanatory practices has not sufficiently addressed a widespread reluctance to recognize narrative's epistemological significances. On many occasions, this reluctance is marked by the derogatory use of the '*just-so story*' label (Gould 1978; Gould and Lewontin 1979) to signify that narrative explanations – or other narrative reason-giving practices (cf. Olmos 2019) – do not meet the epistemic criteria required for scientific appraisal.

In many philosophical *forums*, it is now standard to present a stark opposition between allegedly genuine scientific explanations – invoking a well-established and well-delimited, ideally law-like account, which is amenable to formalization, perhaps including a causal mechanism[1] – and *just-so stories* – reconstructive,

---

[1] Although there are important differences and entrenched discussions between philosophers of science who emphasize the role either of laws, of formalizable statistical relations, or of

typically untestable, conjectures of what *just-in-fact* may or may not have happened to cause a particular phenomenon. However, the problem with the widespread use of this opposition is that it tends to create a strong and somewhat easy association between the noun *story* and the qualification *just-so*, so that every attempt to approach explanatory and justificatory tasks through narrative form within the sciences is easily and cursorily dismissed with the *just-so story* derogatory term. In the worst cases, the use of this summary label even tends to prevent further discussion, acting as a dialectical blockade.[2]

In this chapter, I examine the roots of the *just-so charge* by going back to its now classical source in Stephen J. Gould's original paper (1978) from which it spread in the history and philosophy of science (HPS) field as a negative evaluative term. As it is well known, in choosing this denomination, Gould was inspired by Rudyard Kipling's collected children's tales, *Just So Stories* (1902), containing twelve whimsical etiological fables.[3] Gould tried, thus, to convey to the general public the idea of the unscientific and boldly imaginative nature of his own target examples, namely evolutionary biological accounts.[4]

However, a careful reading of Gould's piece shows that his concerns were neither directed to narrative explanations nor were they, in any case, ineluctably linked to their potential narrative quality. For Gould, assessing an allegedly scientific account as a *just-so story*, was to issue a negative evaluative judgement

describable causal mechanisms in the conformation and appraisal of scientific explanations, all these conceptual possibilities share a ring of respectability within mainstream programmes of study that discussions on narrative models have not yet fully attained. The concerns of the so-called 'new mechanists' (Craver and Tabery 2019) seem closer to narrative science discussions than are the more traditional emphases on nomological and Bayesian models. And yet, important suggestions made from the narrative ranks might, as Crasnow (2017) shows, improve and qualify the mechanistic approach.

[2] As several scholars have noticed, even if this dichotomy and its association with narrative explanatory models does not usually appear as such in published papers on epistemology or philosophy of science, it is still a widespread prejudice that is academically very effective. See, for example, Currie and Sterelny (2017: 16 n. 7): 'These complaints are not often found in the published literature, but both of us have met it *regularly* in conversation, and one of us *regularly* in referee's reports on his narrative-based explanations of hominin evolutionary history' (my emphasis on '*regularly*').

[3] Available at www.gutenberg.org/files/32488/32488-h/32488-h.htm. Kipling's work includes stories such as 'How the Rhinoceros Got His Skin' or 'How the Leopard Got His Spots' and tries to respond to children's typical pressing questions by providing fantastic accounts of how a certain individual of a species (the fable's protagonist) got a particular trait that's now common to all in the tradition of the etiological fable. One of the most renowned tales in the book, 'The Elephant's Child', stands out as interestingly self-referential regarding the book's own theme, as the protagonist child elephant, full of 'satiable curtiosity' (as Kipling's child-like spelling runs), gets its unattractive but very useful trunk precisely for asking questions and being inquisitive.

[4] The success of Gould's felicitous denomination is obviously also due to the coincidence between Kipling's themes and evolutionary biological research. On the narrative difficulties of making particular evolutionary accounts, see J. Beatty's paper (Chapter 20). A recent paper by Hubálek (2021) on *just-so stories* focuses precisely on the central role of the particularities of evolutionary science in the configuration of this topic.

regarding its claim to epistemic relevance (as being a bold and so-far unwarranted conjecture) or to point out its way of presentation as avoiding further discussion and further testing (being an unfalsifiable, self-contained hypothesis).

These conditions, I claim, should not be confused or equated with the discursive and causal narrative quality of an explanatory scheme or excluded from the realm of more classically understood explanations that could also, in the mentioned senses, be *just-so* as well. The narrative quality of many of our scientific reason-giving practices is not, in and of itself, a way to avoid the identification of causal relations (even mechanisms, depending on how we define them; cf. Crasnow 2017) or to exclude further discussion or testing (Al-Shawaf 2019). On the contrary, it might be part of what is hypothesized of certain scientifically interesting phenomena, both in the sense of making them dependent on long-term processes (as witness the timely *historicization* of certain natural or social enquiries, at a certain point in their development) or on a complex, highly contextual and somewhat indeterministic causal web that's *better* rendered in a narrative form.

In what follows, I will carefully examine and analyse Gould's points and then come back to current discussions regarding the use of narratives and narrative reason-giving modes within the sciences where advocates of the epistemic relevance of the topic have felt the need to meet the challenge of the *just-so charge*. I identify two basic defensive approaches. One is the vindication of the use of explanatory narratives in cases where the historical, contingent and causally complex nature of the phenomena involved demands an approach that would avoid the strictures of classical models. When these conditions obtain, scientific narratives might be *less just-so* (i.e., less bold, less self-contained) than their too-narrowly understood mechanistic or easily formalizable rivals. This is basically what is claimed about their case studies by Crasnow (2017) and Currie and Sterelny (2017). Nonetheless, authors working along this line, usually propose some kind of collaboration or integration between these different epistemic modes and tools.

The second kind of vindication[5] of narrative science follows, instead, an unveiling (somewhat genealogical) strategy, hinting at a deeper level of narrativity. Scholars taking this approach (Richards 1992; López Beltrán 1998; Rosales 2017) try to show how there's a narrative – or at least a narrative kind of rationality, in W. Fisher's sense (1989) – behind (or before) each law-like generalization or nomological explanatory formula. This kind of narrative, that depicts and delimits the scenarios in which the particular nomological expression might acquire some sense and specifically become useful for drawing scientific conclusions,[6] is usually obscured and disregarded in its current application as a validated theory. However, it may always re-emerge when the formula comes under scrutiny as an explanatory

---

[5] Which, as can be seen, can be traced back to the early 1990s and, thus, antecedes the current discussions of the 'new mechanists'.

[6] Toulmin's (1953: 51–93) characterization of 'scientific laws' not as traits of nature but as restrictedly applicable and practical inference rules might be of help here.

principle (sometimes, as in Rosales's case study, in comparison with rival theories), which makes it a crucial part of its deep understanding.[7]

As in previous contributions (Olmos 2018; 2019), I approach all these topics with the tools and conceptual framework of argumentation theory, that takes into account the argumentative nature of our discursive, explanatory and justificatory practices in terms of reason-giving, reason-asking and reason-discussing activities. The philosophers of science whose works I examine support their claims with (obviously non-demonstrative) reasons and concentrating on their argumentative moves in these discussions helps clarify their positions.[8] But we must also take into account that their very object of study (scientific explanation and justification) is also of an argumentative, reason-giving nature. The way the grounds of *this* argumentative activity – i.e., scientific justificatory or *forensic* practice, in John Woods's (2017: 143–144) terms – should be assessed is what is finally at stake in philosophical discussions regarding the use of narratives in science and their alleged vulnerability to the *just-so charge*.

A final step that the argumentative approach might help us take is based on the *naturalistic* assumption that scientific argumentative activities are already intrinsically normative and evaluative in nature, so that, even if philosophers might discuss the criteria for the acceptability of the concerned claims and *explanantia*, there is already an intra-scientific evaluative activity going on, whose most basic rule, well beyond the strictures of any aprioristic model, is the *dialectical requirement* of (a posteriori) openness to collective survey and discussion.[9]

A well-understood *just-so charge* will have more to do with possible violations of this basic rule than with the textual and formal characteristics of proposed and supported scientific arguments and explanations.

21.2   The *Just-so Charge*

As already mentioned, the now classical reference for the *just-so charge* is palaeontologist and evolutionary biologist Stephen Jay Gould's (1978) critical article on what he saw as the excesses of certain trends in sociobiology, published in the *New Scientist* under the title 'Sociobiology: The Art of Storytelling'. For Gould, the question was whether and when evolutionary

---

[7] As Mary Morgan has defended (2001: 369), 'the identity of the model is not only given by the structure (or the metaphor), but also the questions we can ask and the stories we can tell with it'.

[8] A genuine *locus classicus* for the argumentative (as opposed to demonstrative) nature of philosophical discourse is Friedrich Waismann's 'How I See Philosophy' (Waismann 1968: 30).

[9] According to Hansson (2017), for example, a minimal criterion of science would be: '*Science is a systematic search for knowledge whose validity does not depend on the particular individual but is open for anyone to check or rediscover*'.

scientists (sociobiologists, particularly) were being excessively speculative and overconfident with their imaginative accounts about the historical origins of the traits they studied.

Gould's target cases included, in the first place, certain explanations of animal behaviour that he analysed as solely based on their 'consistency with natural selection' or 'adaptationism' (Gould 1978: 531). As his first example, Gould picked up David Barash's explanation of the greater aggressiveness of male mountain bluebirds towards other males approaching their nest before rather than after eggs have been laid. According to Gould, Barash's proposed *explanans* to this phenomenon (that he documented, also according to Gould, with rather scarce data) was that this behaviour was advantageous (and so adaptive) as long as it reserved aggression (a costly attitude) to periods where the male was not yet sure to have passed on his genes. This was 'consistent with the expectations of evolutionary theory' (Gould 1978: 531) and that was nearly all there was to it.

Gould's criticism of this case included demands for more data and more tests, exploration of alternative (also testable) explanations and, most significantly, a call for certain control or restraint on the part of the scientist in drawing further conclusions based on his hypothesized *explanans*. Barash was particularly due criticism as he had gone so far as to suggest this was also a way to understand human foibles regarding adultery (Gould 1978: 531). Barash's *hasty jump* from mountain bluebirds to humans brought him near the realm of *sociobiology* and so allowed Gould to introduce his real main target.

This was Gould's illustration of a just-so story where the story-like quality of the negatively evaluated explanans was not really in its narrative nature (which was somewhat missing here), but rather in its far-fetched imaginative play. So Gould's complaints were directed neither against reconstructive *historicized* explanations that, he assumed, were the goal of evolutionary theory in general, nor against the indication of complex (multifactorial), entangled (non-linear) and somewhat indeterministic causal webs behind a target phenomenon. On the contrary, he explicitly opposed panselectionism or panadaptationism and advocated a less rigid version of natural selection that would 'grant a major role to other evolutionary agents (genetic drift, fixation of neutral mutations, for example)' (1978: 531).[10] Although he displayed some irony about the changing styles of evolutionary stories presented in biology – showing the workings of

---

[10] Professor J. Huss (Chapter 3) kindly suggested that a way to understand the place of this piece within Gould's maturing conception of evolutionary theory is as belonging to a transitional stage between his attempts at a nomothetic and computable approach to palaeobiology (Gould et al. 1977) – that Huss (2009) has studied as the 'MBL Model'– and his definitive emphasis on historicity, contingency and causal pluralism exposed in his classical *Wonderful Life* (Gould 1989). See also Turner and Havstad (2019).

a favoured kind of adaptive factor as it gets theoretically trendy – he seemed to assume this feature as a rather inevitable condition of scientific development.

Now, what was exactly Gould's problem with sociobiology in particular? Regarding the aforementioned mountain bluebirds example, Gould said that it presents 'a perfectly plausible story that may well be true. I only wish to criticise its assertion without evidence or test, using consistency with natural selection as the sole criterion for useful speculation' (Gould 1978: 531). This is the first important thing: it is not a question of the stark unsuitability or unacceptability of a certain *methodology*, in the sense of a certain way of supporting a scientific content (a kind of reason) but of its *sufficiency* to establish its conclusion. That, according to Gould, this was happening in sociobiological explanations of human behaviour allegedly *more than* in other areas of evolutionary theory could be attributed to two additional difficulties met by this particular disciplinary approach: the little observational evidence available,[11] and, more significantly for our purposes, the *reductionistic* option for one specific kind of causal explanation (biological adaptive selection) for such highly complex phenomena as human behavioural traits whose etiological history was surely more entangled than that.[12] Along with this last point, the strictly selectionist sociobiological historical explanations of human behaviour Gould had in mind could be charged with the *just-so* label *precisely* for being more reductionistically mechanistic (based on the single principle, 'if adaptive, then genetic') than assumedly and sophisticatedly *narrative*!

The derogatory label, moreover, was the more emphatically attributed by Gould as he perceived that certain ideas taken from the theories advanced by sociobiology were currently being used to uphold practical and political implications as based on what he saw as their *hasty conclusions* (Gould 1978: 532). So here a certain *pragmatist modulation* of what seemed to start as a purely epistemological concern comes to the fore.

To sum up, according to my analysis, Gould suggested three (not completely independent but yet distinguishable) criteria to be taken into account for a *just-so charge* and none of them has to do with either the explanatory *historicization* of phenomena (their being explained not by the workings of constant laws of nature but by the particular detailed history behind them in conditions that may not be repeated) nor with the suggestion that the phenomena involved might be better understood contextually, taking in account the complexities of its entangled causal web, rather than in isolation. Quite the opposite, I would remark.

---

[11] As the saying goes, 'behavior doesn't fossilize' (Kurzban 2012).

[12] Including 'cultural evolution', with rather different causal workings, according to Gould (1978: 533).

These three *criteria* amount to: (a) a charge of *theoretical justificatory insuffi-ciency*; (b) a charge of *unwarranted reductionism*; and (c) a charge of *prescriptive justificatory insufficiency*. Let us analyse these three points separately.

Gould rejects the presentation of a particular historical, reconstructive explan-ation (not necessarily presented through a fully fledged *narrative*) as solely supported by its inner plausibility (understood as 'consistency with evolutionary expectations'). This would be a charge of *theoretical justificatory insufficiency* and can be mitigated by additional future evidence that would still use that initial plausibility as one of the reasons adduced in favour of the supported hypothesis or by a more humble presentation that would consciously advance it as a plausible hypothesis and offer it to the scientific community, assuming that a lot of research is yet necessary to issue a judgement on it – being still a valuable contribution for some reason (e.g., its novelty).

So this criterion (a), demanding additional robustness for establishing scien-tific theories, is not much more than a reminder of the collective rules of scientific research, organized scepticism, public scrutiny and assumed fallibil-ism. This much has been acknowledged by other scholars responding to Gould's piece: 'The goal should not be to expel stories from science, but rather to identify the stories that are also good explanations' (Kurzban 2012). What Gould is asking for here is just 'more evidence'.

Gould's complaint may be, then, argumentatively modelled as requiring for the sought-for conclusion (the assertion of the hypothetical reconstruc-tion) a *conjunction of additional arguments* (Marraud 2013a: 59–62) that, significantly, does not have to drop at all the initial one. Figure 21.1 represents such a conjunction of reasons justifying an evolutionary hypoth-esis, presented this time through a narrative, neither *solely* on the basis of its *narrative coherence* nor disregarding such coherence's contribution in supporting the conclusion.

I have used similar diagrams in previous works (Olmos 2019; 2020a). They are based on Marraud's (2016) interpretation and development of Toulmin's model (1958). In addition to the representation of co-orientated reasons (signalled by the connective 'besides'), they combine the use of justificatory reasons (allied with connective 'so') and explanatory reasons (allied with connective 'that's why') and, whenever needed to clarify both kinds of inferential steps, either justificatory or explanatory *warrants* (in Toulmin's sense) are provided in side boxes. Gould's concern here is structurally similar to Ian Hacking's criticism of the sufficiency of *abduction alone*, i.e., of the explanatory power of a hypothesis including an existential posit, to establish a 'realist claim' regarding the theoretical entity posited by it (Hacking 1983: 271–272). As I have shown elsewhere (Olmos 2018: 50), Hacking's suggestion might be argumentatively modelled as requiring for the sought for conclusion (the assertion of the hypothesis) a *conjunction of*

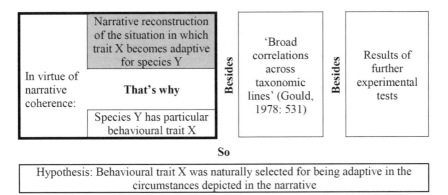

Figure 21.1 **Conjunction of reasons justifying an evolutionary hypothesis**

arguments, specifically including further experimental evidences amounting to the detection and manipulation of the posited entity.[13]

Gould also criticized the concentration on just one kind of causal mechanism (strictly understood natural selection in terms of adaptiveness is, as could be expected, Gould's *usual suspect*) on which to base a historical account of a complex phenomenon. This would be a charge of *unwarranted reductionism*, that is rather more serious as it might prevent rather than encourage research along other lines and easily provide a sense of overconfidence in a particular kind of explanation, precisely for its neat identification of one well-delimited responsible mechanism.

Criterion (b) is Gould's main epistemological point – although criterion (c) may be his main motivation. It is more a caution against *selectionist reductionism* (i.e., *panselectionism*) than any other thing, which is consistent with his well-known position in evolutionary biology (Gould and Lewontin 1979). The caution works this time as the conclusion of an a fortiori argument (Marraud 2013b) based on his own reservations with panselectionism in accounting for biological traits. We could reconstruct this a fortiori rationale behind Gould's case (see Figure 21.2).

This reconstruction, chosen for its clarity at this point in the discussion, could be much more refined if we take into account that it really works

---

[13] Laith Al-Shawaf has recently engaged in a defence of evolutionary psychology (Al Shawaf, Zreik and Buss 2018; Al Shawaf 2019) trying to respond to *just-so stories* charges, along lines rather coincident with my own analysis of criterion (a). The *just-so charge* would be misplaced when evolutionary psychologists do not only concoct and present their storied hypotheses but continue their experimental research and generate and test novel empirical predictions. Al-Shawaf claims that most published research in evolutionary psychology provides evidence and arguments *enough* along these lines and cannot be accused of presenting theories as just-so accounts.

Strict selection alone is insufficient to account for all physiological traits in simpler organisms (Gould's position)	The social behaviour of human beings is a phenomenon with a more complex history than the physiological traits of simpler organisms

The more complex the phenomenon, the more susceptible to be dependent on a multifactorial causal web:

So

Strict selection alone will be insufficient *in a greater degree* to account for the social behaviour of human beings

Figure 21.2 **A fortiori rationale behind the charge of unwarranted reductionism**

meta-argumentatively (Marraud 2013b: 10–12) as a comparison between two reason-giving acts (two explanatory accounts) that are connected by a *scalar topos* of the kind 'the more …, the more …' acting as the warrant of the comparison. This topos, makes possible that the characteristics attributed to one of these reason-giving acts (its being *insufficient* or *inadequate* in a case) be transferred *in an increased measure* to the compared one on the basis of some condition that places them in different positions on the comparison scale.[14]

So the problem this time with sociobiology's 'art of storytelling' is that it has *overconfidently* picked up a scientifically well-defined and well-understood *mechanism* (the 'selection of beneficial traits') and used it as a guide to reconstruct (allegedly too simplistically) the causal origins of some of the most complicated and intractable phenomena available. The *just-so charge* arises here as a charge against misguided scientificism, not against *narrative science*.

Gould finally warns us against a perceived as hasty use of theoretical results from natural science to support practical (even political) decisions. This would be a charge of *prescriptive justificatory insufficiency* that should be weighed in its own merits and according to pragmatic reasons modulated by considerations of risk (among other things).

The significance of criterion (c) (*prescriptive insufficiency*) can be understood (see Figure 21.3) as based on an additional a fortiori line of reasoning: 'if strict

---

[14] The meta-argumentative variety of a fortiori arguments is a scalar version of the meta-argumentative interpretation of analogy (cf. Woods and Hudak 1989) allowing the *simple transfer* (with no increase) of the characteristics attributed to an argument to another argument on the basis of their similarity.

Strict selection alone is insufficient to account for all physiological traits in simpler organisms (Gould's position)	Decisions on social policies should come after a collective value-based deliberation process taking into account people's preferences. Scientific understanding of phenomena plays a secondary role in them.

The more value-related the issue, the greater the demand for collective deliberation:

So

Strict selection alone will be insufficient (and marginal) *in a greater degree* to prescribe social policies

Figure 21.3 **A fortiori rationale behind the charge of prescriptive insufficiency**

selection alone is not enough to account for the actual social behaviour of human beings, how could it be enough to prescribe social policies?' I have again to thank Professor Huss for reminding me of Gould's membership of the Harvard left-wing group 'Science for the People' that held strong positions against the use of scientific results in justifying oppressing policies. This is consistent with my contention that criterion (c) was Gould's main motivation in making his case against sociobiology.

As already said, none of these hints at assessment criteria has really much to do with the *story-telling quality* of assumedly *narrative* models of explanation. After defending the scientific credentials of evolutionary psychology (see n. 13, above), Al-Shawaf asks himself 'why do so many people persist in the notion that evolutionary psychological hypotheses are just-so stories?' (Al-Shawaf 2019; cf. Al-Shawaf, Zreik and Buss 2018: 9). He attributes this mainly to the inescapable fact that evolutionary psychology (as astrophysics, cosmology and geology for that matter) has a central historical component and that historicity tends to be associated with untestability.

So it seems that the noun *story* easily attracts the charge *just-so*. Even if the *historicity* of the phenomenon addressed by many scientific disciplines is not a contested issue, the prejudice against *storied accounts* remains strong enough in many *forums* so as to extract protestations of scientific soundness in those addressed. Scholars interested in narrative science and narrative models of scientific justificatory practice feel, therefore, compelled to answer *just-so charges* even if those charges, when carefully examined, have not much to do with the particular characteristics of narratives and may even be based on just the opposite traits.

21.3   Defenders of Narrative Science Meet the *Just-so Charge*

Among the papers included in the 2017 special issue on 'Narrative Science and Narrative Knowing' of *Studies in History and Philosophy of Science*, edited by Mary S. Morgan and M. Norton Wise (Morgan and Wise 2017), it is significantly Sharon Crasnow's (Crasnow 2017: 6–13) and Adrian Currie and Kim Sterelny's (Currie and Sterelny 2017: 14–21) papers that make an explicit mention of the *just-so story charge* and try somehow to respond to or minimize it.

These two papers stand out in the collection as taking a meta-methodological approach while analysing their respective case studies. Both engage, particularly, in an epistemological appraisal of narratively dense and detailed accounts as opposed to certain efforts to base explanations regarding historically problematic phenomena (the 1898 Fashoda colonial incident, or the evolutionary development of human cooperation, respectively) in too restrictedly understood causal mechanisms or trajectories, amounting to formal models of explanation.

For these authors, narratives help better explore and understand the very causal relations expressed by those allegedly explanatory formal formulae, their contingent nature and the alternatives available at each historical turn. Both make reference to John Beatty's (2016) ideas about *what narratives are good for*, namely dealing with contingencies, alternative possibilities and the particulars of historical turning points, that are still the focus of Beatty's own contribution to the aforementioned special issue, in which he states that: 'Narratives are about not only what actually happened, but also what might have' (Beatty 2017: 31).

But what both papers finally depict in their case studies is not really a situation in which a narrative account of some phenomenon opposes a rival (in the sense of theoretically divergent) *narrowly mechanistic* account of the same phenomenon. The point is rather that certain kinds of identified or hypothesized causal links or mechanisms are *better* understood (explored and discussed) under a narrative rendering than under the crystallized mode of a formal formula or strict inference licence. So, as I will emphasize in the next section, the opposition (or comparison) is not so much *mechanisms* vs *narratives* but *narratives* vs *formalizable laws*.

Sharon Crasnow's paper focuses, in particular, on a case of historical political science, a discipline that, in principle, already accepts its narrative nature. Nevertheless, recent philosophical discussions regarding the requirement for scientific explanations to be based on causal mechanisms, emphasizing, moreover, the use of individual case studies as devices for *causal process tracing*, tend to be read as bringing social scientific disciplines to a point in which narratives might be *dissolved* in favour of a bounded search for discrete pieces of evidence that allow the operation of such allegedly well-identified, well-delimited mechanisms.

This is what Crasnow calls the 'inferential reading of process tracing'[15] that she opposes to the enriched use of narrative accounts of *those same* mechanisms and processes. Her claim is that narratives in political science might, in fact, help tracing causal processes and identifying mechanisms in a *better* way than restricted inferential-type readings, precisely because narratives focus on exploring and discussing contingencies and alternatives (Beatty 2016) and, thus, help making credible and understandable the inner and detailed workings of the very causal connections involved. Contrariwise: 'process tracing as a search for diagnostic pieces of evidence fails to capture the way that a mechanistic account seeks to address the inter-relationship of the parts in a way that narrative elements of case studies can' (Crasnow 2017: 8).

Crasnow acknowledges that narrative approaches to science have been challenged with the *just-so charge* that she equates with the notion of biased *cherry picking* (i.e., suppressing disconfirming evidence or biasedly selecting confirming evidence). Her suggestion to avoid *both* the problem and the charge is finally to *add substance*, detail and discussion of alternatives to narrative accounts, making them, if anything, even *more narrative*:[16]

One worry often raised about the use of case studies is the idea that it may devolve to cherry picking or just-so stories. This is indeed a concern but one that can be addressed by requiring that all of the relevant details of the case be considered and not just those that are relevant to the favored hypothesis. In order to assure that these details are addressed, alternative hypotheses – different ways that the story could have gone, different paths that could have been taken, different mechanisms through which the case can be understood – need to be explored. (Crasnow 2017: 10–11)

The four virtues that Crasnow ultimately associates with the narrative discussion of causal links or *process tracing* (i.e., closure, connectivity, elimination of alternatives and examination of counterfactual options) (Cranow 2017: 10–11) seem to be doing the work of avoiding too simplified accounts based on the biased selection of a restricted kind of evidence. This is more than consistent with Robert Richards's (1992: 41–42) suggestion that narratives (as opposed to *nomological* models of explanation) are the adequate vehicle for ordering and

---

[15] On the concept of 'process tracing', see Andrew Hopkins's and Sharon Crasnow's chapters (Chapters 4 and 11).

[16] Equating the *just-so charge* to a charge of *cherry picking* is a charitable (and dialectically fruitful) choice as it concedes that a narrative account (as well as any other scientific account) might be in need of further and more detailed justification regarding unmentioned or unqualified evidence (i.e., additional arguments). It would be part of a normal scientific evaluative discussion to check any account for *cherry picking*. As I have already said, sometimes the *just-so charge* tends to work in a more prejudiced and dialectically blocking way against certain modes of presenting scientific accounts. Responding to such attempts at blockade by charitably acknowledging that one is being asked to make a better case is a rather reasonable strategic move.

weighing (downgrading and emphasizing) the contributions of possibly many different causal links that could be invoked to account for a historically situated event.

Currie and Sterelny (2017) conduct an even bolder and more committed defence of the benefits of scientific story-telling. Their recipe, though, for obtaining such benefits without incurring *just-so charges* is a bit different. First of all, they are prepared to *defend speculation*, not anymore a vice whenever it yields the appropriate kind of Lakatosian fruitfulness (Currie and Sterelny 2017: 16). It is, precisely, on account of the value attributed to such fruitfulness that they acknowledge that Gould and Lewontin were probably right (or at least consistent) when criticizing strictly adaptionist hypotheses, because, as I have already remarked, these may tend to prevent rather than encourage research along other lines:

> Gould & Lewontin's complaints about adaptationist reasoning is in part clarified by this distinction: the charge of 'just-so' storytelling is in effect the charge of idle speculation: adaptationist hypotheses fail to open new investigative routes and actively discourage them (here is not the place to consider whether such a charge is plausible). (Currie and Sterelny 2017: 17 n. 11)

A second step in Currie and Sterelny's defence regards *coherence* as an epistemic virtue. Mere *internal coherence*, so to say, might be insufficient – although not thereby *negligible* in this respect – to support a historical reconstruction of the causal web leading to an explanandum-phenomenon. But, insofar as such a reconstruction is pressed (by scientific method and community) to cohere with all kinds of constraints, issuing from material discoveries, other reconstructions, general theories, etc. such *extended coherence* becomes a noticeable achievement. This idea may be understood as amounting to appraising *consilience* as a kind of master scientific virtue (Weinstein 2009) or, alternatively, as demanding from us a sufficiently flexible, assumedly multifactorial and open-ended, model of scientific argumentative assessment (Olmos 2020b) in which the contributions of different strategies (some possibly more narrative than others) may be weighed and, at least to a certain point, harmonized.

This last idea is much in line with Currie and Sterelny's final suggestion that their defence of scientific storytelling aims more at integration than substitution. The virtues and benefits of narrative approaches should *combine* with the virtues and benefits of formal models and the possible shortcomings of each of them be compensated by the other. And this is so because, as they try to show (although they do not express it with these words), some kind of *just-so charge* could be attributed to both. There might be *just-so stories* but there are also *just-so formal explanatory models* insofar as they unwarrantedly claim to be self-standing explanations.

This is what purportedly happens in their discussed example of a neatly modelled *threshold-dependent* explanation of the emergence of punishment in early human communities (sociobiology again). Such a clean, self-standing explanation, leaning on the pristine comprehensibility of the mechanism invoked is exposed as a *just-so* attempt, insofar as it is not taking into account enough contextual constraints as the emergence of other factors leading to human cooperation. What's missing here (according to Currie and Sterelny) is a good integrative narrative that's lost in the *decompositional* strategy of formalized (*narrowly mechanistic*) models:

Highly complex explananda like the evolution of human cooperation are resistant to approaches which depend solely on the decomposition and abstraction which enables modellers to probe aspects of constituent dynamics in isolation. For highly complex, multi-factorial, and multi-stage causal trajectories there are no master-models to be had, and so we must instead combine narratives and models, allowing us to navigate between the trade-offs generated by complexity. (Currie and Sterelny 2017: 20)

No wonder that Currie and Sterelny come to agree with Gould and Lewontin's complaints. Their *just-so* criticism, even if it was coupled with the noun *story*, was not directed towards the *storied* character of the accounts they criticized, but to their overconfident self-standing reliance on just one supposedly well-known and well-comprehended natural mechanism.

21.4   The Narrativity behind Nomicity

A somewhat different strategy to appraise narrative models of scientific explanation and justification is the one that exploits a kind of *genealogical* argument based on the idea that there's a narrative (or at least a narrative kind of rationality; cf. Fisher 1989) *behind* (or before) each law-like generalization or nomological explanatory formula that – even if it may be rather opaque and disregarded in its current application as a validated theory – may always re-emerge when the formula comes under scrutiny as an explanatory principle.

The point here is not that there's a *story* behind its *establishment* that may make it more understandable or even be part of its justificatory framework. These kinds of ideas would pertain to either the history of the discovery and acceptance of particular scientific laws and theories or more generally to what I have called the narrative account of scientific experimental and research activities (Olmos 2020a; cf. Meunier's paper in this volume on 'research narratives', Chapter 12). In this sense, there are recent significant case studies of how scientists themselves use a narrative rendering of their *interventions* and experiments (e.g., Mary Terrall's (2017) account of Réamur and Trembley's 'tales of quest and discovery' or M. Norton Wise's (2019) work on Faraday's series). However, this is not what I specifically want to focus on here.

The claim I want to examine is rather that any scientific law-like generalization would somehow depict and delimit the scenario of its own validity and applicability as based on considerations regarding the possibilities of isolating natural phenomena and letting them develop in a controlled setting and making them solely dependent on a relevant set of variables. Such scenarios and the assumptions that make them plausible and assessable would be *narrative* in the sense of describing what can be expected of either a spontaneous or a more or less controlled course of events. Invoking and exposing them in their narrative detail would be just what's needed whenever those generalizations, instead of being *just applied*, are discussed and weighed against alternative ones – which is something scientists involved in original research, as opposed to science teachers and appliers of scientific current theories, are expected to do.

Several authors have defended the interest of approaching and exposing such kind of *narrative* ground that, on the one hand, purportedly gives support to, and, on the other, is somewhat obscured by, scientific nomological formulae. I take Alirio Rosales's (2017) comparison between Ronald A. Fisher and Sewall Wright's mathematical solutions (i.e., nomological models) for certain problems of population genetics in terms of the diverse narratives that not only support them but give them meaning to be fairly understandable along these lines.

An even more theoretically committed contribution in this respect is Carlos López Beltrán's (1998) paper, centred on the combination of narrative and statistical explanations in biology and medicine. As other philosophers interested in *narrative science*, López Beltrán starts with the factual assumption that certain specific scientific areas and practices (his focus is on medicine and biology) make an extensive use of *narrative patterns of explanation* for their very particular, unique and eventful *explananda* (a clinical case or the evolution of a particular trait). But his most thought-provoking point is that the statistical numerical models that these same disciplines also construe still reveal their narrative warp and woof, as issuing from data collection practices whose particulars are more than present in their final presentation and effective use. López Beltrán situates statistical models midway between the particularity of the unique case and the universality of classical nomological generalizations and does so by invoking a fourth intermediary state between the unique and the statistical in the clinical-case based on typicality.[17]

---

[17] López Beltrán uses here extensively the work of Spanish medical doctor P. Laín Entralgo, who, in 1950, published a book on the significance of *clinical stories* that has been invoked as a forerunner of contemporary approaches to *narrative medicine* (Charon 2006). The claim about the *typicality* of a case-narrative may become a claim for its *exemplarity* in the sense that it may allow drawing conclusions thereof that are more based on the saliency and usefulness of its traits than on the statistical probability of its real occurrence (Morgan 2007: 167). However, the

Thus, López Beltrán predicates the genealogical and conceptual *continuity* of narrative and statistic explanatory strategies – against their current alleged rivalry – and even places narrativity, especially narrative cognitive capacities, as grounding both:

> The continuity between these two strategies I want to expose gives certain priority to the narrative one. I want to show that, not just historically but also conceptually, the efficiency of statistical procedures is based on the either explicit or implicit use of cognitive capacities associated to narrativity. That is, the use of statistics implies a (currently nearly always occult) narrative resource and makes the same kind of explanatory work. Both strategies try to establish more or less reliable connections between strictly unrepeatable singular events and the sought for syntheses and generalizations that motivate scientific research. (López Beltrán 1998: 275; my translation)

A major reference for López Beltrán is Robert Richards's (1992) seminal paper, so far probably the most radical defence of the ultimately *narrative* character of scientific explanation *in general*: 'When the barriers are down, we will see, not that historical narrative fails as a scientific explanation, but that much of science succeeds only as historical narrative' (Richards 1992: 40).

The idea of a generalized narrative approach to scientific practice that may be just temporarily and only very superficially *circumvented* by relevant simplifications is very present in Richards's radical proposal. For Richards, the narrative quality of scientific explanatory practice would be, somehow, at the bottom of any explanatory attempt in such a way that it is only when making certain simplifications and taking certain methodological decisions that some disciplines *just apparently* and for a limited range of phenomena succeed in leaving their narrative nature behind:

> [e]volutionists cannot make many predictions of consequence. I should add physicists are not logically better off; their projected systems are usually simpler and, as far as circumstances go, dead. But they cannot more accurately predict the exact trajectory of a falling leaf on a blustery Chicago day than Darwin could have divined the rise and evolutionary development of the HIV virus. (Richards 1992: 36–37)

Richards placed his narrative approach to explanation in opposition to law-based explanatory models, but most especially to the attempt to understand less strict patterns under the epistemological dominance of the nomological

---

functions played by what is supposedly *typical* or *exemplary* in the assessment of the epistemic relevance of narratives may be varied enough (cf. Morgan 2019). For example, in Toker's (2017) study of Gulag's literature, fictional but supposedly sample cases function as representing what really happened many times and may be so discussed in a scientific setting as 'history'. In Meunier's (Chapter 12) 'research narratives', depersonalized accounts of what really happened once (and not exactly so) become epistemically relevant for a community inasmuch as they depict procedures that might be generally implemented.

model. Instead of considering such nomological models as the successful peak from which any degree of divergence would diminish the scientific quality of an account, Richards somehow maintains that keeping in touch with the narrative roots of our scientific explanatory attempts – instead of contemplating and appraising their skeleton-like yields – will in fact *improve* epistemological research.

My second claim goes further: it is that all explanations of events in time are ultimately narrative in structure. This means that Hempel got it just backwards: it is not that history can offer only explanation sketches, but that nomological-deductive accounts [. . .] provide only narrative sketches; the covering law model yields sound explanations only insofar as that skeleton can be fleshed out imaginatively with the sinew and muscle of the corresponding narrative. (Richards 1992: 23)

According to Richards, the problem with Hempel's nomological model as well as other equally *nomologically eager* models is that they assume that currently valid law-like generalizations, first, lay ready at hand and, second, simply match as objective patterns the (pre-determined as) relevant facts of the *explananda* they allegedly cover. However, only in very limited, artificial, textbook-like situations (insofar as the isolation of the phenomenon is ascertained) this seems to be the case. Whenever we want to explain a *real event in time* the explanatory work will not really be done by any prearranged formal relations between selected antecedent conditions and matching laws, but precisely by the detailed investigation of the case that would, among other things, justify their use. And for that, according to Richards, we need narratives, narrative principles and narrative cognition.

López Beltrán's claimed continuity between scientific explanatory methodologies, striving at different ranges of applicability, as based on their common ultimate narrative nature, finally becomes a plea for a reasonable and healthy combination of approaches (1998: 277–278) that is rather in line with Currie and Sterelny's (2017: 20) integrative proposal. Such self-assumed *explanatory pluralism* (cf. Mantzavinos 2016) would avoid the downright dismissal of scientific explanations solely based on their form or mode of presentation and thus be more than compatible with a more nuanced and specifically *argumentative* approach to explanation discussion and assessment.

21.5   Conclusion

The *just-so charge* is a derogatory label, a negative assessment judgement that has been often misinterpreted and hastily attributed to explanatory attempts of a *narrative* nature on account of their form or discursive presentation through the catch-phrase *just-so story*. This is misleading and rather at odds with

S. J. Gould's original introduction of the concept within epistemological discussions.

However, the academic effectiveness of the label, working as a global flaw charge and preventing, in many cases, a more careful analysis of significant epistemological suggestions, has, in many cases, forced defenders of the relevance of narrative science to meet the challenge and try to respond to it.

In this chapter, I have analysed those responses that range from assuming the methodological benefits of narrative formats (or at least of the integration of narratives with other epistemological approaches) whenever the phenomena under scrutiny meet certain conditions to the bold postulation of the ultimate narrative nature of all explanatory endeavour.

These qualified defences of narrative science constitute a contribution to contemporary discussions on *explanatory pluralism* and, together with other suggestions, establish the possibility of analysing scientific reason-giving practices as primarily subject to the *dialectical requirement* of openness to collective survey and discussion rather than to aprioristic predetermined formulae precisely aiming at circumventing it. Nothing could be more *just-so*.[18]

Bibliography

Al-Shawaf, L. (2019) 'Seven Key Misconceptions about Evolutionary Psychology'. *Areo*. https://areomagazine.com/2019/08/20/seven-key-misconceptions-about-evolutionary-psychology/.

Al-Shawaf, L., K. A. Zreik and D. M. Buss (2018). 'Thirteen Misunderstandings about Natural Selection'. In T. K. Shackelford and V. A. Weekes-Shackelford, eds. *Encyclopedia of Evolutionary Psychological Science*. Cham: Springer, 96–102.

Beatty, J. (2016). 'What Are Narratives Good For?' *Studies in History and Philosophy of Science Part C* 58: 33–40.

(2017). 'Narrative Possibility and Narrative Explanation'. *Studies in History and Philosophy of Science Part A* 62: 31–41.

Charon, R. (2006). *Narrative Medicine: Honoring the Stories of Illness*. New York: Oxford University Press.

Crasnow, S. (2017). 'Process Tracing in Political Science: What's the Story?' *Studies in History and Philosophy of Science Part A* 62: 6–13.

Craver, Carl, and James Tabery (2019). 'Mechanisms in Science'. In Edward N. Zalta, ed. *The Stanford Encyclopedia of Philosophy*. https://plato.stanford.edu/archives/sum2019/entries/science-mechanisms/.

---

[18] This work has been funded by the Spanish Ministry of Science and Innovation. Research Project: PGC 2018-095941B-100, 'Argumentative practices and pragmatics of reasons'. *Narrative Science* book: This project has received funding from the European Research Council under the European Union's Horizon 2020 research and innovation programme (grant agreement No. 694732). www.narrative-science.org/.

Currie, A., and K. Sterelny (2017). 'In Defence of Story-Telling'. *Studies in History and Philosophy of Science Part A* 62: 14–21.

Fisher, W. R. (1989). *Human Communication as Narration: Toward a Philosophy of Reason, Value, and Action*. Columbia: University of South Carolina Press.

Gould, S. J. (1978). 'Sociobiology: The Art of Storytelling'. *New Scientist* (16 November): 530–533. http://mountainsandminds.org/wp-content/uploads/201 6/12/Gould-1978-Sociobiology.pdf.

    (1989). *Wonderful Life: The Burgess Shale and the Nature of History*. New York: W. W. Norton.

Gould, S. J., D. M. Raup, J. J. Sepkoski, T. J. M. Schopf and D. S. Simberloff (1977). 'The Shape of Evolution: A Comparison of Real and Random Clades'. *Paleobiology* 3.1: 23–40.

Gould, S. J., and R. C. Lewontin (1979). 'The Spandrels of San Marco and the Panglossian Paradigm: A Critique of the Adaptationist Programme'. *Proceedings of the Royal Society of London. Series B* 205.1161: 581–598.

Hacking, I. (1983). *Representing and Intervening*. Cambridge: Cambridge University Press.

Hansson, S. O. (2017). 'Science and Pseudo-Science'. In Edward N. Zalta, ed., *The Stanford Encyclopedia of Philosophy*. https://plato.stanford.edu/archives/su m2017/entries/pseudo-science/

Hubálek, M. (2021). 'A Brief (Hi)Story of Just-So Stories in Evolutionary Science'. *Philosophy of the Social Sciences* 51.5: 447–468.

Huss, J. (2009). 'The Shape of Evolution: The MBL Model and Clade Shape'. In D. Sepkoski and M. Ruse, eds. *The Paleobiological Revolution: Essays on the Growth of Modern Paleontology*. Chicago: Chicago University Press, 326–345.

Kurzban, R. (2012). 'Just So Stories Are (Bad) Explanations: Functions Are Much Better Explanations'. *Evolutionary Psychology Blog Archive*. http://web.sas.upenn.edu/kurz banepblog/2012/09/24/just-so-stories-are-bad-explanations-functions-are-much-better -explanations/.

Laín Entralgo, P. (1950). *La historia clínica, historia y teoría del relato patográfico*. Madrid: Consejo Superior de Investigaciones Científicas.

López Beltrán, Carlos (1998). 'Explicación narrativa y explicación estadística en medi-cina y biología'. In A. Barahona and S. Martínez, eds. *Historia y explicación en biología*. Mexico City: Fondo de Cultura Económica, 275–288.

Mantzavinos, C. (2016). *Explanatory Pluralism*. Cambridge: Cambridge University Press.

Marraud, H. (2013a). *¿Es lógic@? Análisis y evaluación de argumentos*. Madrid: Cátedra.

    (2013b). 'Variedades de la argumentación *a fortiori*'. *Revista Iberoamericana de Argumentación* 6: 1–17.

    (2016). 'Diagramas y estructuras argumentativas'. Edición revisada y ampliada, noviem-bre de 2019. www.academia.edu/41045963/DIAGRAMAS_Y_ESTRUCTURAS_ ARGUMENTATIVAS_2019_2020.

Morgan, Mary S. (2001). Models, Stories and the Economic World. *Journal of Economic Methodology* 8.3: 361–384.

(2007). 'The Curious Case of the Prisoner s Dilemma: Model Situation? Exemplary Narrative?' In A. N. H. Creager, E. LunbecK and M. Norton Wise, eds. *Science Without Laws: Model Systems, Cases, Exemplary Narratives.* Durham, NC: Duke University Press, 157–185.

Morgan, Mary S., and M. Norton Wise (2017). 'Narrative Science and Narrative Knowing: Introduction to Special Issue on Narrative Science'. *Studies in History and Philosophy of Science Part A* 62: 1–5.

Morgan, N. (2019). 'Exemplification and the Use-Values of Cases and Case Studies'. *Studies in History and Philosophy of Science* 78: 5–13.

Olmos P. (2018). 'La justificación de la abducción en el contexto del debate sobre el realismo científico: una aproximación argumentativa'. *ArtefaCToS* 7.2: 35–57.

(2019). 'Abduction and Comparative Weighing of Explanatory Hypotheses: An Argumentative Approach'. *Logic Journal of the IGPL* 29.4: 523–535.

(2020a). 'Revisiting Accounts of Narrative Explanation in the Sciences: Some Clarifications from Contemporary Argumentation Theory'. *Argumentation* 34: 449–465.

(2020b) 'The Value of Judgemental Subjectivity'. OSSA Conference Archive 12. https://core.ac.uk/download/pdf/323559007.pdf.

Richards, R. J. (1992). 'The Structure of Narrative Explanation in History and Biology'. In M. H. Nitecki and D. V. Nitecki, eds. *History and Evolution.* Albany: State University of New York Press, 19–54.

Rosales, A. (2017). 'Theories that Narrate the World: Ronald A. Fisher's Mass Selection and Sewall Wright's Shifting Balance'. *Studies in History and Philosophy of Science Part A* 62: 22–30.

Terrall, M. (2017). 'Narrative and Natural History in the Eighteenth Century'. *Studies in History and Philosophy of Science Part A* 62: 51–64.

Toker, L. (2017). 'The Sample Convention; Or, When Fictionalized Narratives Can Double as Historical Testimony'. In P. Olmos, ed. *Narration as Argument.* Cham: Springer, 123–140.

Toulmin, S. E. (1953). *The Philosophy of Science.* London: Hutchinson University Library.

(1958). *The Uses of Argument.* Cambridge: Cambridge University Press.

Turner, D., and J. C. Havstad (2019). 'Philosophy of Macroevolution'. In Edward N. Zalta, ed. *The Stanford Encyclopedia of Philosophy.* https://plato.stanford.edu/archives/sum2019/entries/macroevolution/.

Waismann, F. (1968 [1956]). 'How I See Philosophy'. In R. Harré, ed. *How I See Philosophy by F. Waismann.* London: Palgrave Macmillan, 1–38.

Weinstein, M. (2009). 'A Metamathematical Model of Emerging Truth'. In J.-Y. Béziau and A. Costa-Leite, eds. *Dimensions of Logical Concepts.* Campinas: Centro de Lógica, Epistemologia e História da Ciência, 49–64.

Wise, M. Norton (2019). 'Does Narrative Matter? Engendering Belief in Electromagnetic Theory'. Economic History Working Papers, 003. Narrative Science Series, Economic History Department, London School of Economics and Political Science. www.narrative-science.org/uploads/3/1/7/6/31762379/norton_wise_-_wp_003.pdf.

Woods, J. (2017). 'Reorienting the Logic of Abduction'. In L. Magnani and T. Bertolotti, eds. *Springer Handbook of Model-Based Science*. Dordrecht: Springer, 137–150.

Woods, J., and B. Hudak (1989). 'By Parity of Reasoning'. *Informal Logic* 9: 125–139.

VII

Finale

22   Narrative and Natural Language

M. Norton Wise

Abstract

The distinction that has become standard between natural language and formal language, which rests on differentiating what is socially evolved and experiential from what is purposefully planned, suggests that a similar emphasis on experientiality may illuminate the distinction between narrative and formal modes of knowing, which figures prominently in this volume. Support for that perspective comes from developments in both narratology and computational linguistics. A key concept from both specialties – and for this volume – is that of 'scripts', which indicates how even texts that are explicitly formal may be understood as narratives by experienced readers. An explicit example that illuminates these themes comes from James Clerk Maxwell's classic paper 'On Faraday's Lines of Force'. It juxtaposes narrative and formal modes of representation and displays their relative advantages, suggesting that the development of scientific knowledge often depends on continual feedback between natural narrative and formal analysis.

22.1   Introduction

The chapters in this volume all respond to the question, what work does narrative do for practitioners in the sciences? For many authors their answer involves a distinction between narrative modes of knowing and formal modes, even when their aim is to undermine the distinction as a dichotomy. A clear statement appears in Paula Olmos's reassessment of the meaning of *just-so stories*. She seeks a middle way between the attempt to subsume phenomena under the skeleton of formal, lawlike, causal explanations and the fleshed out narrative treatment of 'a complex, highly contextual and somewhat indeterministic causal web' (Olmos, Chapter 21). The qualities of narrative that seem strongest here and throughout the volume include its capacity to capture subtlety, ambiguity, complexity, pattern, temporality, contingency, counterfactuals and, perhaps most centrally, colligation (Morgan, Chapter 1). Formality is weak in these capacities. Its strength lies in simplification, precision, rigour, unification and logic. But why does the distinction between what counts as narrative and what counts as formal seem so commonsensical to so many of us, as though it requires no accounting? Is there not something quite straightforward behind it?[1]

22.2   Natural Language is Evolutionary and Experiential

I take my cue initially from Thomas Piketty whose 1,000-page best-seller *Capitalism and Ideology* (2020) has received widespread acclaim. It is a professional economist's analysis of how income inequality has developed over the past 200 years, based on a massive amount of data assembled from many countries, emphasizing their diverse histories and the multidimensionality of current choices. In methodological remarks 'on the complementarity of natural language and mathematical language', Piketty asserts that such an undertaking has necessarily required that he rely primarily on natural language, for 'there is no substitute for natural language when it comes to expressing social identities or defining political ideologies' (Piketty 2020: 43).

Piketty's appeal to natural language is at the same time an appeal to narrative. It opens the way for him to write economic analysis as narrative history and to make extensive use of literary depictions to give an accurate sense of economic conditions as lived experience. Jane Austen's *Sense and Sensibility*, for example, provides a real-life sense of how capitalism operated around 1800 and what it meant in personal and social terms for a gentrified family in straitened circumstances to have an income from investment capital of 100 pounds a year rather than 4,000 pounds (Piketty 2020: 15, 170).

The great lesson, of course, is that 'Those who believe that we will one day be able to rely on a mathematical formula, algorithm, or econometric model to determine the "socially optimal" level of inequality are destined to be disappointed'. Only natural language, and thus narrative understanding, 'can promise the level of nuance and subtlety necessary to make choices of such magnitude'. Nevertheless, Piketty also relies heavily on formal language, 'the language of mathematics, statistical series, graphs, and tables', which fill many pages and are equally indispensable for social and political reflection (Piketty 2020: 43).

Taking this hint from Piketty, I want to suggest that we think of the easy distinction between narrative and formal as reflecting the distinction as now commonly formulated between natural and formal language. A natural language – also a human or ordinary language – is a naturally evolved product of practical use and repetition. Similarly, a native speaker acquires the capacity for

---

[1] Although a number of chapters in this volume use 'narrative' in the sense once standard among narratologists of an unfolding in time of a causally connected sequence of events, I will use it here in the broader sense of an unfolding of a representation or interpretation, without any necessary reference to temporality but prioritizing experientiality, as in more recent 'natural narratology' (n. 2).

its subtle usage, and thus for the qualities we typically associate with narrative, through many years of lived experience, including sensory experience. In contrast, formal languages – mathematics, logic, programming languages, technical vocabularies – are purposefully designed and purposefully developed, rather than socially and informally evolved. Arguably, there are no native speakers of formal languages, which has much to do with their limited capacity for narrative. But it is easy to overdo this rather static emphasis, since formal languages do develop over time and their experienced readers do inflect them with narrative characteristics.

The experiential perspective on natural language resonates strongly with the recent turn in narratology to 'natural narrative', following the seminal work of Monika Fludernik (1996).[2] 'Natural' here refers to the grounding of narrative in lived experience, so that narrativity is virtually identical with 'experientiality'. Fludernik's model has the advantage of decoupling the concept of narrative from the traditional plot-based requirements of temporal progression and causal connectedness. It also highlights the experience of the reader, and not only the author, in producing the narrativity of a text (Caracciolo 2014). This text–reader interaction will figure importantly below with respect to 'scripts'.

The significance of natural language and natural narrative being interconnected through experientiality finds ready expression in Brian Hurwitz's lovely paper on epistemic switching in medical narratives (Chapter 17). Focusing on their narrative features, he regularly emphasizes the tension between the 'personal experiential' character of medical anecdotes and the 'more formal, impersonal' nature of clinical case reports. 'Unlike case reports, which have become highly regulated medical accounts, anecdotes remain informally patrolled schema, cast in a vernacular language that has less recourse to technical and formal terminology than cases' (Hurwitz, Chapter 17). This is not to say, however, that anecdotes have had little role in medical knowledge. Although much maligned at times as subjective and untrustworthy, they have continued to occupy a prominent place in medical reasoning.

Querying how that happens, Hurwitz highlights another important aspect of the narrative/formal distinction: the 'epistemic switch' that occurs when anecdotal testimony of personal experience gets 'revoiced' as evidence. He offers the striking example of how Pfizer chemists almost serendipitously took up the experience of a few miners who sheepishly reported that a potential medication they were taking in a clinical trial seemed to produce erections. Through

---

[2] My thanks to Kim Hajek for calling my attention to natural narratology and for discussion of the issues involved. For the purposes of this essay I am ignoring the possible problem that, with respect to language, natural is opposed to formal while, with respect to narrative, natural is sometimes opposed to unnatural (meaning impossible in the real world) rather than simply nonnarrative. Fludernik casts doubt on the natural/unnatural distinction.

quantification and standardization in a much larger trial the chemists transformed the rather undefined substance with anomalous side effects into a fully medicalized treatment for erectile dysfunction (sildenafil). 'The miners' natural language testimonies came to be revoiced in the "de-anecdotalized" formal language of Pfizer's subsequent trial participants [. . .] expressed in datapoints', thereby according them objective, scientific status (Hurwitz, Chapter 17). The change in language, from natural to formal, was at the same time a change in speakers and in context, producing an epistemic switch that transformed the very meaning of the miners' testimonies.[3]

22.3   Computational Linguistics

Issues of this kind have taken on new relevance and have led to an explosion of research and development in relation to natural language processing (NLP) and the more sophisticated expectations for natural language generation (NLG) and ultimately natural language understanding (NLU) using artificial intelligence. The questions that arise in this area exhibit so many parallels to those of the present volume that it should perhaps come as no surprise that one commercial company has taken on the same name: 'Narrative Science'. The company specializes in NLG, meaning that its programs convert business data into narrative form, so it advertises itself as 'a data storytelling company, creating products that turn business data into plain-English stories' (Narrative Science 2020).

At the simplest level, NLP has shown considerable success in extracting from narrative texts specific data items that are readable in formal computer programming languages. For example, massive digitization of medical records has made it imperative to be able to extract from patient histories contained in clinical notes and pathology reports the sorts of specific information that would be helpful for continuing care. One study from 2012 looked for temporal expressions of time, date, duration, and sets of these expressions in narrative records of 33,000 individuals. Judged against trained human reviewers, the success rate was a respectable 83 per cent. Still, the false positives and negatives are instructive for just how limited such programs still are. The phrase 'capsule may be opened and sprinkled on applesauce' produced a spurious categorization of 'date', while 'diarrhea daily for about 1–2 months' failed to produce a 'duration' (Reeves et al. 2012). Another NLP study from 2014, using key word searches to extract data, was able to document a striking lack of continuity in the narrative records of patients moving from inpatient to outpatient care and suffering from 'post-intensive care syndrome'. At the same

---

[3] See Paskins (Chapter 13), for an excellent exploration of the epistemic issues involved in shifting between narrative (thick) and formal (thin) language.

time, it showed the severe limitations of its own capabilities, misidentifying the word 'depressed' (mental state) with 'depressed ejection fraction' (heart function) (Sjoding and Liu 2016: 1444).

What makes such attempts at natural language processing so interesting for the reflections on narrative and formal knowledge in this volume is just how difficult it is to formulate in programming language the sorts of subtle distinctions in natural language that humans recognize without thinking. That difficulty reinforces the tendency among many observers to essentialize the narrative/formal distinction as a matter of two dichotomous modes of comprehension. This response may be particularly pressing in the context of computer science education. A thought-provoking paper on priorities in the teaching of programming and assessment of skills stresses that there are 'two quite different mental processes' and summarizes the problem as follows:

- For a formal language, a single and complete meaning is contained entirely within the text, and the understanding process consists of determining that meaning from a close analysis of the text alone.
- For a natural language, an analysis of the text is only part of the task, as this may produce multiple possible meanings. A particular meaning can only be derived by making use of available contextual information for disambiguation. (Cutts et al. 2014: sect. 3)

The authors therefore argue that the two sorts of comprehension should be kept strictly apart, for otherwise confusion will reign. So-called pseudo-code, which blends formal and natural languages and is intended to be used 'for human understanding of algorithms rather than machine understanding' is, according to this view, problematic at best, at least for novices. Interestingly, the authors acknowledge that experienced programmers, reading past ambiguities in the pseudo-code, 'will be able to infer exactly what is meant'. If so, one wonders, then why not teach those skills in the first place and explore with budding programmers how to relate them to machine language?

That is in fact the goal of the sophisticated field of computational linguistics: 'the scientific and engineering discipline concerned with understanding written and spoken language from a computational perspective, and building artefacts [software] that usefully process and produce language' (Schubert 2020: preface). While specialists' views differ on language as a mirror of mind, they agree on the goal of building linguistically competent computers and on the fact that the project faces a myriad of intractable hurdles. Most telling is that natural language is ambiguous at all levels of syntax, semantics and pragmatics and depends for its understanding on a vast store of contextual and world knowledge (or background knowledge).[4] But, given the facility that humans have in

---

[4] On the further issue of value differences expressed in computerized representations, see Dick (Chapter 15).

disambiguating natural language, some analysts have argued that what might be called their narrative or colligatory skills 'more nearly resemble fitting the observed texts or utterances to familiar patterns' than solving complex logical problems (Schubert 2020: sect. 2.4, 5.4).

22.4   Scripts

One early approach of this kind was pioneered by Roger Schank and Philip Abelson, who recognized that understanding and inference in natural language were heavily dependent not simply on a large store of background knowledge but on patterns of belief and expectation that they called 'scripts' (Schank and Abelson 1977). Scripts are 'the prototypical ways in which familiar kinds of complex events [. . .] unfold' (Schubert 2020: sect. 1.2, 4.2). Implementing this perspective in the formal language of machines remains elusive – pattern recognition in general being a notoriously difficult problem – but it caught the interest of people outside computational linguistics, including the well-known cognitive narratologist David Herman (1997: 1047–1048). And that brings me back to the present volume.

In her illuminating chapter, 'Reading Mathematical Proofs as Narratives' (Chapter 19), Line Andersen presents an empirical study of how mathematicians read proofs as narratives. This seems surprising since we normally think of proofs as epitomizing the rigour of formal language and at the farthest remove from narrative. But mathematicians, it turns out, do not necessarily read proofs in a line by line checking of the logical argument being presented. Instead, reading a proof as a telling of how something happened, they often narrativize it by drawing on their own experiential background knowledge, recognizing whole sections of a proof mimetically as the familiar patterns that Schank called scripts. These sections they can skim over and fill in from the scripts. Less familiar parts may throw up surprises, which require close attention and may lead to new scripts. In this way, mathematicians can come to *understand* proofs by reading them as narratives (Andersen, Chapter 19).

To put this a bit differently, a mathematician's understanding of a proof, on Andersen's account, offers an excellent opportunity to reflect on how the experientiality of 'natural narrative' needs to be interrelated with that of 'natural language'. Reading a proof in experiential terms changes what looks to an outsider like a purely formal structure into a natural narrative for the reader; so too the experiential reading enriches the formal language of rigorous proof with the natural language of narrative, for it calls up meanings that the unaided formal language, lacking background and context, cannot convey. On the other hand, without its formal language the proof would not be a proof. This is the sort of conundrum that bedevils computational linguistics. It is meat and potatoes, however, for the narrative science of this volume.

Similar examples from the volume illuminate the point in additional ways. Nina Kranke, in 'The Trees' Tale: Filigreed Phylogenetic Trees and Integrated Narratives' (Chapter 10), argues that phylogenetic tree diagrams, which accompany texts written in the formal language of molecular biology and computer-assisted analysis, are produced and read by biologists as visual narratives of evolutionary history. She observes that biologists very often do not read the entire text of a paper but that 'the informed reader understands the central argument of the paper just by looking at the diagram'. In thus 'reading' the text/diagram, they fill in from their own background knowledge, in the manner of scripts and of natural narrative, much information that is not actually present in the text or even the diagram itself. In this, Kranke's biologists do their reading much like Andersen's mathematicians. But the diagrams introduce an even stronger element of text–reader interaction – one that is perhaps more typical of narrative science – for they are created in the first place not simply as formal diagrams but as visual narratives that already express the author's experience and aim to evoke the experience of the reader. The images of real animals sometimes placed at strategic locations on the more formal diagrams seem to announce this sought-for interaction. Finally, reading the diagrams as visual narratives highlights the sensory character of much natural language.[5]

Yet another aspect of the importance of scripts appears in Andrew Hopkins's discussion, 'The Narrative Nature of Geology and the Rewriting of the Stac Fada Story' (Chapter 4). Hopkins argues that a geologist, in habitually reading professional papers as temporal narratives rather than the non-temporal descriptions they appear to be, relies on an array of scripts that 'derives from a geologist's specific training and experience' and arises in conversation. This emphasis on training and informal communication signals that the scripts are a community affair. Indeed, how could they not be since the natural language of experts is socially evolved as well.

22.5   Narrative and Formal Juxtaposed: A Historical Case

As is apparent already from the chapters in this volume referenced above, it is common to see works in the sciences that employ both narrative and formal modes of knowing, but it is unusual to see the two approaches set side by side and treated quite separately in a single work, thus highlighting their comparison. One such example, however, comes from a canonical paper of James Clerk Maxwell, 'On Faraday's Lines of Force' (Maxwell 1855).[6] Because Maxwell

---

[5] A similar script-like interpretation may apply to the formalized 'storm cards' discussed by Bhattacharyya (Chapter 8).

[6] The discussion below is adapted from Wise (2021), which compares the theories of 'lines of force' and 'action at a distance' in terms of the narrative qualities that make them believable. See references there to illuminating discussions of Maxwell's method of physical analogy.

reflected deeply on the significance of the natural (from 'nature') and the formal ('mathematical') for what he regarded as different 'minds', I will attempt in what follows to extract from his example both a clear expression of the differences and what we might take away from his discussion of their relation.

Published in 1855, 'Faraday's Lines' was Maxwell's first contribution to what was becoming British electromagnetic field theory. In it he took up Michael Faraday's long-running experimental study of electric and magnetic action, which Faraday treated as a mediated action taking place through fields of force in the space surrounding electric or magnetic materials, rather than as the direct unmediated action known as 'action at a distance' (like Newtonian gravitational force). Wilhelm Weber in Germany, working in the action at a distance tradition, had already unified all known phenomena of electromagnetism in a single mathematical formula. It expressed the force acting directly between two electrical particles simply in terms of their distance apart and their relative velocity and acceleration. Faraday instead represented electrical and magnetic phenomena in terms of 'lines of force' distributed in space with an accompanying 'electrotonic state', but just how to conceive the lines of force and the electrotonic state remained rather nebulous and he had no mathematical account of their action.

That is where Maxwell entered the picture. In well-known lines, he expressed his attitude to the two theoretical perspectives of Faraday and Weber: one in the natural language of narrative and the other in formal mathematical language.

What is the use then of imagining an electro-tonic state of which we have no distinctly physical conception, instead of a formula of attraction which we can readily understand? I would answer, that it is a good thing to have two ways of looking at a subject, and to admit that there *are* two ways of looking at it. (Maxwell 1855: 208)

That is the attitude towards the narrative and the formal that informs Maxwell's own representations of Faraday's theory in two quite different ways.

Narrative representation. Maxwell devoted the first half of his long paper to what he famously called a 'physical analogy' between Faraday's lines of force and fluid flow lines, asking his reader to 'consider these curves not as mere lines, but as fine tubes of variable section carrying an incompressible fluid' (Maxwell 1855: 158). Beginning from this simple verbal image, available to anyone who had watched water flowing down a drain, he gradually unfolded a three-dimensional picture of a space full of flowing fluid, including velocity distribution, sources and sinks, a resisting medium, pressure gradients, and changes in the properties of the fluid. The entire account required only the simplest of mathematical relations, remaining almost entirely within the realm of natural language and common imagination.

In elaborating on the virtues of this physical analogy, Maxwell remarked that 'my aim has been to present the mathematical ideas to the mind in an *embodied* form, as systems of lines and surfaces and not as *mere symbols*, which neither convey the same ideas, nor readily adapt themselves to the phenomena to be explained' (Maxwell 1855: 156, 187; emphasis added). In the concept of embodied ideas, he here prefigured a critical concept of natural narratology. Embodiment, Fludernik emphasizes, 'evokes all the parameters of a real-life schema of existence [...] and the motivational and experiential aspects of human actionality likewise relate to the knowledge about one's physical presence in the world' (Fludernik 1996: 30; Caracciolo 2014: sect. 2). Similarly, by embodied mathematics Maxwell did not mean simply that he was giving a physical exemplification of an underlying and more fundamental mathematical structure. It was physical understanding he was after and that did not inhere in 'mere symbols'.

Embodiment here has a literal significance that Maxwell expressed repeatedly through his life. As he would put it in his 'Address to the Mathematical and Physical Sections of the British Association', in 1870: '[many physicists] calculate the forces with which the heavenly bodies pull at one another and they feel their own muscles straining with the effort. To such men momentum, energy, mass are not mere abstract expressions of the results of scientific inquiry. They are words of power, which stir their souls like the memories of childhood' (Maxwell 1870: 220). This highly sensory and emotional aspect of embodiment helps to illuminate Maxwell's presentation of Faraday's lines in narrative form. It was grounded in experience and memory, both conceptual and sensory, and preserved the 'vividness' and 'fertility' of such experience.

It may be helpful also to recognize that Maxwell's presentation of lines of force was explicitly a fictional narrative in which the flowing fluid was an imaginary substance. 'It is not even a hypothetical fluid which is introduced to explain actual phenomena. It is merely a collection of imaginary properties which may be employed for establishing certain theorems in pure mathematics in a way more intelligible to many minds and more applicable to physical problems than that in which algebraic symbols alone are used' (Maxwell 1855: 160).

Having established his basic image in familiar verbal terms, Maxwell employed it to draw together nearly all of the phenomena of electricity and magnetism as conceived by Faraday, including the distribution of magnetic lines around a magnet (Figure 22.1) and the equivalent distribution of magnetic lines produced by electric currents, or electromagnetism. The existence of

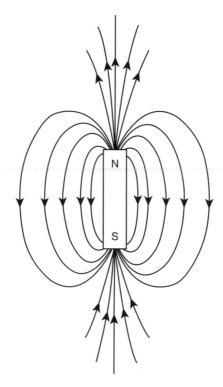

Figure 22.1 **A representation of lines of force
surrounding a bar magnet with north and south poles**

electromagnetism meant that electric current lines and magnetic lines, each conceived separately in terms of flow, had to be interrelated dynamically. Their qualitative relation can be readily understood pictorially with reference to a coil of wire carrying a current (Figure 22.2a), which behaves like a bar magnet with north and south poles, and produces an equivalent distribution of magnetic lines (compare Figure 22.1).

The pattern of the magnetic distribution by itself can be seen as a dynamic balance, which Faraday described as resulting from a tendency of each magnetic line to *contract* along its length and for adjacent lines to *repel* each other laterally. But these effects in the magnetic lines are mirrored reciprocally in the electric lines by the tendency of each electric line (or turn in the coil) to *extend* along its length and for adjacent lines to *attract* laterally. He depicted the reciprocity visually as in Figure 22.2b (Faraday 1855: para. 3265 and plate IV, fig. 1).

Always pursuing the unity of natural powers, Faraday had said of these linked rings and their dynamic balance that it 'probably points to the intimate physical

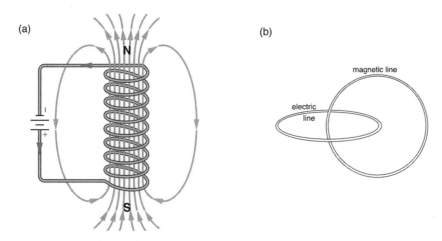

Figure 22.2 (a) and (b) **Current-carrying coil and Faraday's depiction of the relation of electric current lines**
(a) current-carrying coil (dark lines) behaves like a bar magnet. (b) Faraday's depiction of the relation of electric current lines and magnetic lines, which Maxwell called a 'mutual embrace'.

relation, and it may be, to the oneness of condition of that which is apparently two powers or forms of power, electric and magnetic' (Faraday 1855: para. 3268). Maxwell agreed, but with his flair for evoking sensory perception, he labelled their relation a 'mutual embrace' of electricity and magnetism (Maxwell 1855: 184, 194 n.).[7] The importance of this heuristic image can be seen in the fact that it would guide his theorizing through successive versions until he reached his mature theory. For the moment, however, the analogy of lines of force as lines of fluid flow provided no understanding of what the reciprocal dynamics of magnetic and electric lines might consist in physically.

Formal representation. Failing in his quest to understand the mutual embrace physically, Maxwell took up in the second half of his paper an abstract mathematical approach, although still one in which the embrace held a central place. From Faraday and from the flow analogy he had available for mathematical expression the concepts of flow velocity and pressure gradient at any point, or 'quantity' and 'intensity' of flow, which provided his starting point. The reciprocal dynamics of the mutual embrace suggested

---

[7] For Maxwell's continuing use of the metaphor in later papers, see Wise (1979).

further that just as the *quantity* of current passing through a surface surrounded by a magnetic line could be expressed in terms of the *intensity* in the magnetic line, so the *quantity* of magnetic force passing through a surface enclosed by a current line should be expressible in terms of the current's *intensity*. But no such relation of magnetic quantity to current intensity existed experimentally, which is why the mutual embrace remained a suggestive image, an ambiguous symbol of what one might hope to realize physically.

This ambiguity was particularly troubling for Faraday's great discovery of electromagnetic induction, whereby an increase or decrease of the magnetic quantity passing through a surface surrounded by a closed conductor would induce a current in the conductor. Like Faraday, Maxwell thought there must be some corresponding physical condition in the conductor, an 'electrotonic state', whose changing intensity would correspond to the current produced. If so, then this hypothetical electrotonic state might also serve to complete the reciprocal dynamics of the mutual embrace.

Utilizing known laws of electric currents and known theorems of partial differential equations, Maxwell developed his abstract theory of the electrotonic state in a set of six interrelated laws. For the sake of 'seeing' what this formal structure looked like – simply as a formal object – it may be useful to write down four of the laws in modern vector notation, noting three parts (Figure 22.3).

By incorporating the electrotonic state $\mathbf{I}_o$ in this set of equations Maxwell was able to give beautifully coherent expression in the second set of laws to the dynamics of the mutual embrace (electromagnetism) and in the final law to the production of currents in a changing magnetic field (electromagnetic induction). But what was the electrotonic state? It remained a mysterious stranger physically and experimentally. As he put it, 'I have endeavoured to express the idea which I believe to be the mathematical foundation of the modes of thought indicated in [Faraday's] *Experimental Researches*. I do not think that it contains even the shadow of a true physical theory; in fact, its chief merit as a temporary instrument of research is that it does not, even in appearance, *account for* anything' (Maxwell 1855: 207).

22.6   Feedback between Narrative and Formal Representations

In Maxwell's seminal paper, we see juxtaposed two very different representations of the mutual embrace of electric and magnetic lines of force. The first is a narrative unfolding in natural language of a physical analogy, leading to a visual image of the embracing lines and a verbal description of their

1. Two laws (from the flow analogy) expressed the proportionality of quantity **Q** and intensity **I** in each system of lines of force, electric and magnetic,

$$\mathbf{Q}_e \propto \mathbf{I}_e$$

$$\mathbf{Q}_m \propto \mathbf{I}_m$$

2. Two more laws expressed a newly conceived reciprocity in the mutual embrace, now written in terms of the supposed electrotonic state. The total magnetic quantity **Q**$_m$ passing through a surface is given by the electrotonic intensity **I**$_o$ summed around its edge,

$$\iint \mathbf{Q}_m \cdot d\ \sigma = \oint \mathbf{I}_o \cdot d\ \lambda$$

And reciprocally, the total electric quantity **Q**$_e$ passing through a surface is given by the magnetic intensity summed around its edge,

$$\iint \mathbf{Q}_e \cdot d\ \sigma = \oint \mathbf{I}_m \cdot d\ \lambda$$

3. A final law stated that the electric intensity at any point in a conductor is proportional to the rate of change of the electrotonic state at that point.

Figure 22.3 **Maxwell's abstract theory of the electrotonic state**

dynamics. The second is an abstract structure in formal mathematical language, having no necessary relation to the physical analogy. The two sharply contrasting modes of representation are reminiscent not only of the difficulties computational linguists face in their attempts to relate natural language to machine language but of the 'epistemic switch' that Brian Hurwitz observes for the way in which anecdotal knowledge gets 'revoiced' as medical knowledge. They have different meanings in their very different contexts and do not translate one into the other. It was in reference to this sort of epistemic difference that Maxwell remarked that 'mere symbols', in contrast to an embodied analogy, 'neither convey the same ideas, nor readily adapt themselves to the phenomena to be explained'. Reflecting in his 1870 'Address to the Mathematical and Physical Sections' on how fundamental the difference is, he ascribed it to different minds. 'There are [...] some minds which can go on contemplating with satisfaction pure quantities presented to the eye by symbols, and to the mind in a form which none but mathematicians can conceive'. But there are other minds which 'are not content unless they can project their whole physical energies into the scene which they conjure up' (Maxwell 1870: 220).[8] To put that in the terms I am pursuing here, the formal language of the mathematical representation would be poorly described as a 'translation' from the natural language of the physical analogy. For they are different modes of knowing based on different kinds of experience. This key point about experience deserves some development.

With respect to the embodied physical analogy, its creative power depended entirely, both for Maxwell and his readers, on their prior experience of fluid flow – indeed, on their own embodiment and sensory experience – on what natural narratologists call experientiality and on what the linguists call contextual and world knowledge. In his narrative representation this experiential character is explicit. That goes to the heart of the productive work narrative commonly does for scientists, as we see throughout this volume. In contrast, when Maxwell formalized the mutual embrace within a mathematical structure by introducing the electrotonic state, both the embrace and the state were abstracted from physical experience and became mathematical objects defined by the structure. As such they did not, 'even in appearance, account for anything'. Instead, they became well-defined mathematical objects, or, better, mathematical possibilities seeking experimental and conceptual realization. Such creations are of course critically important in the sciences, although as formal representations there is nothing explicitly experiential or narrative-like about them.

But we should not go too fast here and suppose that understanding Maxwell's formal structure was independent of experience. Instead, looking not at the formal laws but at his derivation of them raises the issue of experience in a different manner. The derivation consists in 15 pages of carefully orchestrated mathematical reasoning based on known relations in electromagnetism and known mathematical theorems, known to Maxwell specifically through his friends William Thomson and George Stokes. For him, then, the derivation reflected his personal experience with the mathematics involved, even though that experientiality did not – and could not – appear in the formal language of the text. Similarly, as Line Andersen has made us aware, any reader who shared large portions of that prior knowledge and could therefore see the developing pattern of the derivation might replace much of it with their own experience in getting to the resulting laws. That is, the knowledgeable reader, relying on familiar 'scripts', would read the derivation – and would understand it – more

---

[8] Maxwell inserted an intermediate type who preferred visualization in geometrical forms, drawn or imagined.

like a natural narrative of how the results emerged than as an exercise in logic and would only question the logic if it seemed problematic. Narrativity, it seems, is difficult to escape.

These reflections lead me to a final question about how we should think more generally of the relation between the more narrative and the more formal aspects of scientific reasoning. Are they epistemically different modes of knowing? The refractory character of overcoming the natural/formal distinction in computational linguistics suggests that they are. Maxwell's juxtaposition of narrative and formal representations of the mutual embrace offers a rather stark example to reinforce that view. On the other hand, many of the chapters in this volume show that narrative and formal aspects are not so easily separable and that both play highly creative roles. So again, how should we think of the relation?

Paula Olmos argues for an 'integrative approach'. She cites Sharon Crasnow in support of the view that 'causal links or mechanisms are *better* understood [...] under a narrative rendering than under the crystallized mode of a formal formula'. She also cites Adrian Currie and Kim Sterelny for the view that 'narrative approaches should *combine* with the virtues and benefits of formal models' (Olmos, Chapter 21; Crasnow 2017; Currie and Sterelny 2017). Crasnow's approach might suggest that narratives subsume the formal while Currie and Sterelny's approach would suggest complementarity. Both subsumption and complementarity have attractive qualities, as the cited papers themselves so well attest.

Subsumption would imply that formal modes of knowledge are reductions or abstractions from narrative modes, which are more primitive (in the sense of prior and more basic) and more general. Maxwell made just this point in critiquing the view characteristic of mathematical minds. For them, 'the physical nature of [a] quantity is subordinated to its mathematical form', but this point of view 'stands second to the physical aspect in order of time, because the human mind, in order to conceive of different kinds of quantities, must have them presented to it by nature' (Maxwell 1870: 218). The reduction from nature, or from lived experience, would account for why it is so difficult to encompass the subtleties of natural language in formal language, or why natural language understanding remains rudimentary while natural language generation is making significant strides.

Complementarity, on the other hand, would suggest that neither mode has epistemic priority (at least as a practical matter of use if not a developmental one). They are so distinct that they do not overlap significantly but sit side by side. Once again, Maxwell put it succinctly in terms of modes of knowing: 'For the sake of persons of these different types, scientific truth should be presented in different forms, and should be regarded as equally scientific, whether it appears in the robust form and the vivid colouring of a physical illustration, or in the tenuity and paleness of a symbolical expression' (Maxwell 1870: 220). Thomas Piketty, with whose expression of the complementarity of natural and formal language I began, similarly reminds us of this 'paleness' of mathematical econometric models in comparison with the 'vivid colouring' of narrative history, as well as of the need for both. (See also Paskins, Chapter 13, on 'thin' and 'thick'.)

It seems that Maxwell believed both that formal truths develop as abstractions from narrative truths and that the two are complementary. If he was right then we need a model that encompasses both. Such a model might be found in feedback, in the view that scientists are typically shifting back and forth between narrative and formal modes of representation in a continuous feedback loop, in which each stimulates the other and in which the mutual stimulation is a source of development.[9] From a relatively primitive natural conception an initial formal representation is abstracted, which suggests a more elaborate natural conception, and so on. Maxwell seems to have intended that understanding when he wrote: 'If the skill of the mathematician has enabled the experimentalist [physicist] to see that the quantities which he has measured are connected by necessary relations, the discoveries of physics have revealed to the mathematician new forms of quantities which he could never have imagined for himself' (Maxwell 1870: 218). A bit more history will support that feedback reading for Maxwell's own work.

Prior to 'Faraday's Lines' of 1855, Maxwell had immersed himself in both the narrative papers of Michael Faraday and the mathematical papers of Thomson, who had himself been mathematizing Faraday, with a flow analogy and with Faraday's support. So an ongoing dialectic was already in full swing in the letters that passed between, first, Thomson and Faraday, and then Maxwell and Thomson. It would continue in the series of papers that Maxwell subsequently published, pursuing both more adequate physical analogies and more complete mathematical structures. Already in 1855 he left his reader with the hope that an extended physical analogy would someday complete the picture of electromagnetism with an electrotonic state. 'By a careful study of the laws of elastic solids and of the motions of viscous fluids, I hope to discover a method of forming a mechanical conception of this electro-tonic state adapted to general reasoning' (Maxwell 1855: 188).

Famously, although physical analogies continued to stimulate mathematical formulations, neither Maxwell nor any of the others who tried would find an adequate mechanical conception of an etherial medium in space that would

---

[9] See also Meunier (Chapter 12) on the view that when objects of research, or 'epistemic things', become stabilized the more fluid research narratives drop out.

fully meet the need. Equally famously, as the physical analogies became ever more problematic, the formal structure became ever more dominant, until Heinrich Hertz, discoverer of electromagnetic waves in 1887, famously remarked that 'Maxwell's theory is Maxwell's Equations'. But for Maxwell himself, who had died in 1879, this state of things could only have been temporary. 'We are probably ignorant even of the name of the science which will be developed out of the materials we are now collecting, when the great philosopher next after Faraday makes his appearance' (Maxwell 1873: 360).[10]

References

Caracciolo, M. (2014 [2013]). 'Experientiality', in P. Hühn, J. C. Meister, J. Pier and W. Schmid, eds. *The Living Handbook of Narratology*. Hamburg: Hamburg University Press, sections 3–4. www.lhn.uni-hamburg.de/node/102.html.

Crasnow, S. (2017). 'Process Tracing in Political Science: What's the Story?' *Studies in History and Philosophy of Science Part A* 62: 6–13.

Currie, A., and K. Sterelny (2017). 'In Defence of Story-Telling'. *Studies in History and Philosophy of Science Part A* 62: 14–21.

Cutts, Q., R. Connor, P. Donaldson and G. Michaelson (2014). 'Code or (Not Code): Separating Formal and Natural Language in CS Education'. WiPSCE '14. Proceedings of the 9th Workshop in Primary and Secondary Computing Education: 5–7 November, Berlin, Germany. www.macs.hw.ac.uk/~greg/publications/ccdm.WIPSCE14.pdf.

Faraday, M. (1855). *Experimental Researches in Electricity*, 3 vols. Facsimile reprint. London: Quaritch.

Fludernik, M. (1996). *Towards a 'Natural' Narratology*. London: Routledge.

Herman, D. (1997). 'Scripts, Sequences, and Stories: Elements of a Postclassical Narratology'. *PMLA* 112.5: 1046–1059.

Maxwell, J. C. (1855). 'On Faraday's Lines of Force'. *Transactions of the Cambridge Philosophical Society* 10 (Part I) [Reprinted in Maxwell 1890: 1.155–229] https://archive.org/details/scientificpapers01maxw.

   (1870). 'Address to the Mathematical and Physical Sections of the British Association'. *Report of the British Association for the Advancement of Science* 40 [Reprinted in Maxwell 1890: 2.215–229].

   (1873). 'Scientific Worthies'. *Nature* (18 September): 397–399.

   (1890). *The Scientific Papers of James Clerk Maxwell*. 3 vols. Cambridge: Cambridge University Press.

[10] I thank Mary Morgan and Kim Hajek for detailed commentary on these reflections. Thanks too to Line Andersen, Dominic Berry, Sharon Crasnow, Brian Hurwitz and Paula Olmos. *Narrative Science* book: This project has received funding from the European Research Council under the European Union's Horizon 2020 research and innovation programme (grant agreement No. 694732). www.narrative-science.org/.

Morgan, Mary S., and M. Norton Wise, eds. (2017). 'Narrative Science and Narrative Knowing: Introduction to Special Issue on Narrative Science'. *Studies in History and Philosophy of Science Part A* 62: 1–5.

Narrative Science (2020). https://narrativescience.com.

Piketty, T. (2020). *Capital and Ideology.* Trans. Arthur Goldhammer. Cambridge, MA: Belknap Press.

Reeves, R. M., F. R. Ong, M. E. Matheny, J. C. Denny et al. (2012). 'Detecting Temporal Expressions in Medical Narratives'. *International Journal of Medical Informatics* 82.2: 118–127.

Schank, R. C., and R. P. Abelson (1977). *Scripts, Plans, Goals, and Understanding: An Inquiry into Human Knowledge Structures.* Hillsdale, NJ: Lawrence Erlbaum Associates.

Schubert, L. (2020). 'Computational Linguistics'. In Edward N. Zalta, ed. *The Stanford Encyclopedia of Philosophy.* https://plato.stanford.edu/archives/spr2020/entries/computational-linguistics/.

Sjoding, M. W., and V. X. Liu (2016). 'Can You Read Me Now? Unlocking Narrative Data with Natural Language Processing'. *Annals of the American Thoracic Society* 13.9: 1443–1445.

Wise, M. Norton (1979). 'The Mutual Embrace of Electricity and Magnetism'. *Science* 203.4387: 1310–1318.

(2021). 'Does Narrative Matter? Engendering Belief in Electromagnetic Theory'. In M. Carrier, R. Mertens and C. Reinhardt, eds. *Narratives and Comparisons: Adversaries or Allies in Understanding Science?* Bielefeld: Bielefeld University Press, 29–61.

Index

Abelson, Philip, 452
accident/s, 173
    accident reports, 25
accretionary lapilli, 86, 88, 91, 92, 94, 99
adaptation/adaptationism (biological), 22, 147,
    158, 407, 428, 436
aesthetic, 6, 68, 332, *See* emotions, emotions
affect, 6, 12, 42, 50, 52, 363, 364, 374, 376
Aidinoff, Marc, 310
algebra
    algebraic simplification, 318
    computer algebra, 312
Al-Shawaf, Laith, 431
Alvarez, Walter, 61, 62, 65, 98, 99
ambiguity, 447, 451, 458
American Psychological Association
    (APA), 373
Amherst, Barge (ship), 173–174
Amor, Ken, 89, 91, 96, 239
Amsterdamska, Olga, 288
analogy, physical, 453, 454, 455, 458, 460, 462
ancestor (biological), 212, 215, 405, 408,
    412, 420
Andersen, Line, 23, 52, 156, 207, 216, 274,
    310, 334, 372, 397, 398, 416, 452,
    453, 460
anecdote/s, 19, 35, 242, 351, 352, 353, 354,
    355, 357, 358, 359, 360, 361, 362, 363,
    364, 365, 366
Anekdota, 351, 352
Ankeny, Rachel, 340, 359
Anna O, 45
anomaly, 240
    iridium, 65, 70, 77
Anonimalle Chronicle, 329
anthropology, 7, 14, 18, 125, 208, 268, 291
ant-lions, 25
apprentice, 254
Arber, Agnes, 193, 197
archaeology, 8, 16, 19, 39, 117, 122, 123, 187,
    189, 207, 208, 293
    stratigraphy, 128

theory, 123
time, 131
archive/s, 132, 176, 185, 186, 188, 194, 195,
    197, 201, 208, 229, 292, 328, 342
Argonne National Laboratory, 312
argument, 22, 84, 304, 331, 333
argumentation theory, 424, 427
Aronson, Jeff, 361
artificial intelligence, 50, 317, 318
Asiatic Society of Bengal, 168
asperities (geological), 107, 117
astronomy, 69, 70
astrophysics, 70, 433
Atkinson, Paul, 357
atlases, 192, 200, 201
audience, 212, 220, 290, 335, 358, 371
Austen, Jane, 144, 448
automation, 7, 309, 310, 312, 313, 315, 316,
    317, 325, 329
auxanometer, 149, 151

Bachelard, Gaston, 280
background knowledge, 242, 451, 452, 453,
    See tacit knowledge
back-projection (of seismological data), 117
backstory, 22, 23, 74, 233, 269, 405, 409, 414,
    416, 420
bacteriology, 291, 296
Bal, Mieke, 145, 151
Bar, Moshe, 374, 381
Barash, David, 428
Barrett, Lisa Feldman, 374, 381
Bary, Heinrich Anton de, 195
Bauman, Richard, 373, 375, 379
Baxter, Janella, 338
Bay of Bengal, 165, 167
Bayesian analysis, 372
Beach, Derek, 230
Beadle, George, 247, 248, 255, 257
Beatty, John, 23, 38, 45, 84, 147, 149, 208, 216,
    217, 242, 273, 331, 353, 417
Beer, Gillian, 144